Tectonic Evolution of the Tethys Ocean in the Qinghai–Tibet Plateau

Tectonic Evolution of the Tethys Ocean in the Qinghai–Tibet Plateau

Guest Editor

Ming Wang

MDPI

Basel • Beijing • Wuhan • Barcelona • Belgrade • Novi Sad • Cluj • Manchester

Guest Editor
Ming Wang
College of Earth Sciences
Jilin University
Changchun
China

Editorial Office
MDPI AG
Grosspeteranlage 5
4052 Basel, Switzerland

This is a reprint of the Special Issue, published open access by the journal *Minerals* (ISSN 2075-163X), freely accessible at: https://www.mdpi.com/journal/minerals/special_issues/397Q1RI2S3.

For citation purposes, cite each article independently as indicated on the article page online and as indicated below:

Lastname, A.A.; Lastname, B.B. Article Title. *Journal Name* **Year**, *Volume Number*, Page Range.

ISBN 978-3-7258-6816-2 (Hbk)
ISBN 978-3-7258-6817-9 (PDF)
https://doi.org/10.3390/books978-3-7258-6817-9

Contents

About the Editor

Ming Wang

Ming Wang is a professor and doctoral supervisor in Structural Geology at Jilin University. His primary research focuses on the tectonic evolution of the Tethyan Ocean within the Qinghai–Tibet Plateau. He has led 22 research projects, including grants from the National Natural Science Foundation of China, regional geological surveys, and special and general postdoctoral science funds, all centered on Tethyan evolution in Tibet. To date, he has authored or co-authored 48 papers as the first or corresponding author, of which 35 are SCI-indexed, including 9 papers in Chinese Academy of Sciences Zone I journals and 1 in a Nature Index journal. He has also published one monograph as first author and completed three national 1:50,000 geological survey reports. His first/corresponding author works have been cited over 900 times. Prof. Wang has been recognized as a high-level scientific and technological innovation talent by the Ministry of Natural Resources and a "Geological Mapping Scientist" by the China Geological Survey, and was selected into the Kuang Yaming/Tang Aoqing Talent Program at Jilin University. He serves as a youth Editorial Board Member for *Minerals*, *Continent and Life Evolution*, as well as *Sedimentary Geology and Tethyan Geology*. His awards include the Youth Geological Science and Technology Prize (Silver Hammer Award) from the Geological Society of China, the Qinghai–Tibet Plateau Youth Science and Technology Award, and the Jilin Provincial Youth Geological Science and Technology Award.

minerals MDPI

Editorial

Editorial for Special Issue "Tectonic Evolution of the Tethys Ocean in the Qinghai–Tibet Plateau"

Ming Wang

College of Earth Sciences, Jilin University, Changchun 130061, China; wm609@163.com

1. Introduction

The Tethys Ocean, a geodynamically pivotal transcontinental oceanic system sandwiched between the Laurasian and Gondwanan supercontinents, underwent successive evolutionary phases (Proto-Tethys, Palaeo-Tethys, Neo-Tethys) sculpted by continental drift, convergence, and breakup. Its demise imprinted profound tectonic signatures across the 15,000 km extensive Tethyan tectonic realm, with the Qinghai–Tibet Plateau emerging as a "natural laboratory" archiving a near-complete geochronological record of this epic geodynamic journey. Acclaimed as a premier global hotspot for Earth system science and a strategic repository of mineral and hydrocarbon resources, the Plateau affords important perspectives on continental assembly–disassembly, oceanic opening–closure, and deep Earth geodynamics. Tackling long-standing debates surrounding Tethyan subdivision, geodynamic driving mechanisms, and resource–genetic links, this Special Issue (affiliated with the section "Mineral Geochemistry and Geochronology") brings together ten original contributions spanning the full spectrum of Tethyan evolutionary stages and diverse tectonic settings, pushing the frontiers of our understanding of this iconic geological system.

2. Tethyan Subduction-Driven Magmatism and Tectonic Transition

Six contributions focus on magmatic responses and tectonic transformations across the Neo-Tethyan, Palaeo-Tethyan, and Meso-Tethyan stages. Yang et al. (Contribution 1) documented the first A1-type granite in central Lhasa's Meibaqieqin region (130 Ma), linking its crust–mantle mixed sources to lithospheric extension driven by northward Neo-Tethyan subduction. Hui et al. (Contribution 2) dated the East Kunlun Xingshugou A2-type monzogranite to 247.1 Ma, attributing its formation to the high-temperature partial melting of juvenile crust under extensional conditions triggered by Palaeo-Tethyan slab rollback. He et al. (Contribution 3) investigated Early–Middle Devonian volcanic rocks and granites in the Southern Beishan Orogenic Belt, revealing a tectonic transition from advancing to retreating subduction and constraining ocean–continent transformation in the southern Central Asian Orogenic Belt. Sahoo et al. (Contribution 4) studied amphibole-rich cumulates in the NE Himalayan Mayudia Ophiolite Complex, identifying an island arc root formed by fractional crystallization of super-hydrous magma (10.56–13.61 wt.% H_2O) from sub-arc mantle wedge melting, which constrains late Cretaceous–early Tertiary Neo-Tethyan closure processes. Xia et al. (Contribution 5) documented magma mixing in the North Qilian Menyuan Granodioritic Pluton (ca. 456 Ma), showing that interactions between mantle-derived mafic magma and ancient lower crustal felsic melts in an active continental arc triggered by Proto-Tethyan subduction. Bai et al. (Contribution 6) studied Early Silurian Shandan adakites in the North Qilian Orogen, constraining the 446–440 Ma collision between the Alxa and Qilian blocks and highlighting post-collisional crustal thickening and crust–mantle interactions. These studies collectively capture the diverse magmatic and tectonic dynamics across multiple Tethyan evolutionary stages.

3. Proto-Tethyan and Meso-Tethyan Closure-Related Magmatism

Three contributions address magmatic processes linked to Proto-Tethyan orogeny and Meso-Tethyan closure. Mo et al. (Contribution 7) reported ca. 118.3–120.5 Ma old rhyolites in the Tengchong Terrane, linking their calc–alkaline affinity and crust–mantle mixed sources to Meso-Tethyan subduction, slab rollback, and closure-related Burma-Tengchong terrane collision. Zhang et al. (Contribution 8) analyzed S-type granites from central Tibet's Guomang-Co area, identifying Paleoproterozoic metasedimentary sources and a syn-collisional setting and linking them to Early Paleozoic Andean-type orogeny and southward Proto-Tethyan subduction. Shan et al. (Contribution 9) focused on mafic intrusions in the eastern Himalayan Cuona area, dating diabase porphyrite to 135 Ma and identifying OIB-type geochemical signatures. Their $\varepsilon Nd(t)$ and $\varepsilon Hf(t)$ data link the intrusions to enriched mantle sources influenced by the Kerguelen mantle plume, providing critical constraints on Early Cretaceous East Gondwana rifting and mantle plume activity.

4. Neo-Tethyan Pre-Collisional Magmatism and India–Asia Collision Prelude

Wang et al. (Contribution 10) identified Mazin rhyolite porphyries (58.8–56.1 Ma) as the northernmost extension of the Linzizong volcanic suite, attributing their formation to juvenile lower crust remelting triggered by Neo-Tethyan slab rollback at the onset of the India–Asia collision. This study fills an important gap in understanding the magmatic response to early collisional tectonics in the Tibetan Plateau, bridging Neo-Tethyan closure and continental collision.

This Special Issue brings together research from leading teams worldwide, covering all major Tethyan stages and integrating diverse analytical techniques. The contributions advance key debates on subduction dynamics, crust–mantle interactions, and continental rifting/collision, while providing practical implications for resource exploration. We anticipate these findings will inspire further research into Tethyan evolution and deepen our understanding of Earth's dynamic history.

Acknowledgments: We would like to express our sincere gratitude to all the reviewers for their constructive comments and suggestions, which greatly improved the quality of the manuscripts included in this Special Issue. We also appreciate the professional support provided by the editorial team of *Minerals* throughout the organization and publication process.

Conflicts of Interest: The author declares no conflicts of interest.

List of Contributions:

1. Yang, Y.; Zhao, J.K.; Gao, K.; Zhang, Z.; Ding, S.; Gong, J.S.; Wu, J.Y.; Xu, P.Y.; Li, Y.X. The First Discovery of A1-Type Granite in the Meibaqieqin Region, Central Lhasa Terrane, Xizang. *Minerals* **2025**, *15*, 1093. https://doi.org/10.3390/min15101093.
2. Hui, C.; Sun, F.Y.; Bakht, S.; Yang, Y.Q.; Yan, J.M.; Yu, T.; Chen, X.S.; Zhang, Y.J.; Liu, C.X.; Zhu, X.R.; et al. Petrogenesis of an Anisian A2-Type Monzogranite from the East Kunlun Orogenic Belt, Northern Qinghai–Tibet Plateau. *Minerals* **2025**, *15*, 685. https://doi.org/10.3390/min15070685.
3. He, T.T.; Wang, Y.X.; Yan, J.; Yang, Z.Y.; Li, K.N.; Liu, Z.R.; Wang, Z.X.; Wu, L. The Orogeny Transition of the Southern Beishan Orogenic Belt During the Early–Middle Devonian: Evidence from the Wudaomingshui Volcanic Rocks and Granite. *Minerals* **2025**, *15*, 632. https://doi.org/10.3390/min15060632.

4. Sahoo, S.; Majumdar, A.S.; Anand, R.; Ray, D.; Fuenlabrada, J.M. Petrogenesis of Mafic–Ultramafic Cumulates in the Mayudia Ophiolite Complex, NE Himalaya: Evidence of an Island Arc Root in Eastern Neo-Tethys. *Minerals* **2025**, *15*, 572. https://doi.org/10.3390/min15060572.
5. Xia, S.G.; Qi, Y.; Yu, S.Y.; Jiang, X.C.; Gao, X.Y.; Wang, Y.; Li, C.Z.; Wang, Q.; Wang, L.T.; Peng, Y.B. Magma Mixing Origin for the Menyuan Granodioritic Pluton in the North Qilian Orogenic Belt, China. *Minerals* **2025**, *15*, 391. https://doi.org/10.3390/min15040391.
6. Bai, Z.H.; Yang, Y.; Liu, X.J.; Liu, P.D.; Chen, G.; Liu, X.; Hu, R.G.; Tian, H.; Liu, Y.D.; Huang, W.M.; et al. A Geochemical and Sr–Nd–Hf–O Isotopic Study of the Early Silurian Shandan Adakites in the Longshoushan Area: Implications for the Collisional Setting of the Proto–Tethyan North Qilian Orogen, Northwest China. *Minerals* **2025**, *15*, 352. https://doi.org/10.3390/min15040352.
7. Mo, X.; Gong, C.; Shang, Y.; Wu, J.L.; Wu, J.L.; Qi, R.H.; Wang, X.F.; Guan, Q.; Kong, X. Zircon LA-ICP-MS Dating and Geochemical Characteristics of Rhyolites from the Qushi Area, Tengchong Terrane, Yunnan Province. *Minerals* **2025**, *15*, 315. https://doi.org/10.3390/min15030315.
8. Zhang, Y.H.; Wang, M.; Yu, C.S.; Li, Z.L. S-Type Granites from the Guomang-Co Area in Central Tibet: A Response to Early Paleozoic Andean-Type Orogeny Along the Northern Margin of East Gondwana. *Minerals* **2025**, *15*, 284. https://doi.org/10.3390/min15030284.
9. Shan, C.X.; Kang, Z.Q.; Yang, F.; Ma, C.Y.; Qiao, Z.D.; Liu, Z.H.; Mu, J.Z.; Wu, L.L.; Zhou, L. Mafic Intrusions in the Cuona Area, Eastern Tethyan Himalaya: Early Kerguelen Mantle Plume Activity and East Gondwana Rifting. *Minerals* **2025**, *15*, 281. https://doi.org/10.3390/min15030281.
10. Wang, Z.S.; Wang, N.; Liu, Z.B.; Ma, X.D. The Northernmost Effects of the Neo-Tethys Oceanic Slab Subduction Under the Lhasa Terrane: Evidence from the Mazin Rhyolite Porphyry. *Minerals* **2024**, *14*, 1292. https://doi.org/10.3390/min14121292.

minerals

MDPI

Article

The First Discovery of A1-Type Granite in the Meibaqieqin Region, Central Lhasa Terrane, Xizang

Yi Yang [1], Junkang Zhao [2,*], Ke Gao [2,*], Zhi Zhang [3], Shuai Ding [2], Jiansheng Gong [2], Jianyang Wu [3], Peiyan Xu [3] and Yingxu Li [3]

1 College of Engineering, Tibet University, Lhasa 850000, China; z8642057@126.com
2 Zijin Mining Group Southwest Geological Exploration Co., Ltd., Chengdu 610059, China; ding_shuai1@zijinmining.com (S.D.); gong_jiansheng@zijimining.com (J.G.)
3 Chengdu Center, China Geological Survey (Geoscience Innovation Center of Southwest China), Chengdu 610218, China; zhangz0901@163.com (Z.Z.); wujianyang0208@163.com (J.W.); xu_py1991@163.com (P.X.); pzhliyingxu@126.com (Y.L.)
* Correspondence: zhaojunkang@163.com (J.Z.); gaokely@163.com (K.G.)

Abstract: This study documents the first A1-type granite identified on the southern margin of the central Lhasa terrane: a two-mica syenogranite pluton in the Meibaqieqin region. Because A-type granite provides sensitive records of crustal melting and lithospheric extension, this pluton offers important insights into magmatic processes and tectonic evolution along the southern margin of the Lhasa terrane. We analyzed two sample suites collected from different sites within the same pluton using zircon U–Pb geochronology and Hf isotopes, whole-rock geochemistry and Nd isotope. Zircon U–Pb weighted mean ages were 130.5 ± 0.7 Ma and 130.0 ± 0.7 Ma, placing emplacement in the Early Cretaceous. Zircon $\varepsilon_{Hf}(t)$ values ranged from −11.29 to −9.00 and −11.04 to −7.27, with two-stage Hf model ages (T_{DM2}) of 1.76–1.90 Ga and 1.65–1.89 Ga. Whole-rock $\varepsilon_{Nd}(t)$ values clustered between −11.77 and −11.36, yielding two-stage Nd model ages (T_{NdDM2}) of 1.85–1.88 Ga. Geochemically, the pluton is high-K calc-alkaline. These isotopic signatures indicate derivation predominantly from ancient crustal sources with a little mantle material. Chondrite-normalized REE patterns are overall right-inclined and display a V-shaped profile. Together with trace-element characteristics, these features support classification as A1-type granite. Regional comprehensive data suggest that pluton emplacement was controlled mainly by lithospheric extension related to northward subduction of the Neo-Tethyan oceanic plate, with a lesser contribution from southward subduction along the Bangongco–Nujiang suture. The source characteristics and geodynamic context differ markedly from A2-type granites on the northern margin of the central Lhasa terrane, which reflect distinct magmatic sources and tectonic regimes.

Keywords: geochemistry; Nd-Hf isotope; Zircon U-Pb; A1-type granite; Lhasa terrane

1. Introduction

The Lhasa terrane, a key component of the Tibetan Plateau, underwent northward subduction of the Neo-Tethyan oceanic crust and southward subduction of the Bangongco–Nujiang oceanic crust, followed by collisions with the Indian continent and the Qiangtang terrane [1–7]. These processes produced large-scale Mesozoic–Cenozoic magmatism, forming the Gangdese batholith [5]. The southern batholith is dominated by diorite, granodiorite, and porphyritic biotite granite, with subordinate biotite and two-mica granite, mostly emplaced between ca. 65 and 40 Ma [8]. The Lhasa terrane is subdivided into the north, central, and south Lhasa terranes, separated by the Shiquanhe–Namco ophiolite belt and the Luobadui–Milashan fault [4–6,9].

A-type granite, originally defined as alkaline, relatively anhydrous, and anorogenic [10], covering almost all granite types except typical I- and S-type granites [11,12], commonly forms in rift zone and within stable continental blocks [13,14]. It includes two subtypes: A1, typically associated with intraplate extension or rift zone, and A2, commonly linked to post-collisional plate-margin extension [15,16]. Because of their diagnostic geochemical and tectonic signatures, A-type granites are key to understanding extensional magmatism and regional tectonic evolution [10,15,17–19].

In the northern Lhasa terrane, numerous Early Cretaceous A-type granites have been identified, offering critical constraints on the evolution of the Bangongco–Nujiang Ocean and the Shiquanhe–Namco ophiolite belt. Examples include the Shenza biotite granite [20], the Zhaduding monzogranite in Gaer County (ca. 104 Ma [21]), the Asa porphyritic granite (ca. 107–106 Ma [22]), the Qusanggele granite in Nima County (ca. 101 Ma [23]), and the Kongcuo granite (ca. 105 Ma [24,25]) (Figure 1b). However, these plutons are largely confined to the North Lhasa terrane and the northern margin of the central Lhasa terrane. To date, no A-type granite has been reported along the southern margin of the central Lhasa terrane. Previous studies attribute the northern A-type magmatism mainly to southward subduction of the Bangongco–Nujiang oceanic crust and subsequent post-collisional extension [20,22,23]. In contrast, few extensional granites have been documented in the southern Lhasa terrane, and the petrogenesis and tectonic setting of any A-type granites there remain poorly constrained.

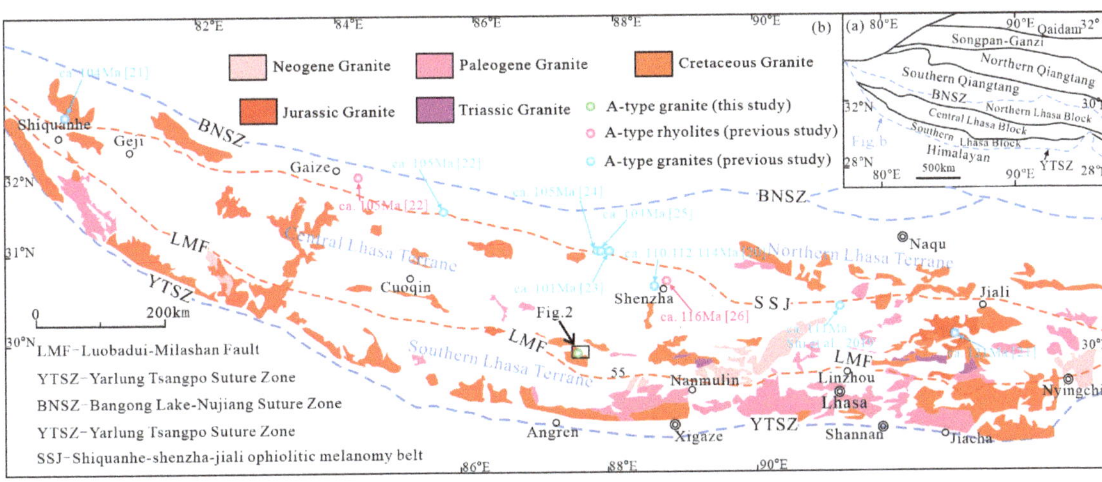

Figure 1. Tectonic Map of the Tibetan Plateau (**a**) and pluton distribution sketch (**b**), Lhasa terrane, Xizang. Age data of A-type granite and rhyolites from [20–26]. Figure 2 illustrates the tectonic location of the geological map of the study area.

During the past two years of geological surveys, we identified an A1-type two-mica syenogranite along the southern margin of the central Lhasa terrane. This study integrates field mapping, petrography, geochronology, whole-rock geochemistry, and isotope analyses to investigate its petrogenesis, source characteristics, and tectonic setting. As the first reported A1-type granite from the central Lhasa terrane, our results provide new evidence for Early Cretaceous magmatism and geodynamics in the region.

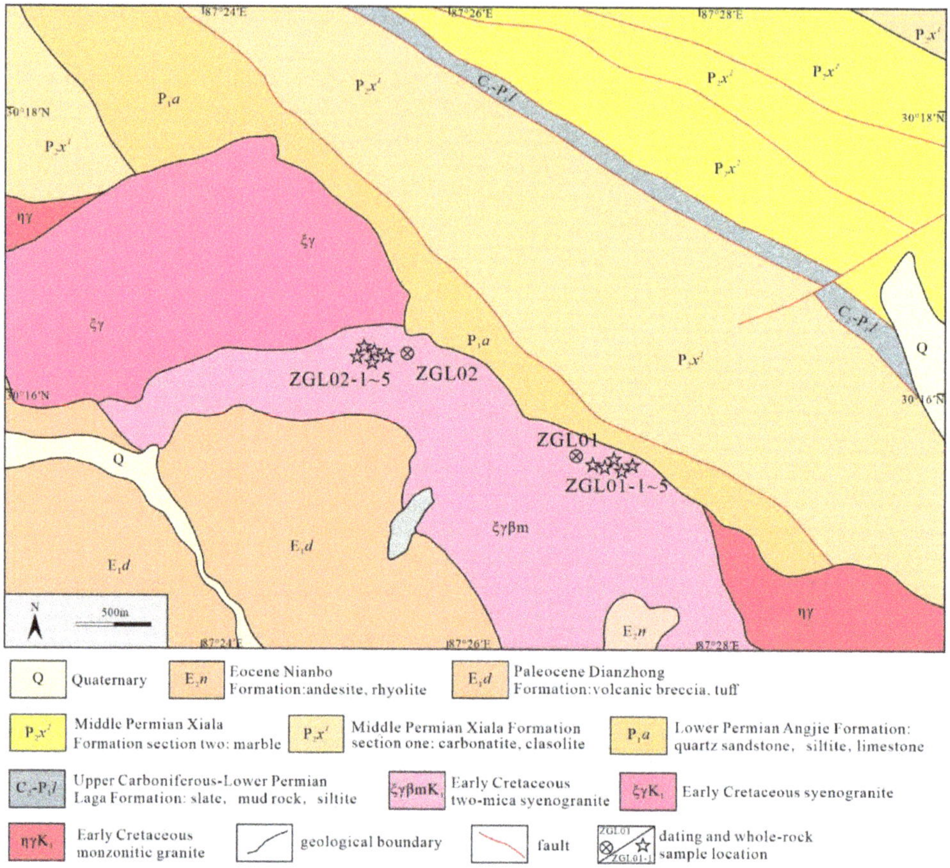

Figure 2. Geological map in the Meibaqieqin region, central Lhasa terrane.

2. Geological Background

The Lhasa terrane is situated between the Yarlung Tsangpo suture zone and the Bangongco-Nujiang suture zone, formed by the northward subduction of the Yarlung Tsangpo oceanic crust, the southward subduction of the Bangongco-Nujiang oceanic crust, and subsequent collision processes. Extensive tectonic evolution of the Lhasa Terrane driven by east–west–trending internal structures and subsidiary faults has been documented by numerous studies [2,27]. Building on the discovery of oceanic subduction–type eclogites in the Sumdo area, Yang et al. (2006) partitioned the terrane into the southern and northern Lhasa terranes [28]. Zhu et al. (2011) subsequently refined this framework, recognizing three regions—the Northern Lhasa terrane, Central Lhasa terrane, and Southern Lhasa terrane—on the basis of variations in the cover and basement, with boundaries defined by the Shiquanhe–Shenzha-Jiali ophiolitic belt (SSJ) and the Luobadui–Milashan fault (LMF) [5]. This study focuses on the southern part of the central Lhasa terrane (Figure 1b).

The study area lies north of Meibaqieqin Township, Angren County, Shigatse City, immediately north of the Luobadui–Milashan fault belt, within the southern margin of the Central Lhasa terrane. The stratum includes: Carboniferous Yongzhu Formation, Upper Carboniferous–Lower Permian Laga Formation, Lower Permian Angjie Formation, Middle Permian Xiala Formation, Paleocene Dianzhong Formation, Eocene Nianbo Formation, Pliocene Wuyu Group, Quaternary deposits.

Magmatic rocks are widespread in the study area and are dominated by intrusions, with a small amount of volcanics. Magmatism spans three main stages—Early Cretaceous, Late Cretaceous, and Paleogene—with the Cretaceous representing the most intense phase and dominated by felsic compositions. Volcanism is best developed in the Paleogene Linzizong Group, followed by the Neogene Wuyu Group. The authors conducted mapping through conventional geological surveys and drew the geological map of the study area.

3. Petrology

The two-mica syenogranite that intrudes the Lower Permian Angjie Formation and is well exposed. The upper part of the pluton is overlain by volcanic strata of the Paleocene Dianzhong and Eocene Nianbo formations (Figure 2). The two-mica syenogranite is in an intrusive contact relationship with syenogranite and monzonitic granite. The exposed area of the pluton is approximately 42 km^2. Two sample sets were collected from the same pluton and show similar petrography. The main mineral assemblage comprises quartz (24%–27%), K-feldspar (50%–53%), plagioclase (13%–15%), muscovite (4%–6%), and biotite (3%–5%), with accessory zircon, monazite, and apatite (Figure 3). K-feldspar locally replaces plagioclase, preserving plagioclase relicts enclosed within K-feldspar. Figure 3a,b show that the mineral composition mainly consists of K-feldspar and plagioclase, with some plagioclase being replaced by K-feldspar; quartz grains mainly exhibit subhedral structure; biotite and muscovite display alteration transitional distribution with K-feldspar.

Figure 3. Hand specimens (**a**,**b**) and microphotograph (**c**,**d**) of the two-mica syenogranite in the Meibaqieqin region, central Lhasa terrane. Qtz-Quartz, Bt-Biotite, Kfs-k-feldspar, Pl-Plagioclase Ms-muscovite. Sample locality for photo (**a**,**c**) is E 87°26′22″ and 30°16′05″ (ZGL01); Sample locality for photo b, d is E 87°25′34″ and 30°16′23″ (ZGL02).

4. Results

4.1. Whole-Rock Geochemistry

Two groups of samples were collected from the same pluton representing two-mica syenogranites, showing consistent petrogeochemical features (Table 1). See Appendix A for the analytical methods. Loss on ignition (LOI) is 0.54–1.51 wt.%, indicating weak alteration and reliable geochemical data. The rocks have SiO_2 = 73.97–76.97 wt.%, $K_2O + Na_2O$ = 8.12–8.72 wt.%, and Al_2O_3 = 12.00–13.64 wt.%, i.e., high-silica, alkali-rich, and relatively low Al. In the TAS diagram, all samples plot in the granite domain and are subalkaline [29] (Figure 4a). The aluminum saturation index (A/CNK) is 0.95–1.14; in the A/CNK–A/NK diagram, samples fall in the metaluminous to peraluminous domain [30] (Figure 4b). The Rittmann index (σ) is 1.97–2.38 (<3.3), consistent with calc-alkaline affinity; the SiO_2–K_2O plot places all samples in the high-K calc-alkaline field [31] (Figure 4c).

Table 1. Major (wt.%), trace and TREE (μg/g) element data of the two-mica syenogranite in the Meibaqieqin region.

Sample	ZGL01-1	ZGL01-2	ZGL01-3	ZGL01-4	ZGL01-5	ZGL02-1	ZGL02-2	ZGL02-3	ZGL02-4	ZGL02-5
SiO_2	73.97	76.37	75.27	76.97	75.88	76.46	75.23	74.84	74.84	76.06
Al_2O_3	12.60	12.00	12.87	12.02	12.11	12.27	13.64	12.99	13.13	12.32
$^{T}Fe_2O_3$	1.26	0.51	1.24	0.37	0.54	1.09	0.65	1.10	0.75	0.99
CaO	1.17	1.06	0.57	0.92	1.00	0.44	0.23	0.32	0.27	0.39
MgO	0.18	0.18	0.14	0.09	0.11	0.13	0.14	0.21	0.13	0.14
K_2O	5.29	5.09	5.27	5.19	5.09	5.00	4.71	5.21	5.19	5.01
Na_2O	3.33	3.03	3.14	3.14	3.03	3.27	3.95	3.21	3.53	3.24
P_2O_5	0.01	0.01	0.02	0.01	0.01	0.01	0.01	0.02	0.01	0.01
MnO	0.05	0.02	0.05	0.02	0.02	0.02	0.04	0.03	0.05	0.02
TiO_2	0.14	0.12	0.15	0.12	0.12	0.12	0.03	0.13	0.09	0.12
LOI	1.51	1.09	0.57	0.77	1.24	0.77	0.58	0.94	0.55	0.54
$K_2O + Na_2O$	8.63	8.12	8.41	8.33	8.12	8.27	8.66	8.42	8.72	8.25
DI	93.38	94.23	94.39	95.33	94.28	95.56	96.01	95.1	96.1	95.7
σ	2.38	1.97	2.18	2.03	1.99	2.04	2.31	2.21	2.37	2.05
10,000 Ga/Al	3.69	3.89	3.53	4.05	3.83	3.57	4.63	3.88	4.51	3.99
A/NK	1.12	1.14	1.18	1.12	1.15	1.14	1.18	1.19	1.15	1.15
A/CNK	0.95	0.97	1.08	0.97	0.99	1.06	1.14	1.14	1.1	1.08
Sr	21.6	37.8	31.3	30.1	34.9	14.9	14.9	22.3	14.4	15.4
Ba	67.6	128	112	66.2	84.2	69.0	46.6	64.1	40.0	55.2
V	2.45	7.16	6.96	1.46	1.38	5.26	4.92	7.78	4.60	4.24
Zr	160	158	175	161	166	171	76.0	141	133	158
As	7.18	3.09	8.52	1.60	3.65	0.63	0.98	1.47	0.93	0.35
Li	17.6	9.38	41.4	8.34	8.35	46.1	129	58.0	155	42.5
Be	3.65	3.62	3.52	3.76	3.44	3.57	5.05	5.73	5.05	4.38
Sc	1.94	3.47	2.43	4.06	3.25	2.89	7.62	4.41	8.02	2.86
Cr	2.41	3.86	3.65	1.13	1.39	1.15	2.38	7.01	1.39	1.09
Co	0.60	0.61	0.87	0.27	0.72	0.17	0.36	0.84	0.39	0.79
Ni	1.26	1.61	2.05	0.40	0.59	0.47	0.86	3.17	0.40	0.29
Cu	7.32	13.8	141	4.85	4.35	1.89	2.97	5.25	21.5	1.75
Zn	31.4	26.3	39.0	21.5	25.7	30.3	22.8	36.7	28.2	27.5
Ga	24.6	24.7	24.0	25.8	24.5	23.2	33.4	26.7	31.3	26.0
Rb	336	514	341	537	445	395	791	491	847	403
Y	22.6	43.5	24.7	48.3	43.4	35.5	23.2	32.7	29.7	33.9
Nb	36.4	43.0	34.5	50.6	43.6	36.2	67.4	34.3	48.0	36.8
Cs	4.21	5.37	5.34	5.55	4.72	5.04	12.1	9.45	14.00	5.28
La	22.5	59.1	19.1	60.9	62.0	34.9	30.9	54.4	51.1	45.0
Ce	93.6	102	94.0	106.8	110	91.1	62.9	106	105	104
Pr	4.71	9.76	3.94	10.3	10.2	7.63	5.47	10.4	9.55	9.98
Nd	15.2	28.9	12.6	30.1	30.7	24.7	15.1	32.3	28.4	32.3
Sm	3.23	5.44	2.75	5.63	5.70	4.57	2.52	5.64	4.87	5.80
Eu	0.34	0.30	0.35	0.28	0.28	0.34	0.17	0.42	0.29	0.39
Gd	2.89	5.24	2.62	5.39	5.31	3.83	2.16	4.58	3.90	4.51
Tb	0.53	0.86	0.58	0.88	0.88	0.68	0.39	0.76	0.63	0.71
Dy	3.63	5.26	3.66	5.56	5.55	4.73	2.87	4.80	4.14	4.44
Ho	0.98	1.31	0.95	1.43	1.38	1.25	0.83	1.18	1.08	1.15
Er	3.31	4.57	3.50	5.14	5.00	4.75	3.48	4.44	4.22	4.33
Tm	0.54	0.75	0.59	0.85	0.78	0.82	0.70	0.72	0.77	0.70
Yb	3.55	5.14	4.80	5.85	5.42	5.55	5.55	4.90	5.66	5.03
Lu	0.57	0.84	0.63	1.08	0.89	0.87	0.96	0.77	0.96	0.75
Ta	3.94	4.87	4.11	5.71	4.40	4.01	9.99	4.34	5.86	4.24
Pb	30.1	30.0	50.0	29.3	36.9	28.4	34.0	30.6	39.9	27.8
Th	56.3	86.1	52.8	84.4	74.7	56.4	28.2	49.1	46.5	67.7
U	3.00	10.0	3.00	18.7	23.0	4.35	5.30	4.45	6.30	3.51
Hf	4.03	3.71	4.44	4.15	4.07	4.51	2.87	3.52	3.84	4.13
Rb/Sr	15.60	13.60	10.91	17.82	12.76	26.59	52.99	21.97	59.03	26.22
Y/Nb	0.62	1.01	0.71	0.95	0.99	0.98	0.34	0.95	0.62	0.92
Rb/Nb	9.24	11.97	9.88	10.59	10.21	10.92	11.75	14.30	17.66	10.95

Table 1. *Cont.*

Sample	ZGL01-1	ZGL01-2	ZGL01-3	ZGL01-4	ZGL01-5	ZGL02-1	ZGL02-2	ZGL02-3	ZGL02-4	ZGL02 5
Ce/Nb	2.57	2.37	2.72	2.11	2.52	2.52	0.93	3.10	2.18	2.83
Nb/Ta	9.24	8.82	8.39	8.87	9.92	9.02	6.74	7.90	8.18	8.68
Ce/Pb	3.11	3.40	1.88	3.65	2.98	3.21	1.85	3.46	2.62	3.74
Zr/Nb	4.41	3.68	5.05	3.19	3.80	4.73	1.13	4.11	2.77	4.30
La/Nb	0.62	1.37	0.55	1.20	1.42	0.96	0.46	1.59	1.06	1.22
Th/Ta	14.3	17.66	12.83	14.77	17.00	14.06	2.82	11.31	7.93	15.96
La/Sm	6.95	10.87	6.95	10.82	10.89	7.63	12.24	9.64	10.49	7.76
La/Yb	6.32	11.50	3.98	10.42	11.45	6.29	5.57	11.10	9.03	8.93
Ce/Ce*	2.23	1.04	2.66	1.05	1.07	1.37	1.19	1.09	1.16	1.20
Eu/Eu*	0.34	0.17	0.40	0.16	0.16	0.25	0.22	0.25	0.20	0.23
ΣREE	155.5	229.4	150.2	240.16	244.25	185.7	134	231.4	220.2	219.1
LREE/HREE	8.72	8.57	7.67	8.17	8.69	7.26	6.91	9.44	9.31	9.14
La_N/Yb_N	4.54	8.25	2.86	7.47	8.21	4.51	3.99	7.96	6.48	6.41

Notes: $Eu/Eu^* = 2 \times Eu_N/(Sm_N + Gd_N)$; $Ce/Ce^* = 2 \times Ce_N/(La_N + Pr_N)$; $\sigma = (Na_2O + K_2O)^2/(SiO_2 - 43)$ (wt.%); Differentiation Index (DI) = Qz + Or + Ab + Ne + Lc + Kp.

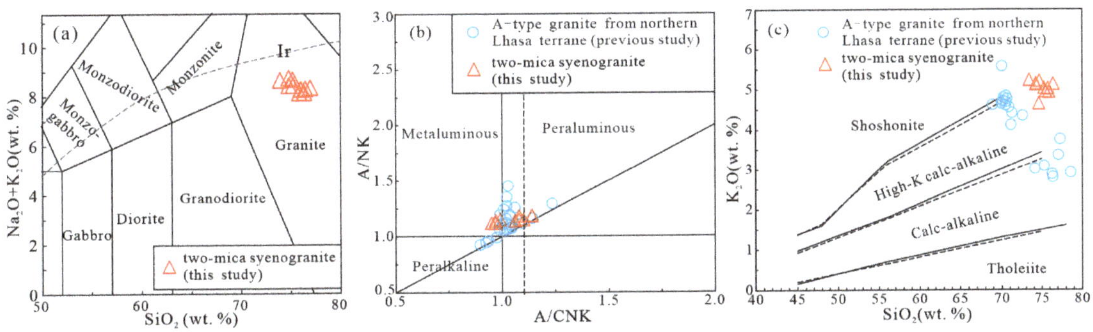

Figure 4. TAS diagram ((**a**), after [29]), A/NK vs. A/CNK diagram ((**b**), after [30]), K_2O vs. SiO_2 diagram ((**c**), after [31,32]) of the two-mica syenogranite in the Meibaqieqin region. A-type granite from central Lhasa terrane from [20–26,33].

Total rare earth elements are 134–244 μg/g, with LREE/HREE = 6.91–9.44 and La_N/Yb_N = 2.86–8.21. Chondrite-normalized REE patterns show LREE enrichment, obviously negative Eu anomalies (Eu/Eu* = 0.16–0.40), and strong LREE–HREE fractionation (Figure 5a). Primitive mantle-normalized spidergram shows enrichment in large ion lithophile elements (e.g., Rb, Zr, Th, U, Hf) and depletion in high field strength elements (e.g., Sr, Ba, Ti, and P), consistent with A-type granite signatures (Figure 5b). In addition, the Rb/Sr ratios of the two-mica syenogranite vary from 10.9 to 53, with average of 25.8.

4.2. Whole-Rock Nd Isotopes

Eight Nd isotope analyses yielded $\varepsilon_{Nd}(t)$ values of −11.77 to −11.36 at 130 Ma, with $(^{143}Nd/^{144}Nd)_i$ = 0.511867–0.511888 and two-stage model ages (T_{NdDM2}) of 1.85–1.88 Ga (Table 2).

Table 2. Nd isotopic data of the two-mica syenogranite in the Meibaqieqin region.

Sample	Sm (μg/g)	Nd (μg/g)	$^{147}Sm/^{144}Nd$	$^{143}Nd/^{144}Nd$	$(^{143}Nd/^{144}Nd)_i$	$\varepsilon_{Nd}(0)$	$\varepsilon_{Nd}(t)$	$f_{Sm/Nd}$	T_{DM}	T_{NdDM2}
ZGL01-1	3.23	15.2	0.129	0.511978	0.511868	−12.87	−11.75	−0.34	2106	1879
ZGL01-2	5.44	28.9	0.1137	0.51197	0.511873	−13.03	−11.65	−0.42	1797	1872
ZGL01-3	2.75	12.6	0.1317	0.512	0.511888	−12.45	−11.36	−0.33	2134	1847
ZGL01-4	5.63	30.1	0.1131	0.511967	0.51187	−13.09	−11.71	−0.43	1791	1876
ZGL02-1	4.57	24.7	0.112	0.511966	0.51187	−13.11	−11.71	−0.43	1773	1876
ZGL02-2	2.52	15.1	0.1011	0.511959	0.511873	−13.25	−11.65	−0.49	1612	1874
ZGL02-3	5.64	32.3	0.1057	0.511975	0.511885	−12.93	−11.42	−0.46	1658	1855
ZGL02-4	4.87	28.4	0.1038	0.511956	0.511867	−13.3	−11.77	−0.47	1655	1881

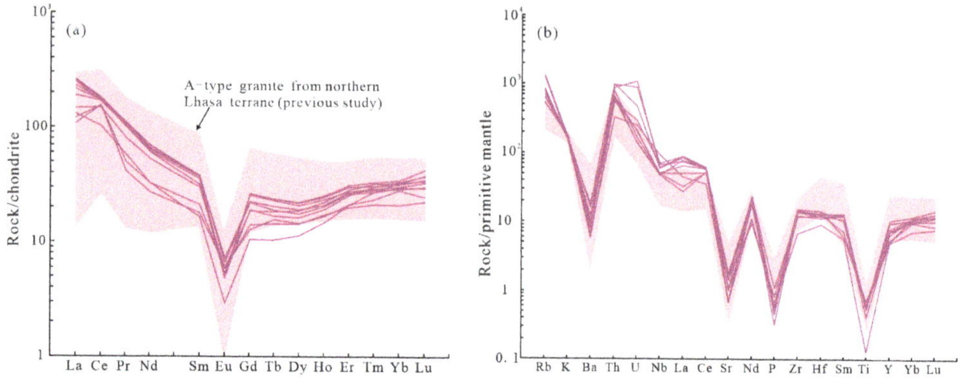

Figure 5. Chondrite-normalized REE patterns (**a**) and primitive mantle-normalized trace element patterns (**b**) of the two-mica syenogranite in the Meibaqieqin region [34]. A-type granite from northern Lhasa terrane from [24,25].

4.3. U-Pb Zircon Geochronology

Zircon U-Pb dating was conducted on two two-mica syenogranite samples (ZGL01 and ZGL02; Table 3). Most zircons are short prismatic grains, 90–210 μm long and 60–130 μm wide, with aspect ratios of 1–2. Cathodoluminescence images show clear oscillatory zoning, consistent with magmatic zircon [35] (Figure 6). Zircons from the two-mica syenogranite exhibit high U and Th contents, and the U-Pb dating results show good concordance, differing from other types of zircon. There are a few detrital zircons in this study. The zircon cores appear dark gray in CL images with relatively low transparency. The margins of the zircons show obvious corrosion rims with embayed morphology. This indicates that the cores underwent thermal metamorphism and metasomatism under magmatic processes. Fractures are well-developed.

Table 3. LA-ICP-MS U-Pb data for zircon from the two-mica syenogranite in the Meibaqieqin region, central Lhasa terrane.

Spot	Pb (μg/g)	Th (μg/g)	U (μg/g)	Th/U	$^{207}Pb/^{206}Pb$ Ratio	1σ	$^{207}Pb/^{235}U$ Ratio	1σ	$^{206}Pb/^{238}U$ Ratio	1σ	$^{207}Pb/^{235}U$ Age	1σ	$^{206}Pb/^{238}U$ Age	1σ	Concordance
ZGL01-1	72	972	716	1.36	0.0484	0.0035	0.1370	0.0087	0.0205	0.0004	130.4	7.8	131.1	2.8	99%
ZGL01-2	33	440	316	1.39	0.0555	0.0060	0.1529	0.0135	0.0206	0.0004	144.5	11.9	131.3	2.4	90%
ZGL01-3	65	915	751	1.22	0.0501	0.0034	0.1408	0.0087	0.0205	0.0003	133.8	7.7	130.8	1.8	97%
ZGL01-4	52	685	510	1.34	0.0467	0.0038	0.1325	0.0090	0.0204	0.0002	126.4	8.1	130.1	1.5	97%
ZGL01-5	169	2334	1833	1.27	0.0492	0.0026	0.1382	0.0068	0.0204	0.0002	131.5	6.1	129.9	1.3	98%
ZGL01-6	83	1090	991	1.10	0.0491	0.0033	0.1379	0.0086	0.0203	0.0003	131.1	7.7	129.8	1.6	98%
ZGL01-7	34	486	352	1.38	0.0492	0.0055	0.1362	0.0137	0.0203	0.0003	129.6	12.2	129.8	1.7	99%
ZGL01-8	40	311	797	0.39	0.0521	0.0037	0.1513	0.0108	0.0208	0.0003	143.1	9.6	132.6	2.2	92%
ZGL01-9	58	822	609	1.35	0.0503	0.0044	0.1395	0.0103	0.0206	0.0004	132.6	9.2	131.2	2.3	98%
ZGL01-10	55	731	563	1.30	0.0470	0.0041	0.1330	0.0103	0.0205	0.0002	126.8	9.2	130.7	1.3	96%
ZGL01-11	51	759	508	1.49	0.0474	0.0043	0.1315	0.0106	0.0204	0.0003	125.5	9.5	130.0	1.8	96%
ZGL01-12	51	739	480	1.54	0.0517	0.0042	0.1465	0.0104	0.0206	0.0003	138.8	9.2	131.1	1.6	94%
ZGL01-13	50	707	484	1.46	0.0486	0.0046	0.1364	0.0120	0.0205	0.0003	129.8	10.7	130.9	2.1	99%
ZGL01-14	69	943	814	1.16	0.0473	0.0038	0.1330	0.0100	0.0204	0.0002	126.8	9.0	130.4	1.4	97%
ZGL01-15	66	926	705	1.31	0.0490	0.0043	0.1378	0.0110	0.0205	0.0003	131.0	9.8	131.0	1.6	99%
ZGL01-16	52	775	418	1.85	0.0506	0.0053	0.1425	0.0147	0.0205	0.0003	135.2	13.1	130.5	1.7	96%
ZGL01-17	95	1262	1164	1.08	0.0463	0.0029	0.1314	0.0082	0.0204	0.0002	125.4	7.4	130.4	1.5	96%
ZGL01-18	32	438	328	1.34	0.0539	0.0058	0.1513	0.0133	0.0206	0.0003	143.1	11.8	131.4	1.7	91%
ZGL01-19	60	828	712	1.16	0.0527	0.0040	0.1461	0.0097	0.0204	0.0002	138.5	8.6	129.9	1.5	93%
ZGL01-20	50	678	509	1.33	0.0506	0.0052	0.1426	0.0144	0.0204	0.0003	135.3	12.8	130.4	1.7	96%
ZGL01-21	28	419	285	1.47	0.0515	0.0064	0.1409	0.0156	0.0204	0.0003	133.8	13.9	130.2	1.9	97%
ZGL01-22	61	862	598	1.44	0.0521	0.0042	0.1480	0.0112	0.0206	0.0003	140.1	9.9	131.1	1.8	93%
ZGL01-23	49	665	537	1.24	0.0482	0.0041	0.1377	0.0110	0.0204	0.0002	131.0	9.8	130.1	1.5	99%
ZGL01-24	50	617	702	0.88	0.0503	0.0038	0.1441	0.0103	0.0205	0.0003	136.7	9.1	130.9	1.9	95%
ZGL02-1	56	833	582	1.43	0.0477	0.0034	0.1354	0.0086	0.0205	0.0003	128.9	7.7	130.6	1.8	98%

Table 3. *Cont.*

Spot	Pb (μg/g)	Th (μg/g)	U (μg/g)	Th/U	$^{207}Pb/^{206}Pb$ Ratio	$^{207}Pb/^{206}Pb$ 1σ	$^{207}Pb/^{235}U$ Ratio	$^{207}Pb/^{235}U$ 1σ	$^{206}Pb/^{238}U$ Ratio	$^{206}Pb/^{238}U$ 1σ	$^{207}Pb/^{235}U$ Age	$^{207}Pb/^{235}U$ 1σ	$^{206}Pb/^{238}U$ Age	$^{206}Pb/^{238}U$ 1σ	Concordance
ZGL02-2	36	506	422	1.20	0.0449	0.0042	0.1264	0.0109	0.0205	0.0003	120.8	9.9	130.5	1.8	92%
ZGL02-3	35	505	367	1.38	0.0516	0.0046	0.1459	0.0120	0.0205	0.0003	138.3	10.6	130.8	1.8	94%
ZGL02-4	36	546	414	1.32	0.0474	0.0040	0.1289	0.0088	0.0202	0.0004	123.1	7.9	129.1	2.4	95%
ZGL02-5	28	377	366	1.03	0.0514	0.0045	0.1459	0.0120	0.0205	0.0003	138.3	10.6	131.0	2.0	94%
ZGL02-6	78	1136	786	1.45	0.0484	0.0030	0.1376	0.0087	0.0203	0.0003	130.9	7.7	129.6	1.7	99%
ZGL02-7	26	377	298	1.27	0.0512	0.0049	0.1420	0.0120	0.0205	0.0004	134.8	10.6	130.5	2.2	96%
ZGL02-8	37	519	425	1.22	0.0513	0.0042	0.1426	0.0102	0.0204	0.0003	135.4	9.0	130.1	2.0	95%
ZGL02-9	52	747	589	1.27	0.0490	0.0036	0.1384	0.0099	0.0204	0.0003	131.6	8.9	130.4	1.8	99%
ZGL02-10	37	630	255	2.47	0.0467	0.0053	0.1290	0.0128	0.0201	0.0003	123.2	11.5	128.1	1.8	96%
ZGL02-11	30	421	340	1.24	0.0447	0.0041	0.1253	0.0101	0.0203	0.0003	119.9	9.1	129.6	1.6	92%
ZGL02-12	74	1159	751	1.54	0.0508	0.0034	0.1422	0.0094	0.0203	0.0003	135.0	8.4	129.7	2.0	95%
ZGL02-13	28	421	300	1.40	0.0530	0.0057	0.1515	0.0147	0.0206	0.0003	143.3	12.9	131.1	2.0	91%
ZGL02-15	77	1174	810	1.45	0.0455	0.0033	0.1273	0.0086	0.0202	0.0003	121.7	7.8	129.2	1.9	94%
ZGL02-16	28	761	554	1.37	0.0441	0.0042	0.1249	0.0113	0.0203	0.0003	119.5	10.2	129.3	1.8	92%
ZGL02-18	87	1223	969	1.26	0.0467	0.0035	0.1304	0.0093	0.0203	0.0002	124.5	8.3	129.3	1.4	96%
ZGL02-19	43	614	430	1.43	0.0493	0.0044	0.1379	0.0112	0.0203	0.0002	131.2	10.0	129.4	1.5	98%
ZGL02-20	48	679	477	1.42	0.0542	0.0045	0.1528	0.0112	0.0206	0.0003	144.4	9.9	131.4	2.1	90%
ZGL02-21	31	419	322	1.30	0.0520	0.0047	0.1459	0.0112	0.0204	0.0003	138.3	9.9	130.0	1.9	93%
ZGL02-22	92	1428	613	2.33	0.0490	0.0040	0.1394	0.0105	0.0204	0.0002	132.5	9.4	130.2	1.4	98%
ZGL02-23	49	665	524	1.27	0.0543	0.0058	0.1486	0.0139	0.0205	0.0003	140.6	12.3	130.6	2.1	92%
ZGL02-24	49	595	672	0.89	0.0512	0.0039	0.1453	0.0105	0.0206	0.0003	137.8	9.3	131.1	1.7	95%
ZGL02-25	73	1069	603	1.77	0.0470	0.0042	0.1318	0.0110	0.0204	0.0003	125.7	9.8	130.3	1.8	96%
ZGL02-26	35	513	370	1.39	0.0483	0.0046	0.1377	0.0115	0.0204	0.0003	131.0	10.3	129.9	1.9	99%
ZGL02-28	39	570	346	1.65	0.0497	0.0051	0.1404	0.0131	0.0203	0.0003	133.4	11.6	129.7	2.0	97%

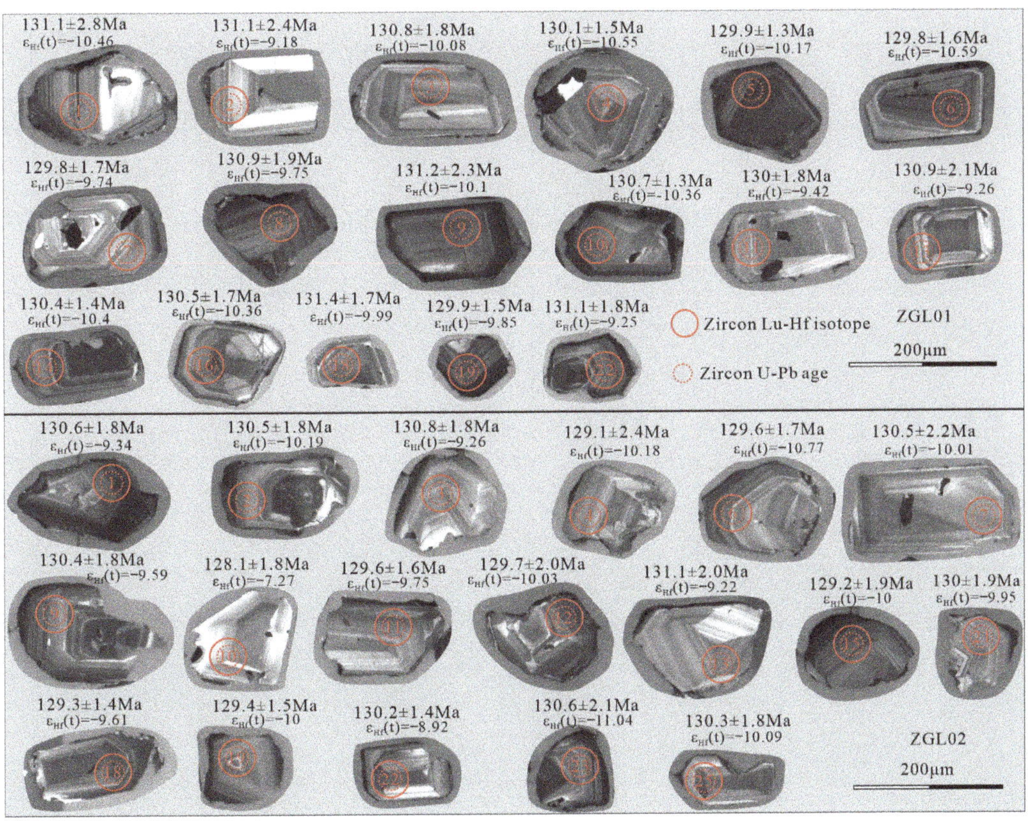

Figure 6. Zircons CL images of the two-mica syenogranite in the Meibaqieqin region.

A total of 24 zircon U-Pb isotope data points were obtained from sample ZGL01, with $^{206}Pb/^{238}U$ ages concentrated between ca. 129.8~132.6 Ma. On the U-Pb age concordia diagram, the data points distribute near the concordia line, yielding a weighted mean age of 130.5 ± 0.7 Ma (MSWD = 1.3) for this sample (Figure 7). Sample ZGL02 yielded 25 U-Pb $^{206}Pb/^{238}U$ ages ranging between 128.1 and 131.4 Ma, and a weighted mean age of 130 ± 0.7 Ma (MSWD = 0.17). These two age values accurately represent the crystallization age of this pluton, indicating large-scale magmatic activity during the Cretaceous in this region.

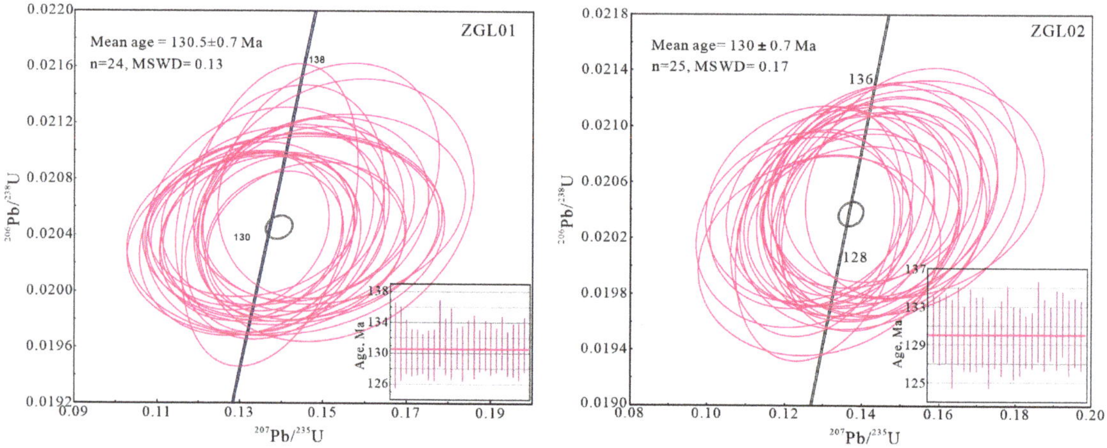

Figure 7. U-Pb concordia diagrams and weighted mean $^{206}Pb/^{238}Pb$ age diagrams for the two-mica syenogranite in the Meibaqieqin region.

4.4. Zircon Hf Isotopes

The zircon crystallization ages were used to calculate $\varepsilon_{Hf}(t)$ values and two-stage model ages. Zircon Hf isotopes were measured at 22 and 23 spots from samples ZGL01 and ZGL02, respectively. Hf spots coincide with the U-Pb analytical sites (Table 4), and both samples show similar isotopic compositions. In total, 45 spots yield initial $^{176}Hf/^{177}Hf$ ratios of 0.282382–0.282446. For ZGL01, $\varepsilon_{Hf}(0)$ values range from −13.79 to −11.53 and $\varepsilon_{Hf}(t)$ from −11.29 to −9.03; single-stage Hf model ages (T_{DM}) are 1.18–1.34 Ga and two-stage model ages (T_{DM2}) are 1.76–1.90 Ga. For ZGL02, $\varepsilon_{Hf}(0)$ ranges from −13.67 to −9.83 and $\varepsilon_{Hf}(t)$ from −11.04 to −7.27; T_{DM} values are 1.13–1.34 Ga and T_{DM2} values are 1.65–1.89 Ga.

Table 4. Hf isotopic compositions of zircons from the two-mica syenogranite in the Meibaqieqin region, central Lhasa terrane.

Spot	$^{176}Yb/^{177}Hf$	1σ	$^{176}Lu/^{177}Hf$	1σ	$^{176}Hf/^{177}Hf$	1σ	Age (Ma)	I_{Hf}	$\varepsilon_{Hf}(0)$	$\varepsilon_{Hf}(t)$	T_{DM}	T_{DM2}	$f_{Lu/Hf}$
ZGL01-1	0.064941	0.000577	0.001670	0.000018	0.282399	0.000013	131.1	0.282395	−13.18	−10.47	1227	1852	−0.95
ZGL01-2	0.086168	0.002356	0.001883	0.000035	0.282436	0.000015	131.3	0.282432	−11.88	−9.16	1181	1770	−0.94
ZGL01-3	0.102841	0.000464	0.002323	0.000018	0.282412	0.000016	130.8	0.282407	−12.72	−10.05	1229	1824	−0.93
ZGL01-4	0.089786	0.000629	0.002247	0.000024	0.282398	0.000014	130.1	0.282392	−13.23	−10.58	1248	1857	−0.93
ZGL01-5	0.097762	0.000681	0.002491	0.000008	0.282410	0.000013	129.9	0.282404	−12.80	−10.15	1238	1830	−0.92
ZGL01-6	0.085560	0.000898	0.002112	0.000002	0.282397	0.000016	129.8	0.282392	−13.26	−10.58	1244	1857	−0.94
ZGL01-7	0.051677	0.001096	0.001266	0.000019	0.282419	0.000014	129.8	0.282416	−12.48	−9.73	1185	1804	−0.96
ZGL01-8	0.094033	0.000602	0.002310	0.000015	0.282420	0.000013	132.6	0.282415	−12.44	−9.76	1217	1806	−0.93
ZGL01-9	0.081256	0.000396	0.002072	0.000002	0.282410	0.000013	131.2	0.282405	−12.79	−10.12	1224	1828	−0.94
ZGL01-10	0.072825	0.000320	0.001833	0.000014	0.282402	0.000014	130.7	0.282398	−13.07	−10.37	1227	1844	−0.94
ZGL01-11	0.090999	0.001807	0.002199	0.000051	0.282430	0.000015	130.0	0.282425	−12.08	−9.41	1199	1784	−0.93
ZGL01-12	0.089261	0.001538	0.002045	0.000031	0.282423	0.000012	131.1	0.282418	−12.33	−9.66	1204	1798	−0.94
ZGL01-13	0.097669	0.000540	0.002185	0.000020	0.282434	0.000015	130.9	0.282429	−11.95	−9.27	1193	1775	−0.93
ZGL01-14	0.076826	0.000325	0.001818	0.000008	0.282401	0.000013	130.4	0.282396	−13.14	−10.44	1230	1849	−0.95
ZGL01-15	0.072102	0.000319	0.001788	0.000010	0.282432	0.000014	131.0	0.282428	−12.02	−9.30	1183	1777	−0.95

Table 4. *Cont.*

Spot	$^{176}Yb/^{177}Hf$	1σ	$^{176}Lu/^{177}Hf$	1σ	$^{176}Hf/^{177}Hf$	1σ	Age (Ma)	I_{Hf}	$\varepsilon_{Hf}(0)$	$\varepsilon_{Hf}(t)$	T_{DM}	T_{DM2}	$f_{Lu/Hf}$
ZGL01-16	0.096401	0.001642	0.002207	0.000028	0.282403	0.000016	130.5	0.282398	−13.04	−10.37	1238	1843	−0.93
ZGL01-17	0.155763	0.004177	0.004179	0.000117	0.282382	0.000014	130.4	0.282372	−13.79	−11.28	1342	1902	−0.87
ZGL01-18	0.084408	0.001249	0.002083	0.000040	0.282413	0.000016	131.4	0.282408	−12.68	−10.01	1220	1821	−0.94
ZGL01-19	0.083414	0.001003	0.002075	0.000013	0.282418	0.000015	129.9	0.282412	−12.54	−9.87	1214	1812	−0.94
ZGL01-20	0.062056	0.000735	0.001516	0.000005	0.282388	0.000016	130.4	0.282385	−13.57	−10.82	1237	1875	−0.95
ZGL01-21	0.182502	0.008026	0.004088	0.000126	0.282446	0.000019	130.2	0.282436	−11.53	−9.02	1241	1759	−0.88
ZGL01-22	0.092732	0.001782	0.002319	0.000049	0.282435	0.000013	131.1	0.28243	−11.91	−9.23	1196	1773	−0.93
ZGL02-1	0.106946	0.000395	0.002503	0.000022	0.282433	0.000012	130.6	0.282427	−11.98	−9.35	1205	1779	−0.92
ZGL02-2	0.067830	0.000183	0.001615	0.000004	0.282407	0.000012	130.5	0.282403	−12.91	−10.20	1214	1834	−0.95
ZGL02-3	0.057646	0.000937	0.001298	0.000011	0.282432	0.000013	130.8	0.282429	−12.03	−9.28	1168	1776	−0.96
ZGL02-4	0.072288	0.000375	0.001717	0.000009	0.282408	0.000016	129.1	0.282403	−12.89	−10.20	1216	1832	−0.95
ZGL02-5	0.094155	0.000463	0.002217	0.000013	0.282430	0.000014	131.0	0.282425	−12.09	−9.42	1200	1785	−0.93
ZGL02-6	0.092970	0.002990	0.002368	0.000082	0.282393	0.000014	129.6	0.282387	−13.41	−10.77	1259	1868	−0.93
ZGL02-7	0.056264	0.000580	0.001317	0.000014	0.282411	0.000013	130.5	0.282407	−12.78	−10.06	1199	1824	−0.96
ZGL02-8	0.064781	0.000244	0.001556	0.000002	0.282430	0.000014	130.1	0.282426	−12.10	−9.39	1179	1782	−0.95
ZGL02-9	0.072894	0.000577	0.001783	0.000005	0.282424	0.000012	130.4	0.282419	−12.32	−9.63	1196	1798	−0.95
ZGL02-10	0.143920	0.001600	0.003122	0.000015	0.282494	0.000014	128.1	0.282486	−9.83	−7.26	1136	1647	−0.91
ZGL02-11	0.083086	0.000530	0.001832	0.000007	0.282420	0.000012	129.6	0.282415	−12.46	−9.77	1202	1805	−0.94
ZGL02-12	0.123662	0.001834	0.003034	0.000046	0.282415	0.000014	129.7	0.282408	−12.63	−10.02	1250	1823	−0.91
ZGL02-13	0.070975	0.000654	0.001555	0.000006	0.282434	0.000013	131.1	0.28243	−11.97	−9.24	1174	1774	−0.95
ZGL02-15	0.078806	0.000724	0.001969	0.000024	0.282414	0.000012	129.2	0.282409	−12.66	−9.99	1215	1819	−0.94
ZGL02-18	0.087561	0.000379	0.001907	0.000010	0.282425	0.000013	129.3	0.282421	−12.26	−9.56	1197	1794	−0.94
ZGL02-19	0.064202	0.000300	0.001584	0.000013	0.282413	0.000017	129.4	0.282409	−12.69	−9.99	1204	1820	−0.95
ZGL02-20	0.085997	0.002079	0.002009	0.000035	0.282473	0.000014	131.4	0.282468	−10.59	−7.90	1132	1688	−0.94
ZGL02-21	0.059703	0.000836	0.001365	0.000007	0.282413	0.000013	130.0	0.28241	−12.69	−9.95	1197	1819	−0.96
ZGL02-22	0.136569	0.002374	0.002993	0.000039	0.282446	0.000015	130.2	0.282439	−11.52	−8.93	1202	1752	−0.91
ZGL02-23	0.098117	0.001078	0.002689	0.000052	0.282386	0.000015	130.6	0.282379	−13.67	−11.05	1281	1887	−0.92
ZGL02-24	0.060680	0.001426	0.001413	0.000024	0.282406	0.000013	131.1	0.282403	−12.93	−10.20	1208	1834	−0.96
ZGL02-25	0.088660	0.001048	0.002110	0.000019	0.282411	0.000018	130.3	0.282406	−12.76	−10.09	1224	1827	−0.94
ZGL02-26	0.062762	0.000351	0.001504	0.000014	0.282388	0.000015	129.9	0.282384	−13.58	−10.87	1237	1875	−0.95

Note: The calculation formula is presented in Appendix A.5.

5. Discussion

5.1. Genetic Type

Granites are commonly grouped into I-, S-, and A-types based on geochemistry and petrogenesis. A-type granites are typically alkaline or peralkaline, rich in high field strength elements (Nb, Ta, Zr, Hf), and formed at relatively high temperatures; they also show elevated Ga/Al and FeO^T/MgO ratios [13,16,17]. Petrologists further subdivide A-type granites into A1 and A2 subtypes that reflect different tectonic environments and chemical features [15]. Because of these attributes, A-type granites are valuable records of extensional magmatism and regional tectonics.

The two-mica syenogranite investigated here is rich in K-feldspar and biotite. Major elements indicate high K, high total alkalis, and Mg depletion. The aluminum saturation index is classified as weakly peraluminous (A/CNK = 0.95–1.14; A/NK = 1.12–1.19; see Figure 4b and Table 1), which aligns with typical A-type compositions [13,36], higher than those of typical I-type granites. The differentiation index (DI) is 93.4–96.1, implying strong magmatic differentiation. Rare earth element patterns show significantly negative Eu anomalies (Eu/Eu* = 0.16–0.40) and positive Ce anomalies (Ce/Ce* = 1.04–2.66), with light REE strongly enriched over heavy REE (LREE/HREE = 6.91–9.44). Chondrite-normalized REE plots display right-inclined, concave-upward patterns. Primitive mantle-normalized spidergrams show enrichment in Rb, Th, U, Zr, Hf and depletion in Sr, Ba, Ti, P. The above geochemical characteristics are consistent with the typical features of A-type granites [13,14]. In addition, they display high (10,000 Ga/Al) ratios (Figure 8a–d). On multiple discrimination diagrams, all samples plot within the A-type field (Figure 8a–d). A1-type granites have very low Y/Nb and Rb/Nb ratios, while A2-type granites are the opposite [15]. The two-mica Syenogranite are consistent with the A1-type granite (Y/Nb ratio < 1.2), with their Y/Nb ratios varying from 0.34 to 1.01. Compared with published Early Cretaceous A-type granites in the Lhasa terrane, the Meibaqieqin two-mica syenogranite has relatively low Y/Nb (0.34–1.01) and Rb/Nb (9.24–17.66). In the Y–Nb–Ce discrimination diagrams of A-type granites proposed by Eby [14] (Figure 8e), the two-mica syenogranites plot in the

A1 subfield. Moreover, the two-mica syenogranites samples plot in the scope of A1 subtype and overlap with the OIB field in the Y/Nb vs. Ce/Nb diagram [14,15] (Figure 8f). All of the above characteristics indicate that the two-mica syenogranite belongs to type A1 granite.

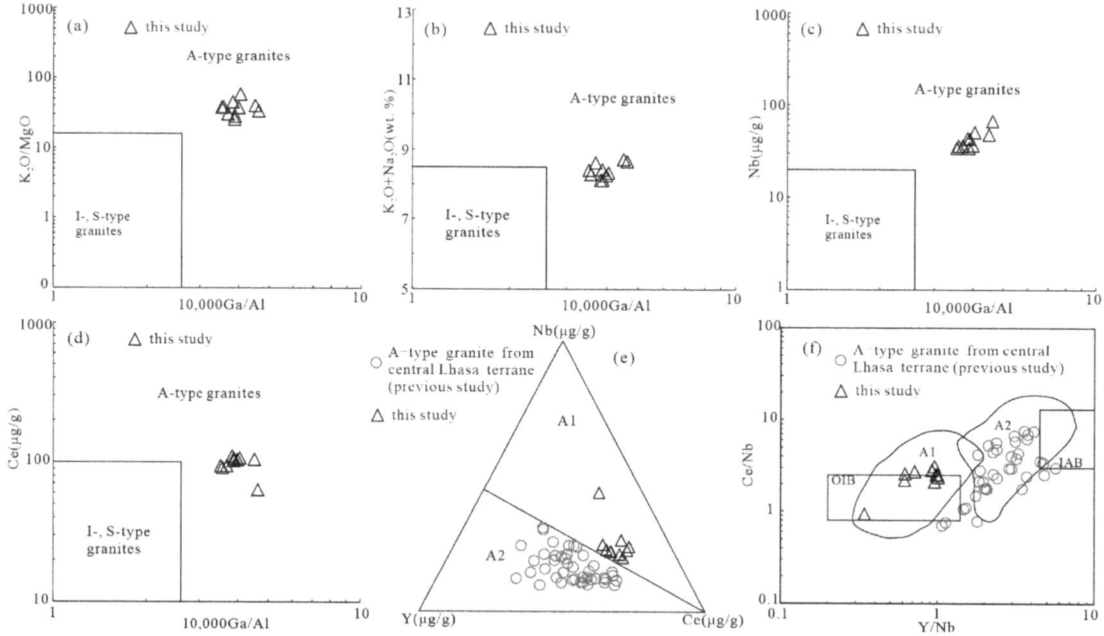

Figure 8. Discrimination diagrams for the two-mica syenogranite in the Meibaqieqin region. Diagram (**a**–**d**), after [13]; Diagram (**e**,**f**), after [15]. The cited data is the same as that in Figure 6.

5.2. Magmatic Sources and Petrogenesis

Several genetic models have been put forward for A-type granites, including partial melting of lower crust [17,28], melting of high-grade (granulite-facies) crustal protoliths [37,38], and mingling or mixing of crustal melts with mantle-derived magmas [28,39]. Despite different origins, A-type granites commonly exhibit high SiO_2, high total alkalis, low Ca and Mg, and relatively high $(Na_2O + K_2O)/Al_2O_3$, FeO^T/MgO, and Ga/Al [13,40–42]. Eby linked A1-type granites to intraplate or rift-related settings, whereas A2-type granites reflect post-collisional reworking and melting of continental crust [14]. In the Lhasa terrane, A2-type granites are commonly explained by interaction between mantle-derived mafic magmas and partially melted juvenile lower crust [21,24,25]. The two-mica syenogranite is the first discovered A1-type granite in the Lhasa terrane. Although it shares some similar geochemical characteristics with A2-type granite, its material source and genesis require further in-depth research.

Low Y/Nb ratios are characteristic of oceanic island, intraplate, and rift zone magmas derived from OIB-type mantle sources [15]. Magmas derived from arc-type sources (lithospheric mantle altered by subduction-related fluids) or dominantly from continental crust typically exhibit high Y/Nb ratios, reflecting the high ratio in continental crust [15]. Therefore, A1-type suites with lower Y/Nb ratios are interpreted as differentiates of mantle-derived melts similar to ocean island basalt (OIB). The two-mica syenogranite exhibits low Y/Nb ratios, possibly indicating a genesis of magmatic differentiation driven by mantle-derived material. The high alkaline characteristics and low Y/Nb and Y/Ta ratios all indicate that the two-mica syenogranite belongs to the A1 type, which is considered to be formed by derivatives of enriched OIB mantle sources [15,16]. However, several trace-element ratios indicate substantial crustal involvement. Nb/Ta = 6.74–9.92 and Ce/Pb = 1.88–3.71 approach continental crust values (Nb/Ta ≈ 11–12; Ce/Pb ≈ 4) and are well below primitive mantle values (Nb/Ta ≈ 17.5; Ce/Pb ≈ 9) [43,44]. High Rb/Sr (10.9–59.0) and low Cr (1.09–7.01 μg /g) also point to evolved, crustal sources. The Sm/Nd ratios of the samples range from 0.17 to 0.22, which are slightly lower than the characteristic values corresponding to the crust (0.3), indicating the magma source region contains material not only from the crust [45]. The Zr/Nb ratio ranges from 1.13 to 5.05, with an average value of 3.72, which is significantly lower than the corresponding ratio of continental crust (16.2) [46]. Nb/U = 5.62–8.22 is far lower than the primitive mantle ratio (~33.6) and typical of continental crust [47]. Strong Eu anomalies and depletion in Sr and Ba (Figure 7) suggest residual or fractionating plagioclase and melting of plagioclase-rich, garnet-poor sources [48]. La/Nb = 0.46–1.58 also supports a crustal component [49]. The strong depletion of Ba and Sr in the two-mica syenogranite (Figure 5b) also reflects that it may be a product of low-degree partial melting of crustal materials [50]. The Th/Ta ratio in granitic rocks is significant for indicating magma–crust interaction, as mantle-derived rocks exhibit Th/Ta ratios of ≈2—lower than those of the upper crust (Th/Ta ≈ 6.9) and lower crust (Th/Ta ≈ 7.9) [51]. Samples of two-mica syenogranites yield an average Th/Ta value of ≈12.86. The two-mica syenogranite was derived from crust-generated magma, possibly. In short, these characteristics indicate that the tectonic affinity of two-mica syenogranite reflects an extensional setting (intraplate and rift zone), while the magma itself is primarily derived from partial melting of ancient continental crust, with the driving force for partial melting possibly originating from shallow mantle magmatism.

Zircon Hf isotopes, owing to zircon's high closure temperature and low Lu/Hf, are robust tracers of magma sources and crustal evolution [39,52]. Positive $\varepsilon_{Hf}(t)$ values generally indicate juvenile crust or depleted mantle sources, whereas negative $\varepsilon_{Hf}(t)$ values point to derivation from older continental crust [53,54]. The A1-type granites studied here yield zircon $\varepsilon_{Hf}(t)$ values of −11.28 to −7.26 (average −9.83) and two-stage Hf model ages (T_{DM2}) of 1.65–1.90 Ga. In the age vs. $\varepsilon_{Hf}(t)$ diagram (Figure 9a), all analyses plot below the chondritic (CHUR) evolution curve, between Paleoproterozoic (ca. 1.6–2.0 Ga) crustal evolution lines, indicating predominantly Paleoproterozoic crustal sources. Whole-rock $\varepsilon_{Nd}(t)$ values of −11.75 to −11.36 further support derivation from ancient crust with limited juvenile input (Figure 9b). To compare source characteristics through time, we compiled published Hf isotope data for A-type granites across Lhasa terrane. Three main emplacement pulses occur at ca. 130 Ma, ca. 114 Ma, and ca. 105 Ma (Figure 9a). The younger A2-type granites (ca. 114 Ma and ca. 105 Ma), chiefly along the northern margin of the central Lhasa terrane, show increased mantle contributions relative to the ca. 130 Ma magmatism, which is dominated by crustal signatures [20–26,33] (Figure 9a). As mentioned above, both the zircon $\varepsilon_{Hf}(t)$ values and whole-rock $\varepsilon_{Nd}(t)$ values of the two-mica syenogranite indicates the involvement of crustal materials in the magma source. The whole-rock geochemical data also indicate characteristics of crustal material participation in petrogenesis.

The characteristics of trace elements such as Th, La, Sm, and Nd can be used to discriminate whether the petrogenetic process is fractional crystallization or partial melting [55]. In the La vs. La/Sm and La vs. La/Yb diagrams (Figure 10a,b), the samples define positive correlations consistent with partial melting. Experimental studies on pelites, greywackes, and mafic protoliths demonstrate contrasting melt chemistries [48,56,57]. The low CaO + FeO^T + TiO_2 contents of our samples resemble melts derived from pelitic sources (Figure 10c).

This evidence indicates that the two-mica syenogranite was generated mainly by partial melting of ancient continental crust, with a pelitic-dominated source and some mantle input. This differs from A2-type granites along the northern margin of the central Lhasa terrane, which reflect stronger crust–mantle interaction.

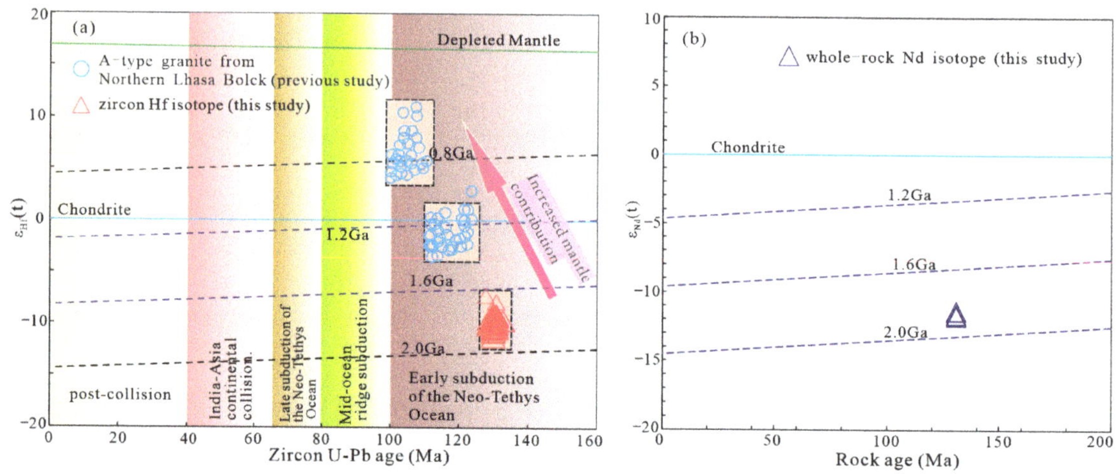

Figure 9. Plot of age vs. $\varepsilon_{Hf}(t)$ ((**a**), after [58]) and age vs. $\varepsilon_{Nd}(t)$ (**b**) for the two-mica syenogranite. The cited data is the same as that in Figure 6.

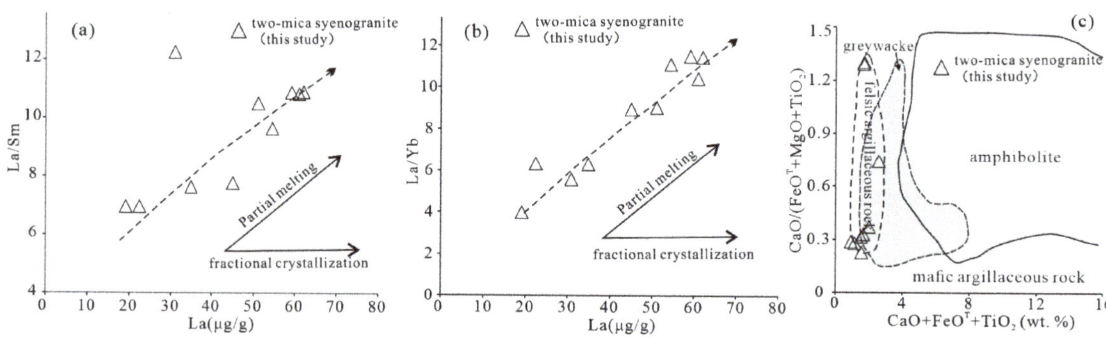

Figure 10. Genetic types and source discrimination diagrams of two-mica syenogranite. (**a**) La vs. La/Sm diagram, (**b**) La vs. La/Yb diagram, (**c**) CaO + FeO + TiO_2 vs. CaO/(FeO^T + MgO + TiO_2) diagram.

5.3. Tectonic Setting

A-type granites typically form in extensional settings, including intraplate, post-collisional, and back-arc environments [13,15]. The A-type granites are usually regarded as a marker of crustal extension [13,17]. In the Lhasa terrane, three A-type episodes are recognized at ca. 130 Ma, ca. 114 Ma, and ca. 105 Ma [20–25]. Discrimination diagrams categorize the Meibaqieqin two-mica syenogranite (ca. 130 Ma) as A1-type, consistent with intraplate extensional magmatism (Figure 11). By contrast, the ca. 114 Ma and ca. 105 Ma A-type granites are A2-type and are linked to southward subduction of the Bangongco–Nujiang oceanic crust and post-collisional extension in northern Lhasa [20,22,23].

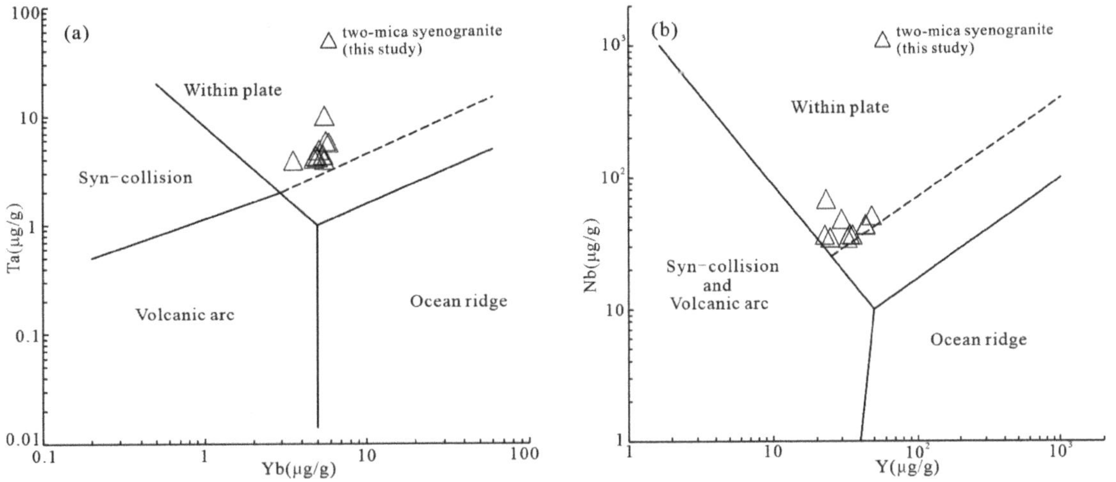

Figure 11. Tectonic discrimination diagrams for the two-mica syenogranite. (**a**) Yb vs. Ta diagram; (**b**) Y vs. Nb diagram.

As analyzed above, the two-mica syenogranite is situated in the southern margin of the central Lhasa terrane, showing clear differences from the continent-continent collision-related A2-type granites in the northern margin of the central Lhasa terrane. Around 130 Ma, the Bangongco-Nujiang oceanic crust was undergoing southward subduction, while the Neo-Tethyan oceanic crust was subducting northward. Which subduction event is more closely related to the samples studied in this research? Recent studies on the Shiquanhe-Namco ophiolitic mélange belt suggest that this belt is a back-arc basin formed during the subduction of the Bangongco-Nujiang Ocean [5,59,60]. Its small scale and rapid evolution imply that the subduction in this basin is not related to the large-scale magmatic activity in the central-northern part of the Lhasa terrane [61]. This paper posits that the northward subduction of the Neo-Tethyan oceanic crust had a stronger influence than the southward subduction of the Bangongco-Nujiang oceanic crust and the Shiquanhe-Namco small oceanic basin crust. The main evidence includes: (1) In the early Early Cretaceous, the Lhasa terrane had a bidirectional subduction system [2], which included the southward subduction of the Bangongco-Nujiang oceanic crust, and the northward subduction of the Neo-Tethyan oceanic crust. This bidirectional subduction led to the formation of the Zenong Group volcanic arc, Coqin-Duowa back-arc basin, Longgar-Nyainqentanglha magmatic arc, and south Gangdese magmatic arc from north to south, respectively [2]. This also created an extensional tectonic environment in the southern Lhasa terrane, distinctly different from the post-collisional A2-type granites in the northern Lhasa terrane because of Shiquanhe-Namco small oceanic basin. (2) The influence of the northward subduction of the Neo-Tethys oceanic plate is comparable to that of the southward subduction of the Bangongco-Nujiang oceanic crust, therefore the distance between the location of plutons and the subduction zone is related to the strength of tectonic influence indication [62]. The A1-type granite discovered in the Meibaqieqin region is situated in the southern area of the central Lhasa terrane, approximately 80 km from the Yarlung Zangbo suture zone and nearly 300 km from the Bangongco-Nujiang suture zone. This proximity suggests that the influence of the southern Neo-Tethyan ocean subduction is more significant. (3) The volcanic rocks from the Duoni Formation and Zenong Group in the central-northern Lhasa terrane to the Early Cretaceous volcanic rocks on the southern side of the central Lhasa terrane also show that the lithologic assemblage transformed from calc-alkaline series to predominantly high-K calc-alkaline and minor shoshonitic series, indicating that the subduction zone depth deepened from north to center, with subduction polarity from north to south, and ages showing a trend of becoming younger from north to center. The volcanic rocks from the Yeba Formation and Sangri Group in the southern Lhasa terrane to the Early Cretaceous volcanic rocks on the southern side of the central Lhasa block show lithologic assemblages that transformed from tholeiitic to calc-alkaline series to high-K calc-alkaline and shoshonitic series, with ages showing a trend from old to young, indicating that the subduction zone depth deepened from south to center, with subduction polarity from south to north, suggesting possible constraints from bidirectional subduction during the Early Cretaceous. The distribution characteristics of different rock series indicate that the northern part of the central Lhasa terrane was more influenced by southward subduction of the Bangongco-Nujiang oceanic crust, while the southern part of the central Lhasa terrane was more influenced by northward subduction of the Neo-Tethyan oceanic crust [62]. Around 130 Ma, the Neo-Tethyan oceanic crust was in a northward subduction stage, forming extensive arc magmatic rock distributions in the southern Lhasa block. At this time, the Bangongco-Nujiang oceanic crust experienced bidirectional subduction. Due to north-south extensional tectonic setting, the Shiquanhe- Namco back-arc ocean basin was formed. The northward subduction of the Neo-Tethyan Ocean did not form a typical back-arc small ocean basin. Instead, its tectonic stress created back-arc rift zones or intraplate hotspots. Lithospheric thinning led to asthenospheric mantle upwelling, providing heat sources for petrogenesis. Magmatic materials metasomatically enriched the lower crust to form primitive magmas, which underwent magmatic evolution and intruded into the shallow crust to form this special type of granite.

Thus, it is evident that the northward subduction of the Neo-Tethyan Ocean played a major role in the extensional tectonic setting in the southern part of the Lhasa terrane. This was the key dynamic factor for the formation of intraplate granites in the Meibaqieqin region, whereas the southward subduction of the Shiquanhe-Namco back-arc basin and the Bangongco-Nujiang oceanic crust had relatively minor influences.

6. Conclusions

1. The crystallization age of the two-mica syenogranite is ca. 130 Ma, representing a large-scale Early Cretaceous magmatism in the Meibaqieqin region and its periphery.
2. Geochemical and isotopic data indicate that the two-mica syenogranite is a high-K calc-alkaline, A-type granite, with magmatic source materials derived from ancient crustal materials, formed through partial melting of pelitic rocks driven by mantle-derived magma.
3. The Meibaqieqin A1-type granite was formed in intraplate extensional tectonic setting, primarily influenced by the northward subduction of the Neo-Tethyan oceanic crust, with relatively minor influence from the southward subduction of the Bangongco-Nujiang oceanic crust. The petrogenesis of two-mica syenogranite is related to the extensional tectonic setting formed by subduction, and this petrogenetic environment is distinctly different from the post-collisional extensional environment of A2-type granites.

Author Contributions: Writing—original draft, Y.Y. and J.Z.; Writing—review & editing, K.G. and Y.Y.; resources, K.G. and Z.Z.; Conceptualization, K.G.; Funding acquisition, J.Z., J.G. and S.D.; investigation, J.W., Y.L. and P.X. All authors have read and agreed to the published version of the manuscript.

Funding: This work was supported by Deep Earth Probe and Mineral Resources Exploration—National Science and Technology Major Project (2024ZD1003205, 2024ZD1003206, 2024ZD1003207), Key Research and Development Program Project of Tibet Science and Technology Department (XZ202501ZY0102), Central Guidance Fund for Local Science and Technology Development Projects (XZ202301YD0030C), China Geological Survey Projects (DD20242518, DD20240069, DD202402028).

Data Availability Statement: Data are contained within the article.

Acknowledgments: The authors are thankful to the reviewers and scientific editor, whose constructive criticism and recommendations helped us to significantly rework and improve the manuscript.

Conflicts of Interest: Junkang Zhao, Ke Gao, Shuai Ding and Jiansheng Gong were employed by the Zijin Mining Group Southwest Geological Exploration Co., Ltd. The remaining authors declare that the research was conducted in the absence of any commercial or financial relationships that could be construed as a potential conflict of interest.

Appendix A. Analytical Methods

Appendix A.1. Whole-Rock Major and Trace Element Analyses

Whole-rock major elements were determined after fusion. The flux consisted of lithium tetraborate, lithium metaborate, and lithium fluoride mixed in a 45:10:5 ratio. Ammonium nitrate and lithium bromide served as the oxidant and release agent, respectively. Fusions were carried out at 1050 °C for 15 min. Major elements were analyzed on a Rigaku ZSX Primus II wavelength-dispersive XRF with a 4.0 kW end-window Rh target X-ray tube, operated at 50 kV and 60 mA. All oxides were measured on Kα lines. Calibration used Chinese national reference materials (rock: GBW07101–14 [63]; soil: GBW07401–08 [64]; stream sediment: GBW07302–12 [65]). Data were corrected using the theoretical α-coefficient method, and the relative standard deviation (RSD) was <2%.

Whole-rock trace elements were measured by ICP-MS (Agilent 7700e) at Wuhan Sample Solution Analytical Technology Co., Ltd. (Wuhan, China). The digestion procedure was: (1) Dry 200-mesh powders at 105 °C for 12 h; (2) Accurately weigh 50 mg into a Teflon bomb; (3)Add 1 mL HNO3 and 1 mL HF slowly; (4)Place the bomb in a stainless-steel jacket and heat at 190 °C in an oven for >24 h; (5) After cooling, open the bomb, evaporate at 140 °C to incipient dryness, add 1 mL HNO_3, and evaporate to dryness again; (6) Add 1 mL HNO_3, 1 mL MQ water, and 1 mL of 1 μg/g In internal standard solution; reseal and heat at 190 °C for >12 h; (7) Transfer the solution to a polyethylene bottle and dilute with 2% HNO_3 to a final mass of 100 g.

Appendix A.2. Nd Isotopic Ratio Measurements

Nd isotope ratios were obtained using a Neptune Plus MC-ICP-MS (Thermo Fisher Scientific, Dreieich, Germany) at Wuhan Sample Solution Analytical Technology Co., Ltd. The double-focusing instrument was fitted with seven fixed electron-multiplier ICs and nine Faraday cups with 1011 Ω resistors. A high-capacity dry interface pump (120 m^3 h^{-1}), an H skimmer cone, and the standard sample cone were used to enhance sensitivity. An Alfa Aesar Nd single-element solution was used for tuning. A 200 μg/L GSB 04-3258-2015 standard was analyzed routinely to assess accuracy and reproducibility; typical $^{142}Nd^+$ intensities exceeded ~2.5 V. Data were collected in static mode at low resolution in ten blocks of ten cycles, with 4.194 s per cycle, for a total of ~7 min per analysis.

Instrumental mass discrimination was corrected using the exponential law [66], with internal normalization to $^{146}Nd/^{144}Nd = 0.7219$ [67]. Samarium was effectively removed by ion exchange; any residual $^{144}Sm^+$ was corrected following Lin et al. [67]. Data reduction for Nd isotope ratios was performed with Iso-Compass software [68]. One GSB 04-3258-2015 standard was measured after every seven unknowns [69]. The measured $^{143}Nd/^{144}Nd$ for

GSB 04-3258-2015 was 0.512440 ± 6 (2SD, n = 31), identical within error to the published value of 0.512438 ± 6 (2SD) [69]. USGS reference materials BCR-2 (basalt) and RGM-2 (rhyolite) yielded 0.512641 ± 11 (2SD, n = 82) and 0.512804 ± 12 (2SD, n = 80), respectively, consistent with published values [70].

Appendix A.3. Zircon U-Pb Dating

Zircon U-Pb ages were acquired by LA–ICP–MS at Wuhan Sample Solution Analytical Technology Co., Ltd. (Wuhan, China). Instrumental parameters for the laser ablation system, ICP–MS, and data reduction followed Zong et al. [71]. Ablation used a GeolasPro platform equipped with a COMPexPro 102 ArF excimer laser (193 nm, maximum energy 200 mJ) and a MicroLas optical system (Agilent Technologies, Bremen, Germany). Ion signals were measured on an Agilent 7900 ICP–MS. The system includes a wire-type signal-smoothing device [72]. In this study, the laser spot diameter was 30 μm and the repetition rate was 6 Hz. Zircon 91500 and NIST SRM 610 glass were used as external standards for U–Pb dating and trace-element calibration, respectively [73]. Off-line background selection/integration, time-drift correction, and quantitative calibration for both trace elements and U–Pb data were performed with the Excel-based ICPMSDataCal 10 software [74,75]. Concordia plots and weighted mean values were calculated using Isoplot/Ex_ver3 [76].

Appendix A.4. Zircon Hf Isotope

Experiments of in situ Hf isotope ratio analysis were conducted using a Neptune Plus MC-ICP-MS (Thermo Fisher Scientific, Dreieich, Germany) in combination with a Geolas HD excimer ArF laser ablation system that was hosted at the Wuhan Sample Solution Analytical Technology Co., Ltd., Wuhan, China. A "wire" signal smoothing device is included in this laser ablation system, by which smooth signals are produced even at very low laser repetition rates down to 1 Hz [73]. Helium was used as the carrier gas within the ablation cell and was merged with argon (makeup gas) after the ablation cell. Small amounts of nitrogen were added to the argon makeup gas flow for the improvement of sensitivity of Hf isotopes [77]. Compared to the standard arrangement, the addition of nitrogen in combination with the use of the newly designed X skimmer cone and Jet sample cone in Neptune Plus improved the signal intensity of Hf, Yb and Lu by a factor of 5.3, 4.0 and 2.4, respectively. All data were acquired on zircon in single spot ablation mode at a spot size of 44 μm. The energy density of laser ablation that was used in this study was ~7.0 J cm^{-2}. Each measurement consisted of 20 s of acquisition of the background signal followed by 50 s of ablation signal acquisition. Detailed operating conditions for the laser ablation system and the MC-ICP-MS instrument and analytical method are the same as description by Hu et al. [77].

The major limitation to accurate in situ zircon Hf isotope determination by LA-MC-ICP-MS is the very large isobaric interference from ^{176}Yb and, to a much lesser extent ^{176}Lu on ^{176}Hf. It has been shown that the mass fractionation of Yb (βYb) is not constant over time and that the βYb that is obtained from the introduction of solutions is unsuitable for in situ zircon measurements [78]. The under- or over-estimation of the βYb value would undoubtedly affect the accurate correction of ^{176}Yb and thus the determined ^{176}Hf/^{177}Hf ratio. We applied the directly obtained βYb value from the zircon sample itself in real-time in this study. The ^{179}Hf/^{177}Hf and ^{173}Yb/^{171}Yb ratios were used to calculate the mass bias of Hf (βHf) and Yb (βYb), which were normalized to ^{179}Hf/^{177}Hf = 0.7325 and ^{173}Yb/^{171}Yb = 1.132685 [79] using an exponential correction for mass bias. Interference of ^{176}Yb on ^{176}Hf was corrected by measuring the interference-free ^{173}Yb isotope and using ^{176}Yb/^{173}Yb = 0.79639 [79] to calculate ^{176}Yb/^{177}Hf. Similarly, the relatively minor interference of ^{176}Lu on ^{176}Hf was corrected by measuring the intensity of the interference-free ^{175}Lu isotope and using the recommended ^{176}Lu/^{175}Lu = 0.02656 [80] to calculate ^{176}Lu/^{177}Hf. We used the mass bias of Yb (βYb) to calculate the mass fractionation of Lu because of their similar physicochemical properties. Off-line selection and integration of analyte signals, and mass bias calibrations were performed using ICPMSDataCal 10 [75].

Appendix A.5. Calculation Formula of Zircon Hf Isotope

$\varepsilon_{Hf}(t) = 10{,}000 \times (((^{176}Hf/^{177}Hf)_S - (^{176}Lu/^{177}Hf)_S \times (e^{\lambda t} - 1))/((^{176}Hf/^{177}Hf)_{CHUR,0} - (^{176}Lu/^{177}Hf)_{CHUR} \times (e^{\lambda t} - 1)) - 1)$; $T_{DM}{}^{C} = T_{DM} - (T_{DM} - t) \times ((f_{cc} - f_s)/(fcc - f_{DM}))$; $T_{DM} = 1/\lambda \times \ln(1 + ((^{176}Hf/^{177}Hf)_S - (^{176}Hf/^{177}Hf)_{DM})/((^{176}Lu/^{177}Hf)_S - (^{176}Lu/^{177}Hf)_{DM}))$; $f_{Lu/Hf} = (^{176}Lu/^{177}Hf)_S/(^{176}Lu/^{177}Hf)_{CHUR} - 1$; $\lambda = 1.867 \times 10^{-11}$ year^{-1} [81]; $(^{176}Lu/^{177}Hf)_S$ and $(^{176}Hf/^{177}Hf)_S$ are the measured values of the samples; $(^{176}Lu/^{177}Hf)_{CHUR} = 0.0336$ and $(^{176}Hf/^{177}Hf)_{CHUR,0} = 0.282785$ [82]; $(^{176}Lu/^{177}Hf)_{mean\ crust} = 0.015$ [83]; $(^{176}Lu/^{177}Hf)_{DM} = 0.0384$ and $(^{176}Hf/^{177}Hf)_{DM} = 0.28325$ [84]; $f_{cc} = ((^{176}Lu/^{177}Hf)_{mean\ crust}/(^{176}Lu/^{177}Hf)_{CHUR}) - 1$; fs= $f_{Lu/Hf}$; $f_{DM} = ((^{176}Lu/^{177}Hf)_{DM}/(^{176}Lu/^{177}Hf)_{CHUR}) - 1$; t = crystallization time of zircon.

References

1. Pan, G.T.; Wang, L.Q.; Li, X.Z.; Wang, J.M.; Xu, Q. The tectonic framework and spatial allocation of the archipelagic arc-basin systems on the Qinghai-Xizang Plateau. *Sediment. Geol. Tethyan Geol.* **2001**, *21*, 1–26. (In Chinese)
2. Pan, G.T.; Mo, X.X.; Hou, Z.Q.; Zhu, D.C.; Wang, L.Q.; Li, G.M.; Zhao, Z.D.; Geng, Q.R.; Liao, Z.L. Spatial-temporal framework of the Gangdese Orogenic Belt and its evolution. *Acta Petrol. Sin.* **2006**, *22*, 521–533. (In Chinese)
3. Hou, Z.Q.; Mo, X.X.; Yang, Z.M.; Wang, A.J.; Pan, G.T.; Qu, X.M.; Nie, F.J. Metallogeneses in the collisional orogen of the Qinghai-Tibet Plateau: Tectonic setting, tempo-spatial distribution and ore deposit types. *Geol. China* **2006**, *33*, 340–351. (In Chinese)
4. Yin, A.; Harrison, T.M. Geologic evolution of the Himalayan-Tibetan orogen. *Earth Planet. Sci.* **2000**, *28*, 211–280. [CrossRef]
5. Zhu, D.C.; Zhao, Z.D.; Niu, Y.; Mo, X.X.; Chung, S.L.; Hou, Z.Q.; Wang, L.Q.; Wu, F.Y. The Lhasa Terrane: Record of a microcontinent and its histories of drift and growth. *Earth Planet. Sci. Lett.* **2011**, *301*, 241–255. [CrossRef]
6. Zhu, D.C.; Zhao, Z.D.; Niu, Y.; Dilek, Y.; Hou, Z.Q.; Mo, X.X. The origin and pre-Cenozoic evolution of the Tibetan Plateau. *Gondwana Res.* **2013**, *23*, 1429–1454. [CrossRef]
7. Zhu, D.C.; Li, S.M.; Cawood, P.A.; Wang, Q.; Zhao, Z.D.; Liu, S.A.; Wang, L.Q. Assembly of the Lhasa and Qiangtang terranes in central Tibet by divergent double subduction. *Lithos* **2016**, *245*, 7–17. [CrossRef]
8. Pan, G.T.; Wang, L.Q.; Yin, F.G.; Geng, Q.R.; Li, G.M.; Zhu, D.C. Researches on geological-tectonic evolution of Tibetan Plateau: A review, recent advances, and directions in the future. *Sediment. Geol. Tethyan Geol.* **2022**, *42*, 151–175. (In Chinese) [CrossRef]
9. Pan, G.T.; Wang, L.Q.; Li, R.S.; Yuan, S.H.; Ji, W.H.; Yin, F.G.; Zhang, W.P.; Wang, B.D. Tectonic evolution of the Qinghai-Tibet Plateau. *J. Asian Earth Sci.* **2012**, *53*, 3–14. [CrossRef]
10. Loiselle, M.C. Characteristics and origin of anorogenic granites. *Geol. Soc. Am. Abstr. Progr.* **1979**, *11*, 468.
11. Wu, S.P.; Wang, M.Y.; Qi, K.J. Present situation of researches on A-type granites:a review. *Acta Petrol. Mineral.* **2007**, *26*, 57–66. (In Chinese)
12. Li, X.W.; Mo, X.X.; Zhao, Z.D.; Zhu, D.C. A discussion on how to discriminate A-type granite. *Geol. Bull. China* **2010**, *29*, 278–285. (In Chinese) [CrossRef]
13. Whalen, J.B.; CurrieBruce, K.L.; Chappell, W. A-type granites: Geochemical characteristics, discrimination and petrogenesis. *Contrib. Mineral. Petrol.* **1987**, *4*, 407–419. [CrossRef]
14. Eby, G.N. The A-type granitoids: A review of their occurrence and chemical characteristics and speculations on their petrogenesis. *Lithos* **1990**, *26*, 115–134. [CrossRef]
15. Eby, G.N. Chemical subdivision of the A-type granitoids:Petrogenetic and tectonic implications. *Geology* **1992**, *20*, 641–644. [CrossRef]
16. Bonin, B. A-type granites and related rocks: Evolution of a concept, problems and prospects. *Lithos* **2007**, *97*, 1–29. [CrossRef]
17. Collins, W.J.; Beams, S.D.; White, A.J.R.; Chappell, B.W. Nature and origin of A-type granites with particular reference to southeastern Australia. *Contrib. Mineral. Petrol.* **1982**, *80*, 189–200. [CrossRef]
18. King, P.L.; White, A.J.R.; Chappell, B.W.; Allen, C.M. Characterization and Origin of Aluminous A-type Granites from the Lachlan Fold Belt, Southeastern Australia. *J. Petrol.* **1997**, *38*, 371–391. [CrossRef]
19. Mushkin, A.; Navon, O.; Halicz, L.; Hartmann, G.; Stein, M. The Petrogenesis of A-type Magmas from the Amram Massif, Southern Israel. *Petrology* **2003**, *44*, 815. [CrossRef]

20. Qu, X.M.; Wang, R.J.; Xin, H.B.; Jiang, J.H.; Chen, H. Age and petrogenesis of A-type granites in the middle segment of the Bangonghu–Nujiang suture, Tibetan plateau. *Lithos* **2012**, *146–147*, 264–275. [CrossRef]
21. Xie, L.; Dun, D.; Zhu, L.D.; Nima, C.R.; Yang, W.G.; Tao, G.; Li, C.; He, B.; He, Y. Zircon U-Pb geochronology, geochemistry and geological significance of the Zhaduding A-type granites in northern Gangdise, Tibet. *Geol. China* **2015**, *42*, 1214–1227. (In Chinese) [CrossRef]
22. Li, H.; Wang, M.; Zeng, X.W.; Luo, A.B.; Zeng, X.J. Slab break-off origin of 105 Ma A-type porphyritic granites in the Asa area of Tibet. *Geol. Mag.* **2020**, *157*, 1281–1298. [CrossRef]
23. Wang, X.X.; Yan, G.Q.; Liu, H.; Huang, H.X.; Lai, Y.; Tian, E.Y.; Ouyang, Y. Genesis of Late Cretaceous Qusang'gele Granitie in Central Lhasa Block, Tibet: Constraints by Geochemistry, Zircon U-Pb Geochronology, and Sr-Nd-Pb-Hf Isotopes. *Earth Sci.* **2021**, *8*, 2832–2849. (In Chinese) [CrossRef]
24. Liu, H.; Li, W.C.; Li, W.C.; Huang, H.X.; Li, Y.G.; Ouyang, Y.; Zhang, X.F.; Zhou, Q. Petrogenesis and tectonic setting of the late Early Cretaceous Kong Co A-type granite in the northern margin of Central Lhasa Subterrane, Tibet. *Acta Petrol. Sin.* **2022**, *38*, 230–252. (In Chinese) [CrossRef]
25. Xiang, A.P.; Liu, H.; Fan, W.X.; Zhou, Q.; Wang, H.; Li, K.Z. Petrogenesis, Geochemistry, and Geological Significance of the Kongco Granitic Porphyry Dykes in the Northern Part of the Central Lhasa Microblock, Tibet. *Minerals* **2025**, *3*, 283. [CrossRef]
26. Chen, Y.; Zhu, D.C.; Zhao, Z.D.; Meng, F.Y.; Wang, Q.; Santosh, M.; Wang, L.Q.; Dong, G.C.; Mo, X.X. Slab breakoff triggered ca. 113Ma magmatism around Xainza area of the Lhasa Terrane, Tibet. *Gondwana Res.* **2014**, *26*, 449–463. [CrossRef]
27. Zhu, D.C.; Mo, X.X.; Zhao, Z.; Niu, Y.L.; Pan, G.T.; Wang, L.Q.; Liao, Z.L. Permian and Early Cretaceous tectonomagmatism in southern Tibet and Tethyan evolution: New perspective. *Earth Sci. Front.* **2009**, *16*, 1–20. (In Chinese)
28. Yang, J.H.; Wu, F.Y.; Chung, S.L.; Wilde, S.A.; Chu, M.F. A hybrid origin for the Qianshan A-type granite, northeast China: Geochemical and Sr–Nd–Hf isotopic evidence. *Lithos* **2006**, *89*, 89–106. [CrossRef]
29. Middlemost, E.A.K. Naming materials in the magma/igneous rock system. *Earth-Sci. Rev.* **1994**, *37*, 215–224. [CrossRef]
30. Maniar, P.D.; Piccoli, P.M. Tectonic discrimination of granitoids. *Geol. Soc. Am. Bull.* **1989**, *101*, 635–643. [CrossRef]
31. Peccerillo, A.; Taylor, S.R. Geochemistry of eocene calc-alkaline volcanic rocks from the Kastamonu area, Northern Turkey. *Contrib. Mineral. Petrol.* **1976**, *58*, 63–81. [CrossRef]
32. Middlemost, E.A.K. Magmas and Magmatic Rocks: An Introduction to Igneous Petrology. *Longman* **1985**, 1–266.
33. Xiao, D.; Yang, X.J.; Teng, C.; Cheng, T.S.; Zhu, N.; Cao, J. Early Cretaceous A-Type Acidic Magmatic Belt in Northern Lhasa Block: Implications for the Evolution of the Bangong–Nujiang Ocean Lithosphere. *Minerals* **2024**, *14*, 681. [CrossRef]
34. Sun, S.S.; McDonough, W.F. Chemical and isotopic systematics of oceanic basalts: Implications for mantle composition and processes. *Geol. Soc. Lond. Spec. Publ.* **1989**, *42*, 313–345. [CrossRef]
35. Wu, Y.B.; Zheng, Y.F. Genesis of zircon and its constraints on interpretation of U-Pb age. *Chin. Sci. Bull.* **2004**, *49*, 1554–1569. [CrossRef]
36. Deng, J.F.; Luo, Z.H.; Su, S.G. *Petrogenesis, Tectonic Environment and Mineralization*; Geological Publishing House: Beijing, China, 2009; p. 381.
37. Creaser, R.A.; Price, R.C.; Wormald, R.J. A-type granites revisited: Assessment of a residual-source model. *Geology* **1991**, *19*, 163–166. [CrossRef]
38. Frost, B.R. A Geochemical Classification for Granitic Rocks. *J. Petrol.* **2001**, *42*, 2033–2048. [CrossRef]
39. Kerr, A.; Fryer, B.J. Nd isotope evidence for crust-mantle interaction in the generation of A-type granitoid suites in Labrador, Canada. *Chem. Geol.* **1993**, *104*, 39–60. [CrossRef]
40. Wu, F.Y.; Sun, D.Y.; Li, H.; Jahn, B.M.; Wilde, S. A-type granites in northeastern China: Age and geochemical constraints on their petrogenesis. *Chem. Geol.* **2002**, *187*, 143–173. [CrossRef]
41. Mukherjee, S. A Review on Out-of-Sequence Deformation in the Himalaya. In *Tectonics of the Himalay*; Mukherjee, S., Carosi, R., vander Beek, P., Mukherjee, B.K., Robinson, D., Eds.; Geological Society: London, UK, 2015; p. 317.
42. Hu, Q.H.; Yu, K.Z.; Liu, Y.S.; Hu, Z.C.; Zong, K.Q. The 131–134Ma A-type granites from northern Zhejiang Province, South China: Implications for partial melting of the Neoproterozoic lower crust. *Lithos* **2017**, *294–295*, 39–52. [CrossRef]
43. Green, T.H. Significance of Nb/Ta as an indicator of geochemical processes in the crust-mantle system. *Chem. Geol.* **1995**, *120*, 347–359. [CrossRef]
44. Hofmann, A.W. Sampling Mantle Heterogeneity through Oceanic Basalts: Isotopes and Trace Elements. *Treatise Geochem.* **2014**, *3*, 67–101.
45. Xu, R.H.; Zhang, Z.Q.; Song, H.B. *New Methods of Geochemistry and Isotopic Geology of Rare Earth Elements*; Geological Publishing House: Beijing, China, 1985; p. 159.
46. Weaver, B.L. The origin of ocean island basalt end-member compositions: Trace element and isotopic constraints. *Earth Planet. Sci. Lett.* **1991**, *104*, 381–397. [CrossRef]
47. Taylor, S.R.; Mclennan, S.M. The continental crust: Its composition and evolution. *J. Geol.* **1985**, *94*, 57–72.

48. Rapp, R.P.; Watson, E.B. Dehydration Melting of Metabasalt at 8–32 kbar: Implications for Continental Growth and Crust-Mantle Recycling. *J. Petrol.* **1995**, *36*, 891–931. [CrossRef]
49. DePaolo, D.J.; Daley, E.E. Neodymium isotopes in basalts of the southwest basin and range and lithospheric thinning during continental extension. *Chem. Geol.* **2000**, *169*, 157–185. [CrossRef]
50. Harris, N.; Inger, S. Trace element modelling of pelite-derived granites. *Contrib. Mineral. Petrol.* **1992**, *110*, 46–56. [CrossRef]
51. Rudnick, R.L.; Gao, S. Composition of the Continental Crust. In *Treatise on Geochemistry*; Holland, H.D., Turekian, K.K., Eds.; Pergamon: Oxford, UK, 2003; Volume 3, pp. 1–64.
52. Wu, F.Y.; Li, X.H.; Zheng, Y.F.; Gao, S. Lu-Hf isotopic systematics and their applications in petrology. *Acta Petrol. Sin.* **2007**, *23*, 185–220. (In Chinese)
53. Jahn, B.M.; Griffin, W.L.; Windley, B. Continental growth in the phanerozoic: Evidence from central Asia. *Tectonophysics* **2000**, *1–2*, 7–10. [CrossRef]
54. Kinny, P.D.; Maas, R. Lu–Hf and Sm–Nd isotope systems in zircon. *Rev. Mineral. Geochem.* **2003**, *53*, 327–341. [CrossRef]
55. Allègre, C.J.; Minster, J.F. Quantitative models of trace element behavior in magmatic processes. *Earth Planet. Sci. Lett.* **1978**, *38*, 1–25. [CrossRef]
56. Patiño Douce, A.E.; Johnston, A.D. Phase equilibria and melt productivity in the pelitic system: Implications for the origin of peraluminous granitoids and aluminous granulites. *Contrib. Mineral. Petrol.* **1991**, *107*, 202–218. [CrossRef]
57. Patiño Douce, A.E. Generation of metaluminous A-type granites by low-pressure melting of calc-alkaline granitoids. *Geology* **1997**, *25*, 743–746. [CrossRef]
58. Chu, M.F.; Chung, S.L.; O'Reilly, S.Y.; Pearson, N.J.; Wu, F.Y.; Li, X.H.; Liu, D.; Ji, J.; Chu, C.H.; Lee, H.Y. India's hidden inputs to Tibetan orogeny revealed by Hf isotopes of Transhimalayan zircons and host rocks. *Earth Planet. Sci. Lett.* **2011**, *307*, 479–486. [CrossRef]
59. Xu, M.J.; Li, C.; Zhang, X.Z.; Wu, Y.W. Nature and evolution of the Neo-Tethys in central Tibet: Synthesis of ophiolitic petrology, geochemistry, and geochronology. *Int. Geol. Rev.* **2014**, *56*, 1072–1096. [CrossRef]
60. Zeng, X.W.; Wang, M.; Fan, J.J.; Li, C.; Xie, C.M.; Liu, Y.M.; Zhang, T.Y. Geochemistry and geochronology of gabbros from the Asa Ophiolite, Tibet: Implications for the early Cretaceous evolution of the Meso-Tethys Ocean. *Lithos* **2018**, *320–321*, 192–206. [CrossRef]
61. Zhu, D.C.; Pan, G.T.; Wang, L.Q.; Mo, X.X.; Zhao, Z.D.; Zhou, C.Y.; Liao, Z.L.; Dong, G.C.; Yuan, S.H. Tempo-spatial variations of Mesozoic magmatic rocks in the Gangdise belt, Tibet, China, with a discussion of geodynamic setting-related issues. *Geol. Bull. China* **2008**, *27*, 1535–1550. (In Chinese)
62. Wang, L.Y.; Zheng, Y.Y.; Gao, S.B.; Li, W.L.; Mao, R.W.; Huang, L.L. The discovery of the Early Cretaceous Zenong Group volcanic rocks and geological significance in Jiwa area in south of the Central Lhasa subterrane. *Acta Petrol. Sin.* **2016**, *32*, 1543–1555. (In Chinese)
63. *GBW07101-GBW07114*; Rock Reference Materials. National Geological Experimental Testing Center: Beijing, China, 2003.
64. *GBW07401-GBW07408*; Soil Reference Materials. Institute of Geophysical and Geochemical Exploration: Langfang, China, 2003.
65. *GBW07302-GBW07312*; Stream Sediment Reference Materials. National Geological Experimental Testing Center: Beijing, China, 2000.
66. Russell, W.A.; Papanastassiou, D.A.; Tombrello, T.A. Ca isotope fractionation on the Earth and other solar system materials. *Geochim. Cosmochim. Acta* **1978**, *42*, 1075–1090. [CrossRef]
67. Lin, J.; Liu, Y.S.; Yang, Y.H.; Hu, Z.C. Calibration and correction of LA-ICP-MS and LA-MC-ICP-MS analyses for element contents and isotopic ratios. *Solid Earth Sci.* **2016**, *1*, 5–27. [CrossRef]
68. Zhang, W.; Hu, Z.C.; Liu, Y.S. Iso-Compass: New freeware software for isotopic data reduction of LA-MC-ICP-MS. *J. Anal. At. Spectrom.* **2020**, *35*, 1087–1096. [CrossRef]
69. Li, J.; Tang, S.H.; Zhu, X.K.; Pan, C.X. Production and Certification of the Reference Material GSB04-3258-2015 as a $^{143}Nd/^{144}Nd$ Isotope Ratio Reference. *Geostand. Geoanalytical Res.* **2017**, *41*, 255–262.
70. Li, C.F.; Li, X.H.; Li, Q.L.; Guo, J.H.; Li, X.H.; Yang, Y.H. Rapid and precise determination of Sr and Nd isotopic ratios in geological samples from the same filament loading by thermal ionization mass spectrometry employing a single-step separation scheme—ScienceDirect. *Anal. Chim. Acta* **2012**, *727*, 54–60. [CrossRef] [PubMed]
71. Zong, K.Q.; Klemd, R.; Yuan, Y.; He, Z.Y.; Guo, J.; Shi, X.; Liu, Y.S.; Hu, Z.C.; Zhang, Z.M. The assembly of Rodinia: The correlation of early Neoproterozoic (ca. 900Ma) high-grade metamorphism and continental arc formation in the southern Beishan Orogen, southern Central Asian Orogenic Belt (CAOB). *Precambrian Res.* **2017**, *290*, 32–48. [CrossRef]
72. Hu, Z.C.; Zhang, W.; Liu, Y.S.; Gao, S.; Li, M.; Zong, K.Q.; Chen, H.H.; Hu, S.H. "Wave" Signal-Smoothing and Mercury-Removing Device for Laser Ablation Quadrupole and Multiple Collector ICPMS Analysis: Application to Lead Isotope Analysis. *Anal. Chem.* **2015**, *87*, 1152–1157. [CrossRef]
73. Hu, Z.C.; Li, X.H.; Luo, T. Tanz zircon megacrysts: A new zircon reference material for the microbeam determination of U–Pb ages and Zr–O isotopes. *J. Anal. At. Spectrom.* **2021**, *36*, 2715–2734. [CrossRef]

74. Liu, Y.S.; Hu, Z.C.; Gao, S.; Gunther, D.; Xu, J.; Gao, C.G.; Chen, H.H. In situ analysis of major and trace elements of anhydrous minerals by LA-ICP-MS without applying an internal standard. *Chem. Geol.* **2008**, *257*, 34–43. [CrossRef]
75. Liu, Y.S.; Gao, S.; Hu, Z.C.; Gao, C.G.; Zong, K.Q.; Wang, D.B. Continental and Oceanic Crust Recycling-induced Melt–Peridotite Interactions in the Trans-North China Orogen: U–Pb Dating, Hf Isotopes and Trace Elements in Zircons from Mantle Xenoliths. *J. Petrol.* **2010**, *52*, 392–399. [CrossRef]
76. Ludwig, K.R. Isoplot/Ex, a geochronological toolkit for Microsoft Excel, Version 3.00. *Berkeley Geochronol. Cent.* **2003**, *4*, 1–43.
77. Hu, Z.C.; Liu, Y.S.; Gao, S. Improved in situ Hf isotope ratio analysis of zircon using newly designed X skimmer cone and jet sample cone in combination with the addition of nitrogen by laser ablation multiple collector ICP-MS. *J. Anal. At. Spectrom.* **2012**, *27*, 1391–1399. [CrossRef]
78. Woodhead, J.; Hergt, J.; Shelley, M.; Eggins, S.; Kemp, R. Zircon Hf-isotope analysis with an excimer laser, depth profiling, ablation of complex geometries, and concomitant age estimation. *Chem. Geol.* **2004**, *209*, 121–135. [CrossRef]
79. Fisher, C.M.; Vervoort, J.D.; Hanchar, J.M. Guidelines for reporting zircon Hf isotopic data by LA-MC-ICPMS and potential pitfalls in the interpretation of these data. *Chem. Geol.* **2014**, *363*, 125–133. [CrossRef]
80. Blichert-Toft, J.; Chauvel, C.; Albarède, F. Separation of Hf and Lu for high-precision isotope analysis of rock samples by magnetic sector-multiple collector ICP-MS. *Contrib. Mineral. Petrol.* **1997**, *127*, 248–260. [CrossRef]
81. Söderlund, U.; Patchett, P.J.; Vervoort, J.D.; Isachsen, C.E. The ^{176}Lu decay constant determined by Lu–Hf and U–Pb isotope systematics of Precambrian mafic intrusions. *Earth Planet. Sci. Lett.* **2004**, *219*, 311–324. [CrossRef]
82. Bouvier, A.; Vervoort, J.D.; Patchett, J. The Lu-Hf and Sm-Nd isotopic composition of CHUR: Constraints from unequilibrated chondrites and implications for the bulk composition of the terrestrial planets. *Earth Planet. Sci. Lett.* **2008**, *280*, 285–295. [CrossRef]
83. Griffin, W.L.; Wang, X.; Jackson, S.E.; Pearson, N.J.; O'Reilly, S.Y.; Xu, X.S.; Zhou, X.M. Zircon chemistry and magma mixing, SE china: In-situ analysis of Hf isotopes, Tonglu and Pingtan igneous complexes. *Lithos* **2002**, *61*, 237–269. [CrossRef]
84. Griffin, W.L.; Pearson, N.J.; Belousova, E.; Jackson, S.E.; Van Achterbergh, E.; O'Reilly, S.; Shee, S.R. The Hf isotope composition of cratonic mantle: LAM-MC-ICPMS analysis of zircon megacrysts in kimberlites. *Geochim. Cosmochim. Acta* **2000**, *64*, 133–147. [CrossRef]

minerals

MDPI

Article

Petrogenesis of an Anisian A_2-Type Monzogranite from the East Kunlun Orogenic Belt, Northern Qinghai–Tibet Plateau

Chao Hui [1], Fengyue Sun [1,2,*], Shahzad Bakht [1], Yanqian Yang [3,4], Jiaming Yan [1], Tao Yu [5], Xingsen Chen [1], Yajing Zhang [1], Chengxian Liu [1], Xinran Zhu [1], Yuxiang Wang [1], Haoran Li [1], Jianfeng Qiao [1], Tao Tian [1], Renyi Song [1], Desheng Dou [1], Shouye Dong [1] and Xiangyu Lu [1]

1 College of Earth Sciences, Jilin University, Changchun 130061, China; huichao20@mails.jlu.edu.cn (C.H.); shahzadbk94@gmail.com (S.B.); yjm19910219@163.com (J.Y.); 13578713517@163.com (D.D.)
2 Key Laboratory of Mineral Resources Evaluation in Northeast Asia, Ministry of Natural Resources of China, Changchun 130061, China
3 Qinghai Geological Survey, Xining 810000, China
4 Technology Innovation Center for Exploration and Exploitation of Strategic Mineral Resources in Plateau Desert Region, Ministry of Natural Resources, Xining 810000, China
5 Northeast Oil Gas Branch of SINOPEC, Changchun 130062, China; yutao19@mails.jlu.edu.cn
* Correspondence: fengyue_sun@163.com

Abstract: Late Paleozoic to Early Mesozoic granitoids in the East Kunlun Orogenic Belt (EKOB) provide critical insights into the complex and debated relationship between Paleo–Tethyan magmatism and tectonics. This study presents integrated bulk-rock geochemical and zircon isotopic data for the Xingshugou monzogranite (MG) to address these controversies. LA-ICP-MS zircon U-Pb dating constrains the emplacement age of the MG to 247.1 ± 1.5 Ma. The MG exhibits a peraluminous and low Na_2O A_2-type granite affinity, characterized by high K_2O (4.69–6.80 wt.%) and Zr + Nb + Ce + Y (>350 ppm) concentrations, coupled with high Y/Nb (>1.2) and A/CNK ratios (1.54–2.46). It also displays low FeO^T, MnO, TiO_2, P_2O_5, and $Mg^\#$ values (26–49), alongside pronounced negative Eu anomalies (Eu/Eu* = 0.37–0.49) and moderately fractionated rare earth element (REE) patterns ($(La/Yb)_N$ = 3.30–5.11). The MG exhibits enrichment in light rare earth elements (LREEs) and large ion lithophile elements (LILEs; such as Sr and Ba), and depletion in high field strength elements (HFSEs; such as Nb, Ta, and Ti), collectively indicating an arc magmatic affinity. Zircon saturation temperatures (T_{Zr} = 868–934 °C) and geochemical discriminators suggest that the MG was generated under high-temperature, low-pressure, relatively dry conditions. Combined with positive zircon $\varepsilon_{Hf}(t)$ (1.8 to 4.7) values, it is suggested that the MG was derived from partial melting of juvenile crust. Synthesizing regional data, this study suggests that the Xingshugou MG was formed in an extensional tectonic setting triggered by slab rollback of the Paleo-Tethys Oceanic slab.

Keywords: East Kunlun Orogenic Belt; Paleo-Tethys Ocean; A_2-type granitoid; petrogenesis; Qinghai–Tibet Plateau

1. Introduction

Since the original definition of A-type granites by Loiselle and Wones [1] as alkaline, anhydrous, and anorogenic rocks distinct from M-, S-, and I-type granites [2–7], their diagnostic criteria have been progressively refined [8,9]. Contemporary classification emphasizes the characteristic high contents of SiO_2, FeO^T, $Na_2O + K_2O$, and Zr + Nb + Ce + Y (>350 ppm), as well as distinctive 10,000 Ga/Al ratios (>2.6) [9,10]. These granites are typically generated under high temperatures and low-pressure conditions in extensional settings [4,6,9,10], and can be aluminous [11,12]. A-type granites can be subdivided into A_1 and A_2 subtypes based on their characteristic geodynamic environments [6,8]. The A_1 subtype (Y/Nb < 1.2) is derived from oceanic inland basalts in continental rift or intraplate settings, whereas the A_2 subtype (Y/Nb > 1.2) originates from continental or underplated crust in post-collisional extensional regimes or during slab rollback events [6,8]. A_2-type granites are widespread in the subducted slab rollback and post-collisional settings, such as those found in the Central Asian Orogenic Belt, NW China [13], Gan-Hang Belt, SE China [14], Sakarya Zone, and NE Turkey [15].

The East Kunlun Orogenic Belt (EKOB), which forms the western portion of China's Central Orogenic Belt, records multiple tectono-magmatic events [16]. Previous research has predominantly focused on the evolution of the Precambrian supercontinents [17,18], as well as the Cambrian–Devonian Proto-Tethys Ocean [19–21], establishing relatively comprehensive evolutionary frameworks. In contrast, the evolutionary processes of the Carboniferous–Triassic Paleo-Tethys Ocean (PTO) remain subject to considerable debate [22–28]. While the Early Carboniferous opening of the PTO is widely accepted, as evidenced by ophiolitic units such as the Haerguole gabbro [29] and Dur'ngoi basalts [30,31], northeast-directed subduction is also recorded in the Xiyingzhaogou gabbro at ca. 280 Ma [26]. Despite this, three contentious tectonic models persist, as follows: (i) the prevailing model suggests that subduction persisted between 278 Ma and 240 Ma, transitioning into collisional and post-collisional stages during 230–190 Ma [26,32–35]; (ii) an early collision hypothesis proposes pre-Permian collision with post-collisional extension by ca. 247 Ma [36–39]; (iii) another hypothesis suggests that subduction continued until the Late Triassic, followed by a transition into a collisional regime [40–42].

Granitoids are fundamental constituents of the continental crust and serve as a key probe for deciphering geodynamic regimes [9,43–48]. Previous studies have focused on I-/S-type granites [49,50], adakitic rocks [51,52], mafic rocks [26,53,54], and volcanic rocks [55], employing predominantly single-rock geochemical approaches to constrain tectonic evolution in the EKOB. In contrast, few A_2-type granites related to the evolution of the PTO in the EKOB have been discovered, such as Balugou granite (ca. 244 Ma; [56]) and Dalijigetang syenogranite (ca. 259 Ma; [57]), which are important for identifying the tectonic setting.

This study presents integrated petrological, geochemical, and zircon U-Pb-Lu-Hf isotopic analyses of a monzogranite sample from the Xingshugou area in the EKOB, which shows A_2-type granite affinity. Combined with published literature, this research aims to (1) delineate the petrogenetic processes and magma sources of the Xingshugou MG; and (2) identify its tectonic setting.

2. Geological Setting and Petrography

2.1. Geological Setting

The EKOB occupies the northern margin of the Qinghai–Tibet Plateau (Figure 1a). Its boundaries comprise (1) the Qaidam Block (north); (2) the Altyn Tagh strike–slip fault system (west); (3) the Bayan Har-Songpan Ganzi Block (south); and (4) the Wenquangou–Wahongshan fault (east; Figure 1b; [49]). This E-W trending orogen extends ~1500 km with a variable width of 50–200 km [24]. Structural analysis reveals four distinct belts, as follows: (1) the Paleo-Proterozoic Northern East Kunlun Belt (NKB); (2) the Paleo-Proterozoic Central East Kunlun Belt (CKB); (3) the Meso- to Neoproterozoic Southern East Kunlun Belt (SKB); and (4) the Paleozoic A'nyemaqen Ophiolitic Belt (AOB; Figure 1b). These belts are separated by three major faults (from north to south), as follows: the Northern East Kunlun Fault, the Central East Kunlun Fault, and the Southern East Kunlun Fault (Figure 1b) [22,24]. Three parallel ophiolitic suites further delineate the EKOB's tectonic architecture, which are the Qimantage–Xiangride (486–423 Ma), Aqikelulehu–Kunzhong (555–243 Ma), and Muztage–Buqingshan–A'nyemaqen ophiolite belts (535–260 Ma), respectively [24,30,31].

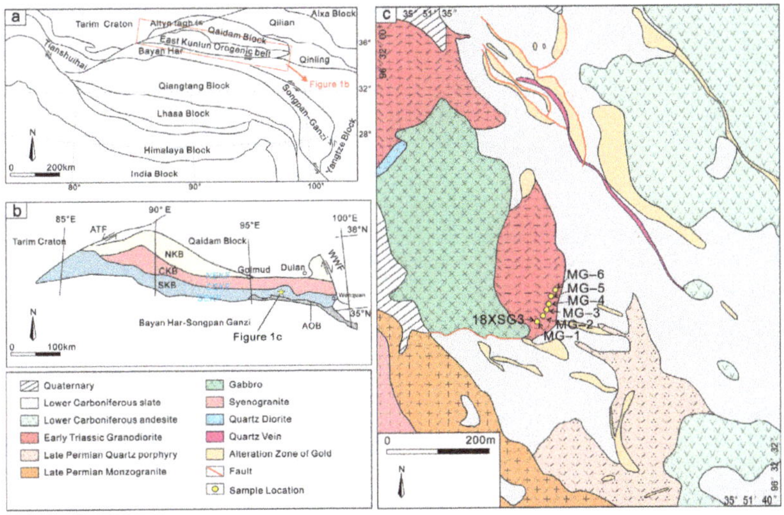

Figure 1. (**a**) Tectonic framework of the Qinghai–Tibetan Plateau (after [28]). (**b**) Structural architecture of the EKOB (modified from [22,24]). (**c**) Geological map of the Xingshugou district. Major tectonic units: Altyn Tagh strike–slip fault (ATF), NKB: Northern East Kunlun Belt. For full terms, see the Abbreviations section.

Precambrian basement rocks exhibit pronounced spatial heterogeneity in the EKOB: (1) Paleoproterozoic schist and gneiss of the Jinshuikou Group dominate the NKB and CKB; and (2) the SKB incorporates Meso- to Neoproterozoic lithologies of the Wanbaogou Group, primarily consisting of basaltic plateau sequences overlain by metavolcanic–metasedimentary units [22]. The overlying sedimentary strata span from the Mesoproterozoic to the Quaternary, and are dominated by the following: (1) the Meso- to Neoproterozoic Binggou Group, consisting of metamorphosed carbonate sequences (marble); (2) the Ordovician Tanjianshan and Nachitai Group, comprising low-grade metasedimentary rocks (schist and slate); (3) the Upper Devonian Maoniushan Formation, containing volcanic rocks (andesite, rhyolite) with minor tuff; and (4) Triassic units, which include continental clastic rocks (sandstone) and localized volcanic layers [22]. The EKOB preserves a polyphase tectonic history, including Precambrian evolution [17,18] and the oceanic cycles of the Proto-Tethyan (Cambrian–Devonian) and Paleo-Tethyan (Carboniferous–Triassic) [22,24,31,33,38,42]. The region exhibits extensive magmatic activity, predominantly associated with the tectonic evolution of the Proto- and Paleo-Tethys Oceans [24]. Granitic rocks represent the most widespread lithological units, while subordinate mafic magmatic activities also occur [24]. Intensive mantle–crust interactions during these episodes have generated widespread mineralization, including skarn Cu-Pb-Zn, hydrothermal vein-type Au-Ag, and porphyry Cu deposits [58–60].

In the Xingshugou (XSG) sector of the SKB (70 km south of Nuomuhong; Figure 1b), the Halaguole Formation (Lower Carboniferous) comprises interbedded andesite flows and carbonaceous slate (Figure 1c; [49]). Voluminous intrusions are identified in the XSG area, including the Late Permian monzogranite and quartz porphyry, the Early Triassic monzogranite, and other associated intrusions (Figure 1c). Two dominant fault assemblages are identified in the area, that is, primary NW–SE to E–W-trending structures and secondary N–S-oriented fracture sets (Figure 1c). The XSG area hosts an epithermal low-sulfidation

Au deposit, with gold-bearing veins (striking northwestward and dipping southwest), spatially associated with alteration halos comprising silicified, kaolinized, carbonatized, and pyritized zones.

2.2. Petrography

Petrographic analysis reveals the monzogranite (MG) as light gray (Figure 2b) with a fine-medium massive granitic texture (Figure 2). Subhedral plagioclase (~30 vol.%) exhibits distinct polysynthetic twinning under cross-polarized light (XPL) with maximum interference colors reaching first-order gray–white (Figure 2c,d). Subhedral K-feldspar (~25 vol.%) appears colorless in plane-polarized light (PPL), displaying Carlsbad twinning in some grains, attaining first-order gray–white interference colors under XPL (Figure 2c–f). Euhedral to subhedral biotite (~5 vol.%) shows pronounced brown pleochroism under PPL, with well-developed {001} cleavage and third-order green interference colors under XPL (Figure 2c–f). Anhedral quartz (~40 vol.%) is optically clear (colorless) and displays first-order yellow–white interference colors (Figure 2c–f). Zircon (<5 vol.%) is the dominant accessory, characterized by high relief, the absence of cleavage, and higher-order interference colors (Figure 2d).

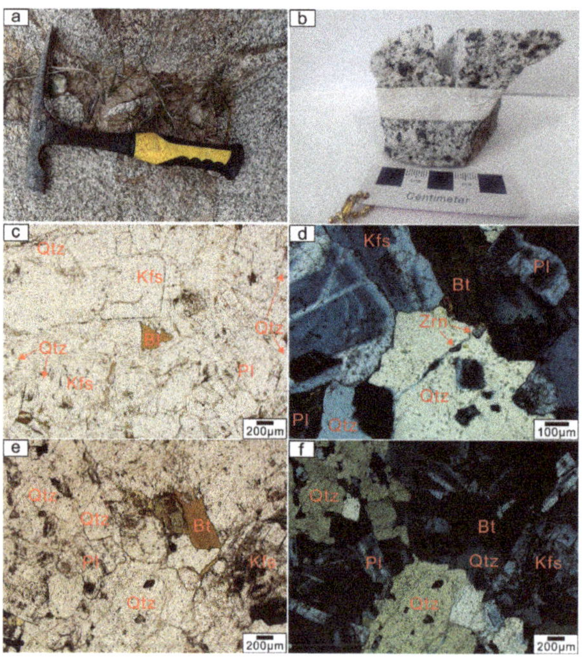

Figure 2. Xingshugou monzogranite textural and mineralogical characteristics. (**a**) Field photographs. (**b**) Hand specimen. (**c**,**e**) Plane-polarized light (PPL) microphotographs. (**d**,**f**) Cross-polarized light (XPL) microphotographs. Qtz: quartz. For full terms, see the Abbreviations section.

3. Sampling and Analytical Methods

3.1. Sampling

The MG was collected from the XSG area (96°32′14″ E, 35°51′30″ N). A total of 10 kg of fresh MG (labeled 18XSG3) was collected for zircon separation. Additionally, six fist-sized, unweathered samples (labeled MG-1 to MG-6) were collected at 10 m intervals from the same outcrop. Major and trace elements were analyzed for samples MG-1 to MG-5, while MG-6 was processed for a thin-sectioned petrographic study (Figure 1c).

3.2. Zircon U-Pb-Lu-Hf Isotope Analyses

At Tuoxuan Ltd. (Langfang, China), zircons were concentrated from crushed samples via conventional density separation, with final selection performed through manual picking under binocular microscopy. Selected grains were epoxy-mounted, polished to expose interiors, and imaged using transmitted-reflected light and cathodoluminescence (CL). Zircon U-Pb isotopes and trace elements were analyzed at Yanduzhongshi Ltd. (Langfang, China), using a NWR193 laser-ablation system (Electro Scientific Industries New Wave Research Division, Fremont, CA, USA) coupled to an Analytikjena PlasmaQuant MS Q-ICP-MS (Analytik Jena, Jena, Germany). Analyses utilized a 32 μm ablation spot. External calibration was applied using the 91500 standard [61] for U-Pb dating and SRM610 [62] for trace elements. Plešovice zircon [63] was analyzed after every 10 unknowns for quality control. Data processing was as follows: (1) isotope calculation via ICP-MS-DATACAL [64,65]; (2) common Pb correction [66]; and (3) weighted average age calculation and Concordia diagram construction using ISOPLOT [67]. Methodological details are referenced [16].

In situ Lu-Hf isotope analyses employed the same zircon domains previously dated for U-Pb, utilizing a Neptune MC-ICP-MS (Thermo Fisher Scientific, Waltham, MA, USA) coupled to a NWR193 laser microprobe (Electro Scientific Industries New Wave Research Division, Fremont, CA, USA) at Yanduzhongshi Ltd. (Beijing, China). A 38 μm laser spot was used for the analyses; the methodology is detailed in [16,68].

3.3. Whole-Rock Geochemical Analyses

Fresh samples were powdered (less than 200 mesh) via an agate mill. Analyses conducted at Key Laboratory of Mineral Resources Evaluation in Northeast Asia, Ministry of Natural Resources of China (KLMRENAMNR; Changchun, China) employed the following: major element analysis was performed using ZSX Primus II X-ray fluorescence (XRF; Rigaku, Tokyo, Japan) on fused glass disks. The flux mixture ($Li_2B_4O_7$:$LiBO_2$:LiF = 45:10:5) incorporated NH_4NO_3 as an oxidant and LiBr as a release agent. Fusion proceeded at 1050 °C for 15 min. The X-ray tube operated at a power of 4 kW with a Rh target, and during the sample measurement, the operating voltage and current were set to 50 kV and 60 mA respectively, with a test diameter of 30 mm. Trace elements were analyzed by Agilent 7500a ICP-MS (Agilent Technologies, Santa Clara, CA, USA) following HF + HNO_3 digestion in Teflon bombs. Precision assessment using BHVO-1 (basalt), BCR-2 (basalt), and AGV-1 (andesite) standards yielded the following: (1) major elements, <1% RSD; and (2) trace elements, <5% for >10 ppm and <10% RSD for <10 ppm. Methodological details follow [69].

4. Results

4.1. Zircon U-Pb Geochronology

The zircon U-Pb isotopic data for the MG are presented in Table 1. Zircon grains selected from the MG exhibited transparent, light brown, euhedral crystals with well-developed oscillatory zoning (Figure 3a). Their high Th/U ratios (0.28–2.17; Table 1) further supported their magmatic origin [70]. Twenty concordant analyses yielded a weighted mean $^{206}Pb/^{238}U$ age of 247.1 ± 1.5 Ma (MSWD = 0.13, n = 20; Figure 3c), which is interpreted as the emplacement age of the MG.

Table 1. LA-ICP-MS zircon U-Pb geochronology of the Xingshugou monzogranite.

Sample Name	Content (ppm)			Isotopic Ratios						Isotopic Ages (Ma)					
	U	Th	Th/U	$^{207}Pb/^{206}Pb$	1σ	$^{207}Pb/^{235}U$	1σ	$^{206}Pb/^{238}U$	1σ	$^{207}Pb/^{206}Pb$	1σ	$^{207}Pb/^{235}U$	1σ	$^{206}Pb/^{238}U$	1σ
XSG-monzogranite, weighted mean age: 247.1 ± 1.5 Ma, MSWD = 0.13															
18XSG31-1	204	420	2.06	0.05247	0.00248	0.27854	0.01132	0.03876	0.00063	306.0	107.8	249.5	9.0	245.1	3.9
18XSG31-2	184	285	1.54	0.05163	0.00216	0.27503	0.01169	0.03880	0.00058	268.9	95.8	246.7	9.3	245.4	3.6
18XSG31-3	276	525	1.90	0.05134	0.00170	0.27432	0.00964	0.03881	0.00062	256.3	75.9	246.1	7.7	245.5	3.8
18XSG31-4	206	334	1.62	0.05182	0.00187	0.27737	0.01019	0.03883	0.00050	277.6	82.8	248.6	8.1	245.6	3.1
18XSG31-5	297	602	2.03	0.05106	0.00137	0.27548	0.00771	0.03900	0.00043	243.6	61.9	247.1	6.1	246.6	2.7
18XSG31-6	215	358	1.67	0.05153	0.00180	0.27491	0.00891	0.03903	0.00049	264.4	80.1	246.6	7.1	246.8	3.0
18XSG31-7	198	262	1.32	0.05064	0.00171	0.27299	0.00972	0.03906	0.00048	224.6	78.2	245.1	7.8	247.0	3.0
18XSG31-8	160	210	1.31	0.05143	0.00227	0.27475	0.01254	0.03906	0.00073	260.0	101.4	246.5	10.0	247.0	4.6
18XSG31-9	153	180	1.18	0.05138	0.00214	0.27492	0.01097	0.03908	0.00051	257.9	95.8	246.6	8.7	247.1	3.2
18XSG31-10	375	858	2.29	0.05038	0.00132	0.27265	0.00803	0.03909	0.00049	212.7	60.5	244.8	6.4	247.2	3.0
18XSG31-11	311	621	2.00	0.05141	0.00124	0.27687	0.00717	0.03911	0.00043	259.5	55.2	248.2	5.7	247.3	2.7
18XSG31-12	102	110	1.07	0.05282	0.00247	0.28371	0.01262	0.03911	0.00054	321.0	106.3	253.6	10.0	247.3	3.4
18XSG31-13	234	398	1.70	0.05094	0.00170	0.27536	0.00927	0.03911	0.00048	238.3	76.7	247.0	7.4	247.3	3.0
18XSG31-14	228	342	1.50	0.05105	0.00171	0.27489	0.00936	0.03918	0.00051	242.9	77.1	246.6	7.5	247.7	3.2
18XSG31-15	182	253	1.39	0.05164	0.00212	0.27846	0.01184	0.03919	0.00059	269.5	94.1	249.4	9.4	247.8	3.7
18XSG31-16	157	236	1.50	0.05108	0.00211	0.27526	0.01210	0.03919	0.00065	244.6	95.2	246.9	9.6	247.8	4.1
18XSG31-17	207	349	1.69	0.05226	0.00236	0.27936	0.01280	0.03926	0.00082	296.6	102.9	250.2	10.2	248.2	5.1
18XSG31-18	181	193	1.06	0.05165	0.00300	0.27342	0.01474	0.03930	0.00057	270.1	133.2	245.4	11.8	248.5	3.6
18XSG31-19	103	108	1.05	0.05141	0.00291	0.27813	0.01506	0.03939	0.00069	259.4	129.8	249.2	12.0	249.1	4.3
18XSG31-20	429	863	2.01	0.05178	0.00181	0.28273	0.01083	0.03978	0.00073	275.7	80.2	252.8	8.6	251.5	4.5
91500	65	23	0.36	0.07336	0.00161	1.80139	0.04462	0.17792	0.00193	1024.0	44.5	1046.0	16.2	1055.6	10.6
91500	66	23	0.36	0.07412	0.00161	1.83355	0.04620	0.17953	0.00224	1044.8	43.9	1057.6	16.6	1064.4	12.2
91500	64	23	0.35	0.07366	0.00150	1.79364	0.04153	0.17689	0.00218	1032.1	41.2	1043.1	15.1	1050.0	11.9
91500	63	23	0.36	0.07563	0.00142	1.86696	0.04028	0.17935	0.00223	1085.2	37.7	1069.5	14.3	1063.4	12.2

Table 1. *Cont.*

Sample Name	Content (ppm)			Isotopic Ratios						Isotopic Ages (Ma)					
	U	Th	Th/U	$^{207}Pb/^{206}Pb$	1σ	$^{207}Pb/^{235}U$	1σ	$^{206}Pb/^{238}U$	1σ	$^{207}Pb/^{206}Pb$	1σ	$^{207}Pb/^{235}U$	1σ	$^{206}Pb/^{238}U$	1σ
XSG-monzogranite, weighted mean age: 247.1 ± 1.5 Ma, MSWD = 0.13															
91500	64	23	0.36	0.07395	0.00142	1.82994	0.03801	0.17960	0.00222	1040.0	38.8	1056.3	13.6	1064.8	12.1
91500	64	23	0.36	0.07560	0.00150	1.84876	0.03991	0.17716	0.00210	1084.5	39.8	1063.0	14.2	1051.5	11.5
91500	66	24	0.36	0.07530	0.00137	1.83971	0.03696	0.17734	0.00219	1076.6	36.4	1059.8	13.2	1052.4	12.0
91500	64	23	0.36	0.07463	0.00142	1.83928	0.03800	0.17871	0.00210	1058.5	38.2	1059.6	13.6	1059.9	11.5
Plešovice	609	52	0.09	0.05387	0.00082	0.40034	0.00709	0.05387	0.00047	365.7	34.5	341.9	5.1	338.3	2.9
Plešovice	690	58	0.08	0.05327	0.00075	0.39886	0.00742	0.05417	0.00058	340.4	32.0	340.8	5.4	340.0	3.5
Plešovice	397	34	0.09	0.05329	0.00092	0.39870	0.00762	0.05424	0.00062	341.0	39.0	340.7	5.5	340.5	3.8
Plešovice	495	42	0.09	0.05317	0.00082	0.39814	0.00727	0.05411	0.00057	336.0	34.8	340.3	5.3	339.7	3.5
SRM610	456	453	1.00	0.90832	0.00974	26.81986	0.37107	0.21398	0.00174	5101.9	15.2	3376.9	13.5	1250.0	9.2
SRM610	459	457	1.00	0.91541	0.00718	26.99940	0.37404	0.21317	0.00209	5112.9	11.1	3383.4	13.6	1245.7	11.1
SRM610	462	456	0.99	0.93054	0.00795	26.78482	0.24975	0.20837	0.00184	5136.0	12.1	3375.6	9.1	1220.1	9.8
SRM610	468	462	0.99	0.92115	0.00442	26.85824	0.22928	0.21065	0.00176	5121.7	6.8	3378.3	8.4	1232.3	9.4

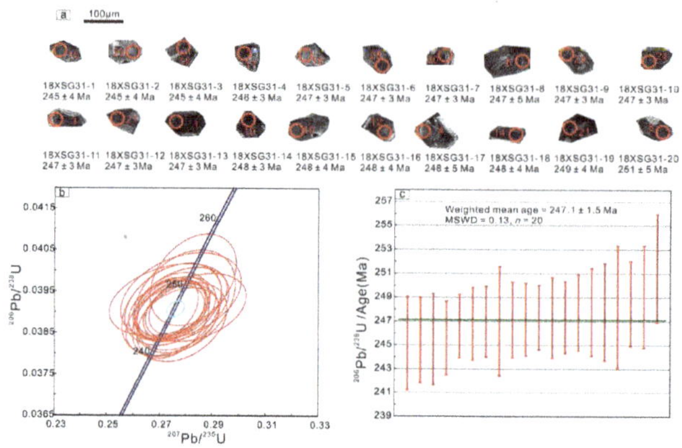

Figure 3. (**a**) Representative zircon CL images for the Xingshugou monzogranite; (**b**) zircon U-Pb Concordia diagrams; and (**c**) weighted mean age diagrams.

4.2. Major and Trace Elements

The major and trace elements of the MG are presented in Table 2. The MG displays high SiO_2 (63.91–74.67 wt.%) and K_2O (4.69–6.80 wt.%) contents, and low Na_2O (0.17–0.82 wt.%), characteristics of granitoids with a subalkaline affinity (Figure 4b) and high K composition (Figure 4c). It also displays peraluminous characteristics, with high A/CNK ratios (A/CNK = molar $Al_2O_3/(CaO + Na_2O + K_2O)$; 1.54–2.46; Figure 4d). The MG exhibits weak fractionation of rare earth elements (REEs; Figure 5a). It also shows significant negative Eu anomalies (Eu/Eu* = 0.37–0.49; Figure 5a). It is enriched in large ion lithophile elements (LILEs, including Sr and Ba; Figure 5b), but depleted in high field strength elements (HFSEs, including Nb, P, and Ti; Figure 5b).

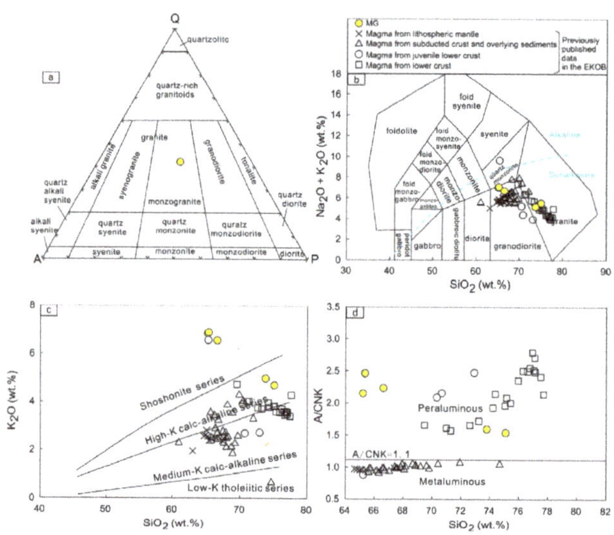

Figure 4. Diagrams of classification for the Xingshugou monzogranite. (**a**) QAP [71]; (**b**) TAS [72]; (**c**) K_2O-SiO_2 [73]; (**d**) A/CNK-SiO_2. All oxides in wt.%. Q represents quartz, P represents plagioclase, and A represents alkaline feldspar in (**a**). Magma from the lithospheric mantle [74], magma from the subducted crust and overlying sediments [36,75], magma from the juvenile lower crust [56], and magma from the lower crust [50]. Supplementary Table S1 presents published whole-rock geochemical data from the EKOB.

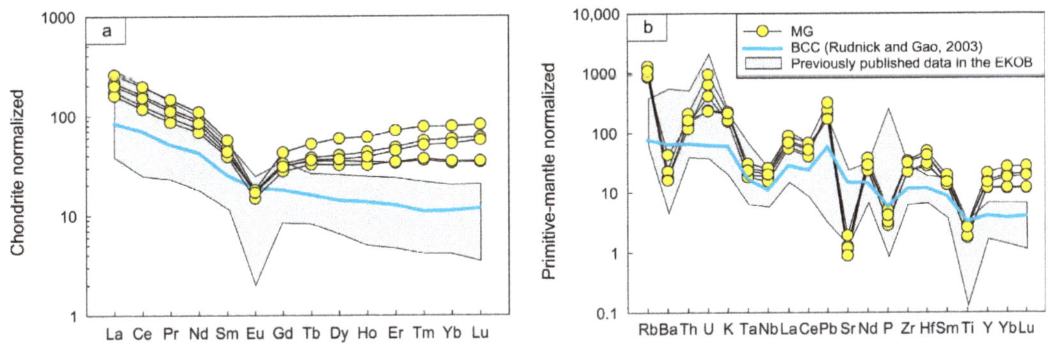

Figure 5. Normalized patterns for the Xingshugou monzogranite. (**a**) Chondrite-normalized REE patterns; (**b**) primitive mantle-normalized trace element patterns. Normalization values are from [76], BCC (bulk crust composition [77]). Supplementary Table S1 presents published whole-rock geochemical data from the EKOB.

Table 2. Whole-rock geochemistry of the Xingshugou monzogranite: major oxide concentrations (wt.%) and trace element abundances (ppm).

Sample No.	MG-1	MG-2	MG-3	MG-4	MG-5
Major elements (wt.%)					
SiO_2	74.67	63.91	72.37	64.43	65.23
TiO_2	0.38	0.59	0.40	0.61	0.57
Al_2O_3	13.61	21.06	14.85	21.54	20.23
Fe_2O_3	1.89	0.98	0.72	1.14	0.93
FeO	1.40	1.81	1.62	1.74	1.84
MnO	0.08	0.11	0.15	0.07	0.11
MgO	0.61	1.40	0.89	1.45	1.42
CaO	1.30	1.18	2.01	0.56	0.98
Na_2O	0.82	0.21	0.18	0.21	0.17
K_2O	4.69	6.72	4.89	6.80	6.43
P_2O_5	0.06	0.10	0.07	0.11	0.09
LOI	0.95	1.15	1.20	1.10	1.20
Total	100.46	99.23	99.36	99.77	99.19
FeO^T	3.11	2.74	2.31	2.80	2.73
$Mg^{\#}$	26.24	48.41	41.46	48.59	48.79
A/CNK	1.54	2.15	1.60	2.46	2.24
A/NK	2.11	2.76	2.65	2.79	2.79
Trace elements (ppm)					
Li	14.3	35.4	24.4	34.0	33.1
Be	6.20	19.4	10.7	14.4	18.4
Sc	5.63	6.50	5.66	9.05	7.18
Ti	2276	3557	2405	3637	3402
V	17.5	26.2	18.8	30.9	25.5
Cr	4.99	3.50	3.20	8.99	2.57
Co	5.96	3.23	3.52	4.52	4.03
Ni	8.63	2.69	10.1	14.2	4.9
Cu	12.5	20.0	8.50	19.8	10.8
Ga	18.2	29.0	20.3	26.4	27.8
As	7.7	35.7	20.2	24.2	20.6
Rb	546	736	569	830	699
Sr	40.7	23.5	26.8	25.3	18.8
Y	53.7	73.0	54.6	97.7	67.3

Table 2. *Cont.*

Sample No.	MG-1	MG-2	MG-3	MG-4	MG-5
Zr	250	368	250	379	358
Nb	13.7	16.1	11.1	18.6	14.6
Mo	4.48	2.53	3.27	4.40	2.84
Ag	0.13	0.14	0.15	0.13	0.1
Cd	0.28	0.07	0.27	1.3	0.11
Sb	1.17	1.33	1.33	1.75	1.42
Ba	306	125	120	163	114
La	41.2	50.6	37.6	61.2	47.3
Ce	77.1	98.3	71.6	120	92.5
Pr	9.07	11.2	8.21	13.8	10.6
Nd	35.2	42.6	31.9	51.2	39.5
Sm	6.30	7.29	5.98	8.76	6.73
Eu	1.02	1.02	0.86	1.05	0.99
Gd	6.45	6.84	5.74	8.83	6.38
Tb	1.31	1.43	1.22	1.95	1.35
Dy	8.97	10.2	8.17	15.1	9.29
Ho	1.96	2.45	1.82	3.46	2.19
Er	5.60	7.98	5.70	11.7	7.40
Tm	0.92	1.43	0.95	1.98	1.30
Yb	5.78	10.0	5.97	13.3	8.93
Lu	0.88	1.55	0.90	2.07	1.45
Hf	8.68	13.1	9.59	15.3	13.1
Ta	0.81	1.06	0.75	1.26	0.98
Tl	2.93	3.45	2.72	3.67	3.11
Pb	39.1	35.1	44.5	58.6	31.5
Th	11.8	12.3	9.95	17.8	13.9
U	8.62	5.01	13.3	19.8	4.87
ΣREEs	201.63	253.04	186.54	314.56	235.84
ΣLREEs	169.78	211.12	156.07	256.16	197.57
ΣHREEs	31.86	41.92	30.47	58.40	38.27
$(La/Yb)_N$	5.11	3.63	4.52	3.29	3.80
Eu/Eu*	0.49	0.44	0.45	0.37	0.46
T_{Zr} (°C)	868	922	869	934	923

LOI = loss on ignition; $Mg^{\#}$ = 100 × molar MgO/(MgO + FeO^T); A/CNK = molar Al_2O_3/(CaO + Na_2O + K_2O); A/NK = molar Al_2O_3/(Na_2O + K_2O); FeO^T = FeO + Fe_2O_3 × 0.8998; $(La/Yb)_N$ = ($La_{analyzed}/La_{CI}$)/($Yb_{analyzed}/Yb_{CI}$); Eu/Eu* = ($Eu_{analyzed}/Eu_{CI}$)/{[($Sm_{analyzede}/Sm_{CI}$) × ($Gd_{analyzed}/Gd_{CI}$]^(1/2)}; T_{Zr} (°C) was determined following [78]. Subscripts denote: analyzed = measured sample concentrations, CI = chondritic reference values from [76].

4.3. *Zircon Lu-Hf Isotopes*

The Lu-Hf data are listed in Table 3. All the $^{176}Lu/^{177}Hf$ ratios (< 0.002) indicate a negligible radiogenic ^{176}Hf contribution from ^{176}Lu decay, suggesting that the measured $^{176}Hf/^{177}Hf$ values reflect the initial composition of the magma system [79,80]. For the MG, the initial $^{176}Hf/^{177}Hf$ values at 247 Ma range from 0.282677 to 0.282765. Analyses carried out on the same magmatic grains that defined the crystallization age provided positive $\varepsilon_{Hf}(t)$ values (1.8–4.7; Figure 6) and Tonian two-stage model ages (T_{DM2}) (832 to 983 Ma).

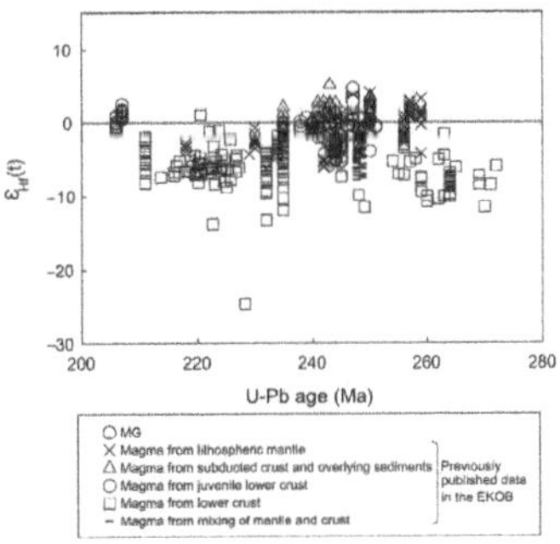

Figure 6. Diagrams of ε_{Hf} (t) values vs. zircon ^{206}Pb-^{238}U ages for the Xingshugou monzogranite. Magma from the lithospheric mantle [33,54,74,81–84], magma from the subducted crust and overlying sediments [32,36,75], magma from the juvenile lower crust [56,84,85], magma from the lower crust [25,50,86,87], magma from the mantle–crust mixing [38]. Supplementary Table S2 presents in situ zircon Lu-Hf isotope data from the EKOB.

Table 3. In situ Lu-Hf isotope analyses of zircon for the Xingshugou monzogranite.

Sample Name	t (Ma)	$^{176}Yb/^{177}Hf$	2σ	$^{176}Lu/^{177}Hf$	2σ	$^{176}Hf/^{177}Hf$	2σ	$\varepsilon_{Hf}(0)$	$\varepsilon_{Hf}(t)$	2σ	T_{DM1}	T_{DM2}	$f_{Lu/Hf}$
18XSG31-1	247	0.073503	0.001425	0.002697	0.000045	0.282756	0.000018	−0.6	4.4	0.6	736	992	−0.92
18XSG31-2	247	0.077179	0.000443	0.002852	0.000017	0.282729	0.000019	−1.5	3.5	0.7	779	1054	−0.91
18XSG31-3	247	0.059668	0.001238	0.002182	0.000035	0.282741	0.000018	−1.1	4.0	0.6	748	1021	−0.93
18XSG31-4	247	0.047300	0.001338	0.001816	0.000053	0.282677	0.000017	−3.4	1.8	0.6	833	1162	−0.95
18XSG31-5	247	0.075752	0.000578	0.002811	0.000020	0.282728	0.000020	−1.6	3.4	0.7	780	1057	−0.92
18XSG31-6	247	0.055265	0.000690	0.002080	0.000026	0.282758	0.000020	−0.5	4.6	0.7	721	981	−0.94
18XSG31-7	247	0.049305	0.000641	0.001948	0.000017	0.282726	0.000023	−1.6	3.5	0.8	765	1052	−0.94
18XSG31-8	247	0.042402	0.000381	0.001624	0.000016	0.282720	0.000018	−1.8	3.3	0.6	766	1062	−0.95
18XSG31-9	247	0.055569	0.000454	0.002136	0.000014	0.282711	0.000023	−2.2	2.9	0.8	790	1087	−0.94
18XSG31-10	247	0.081582	0.000413	0.003029	0.000014	0.282765	0.000019	−0.2	4.7	0.7	729	975	−0.91
18XSG31-11	247	0.067793	0.000852	0.002531	0.000018	0.282708	0.000017	−2.3	2.7	0.6	804	1099	−0.92
18XSG31-12	247	0.069428	0.000637	0.002589	0.000030	0.282711	0.000019	−2.2	2.8	0.7	801	1093	−0.92
18XSG31-13	247	0.045726	0.000561	0.001838	0.000025	0.282705	0.000016	−2.4	2.7	0.6	793	1099	−0.94

The parameters used in our calculations: $(^{176}Lu/^{177}Hf)_{CHUR}$ = 0.0332, $(^{176}Hf/^{177}Hf)_{CHUR}$ = 0.282772, the CHUR value is from [88]; $(^{176}Lu/^{177}Hf)_{DM}$ = 0.0384, $(^{176}Hf/^{177}Hf)_{DM}$ = 0.28325, the DM value is from [89]; $\lambda(^{176}Lu) = 1.867 \times 10^{-11}$ a^{-1} [90]. The $^{176}Lu/^{177}Hf$ (C) = 0.015 [89]. Applied to Xingshugou monzogranite at t = 247 Ma.

5. Discussion

5.1. Petrogenesis of the Xingshugou Monzogranite

Peraluminous indicators, including garnet, cordierite, andalusite, and muscovite, are systematically absent (Figure 2), along with a negative P_2O_5–SiO_2 relationship (Figure 7a), suggesting that the MG belongs to the I-type granite category [2,91,92]. However, the MG shows A-type granite affinity characterized by high Zr + Nb + Ce + Y (287–615 ppm) contents (Figure 7b,c; [3]), and it plots into the peraluminous A-type granite field (Figure 7d; [93]). The lower 10,000 × Ga/Al ratios (2.4–2.7; Figure 7e,f) may be caused by the lower instrument detection capability of Ga or the low contents of Ga in the source [56]. Fractional crystallization of ilmenite, apatite, plagioclase, and mafic minerals is indicated by coherent depletions in Eu (Eu/Eu* = 0.37–0.49; Figure 5a), Ba, P, Ti, and Sr (Figure 5b), alongside negative correlations of P_2O_5-TiO_2-FeO^T-$Mg^\#$ versus SiO_2 (Figure 7g–i). The A-type granites can be distinguished from fractionated I-type granites according to high-temperature, low-pressure, and anhydrous conditions [1,3,94,95]. The zircon saturation temperature (T_{Zr} = 868–934 °C, Table 2) can provide a minimum magma liquidus temperature because of the absence of inherited zircons in the samples [78]. Therefore, the MG originated from a high-temperature condition (greater than 868 °C), which is higher than typical I-type granites [96–98]. The low $CaO/(FeO^T + MgO + TiO_2)$ and $(CaO + FeO^T + MgO + TiO_2)$ values suggest that the magma forms under low-pressure conditions (Figure 8a; [99]). Low La/Yb and Sr/Y ratios also imply that the MG originates from a plagioclase stable area (Figure 8b; [100]). In addition, increasing La_N/Sm_N and Th/Nd ratios with progressive La_N and Th enrichment (Figure 8c,d) indicate predominant partial melting over fractional crystallization during magma evolution. Therefore, the depletion of Eu, Sr, and Ba further indicates that plagioclase is a residual mineral in the source region, rather than the result of strong fractional crystallization of plagioclase [94]. Furthermore, the high HFSE contents and high T_{Zr} value imply that the MG is derived from a relatively dry condition [101]. To summarize, the MG, which forms under high-temperature, low-pressure, and dry conditions, exhibits possible A-type granite affinity with low Na_2O, high K_2O, and a peraluminous composition.

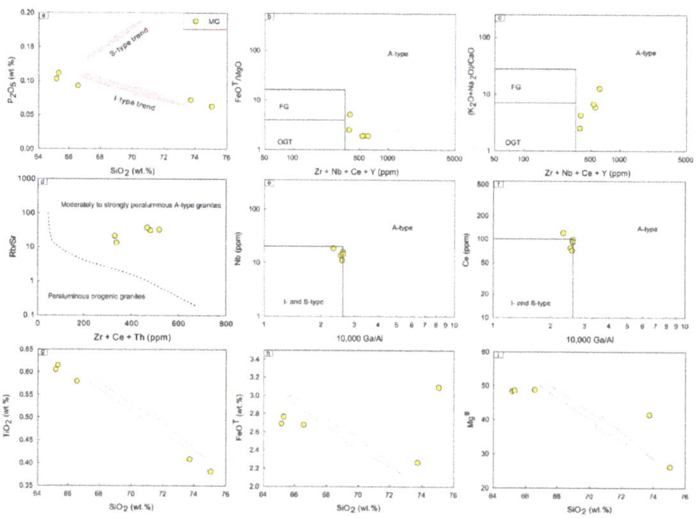

Figure 7. Geochemical discrimination diagrams for the Xingshugou monzogranite. (**a**) P_2O_5–SiO_2; (**b**) FeO^T/MgO–Zr + Nb + Ce + Y (ppm); (**c**) $(K_2O + Na_2O)/CaO$–Zr + Nb + Ce + Y; (**d**) Rb/Sr–Zr + Ce + Th; (**e**) Nb–10,000 Ga/Al; (**f**) Ce–10,000 Ga/Al; (**g**) TiO_2–SiO_2 (wt.%); (**h**) FeO^T–SiO_2; (**i**) $Mg^\#$–SiO_2. (**b**,**c**) are from [3]; (**e**,**f**) are from [6]; (**d**) is from [93]. The yellow dots represent MG in the Xingshugou area. Pink arrows (**a**): I/S type granites trend; gray arrows (**g**–**i**): fractional crystallization vectors. FT and OGT mean fractionated granites and unfractionated granites, respectively. $Mg^\# = 100 \times$ molar $MgO/(MgO + FeO^T)$.

Current petrogenetic hypotheses for A-type granites propose three distinct mechanisms, as follows: (1) partial melting of crustal protoliths [3,4,102]; (2) crust–mantle hybridization [103]; and (3) extreme crystal fractionation of mantle-derived basaltic magmas [104]. For the MG, crust–mantle mixing can be discounted based on the absence of mafic microgranular enclaves (MMEs) and limited variation of $\varepsilon_{Hf}(t)$ values (1.8–4.7; Figure 6) [105,106]. Similarly, mantle-derived fractionation is inconsistent with its high SiO_2 contents (63.91–74.67 wt.%) and low compatible element concentrations such as Cr (2.57–8.99 ppm), Ni (2.69–14.2 ppm), and Co (3.23–5.96 ppm), effectively refuting Model (3) [104,107].

The low Nb/U (0.83–3.20) and Ce/Y (1.61–2.94) ratios of the MG imply a crustal affinity, as the primitive mantle displays higher Nb/U (33.95) and Ce/Pb (9.59) ratios [76], whereas the bulk continental crust exhibits lower Nb/U (6.15) and Ce/Y (3.91) ratios [77]. Source inheritance is further evidenced by low Nb^* (= $[Nb/Th]_{Analyzed}/[Nb/Th]_{Primitive\ mantle}$)

and Ta* (= $[\text{Ta}/\text{U}]_{\text{Analyzed}}/[\text{Ta}/\text{U}]_{\text{Primitive mantle}}$) values (Figure 8e), which remain source-inherited during differentiation [108]. The low ratios of $Al_2O_3/(FeO^T + MgO + TiO_2)$ and $(Na_2O + K_2O)/(FeO^T + MgO + TiO_2)$ of the MG (Figure 8f,g) further indicate that the source of the MG is mafic pelites [99]. Moreover, compared with the alkaline A-type granite, the aluminous A-type granite originates from the crust [12]. The positive $\varepsilon_{Hf}(t)$ (1.8–4.7) values of the MG are overlapped with the granitoids derived from subducted oceanic crust (Figure 6; [32,36,75]) in the EKOB. However, the MG cannot be derived from oceanic crust because of the lower Sr/Y (0.26–0.76) and $(La/Yb)_N$ (3.29–5.11) ratios, and older T_{DM2} (983 to 832 Ma). Furthermore, the MG does not include MMEs, in contrast to magmatic rocks derived from oceanic crust in the EKOB [32,36,75]. Therefore, the Tonian juvenile crustal sediments may be the source of the MG. The MG was generated under high-temperature conditions (more than 934 °C), so additional heat flux will be needed to make the juvenile crust melt. Widespread crust–mantle interaction is evidenced by the abundant presence of MMEs within granitoids in the EKOB [24,32,36–38,51], as well as by the occurrence of a small volume of coeval mafic rocks [82,109], both of which are attributed to asthenospheric upwelling. The MG likely formed through partial melting of the Tonian juvenile crust, driven by thermal input from asthenospheric upwelling.

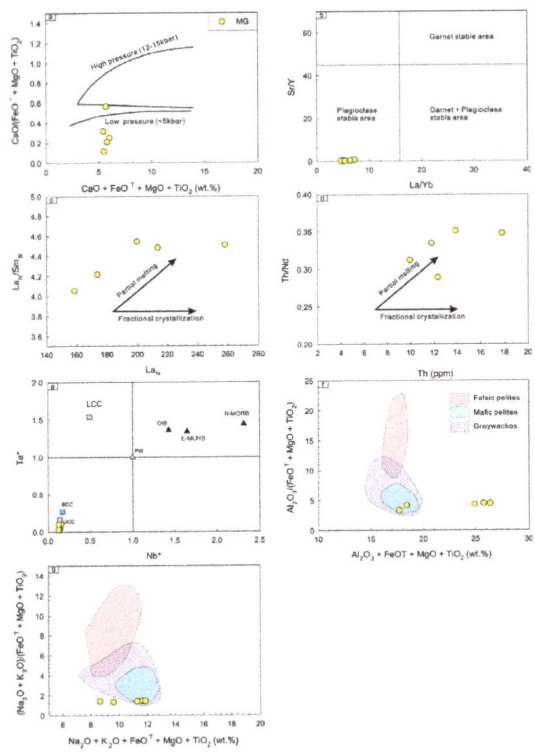

Figure 8. Diagrams of the magmatic source and petrogenesis discrimination of the Xingshugou monzogranite. (**a**) $CaO/(FeO^T + MgO + TiO_2)$ vs. $(CaO + FeO^T + MgO + TiO_2)$ (wt.%) [99]; (**b**) Sr/Y vs. La/Yb [100]; (**c**) La_N vs. La_N/Sm_N [110]; (**d**) Th (ppm) vs. Th/Nd [110] (**e**) Nb* vs. Ta* [108]; (**f**) $Al_2O_3/(FeO^T + MgO + TiO_2)$ vs. $Al_2O_3 + FeO^T + MgO + TiO_2$ (wt.%) [99]; (**g**) $(Na_2O + K_2O)/(FeO^T + MgO + TiO_2)$ vs. $Na_2O + K_2O + FeO^T + MgO + TiO_2$ (wt.%) [99]. OIB, E-MORB, and N-MORB in (**e**) are from [76], and BCC, LCC, and UCC in (**e**) are from [77]. Nb* and Ta* mean $[\text{Nb}/\text{Th}]_{\text{Analyzed}}/[\text{Nb}/\text{Th}]_{\text{Primitive mantle}}$ and $[\text{Ta}/\text{U}]_{\text{Analyzed}}/[\text{Ta}/\text{U}]_{\text{Primitive mantle}}$, respectively.

5.2. Implications for the Geodynamic Evolution

Zircons selected from the MG display magmatic crystallization features, exhibiting transparent, light brown euhedral crystals with prominent growth banding in CL images (Figure 3a) and high Th/U ratios (0.28–2.17) [70]. The zircon $^{206}Pb/^{238}U$ weighted mean age of the MG is ca. 247 Ma (Figure 3b,c). The MG exhibits an arc-related geochemical signature, that is, LILE enrichment coupled with HFSE depletion (Figure 5), and plots within the within-plate granite (WPG) field and oceanic ridge granite (ORG) field on the tectonic discrimination diagrams (Figure 9; [43]). All of the above suggests that the MG was emplaced in an active continental margin setting at ca. 247 Ma.

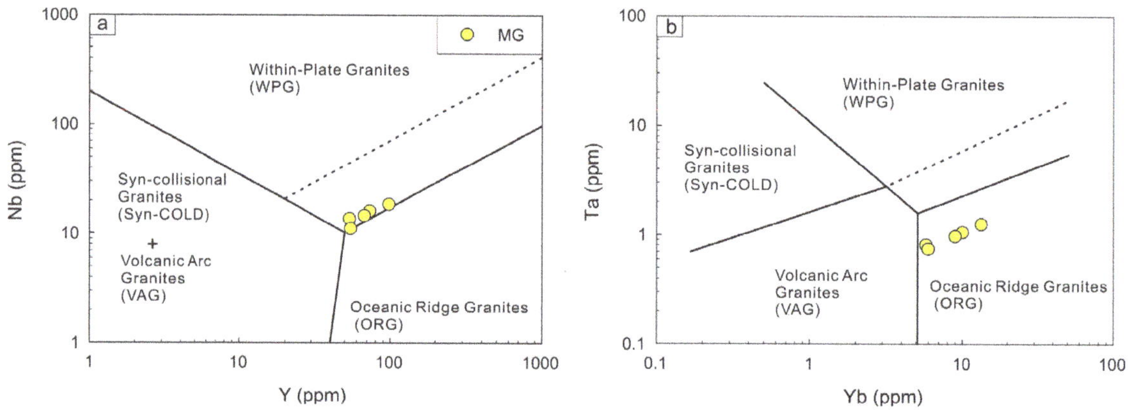

Figure 9. Diagrams of tectonic affinity assessment [43] for the Xingshugou monzogranite. (**a**) Nb–Y; (**b**) Ta–Yb.

A-type granites were initially thought to form within an anorogenic environment [1]. However, Condie et al. (2023) [8] demonstrated that over 85% of A-type granites actually form in orogenic environments. A-type granites can be subdivided into A_1 and A_2 subtypes based on characteristic geodynamic environments [6,8]. The A_1 subtype (Y/Nb < 1.2, enriched in Nb, Ta, Zr, and Hf, and depleted in Ba, Sr, P, Eu and Ti) is derived from oceanic inland basalts in continental rift or intraplate settings, whereas the A_2 subtype (Y/Nb > 1.2, enriched in Tb, Th, U, K and depleted in Ba, Nb, Ta, Sr, P, Eu and Ti) originates from continental or underplated crust in post-collisional extensional regimes or during slab rollback events [6,8]. The MG shows a clear A_2-type feature with high Y/Nb ratios (>1.2), plotting within the A_2-type field in the triangular diagrams of Ce-Nb-Y (Figure 10a; [6]) and 3Ga-Nb-Y (Figure 10b; [6]), and is depleted in Nb, Ta, Ba, Ti, P, and Eu (Figure 5b; [6]). Furthermore, minor A-type granitoids formed in a subduction-related setting in the EKOB have been discovered during the Permian–Triassic, such as Balugou granite (ca. 244 Ma; [56]) and Dalijigetang syenogranite (ca. 259 Ma; [57]). In summary, the MG was likely emplaced in an extensional setting during the northward subduction of the PTO around 247 Ma.

The EKOB is a typical subduction–accretion orogenic belt, and records multiphase tectono-magmatic events, including the Precambrian evolution of the Columbia and Rodinia supercontinents [17,18], the Cambrian–Devonian Proto-Tethys Ocean [19–21], the Carboniferous–Triassic PTO [22–28], and post-Triassic uplift [111–113]. The following contrasting geodynamic models have been proposed to unravel its complex evolution: (1) the accordion-style opening–closing structure, proposed by [39]; (2) a model involving multiple islands and small oceanic basins in the pre-orogenic stage and soft-collision in the orogenic stage (not a typical Wilson Cycle) proposed by [42]; (3) with the identification of the Mesoto Neoproterozoic Wanbaogou Basaltic Plateau (WBP), a new model was presented [22]. All studies agree that the EKOB underwent the evolution of PTO from the Carboniferous to the Triassic, but the detailed evolutionary process remains controversial [22,24,36,37,53,69].

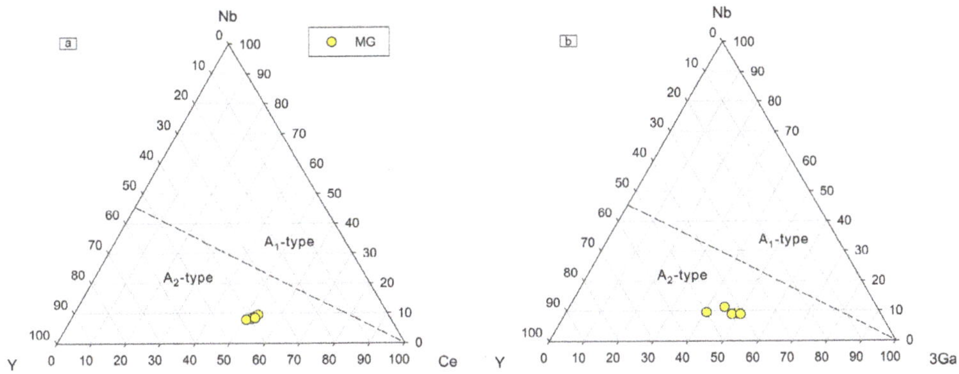

Figure 10. A_1- and A_2-type granitoid discrimination diagrams for the Xingshugou monzogranite [6]. (**a**) Ce–Nb–Y diagram; (**b**) 3 Ga–Nb–Y diagram. The dashed lines represent the Y/Nb ratio = ½; A_1-type granitoids are from anorogenic environments, such as continental rifts or intraplate settings, A2-type granitoids are from post-collisional environments, such as post-orogenic or slab rollback.

There is now broad agreement on the key processes that governed the evolution of the PTO in the EKOB: (1) oceanic opening prior to the Early Carboniferous, as evidenced by ophiolite suites such as the Haerguole gabbro [29] and Dur'ngoi basalts [30,31]; (2) and the initiation of northeastward subduction, recorded in the Xiyingzhaogou gabbro at ca. 280 Ma [26]. However, substantial controversies persist regarding the timing and duration: (i) Some studies propose continuous oceanic subduction from ca. 278–240 Ma, followed by collision (240–225 Ma) [16,22,24,28,34]; (ii) others argue for prolonged subduction lasting into the Late Triassic [40–42,114]; (iii) a different view suggests pre–Late Permian collisional onset [36,37].

To resolve these discrepancies, this study integrates stratigraphic architecture, petrology, geochemistry, and regional tectonic settings with newly published data to reconstruct a refined evolutionary framework for the PTO in the EKOB. Between 280 and 240 Ma, the subduction regime evolved from flat-slab underthrusting, accompanied by magmatic quiescence (Figures 11 and 12c,d), to widespread crust–mantle hybridization (263–240 Ma; Figure 12e,f). The transition culminated in a high-flux igneous episode characterized by diverse I-/A_2-type granites, adakitic, and mafic rocks, which resulted from the upwelling asthenospheric mantle driven by slab rollback [87,109]. Coeval Cu-Ni mineralization hosted in ultramafic rocks [115] also implies this extensional setting. The subsequent collision regime (240–230 Ma) was recorded by a microangular unconformity within the Middle Triassic strata [116]. This syn-collisional phase coincided with a magmatic lull (Figure 11) and localized uplift [116] (Figure 12g,h). The post-collisional stage (230–195 Ma) involved large-scale crustal shortening, evidenced by angular stratigraphic discordance between the Babaoshan Formation and underlying strata [116]. Geophysical evidence delineates orogen-scale crustal attenuation coeval with lithospheric foundering [117,118], instigating asthenospheric influx and late-stage A_2-type/adakitic magmatic resurgence [34,37,56,119] (Figures 11 and 12i,j).

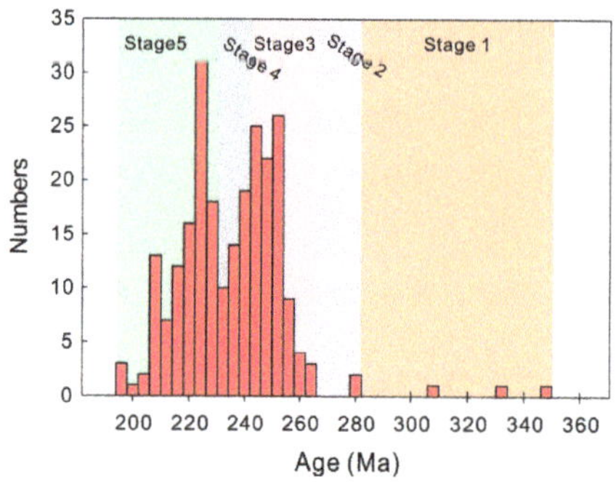

Figure 11. Temporal spectrum of U-Pb geochronology for magmatic rocks (345–195 Ma) in the EKOB. Stage 1: oceanic spreading initiation of the PTO during 345–280 Ma; Stage 2: low-angle subduction; Stage 3: slab rollback-driven flare-up; Stage 4: syn-collisional tectonic regime; Stage 5: post-collisional tectonic regime. Supplementary Table S3 compiles the data and related references.

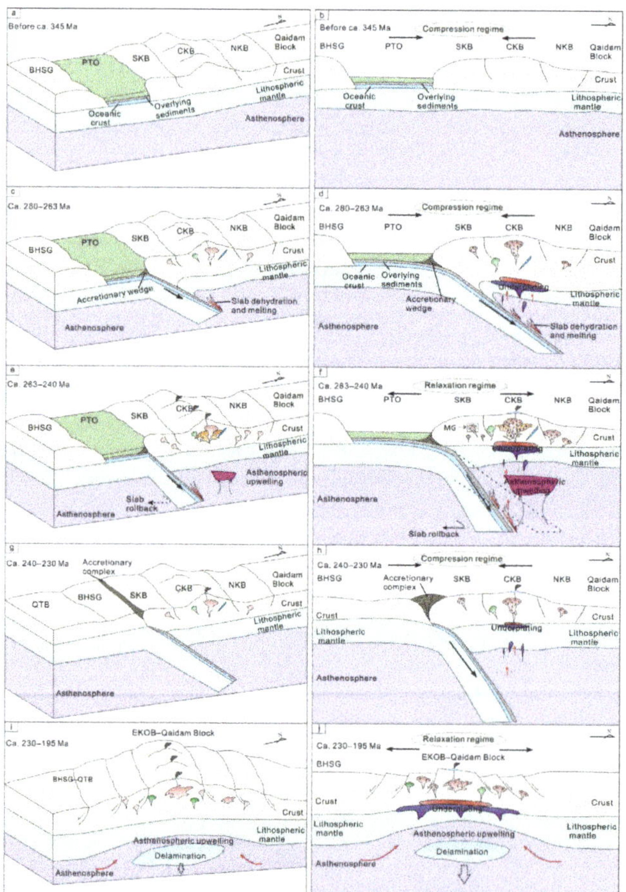

Figure 12. Diagrams of the geodynamic evolution of the PTO in the EKOB [49]. (**a**,**b**) Opening of the PTO before 345 Ma; (**c**,**d**) shallow-dipping slab subduction (280–263 Ma); (**e**,**f**) subducted slab rollback (263–240 Ma); (**g**,**h**) syn-collisional compression (240–230 Ma); (**i**,**j**) post-collisional relaxation (230–195 Ma). PTO: Paleo-Tethys Ocean. For full terms, see the Abbreviations section.

A five-phase evolutionary model of the PTO in the EKOB was established as follows: initial oceanic spreading commenced before the Early Carboniferous (ca. 345 Ma), as evidenced by ophiolitic remnants. Northeastward subduction was initiated by ca. 280 Ma, transitioning from low-angle slab descent (280–263 Ma) to rollback-induced steep subduction (263–240 Ma). Terminal convergence between the EKOB and Bayan Har Block occurred during 240–230 Ma. The EKOB ultimately experienced lithospheric foundering between 230 and 195 Ma.

6. Conclusions

The Xingshugou MG consists of distinctive mineral phases, including quartz, K-feldspar, plagioclase, biotite, and minor zircon. Zircon U-Pb dating constrains its emplacement to 247.1 ± 1.5 Ma. Geochemical data indicate an A_2-type granite affinity, characterized by high K_2O, low Na_2O, elevated A/CNK ratios, high zircon saturation temperatures, and positive $\varepsilon_{Hf}(t)$ values. The MG is interpreted to have originated from the partial melting of Tonian juvenile crust under low-pressure, fluid-absent, and high-temperature conditions. These granites formed in an extensional tectonic setting triggered by slab rollback of the Paleo-Tethys Oceanic slab.

Supplementary Materials: The following supporting information can be downloaded at: https://www.mdpi.com/article/10.3390/min15070685/s1, Table S1. Summary of the published whole-rock major (wt.%) and trace elements (ppm) data of granites in the East Kunlun Orogenic Belt. Table S2. Summary of the published zircon Hf isotopic data of granites in the East Kunlun Orogenic Belt. Table S3. Geochronological summary of the Permian–Triassic igneous rocks in the East Kunlun Orogenic Belt. References [120–174] are cited in Supplementary Materials.

Author Contributions: Conceptualization, C.H. and F.S.; methodology, C.H.; software, S.B.; validation, J.Y. and S.B.; formal analysis, T.Y., S.B., Y.Z., C.L., X.Z. and Y.W.; investigation, J.Q., T.T., X.C. and X.Z.; writing—original draft preparation, C.H.; writing—review and editing, F.S.; visualization, H.L., R.S., D.D., S.D. and X.L.; supervision, F.S.; funding acquisition, Y.Y. All authors have read and agreed to the published version of the manuscript.

Funding: This research was funded by the Qinghai Geological Survey Project, grant numbers 2022012005ky005, 2023085026ky001; the National Natural Science Foundation of China (NSFC) Project (grant number 42302078).

Data Availability Statement: The original contributions presented in the study are included in the article/Supplementary Materials.

Acknowledgments: We would like to thank the Key Laboratory of Mineral Resources Evaluation in Northeast Asia, the Ministry of Natural Resources of China, and Jilin University, as well as the Langfang Tuoxuan Rock and Mineral Testing Service Co., Ltd., for their assistance with the analyses. We thank all the editors and anonymous reviewers for their constructive comments, which improved this manuscript.

Conflicts of Interest: Tao Yu is an employee of the Northeast Oil & Gas Branch of SINOPEC. This paper reflects the views of the scientists and not the company. The other authors declare no conflicts of interest.

Abbreviations

The following abbreviations are used in this manuscript:

EKOB	East Kunlun Orogenic Belt
PTO	Paleo-Tethys Ocean;

AOB	A'nyemaqen Ophiolitic Belt
NKB	Northern East Kunlun Belt
CKB	Central East Kunlun Belt
SKB	Southern East Kunlun Belt
NEKF	Northern East Kunlun Fault
CEKF	Central East Kunlun Fault
SEKF	Southern East Kunlun Fault
ATF	Altyn Tagh Strike–Slip Fault
WWF	Wenquangou–Wahongshan Fault
XSG	Xingshugou
MG	Monzogranite
Qtz	quartz
Pl	plagioclase
Kfs	potassium feldspar
Zrn	zircon
Bt	biotite
CL	cathodoluminescence
REEs	rare earth elements
LILEs	large ion lithophile elements
HFSEs	high field strength elements
TDM2	two-stage model age
MMEs	mafic microgranular enclaves
BHSG	Bayan Har–Songpan Ganzi;
QTB	Qiangtang Block;

References

1. Loiselle, M.C.; Wones, D.R. Characteristics and origin of anorogenic granites. *Geol. Soc. Am* **1979**, *11*, 468.
2. Chappell, B.W.; White, A.J.R. Two contrasting granite types: 25 years later. *Aust. J. Earth Sci.* **2001**, *48*, 489. [CrossRef]
3. Whalen, J.B.; Currue, K.L.; Chappell, B.W. A-type granites: Geochemical characteritics, discrimination and petrogenesis. *Contrib. Mineral. Petrol.* **1987**, *95*, 407–419. [CrossRef]
4. Collins, W.J.; Beams, S.D.; White, A.J.R.; Chappell, B.W. Nature and Origin of A-Type Granites with Particular Reference to Southeastern Australia. *Contrib. Mineral. Petrol.* **1982**, *80*, 189–200. [CrossRef]
5. Defant, M.J.; Drummond, M.S. Derivation of some modern arc magmas by melting of young subducted lithosphere. *Nature* **1990**, *347*, 662–665. [CrossRef]
6. Eby, G.N. Chemical subdivision of the A-type granitoids:Petrogenetic and tectonic implications. *Geology* **1992**, *20*, 641–644. [CrossRef]
7. Didier, J.; Duthou, J.L.; Lameyre, J. Mantle and crustal granites: Genetic classification of orogenic granites and the nature of their enclaves. *J. Volcanol. Geotherm. Res.* **1982**, *14*, 125–132. [CrossRef]
8. Condie, K.C.; Pisarevsky, S.A.; Puetz, S.J.; Roberts, N.M.W.; Spencer, C.J. A-type granites in space and time: Relationship to the supercontinent cycle and mantle events. *Earth Planet. Sci. Lett.* **2023**, *610*, 118125. [CrossRef]
9. Whalen, J.B.; Hildebrand, R.S. Trace element discrimination of arc, slab failure, and A-type granitic rocks. *Lithos* **2019**, *348–349*, 105179. [CrossRef]
10. Frost, C.D.; Frost, B.R. On Ferroan (A-type) Granitoids: Their Compositional Variability and Modes of Origin. *J. Petrol.* **2011**, *52*, 39–53. [CrossRef]
11. Zhang, Y.W.; Zhang, X.F.; Chen, L.X.; Pang, Z.S.; Chen, H.; Xue, J.L.; Zhou, Y.; Teng, C.; Chen, G.C. Geochronology and Geochemistry of Early Cretaceous A-type Granites in Central–Eastern Inner Mongolia, China: Implications for Late Mesozoic Tectonic Evolution of the Southern Great Xing'an Range. *Acta Geol. Sin. (Engl. Ed.)* **2023**, *97*, 1094–1111. [CrossRef]
12. King, P.L.; White, A.J.R.; Chappell, B.W.; Allen, C.M. Characterization and Origin of Aluminous A-type-Granites from the Lachlan Fold Belt, Southeastern Australia. *J. Petrol.* **1997**, *38*, 371–391. [CrossRef]
13. Yin, J.; Chen, W.; Xiao, W.; Yuan, C.; Windley, B.F.; Yu, S.; Cai, K. Late Silurian–early Devonian adakitic granodiorite, A-type and I-type granites in NW Junggar, NW China: Partial melting of mafic lower crust and implications for slab roll-back. *Gondwana Res.* **2017**, *43*, 55–73. [CrossRef]

14. Yang, S.-Y.; Liang, J.-P.; Jiang, S.-Y.; Zhang, X.; Zhang, R.-X. Late Jurassic–Early Cretaceous irregular slab rollback of paleo-Pacific plate beneath southeastern China: Insights from the petrogenesis of volcanic rocks of Moshishan Group in Dazhou volcanic basin, Gan-Hang Belt. *Lithos* **2021**, *392–393*, 106137. [CrossRef]
15. Karsli, O.; Aydin, F.; Uysal, I.; Dokuz, A.; Kumral, M.; Kandemir, R.; Budakoglu, M.; Ketenci, M. Latest Cretaceous "A2-type" granites in the Sakarya Zone, NE Turkey: Partial melting of mafic lower crust in response to roll-back of Neo-Tethyan oceanic lithosphere. *Lithos* **2018**, *302–303*, 312–328. [CrossRef]
16. Pan, Z.C.; Sun, F.Y.; Cong, Z.C.; Tian, N.; Xin, W.; Wang, L.; Zhang, Y.J.; Wu, D.Q. Petrogenesis and Tectonic Implications of the Triassic Granitoids in the Ela Mountain Area of the East Kunlun Orogenic Belt. *Minerals* **2022**, *12*, 880. [CrossRef]
17. Wu, D.Q.; Sun, F.Y.; Pan, Z.C.; Yu, L.; Li, L.; Gao, H.C.; Tian, N.; Xu, C.H. Neoproterozoic magmatic and metamorphic imprints in the East Kunlun Orogenic Belt, North Tibetan Plateau, NW China: Implications for the assembly and initial breakup of the Rodinia supercontinent. *Precambrian Res.* **2021**, *354*, 106076. [CrossRef]
18. He, D.F.; Dong, Y.P.; Hauzenberger, C.A.; Yue, Y.G.; Hui, B.; Zhou, B.; Ren, X.; Zhang, B.; Chong, F.B. 1.38 Ga magmatism and the extension tectonics in East Kunlun, northern Tibetan Plateau. *Precambrian Res.* **2024**, *412*, 107551. [CrossRef]
19. Fu, L.B.; Bagas, L.; Wei, J.H.; Chen, Y.; Chen, J.J.; Zhao, X.; Zhao, Z.X.; Li, A.B.; Zhang, W.K. Growth of early Paleozoic continental crust linked to the Proto-Tethys subduction and continental collision in the East Kunlun Orogen, northern Tibetan Plateau. *GSA Bull.* **2023**, *135*, 1709–1733. [CrossRef]
20. Zhao, X.; Fu, L.B.; Santosh, M.; Wei, J.H.; Chen, J.J. The growth and evolution of continental crust contributed by multiple sources in the East Kunlun Orogen during Early Paleozoic. *Earth-Sci. Rev.* **2022**, *233*, 104190. [CrossRef]
21. Ma, L.T.; Sun, G.C.; Dai, L.Q.; Zhao, Z.F.; Hu, Y.D.; Yang, Q.C.; Qin, L.; Gong, B. Syn- and post-collisional mafic igneous rocks from the East Kunlun Orogenic Belt: Implications for recycling of subducting continental crust during the Silurian-Devonian. *Lithos* **2024**, *474–475*, 107599. [CrossRef]
22. Sun, F.Y. (Geological Survey Institute of Jilin University, Changchun, China); Li, B.L. (Geological Survey Institute of Jilin University, Changchun, China). Research on Major Prospecting Challenges in the East Kunlun Metallogenic Belt. 2009; (Unpublished work; In Chinese)
23. Bian, Q.T.; Li, D.H.; Pospelov, I.; Yin, L.M.; Li, H.S.; Zhao, D.S.; Chang, C.F.; Luo, X.Q.; Gao, S.L.; Astrakhantsev, O.; et al. Age, geochemistry and tectonic setting of Buqingshan ophiolites, North Qinghai-Tibet Plateau, China. *J. Asian Earth Sci.* **2004**, *23*, 577–596. [CrossRef]
24. Dong, Y.P.; He, D.F.; Sun, S.S.; Liu, X.M.; Zhou, X.H.; Zhang, F.F.; Yang, Z.; Cheng, B.; Zhao, G.C.; Li, J.H. Subduction and accretionary tectonics of the East Kunlun orogen, western segment of the Central China Orogenic System. *Earth-Sci. Rev.* **2018**, *186*, 231–261. [CrossRef]
25. Fan, X.Z.; Sun, F.Y.; Xu, C.H.; Wu, D.Q.; Yu, L.; Wang, L.; Yan, C.; Bakht, S. Volcanic rocks of the Elashan Formation in the Dulan-Xiangride Basin, East Kunlun Orogenic Belt, NW China: Petrogenesis and implications for Late Triassic geodynamic evolution. *Int. Geol. Rev.* **2021**, *64*, 1270–1293. [CrossRef]
26. Hu, C.B.; Feng, C.Y.; Li, M.; Zha, X.F.; Gao, X.F.; Gao, Z.K.; Li, M. Subduction and retreat of Permian oceanic plates in the East Kunlun Orogenic Belt: Evidence from mafic-ultramafic intrusive rocks. *Int. Geol. Rev.* **2024**, *66*, 2717–2740. [CrossRef]
27. Liu, B.; Ma, C.Q.; Huang, J.; Wang, L.X.; Zhao, S.Q.; Yan, R.; Sun, Y.; Xiong, F.H. Petrogenesis and tectonic implications of Upper Triassic appinite dykes in the East Kunlun orogenic belt, northern Tibetan Plateau. *Lithos* **2017**, *284–285*, 766–778. [CrossRef]
28. Xia, R.; Deng, J.; Qing, M.; Li, W.L.; Guo, X.D.; Zeng, G.Z. Petrogenesis of ca. 240 Ma intermediate and felsic intrusions in the Nan'getan: Implications for crust–mantle interaction and geodynamic process of the East Kunlun Orogen. *Ore Geol. Rev.* **2017**, *90*, 1099–1117. [CrossRef]
29. Liu, Z.Q.; Pei, X.Z.; Liu, R.B.; Li, Z.C.; Zhang, X.F.; Liu, Z.G.; Chen, G.C.; Chen, Y.X.; Ding, S.P.; Guo, J.F. LA-ICP-MS Zircon U-Pb Geochronology of the Two Suites of Ophiolites at the Buqingshan Area of the A'nyemaqen Orogenic Belt in the Southern Margin of East Kunlun and Its Tectonic Implication. *Acta Geol. Sin.* **2011**, *85*, 185–194, (In Chinese with English Abstract)
30. Chen, L.; Sun, Y.; Pei, X.Z.; Gao, M. Northernmost paleo-tethyan oceanic basin in Tibet: Geo-chronological evidence from $^{40}Ar/^{39}Ar$ age dating of Dur'ngoi ophiolite. *Chin. Sci. Bull.* **2001**, *46*, 1203–1205. [CrossRef]
31. Yang, J.S.; Shi, R.D.; Wu, C.L.; Wang, X.B.; Robinson, P.T. Dur'ngoi ophiolite in East Kunlun, Northeast Tibetan plateau: Evidence for paleo-Tethyan suture in Northwest China. *J. Earth Sci.* **2009**, *20*, 303–331. [CrossRef]
32. Chen, J.J.; Wei, J.H.; Fu, L.B.; Li, H.; Zhou, H.Z.; Zhao, X.; Zhan, X.F.; Tan, J. Multiple sources of the Early Mesozoic Gouli batholith, Eastern Kunlun Orogenic Belt, northern Tibetan Plateau: Linking continental crustal growth with oceanic subduction. *Lithos* **2017**, *292–293*, 161–178. [CrossRef]
33. Wu, D.Q.; Sun, F.Y.; Pan, Z.C.; Tian, N. Geochronology, geochemistry, and Hf isotopic compositions of Triassic igneous rocks in the easternmost segment of the East Kunlun Orogenic Belt, NW China: Implications for magmatism and tectonic evolution. *Int. Geol. Rev.* **2020**, *63*, 1011–1029. [CrossRef]

34. Xiong, F.H.; Ma, C.Q.; Zhang, J.Y.; Liu, B.; Jiang, H.A. Reworking of old continental lithosphere: An important crustal evolution mechanism in orogenic belts, as evidenced by Triassic I-type granitoids in the East Kunlun orogen, Northern Tibetan Plateau. *J. Geol. Soc.* **2014**, *171*, 847–863. [CrossRef]
35. Dai, Q.; Dong, Y.; He, D.; Sun, S.; Hui, B.; Zhang, B.; Zuo, Z.; Chong, F.; Luo, Q.; Xiao, J. Mesozoic crust-mantle interaction in the East Kunlun Orogenic Belt, northern Tibetan Plateau: Constraints from the Tuolahai granodiorite and MMEs. *Lithos* **2025**, *508–509*, 108093. [CrossRef]
36. Huang, H.; Niu, Y.; Nowell, G.; Zhao, Z.; Yu, X.; Zhu, D.C.; Mo, X.; Ding, S. Geochemical constraints on the petrogenesis of granitoids in the East Kunlun Orogenic belt, northern Tibetan Plateau: Implications for continental crust growth through syn-collisional felsic magmatism. *Chem. Geol.* **2014**, *370*, 1–18. [CrossRef]
37. Kong, J.J.; Niu, Y.L.; Hu, Y.; Zhang, Y.; Shao, F.L. Petrogenesis of the Triassic granitoids from the East Kunlun Orogenic Belt, NW China: Implications for continental crust growth from syn-collisional to post-collisional setting. *Lithos* **2020**, *364–365*, 105513. [CrossRef]
38. Li, R.B.; Pei, X.Z.; Pei, L.; Li, Z.C.; Chen, G.C.; Chen, Y.X.; Liu, C.J.; Wang, M. The Early Triassic Andean-type Halagatu granitoids pluton in the East Kunlun orogen, northern Tibet Plateau: Response to the northward subduction of the Paleo-Tethys Ocean. *Gondwana Res.* **2018**, *62*, 212–226. [CrossRef]
39. Jiang, C.F.; Yang, J.S.; Feng, B.G.; Zhu, Z.Z.; Zhao, M.; Chai, Y.C.; Shi, X.D.; Wang, H.D.; Hu, J.Q. *Opening–Closing Tectonicns of Kunlun Mountains*; Geological Press: Beijing, China, 1992; p. 222, (In Chinese with English Abstract)
40. Yuan, C.; Sun, M.; Xiao, W.J.; Wilde, S.; Li, X.H.; Liu, X.H.; Long, X.P.; Xia, X.P.; Ye, K.; Li, J.L. Garnet-bearing tonalitic porphyry from East Kunlun, Northeast Tibetan Plateau: Implications for adakite and magmas from the MASH Zone. *Int. J. Earth Sci.* **2009**, *98*, 1489–1510. [CrossRef]
41. Wang, Q.; Li, Z.X.; Chung, S.L.; Wyman, D.A.; Sun, Y.L.; Zhao, Z.H.; Zhu, Y.T.; Qiu, H.N. Late Triassic high-Mg andesite/dacite suites from northern Hohxil, North Tibet: Geochronology, geochemical characteristics, petrogenetic processes and tectonic implications. *Lithos* **2011**, *126*, 54–67. [CrossRef]
42. Yin, A.; Harrison, T.M. Geologic Evolution of the Himalayan-Tibetan Orogen. *Annu. Rev. Earth Planet. Sci.* **2000**, *28*, 211–280. [CrossRef]
43. Pearce, J.A.; Harris, N.B.W.; Tindle, A.G. Trace Element Discrimination Diagrams for the Tectonic Interpretation of Granitic Rocks. *J. Petrol.* **1984**, *25*, 956–983. [CrossRef]
44. Dong, P.; Dong, G.; Santosh, M.; Mo, X.; Sun, Z.; Ketchaya, Y.B.; Pan, Y.; Lemdjou, Y.B. Eocene magmatism in the western Tengchong Block: Implications for crust-mantle interaction associated with the slab rollback of the Neo-Tethys Ocean. *Gondwana Res.* **2022**, *106*, 259–280. [CrossRef]
45. Turner, S.J.; Langmuir, C.H. Sediment and ocean crust both melt at subduction zones. *Earth Planet. Sci. Lett.* **2022**, *584*, 117424. [CrossRef]
46. Amani, K.; Delavari, M.; Amini, S.; Azizi, H.; Asahara, Y.; Furman, T.; Shabani, A.A.T.; Asiabanha, A.; Mohammadi, A. Geochemistry, Sr-Nd isotopes and zircon U-Pb dating of magmatic rocks from the Talesh range, western Alborz: New insights into Late Cretaceous evolution of the southern Eurasian margin. *Geochemistry* **2024**, *84*, 126042. [CrossRef]
47. Malaviarachchi, S.P.K.; Zhao, L.; Guo, J.; Li, X.; Zhai, M.; Dharmapriya, P.L. Insights for intraplate extensional settings revealed from geochemistry of metamorphosed magmatic rocks of the central Gondwana supercontinent. *Precambrian Res.* **2025**, *425*, 107801. [CrossRef]
48. Sun, H.; Song, Y.; Wilkinson, J.J.; Liu, Z.; Zheng, M.; Wang, B. Petrogenesis of Early Cretaceous Duorenlieqian igneous rocks (113∼117 Ma) in the western Bangong-Nujiang metallogenic belt, Tibet, China: Implications for tectono-magmatic evolution and porphyry Cu-Au mineralization. *Ore Geol. Rev.* **2025**, *182*, 106650. [CrossRef]
49. Hui, C.; Sun, F.; Wang, T.; Yang, Y.; Chai, Y.; Yan, J.; Shahzad, B.; Li, B.; Zhang, Y.; Yu, T.; et al. Petrogenesis and Tectonic Setting of Late Permian Granitoids in the East Kunlun Orogenic Belt, NW China: Constraints from Petrology, Geochemistry and Zircon U-Pb-Lu-Hf Isotopes. *Minerals* **2025**, *15*, 381. [CrossRef]
50. Kamaunji, V.D.; Wang, L.-X.; Ma, C.-Q.; Liu, J.; Zhu, Y.-X. Petrogenesis and tectonic implication of the Permian-Triassic syenogranites from the eastern segment of the East Kunlun Orogen, China. *Lithos* **2021**, *402–403*, 105932. [CrossRef]
51. Gan, J.; Xiong, F.H.; Xiao, Q.R.; Wang, W.; Yan, D.D. Petrogenesis and Geodynamic Implications of Late Triassic Mogetong Adakitic Pluton in East Kunlun Orogen, Northern Tibet: Constraints from Zircon U–Pb–Hf Isotopes and Whole-Rock Geochemistry. *Front. Earth Sci.* **2022**, *10*, 845763. [CrossRef]
52. Norbu, N.; Tang, L.; Li, J.; Kong, H.; Li, Y.; Jia, Q.; Xu, Y. Petrogenesis of Middle Triassic Adakite-like Intrusions in the Asiha Orogenic Gold Deposit, East Kunlun Orogenic Belt, China. *Minerals* **2023**, *13*, 74. [CrossRef]
53. Du, W.; Pei, L.; Li, Z.C.; Li, R.B.; Chen, Y.X.; Liu, C.J.; Chen, G.C.; Pei, X.Z. The Middle Triassic Intermediate to Basic Rocks in the Eastern Kunlun Orogenic Belt, Northeast Tibet: Implication for the Paleo-Tethyan Ocean Closure. *Minerals* **2024**, *14*, 667. [CrossRef]

54. Hu, Y.; Niu, Y.L.; Li, J.Y.; Ye, L.; Kong, J.J.; Chen, S.; Zhang, Y.; Zhang, G.R. Petrogenesis and tectonic significance of the late Triassic mafic dikes and felsic volcanic rocks in the East Kunlun Orogenic Belt, Northern Tibet Plateau. *Lithos* **2016**, *245*, 205–222. [CrossRef]
55. Pan, Z.; Sun, F.; Cong, Z. Petrogenesis, Sources, and Tectonic Settings of Triassic Volcanic Rocks in the Ela Mountain Area of the East Kunlun Orogen: Insights from Geochronology, Geochemistry and Hf Isotopic Compositions. *Minerals* **2022**, *12*, 1085. [CrossRef]
56. Ding, Q.F.; Jiang, S.Y.; Sun, F.Y. Zircon U–Pb geochronology, geochemical and Sr–Nd–Hf isotopic compositions of the Triassic granite and diorite dikes from the Wulonggou mining area in the Eastern Kunlun Orogen, NW China: Petrogenesis and tectonic implications. *Lithos* **2014**, *205*, 266–283. [CrossRef]
57. Peng, B.; Li, B.L.; Zhao, T.F.; Wang, C.; Chang, J.J.; Wang, G.Z.; Yang, W.L. Identification of A-type granite in the southeastern Kunlun Orogen, Qinghai Province, China: Implications for the tectonic framework of the Eastern Kunlun Orogen. *Geol. J.* **2017**, *52*, 454–469. [CrossRef]
58. Qu, H.Y.; Friehauf, K.; Santosh, M.; Pei, R.F.; Li, D.X.; Liu, J.N.; Zhou, S.M.; Wang, H. Middle–Late Triassic magmatism in the Hutouya Fe–Cu–Pb–Zn deposit, East Kunlun Orogenic Belt, NW China: Implications for geodynamic setting and polymetallic mineralization. *Ore Geol. Rev.* **2019**, *113*, 103088. [CrossRef]
59. Yin, S.; Ma, C.Q.; Xu, J.N.; Fu, J.L.; Zhang, X.N. The role of crystal mush in porphyry systems: A case study from the Baishiya ore field, East Kunlun orogenic belt, northern Qinghai-Tibet plateau. *Ore Geol. Rev.* **2022**, *146*, 104962. [CrossRef]
60. Yu, M.; Zeng, Q.H.; Wang, H.; Zhang, J.D.; Mao, J.W.; Feng, C.Y. Lithospheric influence on metallogenesis in the East Kunlun Orogen: Insights from isotopic and geochemical mapping. *J. Geochem. Explor.* **2024**, *263*, 107515. [CrossRef]
61. Wiedenbeck, M.; Alle, P.; Corfu, F.; Griffin, W.L.; Meier, M.; Oberli, F.; Quadt, A.V.; Roddick, J.C.; Spiegel, W. Three Natural Zircon Standards for U-Th-Pb, Lu-Hf, Trace element and REE Analyses. *Geostand. Newsl.* **1995**, *19*, 1–23. [CrossRef]
62. Liu, Y.; Hu, Z.; Zong, K.; Gao, C.; Gao, S.; Xu, J.; Chen, H. Reappraisement and refinement of zircon U-Pb isotope and trace element analyses by LA-ICP-MS. *Chin. Sci. Bull.* **2010**, *55*, 1535–1546. [CrossRef]
63. Sláma, J.; Košler, J.; Condon, D.J.; Crowley, J.L.; Gerdes, A.; Hanchar, J.M.; Horstwood, M.S.A.; Morris, G.A.; Nasdala, L.; Norberg, N.; et al. Plešovice zircon—A new natural reference material for U–Pb and Hf isotopic microanalysis. *Chem. Geol.* **2008**, *249*, 1–35. [CrossRef]
64. Liu, Y.; Gao, S.; Hu, Z.; Gao, C.; Zong, K.; Wang, D. Continental and Oceanic Crust Recycling-induced Melt-Peridotite Interactions in the Trans-North China Orogen: U-Pb Dating, Hf Isotopes and Trace Elements in Zircons from Mantle Xenoliths. *J. Petrol.* **2010**, *51*, 537–571. [CrossRef]
65. Liu, Y.S.; Hu, Z.C.; Gao, S.; Günther, D.; Xu, J.; Gao, C.G.; Chen, H.H. In situ analysis of major and trace elements of anhydrous minerals by LA-ICP-MS without applying an internal standard. *Chem. Geol.* **2008**, *257*, 34–43. [CrossRef]
66. Andersen, T. Correction of common lead in U–Pb analyses that do not report 204Pb. *Chem. Geol.* **2002**, *192*, 59–79. [CrossRef]
67. Ludwig, K.R. *User's Manual for Isoplot 3.00—A Geochronological Toolkit for Microsoft Excel*; Berkeley Geochronology Center Special Publication: Berkeley, CA, USA, 2003; Volume 4, pp. 1–4.
68. Wu, F.Y.; Yang, Y.H.; Xie, L.W.; Yang, J.H.; Xu, P. Hf isotopic compositions of the standard zircons and baddeleyites used in U–Pb geochronology. *Chem. Geol.* **2006**, *234*, 105–126. [CrossRef]
69. Li, H.R.; Qiao, J.F.; Sun, F.Y.; Qian, Y.; Wang, Y.Z.; Bakht, S.; Liu, H.T.; Gu, Y. Three episodes of Triassic volcanism in the Eastern Kunlun Orogen, NW China: Constraints for evolution of the Palaeo-Tethys Ocean. *Int. Geol. Rev.* **2024**, *66*, 1815–1837. [CrossRef]
70. Hoskin, P.W.O. The Composition of Zircon and Igneous and Metamorphic Petrogenesis. *Rev. Mineral. Geochem.* **2003**, *53*, 27–62. [CrossRef]
71. Streckeisen, A.L. To each plutonic rock its proper name. *Earth-Sci. Rev.* **1976**, *12*, 1–33. [CrossRef]
72. Middlemost, E.A.K. Naming materials in the magma/igneous rock system. *Earth-Sci. Rev.* **1994**, *37*, 215–224. [CrossRef]
73. Rickwood, P.C. Boundary lines within petrologic diagrams which use oxides of major and minor elements. *Lithos* **1989**, *22*, 247–263. [CrossRef]
74. Xia, R.; Wang, C.; Qing, M.; Li, W.; Carranza, E.J.M.; Guo, X.; Ge, L.; Zeng, G. Zircon U–Pb dating, geochemistry and Sr–Nd–Pb–Hf–O isotopes for the Nan'getan granodiorites and mafic microgranular enclaves in the East Kunlun Orogen: Record of closure of the Paleo-Tethys. *Lithos* **2015**, *234–235*, 47–60. [CrossRef]
75. Shao, F.L.; Niu, Y.L.; Liu, Y.; Chen, S.; Kong, J.J.; Duan, M. Petrogenesis of Triassic granitoids in the East Kunlun Orogenic Belt, northern Tibetan Plateau and their tectonic implications. *Lithos* **2017**, *282–283*, 33–44. [CrossRef]
76. Sun, S.s.; McDonough, W.F. Chemical and isotopic systematics of oceanic basalts: Implications for mantle composition and processes. *Geol. Soc. Lond. Spec. Publ.* **1989**, *42*, 313–345. [CrossRef]
77. Rudnick, R.L.; Gao, S. *Composition of the Continental Crust*; Pergamon: Oxford, UK, 2003; p. 64.
78. Miller, C.F.; McDowell, S.M.; Mapes, R.W. Hot and cold granites: Implications of zircon saturation temperatures and preservation of inheritance. *Geology* **2003**, *31*, 529–532. [CrossRef]

79. Kinny, P.D.; Compston, W.; Williams, I.S. A reconnaissance ion-probe study of hafnium isotopes in zircons. *Geochim. Cosmochim. Acta* **1991**, *55*, 849–859. [CrossRef]
80. Wu, F.Y.; Li, X.H.; Zheng, Y.F.; Gao, S. Lu-Hf Isotopic Systematics and their Applicaitons in Petrology. *Acta Petrol. Sin.* **2007**, *23*, 185–220, (In Chinese with English Abstract).
81. Yan, D.; Xiong, F.; Ma, C.; Hou, M.; Zhao, H. Petrogenesis of Middle Triassic intermediate-mafic igneous rocks in East Kunlun, Northern Tibet: Implications for the crust growth and Paleo-Tethyan orogeny. *Geosyst. Geoenviron.* **2024**, *3*, 100096. [CrossRef]
82. Xin, W.; Sun, F.Y.; Zhang, Y.T.; Fan, X.Z.; Wang, Y.C.; Li, L. Mafic–intermediate igneous rocks in the East Kunlun Orogenic Belt, northwestern China: Petrogenesis and implications for regional geodynamic evolution during the Triassic. *Lithos* **2019**, *346–347*, 105159. [CrossRef]
83. Xiong, F.; Ma, C.; Chen, B.; Ducea, M.N.; Hou, M.; Ni, S. Intermediate-mafic dikes in the East Kunlun Orogen, Northern Tibetan Plateau: A window into paleo-arc magma feeding system. *Lithos* **2019**, *340–341*, 152–165. [CrossRef]
84. Yao, L.; Dong, S.Y.; Lü, Z.C.; Zhao, C.S.; Pang, Z.S.; Yu, X.F.; Xue, J.L.; Geng, L.; Zhang, Z.H.; Liu, Y. Origin of the Late Permian gabbros and Middle Triassic granodiorites and their mafic microgranular enclaves from the Eastern Kunlun Orogen Belt : Implications for the subduction of the Palaeo-Tethys Ocean and continent–continent collision. *Geol. J.* **2018**, *55*, 147–172. [CrossRef]
85. Yan, D.D.; Zhou, H.; Li, C.X.; Zhang, X.M.; Ma, C.Q.; Hou, M.C.; Huang, H.; Wang, W.; Xiong, F.H. Petrogenesis of Late Triassic adakitic plutons in the East Kunlun Orogen, Northern Tibet: Geodynamic implications for the Paleo-Tethyan orogeny and crustal evolution. *J. Asian Earth Sci.* **2024**, *268*, 106165. [CrossRef]
86. Ren, X.; Dong, Y.P.; He, D.F.; Sun, S.S.; Hauzenberger, C.A.; Zhou, B.; Yue, Y.G.; Hui, B.; Zhang, B. Petrogenesis and tectonic implications of Late Permian S-type granites in the South Kunlun Belt, northern Tibetan Plateau. *J. Asian Earth Sci.* **2022**, *230*, 105204. [CrossRef]
87. Zhao, X.; Wei, J.H.; Fu, L.B.; Huizenga, J.M.; Santosh, M.; Chen, J.J.; Wang, D.Z.; Li, A.B. Multi-stage crustal melting from Late Permian back-arc extension through Middle Triassic continental collision to Late Triassic post-collisional extension in the East Kunlun Orogen. *Lithos* **2020**, *360–361*, 105446. [CrossRef]
88. Blichert-Toft, J.; Chauvel, C.; Albarède, F. Separation of Hf and Lu for high-precision isotope analysis of rock samples by magnetic sector-multiple collector ICP-MS. *Contrib. Mineral. Petrol.* **1997**, *127*, 248–260. [CrossRef]
89. Griffin, W.L.; Pearson, N.J.; Belousova, E.; Jackson, S.E.; van Achterbergh, E.; O'Reilly, S.Y.; Shee, S.R. The Hf isotope composition of cratonic mantle: LAM-MC-ICPMS analysis of zircon megacrysts in kimberlites. *Geochim. Cosmochim. Acta* **2000**, *64*, 133–147. [CrossRef]
90. Sisson, T.W.; Ratajeski, K.; Hankins, W.B.; Glazner, A.F. Voluminous granitic magmas from common basaltic sources. *Contrib. Mineral. Petrol.* **2004**, *148*, 635–661. [CrossRef]
91. Chappell, B.W. Aluminium saturation in I- and S-type granites and the characterization of fractionated haplogranites. *Lithos* **1999**, *46*, 535–551. [CrossRef]
92. Chappell, B.W.; White, A.J.R. I- and S-type granites in the Lachlan Fold Belt. *Earth Environ. Sci. Trans. R. Soc. Edinb.* **2011**, *83*, 1–26. [CrossRef]
93. Parvez, K.; Mondal, M.E.A.; Amal Dev, J.; Ahmad, I.; Khan, W.M.; Tomson, J.K. Geochemistry and geochronology of 1.7–1.8 Ga peraluminous A-type granites and granite-gneiss from the Mahakoshal Basin, Central Indian Tectonic Zone (CITZ): Implications for an accretionary orogen for the evolution of CITZ. *Int. Geol. Rev.* **2025**, *67*, 675–693. [CrossRef]
94. Wu, F.Y.; Li, X.H.; Yang, J.H.; Zheng, Y.F. Discussions on the petrogenesis of granites. *Acta Petrol. Sin.* **2007**, *23*, 1217–1238, (In Chinese with English Abstract)
95. Jiang, N.; Zhang, S.Q.; Zhou, W.G.; Liu, Y.S. Origin of a Mesozoic granite with A-type characteristics from the North China craton: Highly fractionated from I-type magmas? *Contrib. Mineral. Petrol.* **2009**, *158*, 113–130. [CrossRef]
96. Patiño Douce, A.E. Generation of metaluminous A-type granites by low-pressure melting of calc-alkaline granitoids. *Geology* **1997**, *25*, 743–746. [CrossRef]
97. King, P.L.; Chappell, B.W.; Allen, C.M.; White, A.J.R. Are A-type granites the high-temperature felsic granites? Evidence from fractionated granites of the Wangrah Suite. *Aust. J. Earth Sci.* **2001**, *48*, 501–514. [CrossRef]
98. Clemens, J.D.; Holloway, J.R.; White, A.J.R. Origin of an A-type granite; experimental constraints. *Am. Mineral.* **1986**, *71*, 317–324.
99. Patiño Douce, A.E. What do experiments tell us about the relative contributions of crust and mantle to the origin of granitic magmas? *Geol. Soc. Lond. Spec. Publ.* **1999**, *168*, 55–75. [CrossRef]
100. Wang, Z.S.; Wang, N.; Liu, Z.B.; Ma, X.D. The Northernmost Effects of the Neo-Tethys Oceanic Slab Subduction Under the Lhasa Terrane: Evidence from the Mazin Rhyolite Porphyry. *Minerals* **2024**, *14*, 1292. [CrossRef]
101. Yadav, B.S.; Ahmad, T.; Kaulina, T.; Bayanova, T.; Bhutani, R. Origin of post-collisional A-type granites in the Mahakoshal Supracrustal Belt, Central Indian Tectonic Zone, India: Zircon U-Pb ages and geochemical evidences. *J. Asian Earth Sci.* **2020**, *191*, 104247. [CrossRef]

102. Jiang, Y.H.; Zhao, P.; Zhou, Q.; Liao, S.Y.; Jin, G.D. Petrogenesis and tectonic implications of Early Cretaceous S- and A-type granites in the northwest of the Gan-Hang rift, SE China. *Lithos* **2011**, *121*, 55–73. [CrossRef]
103. Yang, J.H.; Wu, F.Y.; Chung, S.L.; Wilde, S.A.; Chu, M.F. A hybrid origin for the Qianshan A-type granite, northeast China: Geochemical and Sr–Nd–Hf isotopic evidence. *Lithos* **2006**, *89*, 89–106. [CrossRef]
104. Turner, S.P.; Foden, J.D.; Morrison, R.S. Derivation of some A-type magmas by fractionation of the basaltic magma: An example from the Padthaway, South Australia. *Lithos* **1992**, *28*, 151–179. [CrossRef]
105. Griffin, W.L.; Wang, X.; Jackson, S.E.; Pearson, N.J.; O'Reilly, S.Y.; Xu, X.S.; Zhou, X.M. Zircon chemistry and magma mixing, SE Chian: In-suit analysis of Hf isotopes, Tonglu and Pingtan igneous complexes. *Lithos* **2002**, *61*, 237–269. [CrossRef]
106. Kemp, A.I.S.; Hawkesworth, C.J.; Foster, G.L.; Paterson, B.A.; Woodhead, J.D.; Hergt, J.M.; Gray, C.M.; Whitehouse, M.J. Magmatic and Crustal Differentiation History of Granitic Rocks from Hf-O Isotopes in Zircon. *Science* **2007**, *315*, 980–983. [CrossRef]
107. Baker, M.B.; Hirschmann, M.M.; Ghiorso, M.S.; Stolper, E.M. Compositions of near-solidus peridotite melts from experiments and thermodynamic calculations. *Nature* **1995**, *375*, 308–311. [CrossRef]
108. Niu, Y.; Batiza, R. Trace element evidence from seamounts for recycled oceanic crust in the Eastern Pacific mantle. *Earth Planet. Sci. Lett.* **1997**, *148*, 471–483. [CrossRef]
109. Hu, C.B.; Li, M.; Zha, X.F.; Gao, X.F.; Li, T. Genesis and Geological Significance of Late Paleozoic Mantle-Derived Magmatism in Qimantag, East Kunlun: A Case Study of Intrusion in Yingzhuagou. *Earth Sci.* **2018**, *43*, 4334–4349, (In Chinese with English Abstract)
110. Sami, M.; Adam, M.M.A.; Lv, X.; Lasheen, E.S.R.; Ene, A.; Zakaly, H.M.H.; Alarifi, S.S.; Mahdy, N.M.; Abdel Rahman, A.R.A.; Saeed, A.; et al. Petrogenesis and Tectonic Implications of the Cryogenian I-Type Granodiorites from Gabgaba Terrane (NE Sudan). *Minerals* **2023**, *13*, 331. [CrossRef]
111. Wang, X.L.; Yuan, W.M.; Hu, Z.Q.; Ji, C.G. Evidence from zircon and apatite thermochronology provides evidence for the tectonic-thermal evolution and denudational processes in Dulan, Eastern Kunlun Mountains, China. *J. Asian Earth Sci.* **2025**, *278*, 106417. [CrossRef]
112. Feng, Z.R.; Yuan, W.M.; Zhao, Z.D.; Dong, G.C.; Li, X.W.; Sun, W.L.; Yang, L.; Hong, S.J.; Zhao, M.M.; Hu, C.X.; et al. Mesozoic–Cenozoic cooling, exhumation and tectonic implications of Chaqiabeishan–Shaliuquan LiBe ore district in the northeastern Qinghai–Tibet Plateau. *Tectonophysics* **2023**, *866*, 230040. [CrossRef]
113. Li, C.P.; Zheng, D.W.; Yu, J.X.; Wang, Y.Z.; Pang, J.Z.; Wang, Y.; Ma, Y.; Hao, Y.Q.; Xu, Y.G. Late Oligocene Orogen-Scale Tilting in Northern Tibet: A Response to Northward Injection of the Tibetan Lower Crust? *Geophys. Res. Lett.* **2023**, *50*, e2022GL102700. [CrossRef]
114. Liu, H.T. Petrology, geochemistry and geochronology of late Triassic volcanics, Kunlun orogenic belt, western China: Implications for tectonic setting and petrogenesis. *Geochem. J.* **2005**, *39*, 1–20. [CrossRef]
115. Sun, Q. Sudy on Geological Characteristics and Genesis of Xiwanggou Copper-Nickel Deposit in Eastern Kunlun, Qinghai. Master's Thesis, Jilin University, Changchun, China, 2021. (In Chinese with English Abstract)
116. Li, R.B.; Pei, X.Z.; Li, Z.C.; Liu, Z.Q.; Chen, G.C.; Chen, Y.X.; Wei, F.H.; Gao, J.M.; Liu, C.J.; Pei, L. Geological characteristics of Late Palaeozoic-Mesozoic unconformities and their response to some signigicant tectonin events in eastern part of Eastern Kunlun. *Earth Sci. Front.* **2012**, *19*, 244–254, (In Chinese with English Abstract).
117. Xu, Z.Q.; Jiang, M.; Yang, J.S.; Xue, G.Q.; Su, H.P.; Li, H.B.; Cui, J.W.; Wu, C.L.; Liang, F.H. Mantle structure of Qinghai-Tibet Plateau: Mantle plume, mantle shear zone and delamination of lithospheric slab. *Earth Sci. Front.* **2004**, *11*, 329–343, (In Chinese with English Abstract) [CrossRef]
118. Deng, J.F.; Wu, Z.X.; Yang, J.J.; Zhao, H.L.; Liu, H.X.; Lai, S.C.; Di, Y.J. Crust-mantle petrological strcucture and deep processes along the Golmud–Ejin Qi geoscience section. *Acta Geophys. Sin.* **1995**, *38*, 130–144, (In Chinese with English Abstract)
119. Yu, M.; Feng, C.Y.; Zhao, Y.M.; Li, D.X. Genesis of post-collisional calc-alkaline and alkaline granitoids in Qiman Tagh, East Kunlun, China. *Lithos* **2015**, *239*, 45–59. [CrossRef]
120. Liu, W.D. Study on rock age and formation mechanism of intrusive rocks in Early Triassic in Xiangride Area in East Kunlun Orogenic. *Shandong Land Resour.* **2014**, *32*, 1–7, (In Chinese with English abstract)
121. Kong, H.L.; Li, Y.Z.; Li, Y.Z.; Jia, Q.Z.; Guo, X.Z.; Zhang, B. LA-ICP-MS zircon U-Pb dating and geochemical characteristics of the Xiwanggou olivine pyroxenolite in East Kunlun. Journal of Geomechanics. *J. Geomech.* **2019**, *25*, 440–452, (In Chinese with English abstract)
122. Ni, J.Y. Zircon U-Pb age and tectonic setting of Permian-Triassic Volcanic Rock in East Kunlun Orogenic Belt. Master's Thesis, Chinese Academy of Geological Sciences, Beijing, China, 2010. (In Chinese with English abstract)
123. Luo, M.; Mo, X.; Yu, X.; Li, X.; Huang, X. Zircon U-Pb geochronology, petrogenesis and implication of the Late Permian granodiorite from the Wulonggou area in East Kunlun, Qinghai Province. *Earth Sci. Front.* **2015**, *22*, 182–195, (In Chinese with English abstract) [CrossRef]

124. Chen, X.; Gehrels, G.; Yin, A.; Li, L.; Jiang, R. Paleozoic and Mesozoic Basement Magmatisms of Eastern Qaidam Basin, Northern Qinghai-Tibet Plateau:LA-ICP-MS Zircon U-Pb Geochronology and its Geological Significance. *Acta Geol. Sin. (Engl. Ed.)* **2012**, *86*, 350–369. [CrossRef]
125. Sun, Y.; Pei, X.Z.; Ding, S.P.; Li, R.B.; Feng, J.Y.; Zhang, Y.F.; Li, Z.C.; Chen, Y.X.; Zhang, X.F.; Chen, G.C. Halagatu magma mixing granite in the East Kunlun Mountains–evidence from zircon U-Pb dating. *Acta Geol. Sin.* **2009**, *83*, 1000–1010, (In Chinese with English abstract)
126. Chen, G.C.; Pei, X.Z.; Li, R.B.; Li, Z.C.; Pei, L.; Liu, C.J.; Chen, Y.X.; Li, X.B. Triassic magma mixing and mingling at the the eastern section of Eastern Kunlun: A case study from Xiangjiananshan granitic batholith. *Acta Petrol. Sin.* **2018**, *34*, 2441–2480, (In Chinese with English abstract)
127. Zhang, J.-Y.; Ma, C.-Q.; Xiong, F.-H.; Liu, B.I.N. Petrogenesis and tectonic significance of the Late Permian–Middle Triassic calc-alkaline granites in the Balong region, eastern Kunlun Orogen, China. *Geol. Mag.* **2012**, *149*, 892–908. [CrossRef]
128. Ren, H.D.; Wang, T.; Zhang, L.; Wang, X.X.; Huagn, H.; Feng, C.Y.; Teschner, C.; Song, P. Ages, Sources and Tectonic Settings of the Triassic Igneous Rocks in the Easternmost Segment of the East Kunlun Orogen, Central China. *Acta Geol. Sin. (Engl. Ed.)* **2016**, *90*, 641–668. [CrossRef]
129. Xiong, F.H.; Ma, C.Q.; Zhang, J.Y.; Liu, B.; Jiang, H.A.; Huang, J. Zircon LA-ICP-MS U-Pb dating of Bairiqili gabbro pluton in East Kunlun orogenic belt and its geological significance. *Geol. Bull. China* **2011**, *30*, 1196–1202, (In Chinese with English abstract)
130. Xiong, F.H.; Ma, C.Q.; Zhang, J.Y.; Liu, B. LA-ICP-MS zircon U-Pb dating, elements and Sr-Nd-Hf isotope geochemistry of the Early Mesozoic mafic dyke swarms in East Kunlun orogenic belt. *Acta Petrol. Sin.* **2011**, *27*, 3350–3364, (In Chinese with English abstract)
131. Kong, H.L.; Li, J.C.; Li, Y.Z.; Jia, Q.Z.; Guo, X.Z.; Zhang, B. Zircon U-Pb dating and geochemistry of the Jiadang olivine gabbro in the Eastern section of East Kunlun, Qinghai province and their geological significance. *Acta Grologica Sin.* **2018**, *92*, 964–978, (In Chinese with English abstract)
132. Li, R.B.; Pei, X.Z.; Li, Z.C.; Pei, L.; Chen, G.C.; Chen, Y.X.; Liu, C.J.; Wang, S.M. Paleo-Tethys Ocean subduction in eastern section of East Kunlun Orogen: Evidence from the geochronology and geochemistry of the Wutuo pluton. *Acta Petrol. Sin.* **2018**, *34*, 3399–3412, (In Chinese with English abstract)
133. Chen, X.-D.; Li, B.; Sun, C.-B.; Zhou, H.-B. Protracted Storage for Calc-Alkaline Andesitic Magma in Magma Chambers: Perspective from the Nageng Andesite, East Kunlun Orogen, NW China. *Minerals* **2021**, *11*, 198. [CrossRef]
134. Gao, H.; Sun, F. Middle to Late Triassic granitic magmatism in the East Kunlun Orogenic Belt, NW China: Petrogenesis and implications for a transition from subduction to post-collision setting of the Palaeo-Tethys Ocean. *Geol. J.* **2021**, *56*, 3378–3395. [CrossRef]
135. Song, K.; Ding, Q.F.; Zhang, Q.; Cheng, L.; Han, Y.; Liu, F.; Liu, Y. Zircon U–Pb geochronology, Hf isotopes, and whole-rock geochemistry of Hongshuihe Early to Middle Triassic quartz diorites and granites in the Eastern Kunlun Orogen, NW China: Implication for petrogenesis and geodynamics. *Geol. J.* **2019**, *55*, 1507–1528. [CrossRef]
136. Chen, G.; Pei, X.; Li, R.; Li, Z.; Liu, C.; Chen, Y.; Pei, L. Paleo-tethyan oceanic crust subduction in the Eastern Section of the East Kunlun Orogenic Belt: Geochronology and Petrogenesis of the Qushi'ang Granodiorite. *Acta Geol. Sin. (Engl. Ed.)* **2017**, *91*, 565–580. [CrossRef]
137. Li, X.; Huang, X.; Luo, M.; Dong, G.; Mo, X. Petrogenesis and geodynamic implications of the Mid-Triassic lavas from East Kunlun, northern Tibetan Plateau. *J. Asian Earth Sci.* **2015**, *105*, 32–47. [CrossRef]
138. Tian, N.; Sun, F.; Pan, Z.; Li, L.; Yan, J.; Wu, D.; Gu, Y.; Deng, J.; Liu, Z.; Wang, L.; et al. Petrogenesis and tectonic setting of Mid-Triassic volcanic rocks in the East Kunlun orogenic belt, NW China: Insights from geochemistry, zircon U–Pb dating, and Hf isotopes. *Geol. J.* **2021**, *56*, 3257–3274. [CrossRef]
139. Tian, N.; Sun, F.-Y.; Pan, Z.-C.; Li, L.; Gu, Y.; Wu, D.-Q.; Deng, J.-F.; Liu, Z.-D.; Wang, L.; Zhang, Y.-J. Triassic igneous activities in the east flank of the East Kunlun orogenic belt: The Daheba complex example. *Int. Geol. Rev.* **2021**, *65*, 1077–1104. [CrossRef]
140. Wang, K.; Wang, L.X.; Ma, C.Q.; Zhu, Y.X.; Gao, L.Y. Petrogenesis and Geological Implications of the Middle Triassic Garnet-Bearing Two-Mica Granite from Jialuhe Region, East Kunlun. *Earth Sci.* **2020**, *45*, 400–418, (In Chinese with English abstract)
141. Xu, Q.L.; Sun, F.Y.; Li, B.L.; Qian, Y.; Li, L.; Yang, Y.Q. Geochronological dating, geochemical characteristics and tectonic setting of the granite-porphyry in the Mohexiala silver polymetallic deposit, Eastern Kunlun Orogenic Belt. *Geotecton. Et Metallog.* **2014**, *38*, 421–433, (In Chinese with English abstract) [CrossRef]
142. Chen, G.; Pei, X.Z.; Li, Z.C.; Li, R.B.; Chen, Y.X.; Liu, C.J.; Chen, G.C.; Wang, X.B.; Sang, J.Z.; Yang, S.; et al. Zircon U-Pb geochronology, geochemical characteristics and geological significance of Chaohuolutaolegai granodiorite in Balong area, East Kunlun Mountains. *Geol. Bull. China* **2016**, *35*, 1990–2005, (In Chinese with English abstract)
143. Zhang, Y.; Pei, X.Z.; Li, R.B.; Liu, C.J.; Chen, Y.X.; Li, Z.C.; Wang, X.; Hu, C.G. Zircon U-Pb geochronology, geochemistry of the Alasimu gabbro in eastern section of East Kunlun Mountains and the closing time of Paleoocean basin. *Geol. China* **2017**, *44*, 526–540, (In Chinese with English abstract)

144. Yao, X.G.; Zhou, S.A.; Jia, Q.Z.; Li, J.C.; Kong, H.L.; Guo, X.Z.; Wang, Y. Zircon U-Pb dating and geological characteritics of Monzogranite in the Shengli Iron-Copper deposit from East Kunlun Mountains and its prospecting significance. *Geol. Sci. Adn Technol. Inf.* **2018**, *37*, 11–20, (In Chinese with English abstract) [CrossRef]
145. Deng, W.B.; Pei, X.Z.; Liu, C.J.; Li, Z.C.; Li, R.B.; Chen, Y.X.; Chen, G.C.; Yang, S.; Chen, G.; Sang, J.Z.; et al. LA-ICP-MS zircon U-Pb dating of the Chahantaolegai syenogranites in Xiangride area of East Kunlun and its geological significance. *Geol. Bull. China* **2016**, *35*, 687–699, (In Chinese with English abstract)
146. Guo, X.Z.; Jia, Q.; Li, J.; Kong, H.; Yao, X.; Li, Y. Geochronology and geochemical characteristics of syenogranites from Zhamaxiuma area in east Kunlun and their tectonic significance. *Acta Geol. Sin.* **2019**, *93*, 830–842, (In Chinese with English abstract) [CrossRef]
147. He, C.; Wang, L.Y.; Tian, L.M.; Xu, J. Petrogenesis and geological implications of granitoids from Halasen, East Kunlun. *Earth Sci.* **2018**, *43*, 1207–1221, (In Chinese with English abstract)
148. Li, H.-R.; Qian, Y.; Sun, F.-Y.; Li, L. Geochemistry, zircon geochronology, and isotopic systematics of the Zhanbuzhale granites in the East Kunlun, Qinghai Province, northwestern China: Implications for the tectonic setting. *Can. J. Earth Sci.* **2020**, *57*, 275–291. [CrossRef]
149. Xiong, F.; Ma, C.; Jiang, H.A.; Zhang, H. Geochronology and petrogenesis of Triassic high-K calc-alkaline granodiorites in the East Kunlun orogen, West China: Juvenile lower crustal melting during post-collisional extension. *J. Earth Sci.* **2016**, *27*, 474–490. [CrossRef]
150. Xia, R.; Qin, Y.; Wang, C.M.; Li, W.L. The genesis of the Ore-bearing porphyry of the Tuoketuo porphyry Cu-Au-(Mo) deposit in the East Kunlun, Qinghai Province: Constraints from zircon U-Pb geochronological and geochemistry. *J. Jilin Univ. Earth Sci. Ed.* **2014**, *44*, 1502–1524, (In Chinese with English abstract) [CrossRef]
151. Chen, G.C.; Pei, X.Z.; Li, R.B.; Li, Z.C.; Liu, C.J.; Chen, Y.X.; Pei, L.; Li, X.B. Age and lithogenesis of Keri syenogranite from eastern part of East Kunlun Orogenic Belt: Constraint on the Middle Triassic tectonic evolution of East Kunlun. *Acta Petrol. Sin.* **2018**, *34*, 567–585, (In Chinese with English abstract)
152. Wu, Z.N.; Ji, W.H.; He, S.P.; Chen, S.J.; Yu, P.S.; Shi, C.; Chen, F.N.; Zhang, H.S.; Peng, Y. LA-ICP-MS zircon U-Pb dating and geochemical characteristics of granodiorite in Rilonggou area, Xinghai County, Qinghai Province. *Geol. Bull. China* **2015**, *34*, 1677–1688, (In Chinese with English abstract)
153. Xin, W.; Ding, Z.-J.; Meng, Y.-K.; Bo, J.-W.; Li, L.; Mao, G.-Z. Late Triassic granites with mafic microenclaves in the East Kunlun Orogenic Belt, northwestern China: Petrogenesis and implications for continental crust evolution and geodynamic evolution. *Can. J. Earth Sci.* **2023**, *60*, 672–691. [CrossRef]
154. Wen, G.; Sun, G.-C.; Zhao, Z.-F.; Dai, L.-Q.; Zhou, Y. Post-collisional reworking of juvenile mafic lower crust for the petrogenesis of late Triassic adakitic rocks in the East Kunlun Orogen. *Lithos* **2024**, *482–483*, 107730. [CrossRef]
155. Zhao, F.F. *Metallogenesis of Precious-Nonferrous Metal Deposits in Gerizhuotuo Area, Dulan Qinghai Province*; Jilin of University: Changchun, China, 2017; (In Chinese with English abstract)
156. Zhang, Y.; Zhang D, M.; Liu, G.Y.; Li, Z.F.; Zhao, Y.L.; Li, H.H.; Wang, S.M. Zircon U-Pb dating of porphyroid monzonitic granitein the Kaerqueka copper polymetallic deposit of East Kunlun Mountains and its geological significance. *Geol. Bull. China* **2017**, *36*, 270–274, (In Chinese with English abstract)
157. Li, Z.C.; Pei, X.Z.; Liu, Z.Q.; Li, R.B.; Pei, L.; Chen, G.C.; Wu, S.K. Geochronology and geochemistry of the Gerizhuotuo diorites from the Buqingshan Tectonic Mélange Belt in the southern margin of East Kunlun and their geologic implications. *Acta Geol. Sin.* **2013**, *87*, 1089–1103, (In Chinese with English abstract)
158. Chen, G.C.; Pei, X.Z.; Li, R.B.; Li, Z.C.; Pei, L.; Liu, Z.Q.; Chen, Y.X.; Liu, C.J. Geochronology and Genesis of the Helegang Xilikete Granitic Plutons from Southern Margin of the Eastern Kunlun Orogenic Belt and Their Tectonic Significance. *Geol. China* **2013**, *40*, 1044–1065, (In Chinese with English abstract)
159. Zhou, H.; Zhang, D.; Wei, J.; Wang, D.; Santosh, M.; Shi, W.; Chen, J.; Zhao, X. Petrogenesis of Late Triassic mafic enclaves and host granodiorite in the Eastern Kunlun Orogenic Belt, China: Implications for the reworking of juvenile crust by delamination-induced asthenosphere upwelling. *Gondwana Res.* **2020**, *84*, 52–70. [CrossRef]
160. Luo, M.; Mo, X.; Yu, X.; Huang, X.; Yu, J. Zircon LA-ICP-MS dating, petrogenesis and tectonic implications of the Late Triassic granites from Xiangride area, East Kunlun. *Acta Petrol. Sin.* **2014**, *30*, 3229–3241, (In Chinese with English abstract)
161. Xia, R.; Wang, C.; Deng, J.; Carranza, E.J.M.; Li, W.; Qing, M. Crustal thickening prior to 220Ma in the East Kunlun Orogenic Belt: Insights from the Late Triassic granitoids in the Xiao-Nuomuhong pluton. *J. Asian Earth Sci.* **2014**, *93*, 193–210. [CrossRef]
162. Yu, J.Z.; Zheng, Y.Y.; Xu, R.K.; Hou, W.D.; Cai, P.J. Zircon U-Pb Chronology, Geochemistry of Jiangjunmu Ore-Bearing Pluton, Eastern Part of East Kunlun and Their Geological Significance. *Earth Sci.* **2020**, *45*, 1151–1167, (In Chinese with English abstract)
163. Guo, X.; Jia, Q.; Lü, X.; Li, J.; Kong, H.; Yao, X. The Permian Sn metallogenic event and its geodynamic setting in East Kunlun, NW China: Evidence from zircon and cassiterite geochronology, geochemistry, and Sr–Nd–Hf isotopes of the Xiaowolong skarn Sn deposit. *Ore Geol. Rev.* **2020**, *118*, 103370. [CrossRef]

164. Ding, S.; Huang, H.; Niu, Y.L.; Zhao, Z.D.; Yu, X.H.; Mo, X.X. Geochemistry, geochronology and petrogenesis of East Kunlun high Nb-Ta rhyolites. *Acta Petrol. Sin.* **2011**, *27*, 3603–3614, (In Chinese with English abstract)
165. Zhang, B.; Dong, Y.; Sun, S.; He, D.; Hui, B.; Yue, Y.; Ren, X.; He, W. Late Triassic extension of thickened lithosphere of the East Kunlun orogenic Belt, northern Tibetan Plateau: Evidence from the geochemistry and geochronology of mafic magmatism. *Gondwana Res.* **2025**, *137*, 99–116. [CrossRef]
166. Zhang, Z.C.; Zhang, X.J.; Gao, W.L.; Hu, D.G.; Lu, L. Evidence of zircon U-Pb ages for the formation time of the East Kunlun left-lateral ductile shear belt. *J. Geomech.* **2010**, *16*, 51–58, (In Chinese with English abstract)
167. Feng, L.Q.; Gu, X.X.; Zhang, Y.M.; He, G.; Kang, J.Z. Age and structural deformation of ductile shear zones on the southern margin of the East Kunlun Mountains. *Geol. Bull. China* **2017**, *36*, 987–1000, (In Chinese with English abstract)
168. Yan, J.; Sun, F.; Qian, Y.; Tian, N.; Yan, Z.; Zhang, Y.; Yang, X. Petrogenesis and tectonic implications of the Late Triassic Nangou granodiorite porphyry in the Eastern Kunlun Orogenic Belt, northern Tibetan Plateau. *Can. J. Earth Sci.* **2020**, *57*, 801–813. [CrossRef]
169. Zhu, Y.-X.; Wang, L.-X.; Ma, C.-Q.; He, Z.-X.; Deng, X.; Tian, Y. Petrogenesis and tectonic implication of the Late Triassic A1-type alkaline volcanics from the Xiangride area, eastern segment of the East Kunlun Orogen (China). *Lithos* **2022**, *412–413*, 106595. [CrossRef]
170. Xu, B.; Li, H.B.; Nan, Y.Y.; Wang, C.Y.; Yue, T.; Zhao, M.F. LA-MC-ICP-MS zircon U-Pb ages, geochemical characteristics and tectonic significance of the Late Triassic igneous rocks in Ageteng area, Qimantagh Mountains. *Geol. Rev.* **2019**, *65*, 353–369, (In Chinese with English abstract) [CrossRef]
171. Wang, P.; Zhao, G.C.; Liu, Q.; Yao, J.L.; Han, Y.G. Evolution of the Paleo-Tethys Ocean in Eastern Kunlun, North Tibetan Plateau: From continental rift-drift to final closure. *Lithos* **2022**, *422*, 106717. [CrossRef]
172. Yang, X.M.; Sun, F.Y.; Zhao, T.F.; Liu, J.L.; Peng, B. Zircon U-Pb dating, geochemistry and tectonic implications of Akechukesai gabbro in East Kunlun orogenic belt. *Geol. Bull. China* **2018**, *37*, 1842–1852, (In Chinese with English abstract)
173. Li, Z.; Pei, X.; Pei, L.; Liu, C.; Xu, L.; Li, R.; Lin, H.; Wang, M.; Ji, S.; Qin, L.; et al. Petrogenesis and Geochronology of A1-Type Rhyolites in the Late Late Triassic of the East Kunlun Orogenic Belt: Constraints on the End of the Paleo-Tethys Orogenic Event. *Minerals* **2023**, *13*, 290. [CrossRef]
174. Ren, X.; Dong, Y.; He, D.; Sun, S.; Zhang, B.; Zhou, B.; Hui, B.; Yue, Y. Late Triassic magmatic rocks in the southern East Kunlun Orogenic Belt, northern Tibetan Plateau: Petrogenesis and tectonic implications. *Int. Geol. Rev.* **2023**, *65*, 2980–3003. [CrossRef]

minerals

MDPI

Article

The Orogeny Transition of the Southern Beishan Orogenic Belt During the Early–Middle Devonian: Evidence from the Wudaomingshui Volcanic Rocks and Granite

Tongtong He [1,2,3], Yuxi Wang [1,2,4], Jing Yan [1,2], Zhiyong Yang [1,2], Kangning Li [1,2,3], Zirui Liu [3,5], Zixuan Wang [1,2] and Lei Wu [6,*]

1 The Third Geological and Mineral Exploration Institute of Gansu Province Bureau of Geology and Mineral Resources, Lanzhou 730000, China; hett2024@lzu.edu.cn (T.H.); wangyx15@lzu.edu.cn (Y.W.); mn893@126.com (J.Y.); zhiyong011054@163.com (Z.Y.); likn2023@lzu.edu.cn (K.L.); 19993072136@163.com (Z.W.)
2 Gold Mine Resource Exploration and Utilization Technology Innovation Center of Gansu Province, Lanzhou 730050, China
3 College of Earth and Environmental Sciences, Lanzhou University, Lanzhou 730000, China; liuzr851985@163.com
4 Geological Survey of Gansu Province, Lanzhou 730000, China
5 Key Laboratory of Mineral Resources in Western China (Gansu Province), School of Earth Science, Lanzhou University, Lanzhou 730000, China
6 Geological Hazards Prevention Institute, Gansu Academy of Sciences, Lanzhou 730000, China
* Correspondence: wul20@lzu.edu.cn

Abstract: The Southern Beishan Orogenic Belt (SBOB), an integral part of the Southern Central Asian Orogenic Belt (CAOB), is characterized by extensive Late Paleozoic magmatism. These igneous rocks are the key to studying the tectonic evolution process and the ocean–continent tectonic transformation in the southern margin of the CAOB and Paleo-Asian Ocean. We present zircon U-Pb chronology, in situ Lu-Hf isotopes, and whole-rock geochemistry data for Early–Middle Devonian volcanic rocks in the Sangejing Formation and granites from the Shuangyingshan-Huaniushan (SH) unit in the SBOB. The Wudaomingshiu volcanic rocks (Ca. 411.5 Ma) are calc-alkaline basalt-basaltic andesites with low SiO_2 (47.35~55.59 wt.%) and high TiO_2 (1.46~4.16 wt.%) contents, and are enriched in LREEs and LILEs (e.g., Rb, Ba, and Th), depleted in HREEs and HFSEs (Nb, Ta, and Ti), and weakly enriched in Zr-Hf. These mafic rocks are derived from the partial melting of the depleted lithosphere metasomatized by subduction fluid and contaminated by the lower crust. Wudaomingshui's high-K calc-alkaline I-type granite has a crystallization age of 383.6 ± 2.2 Ma (MSWD = 0.11, n = 13), high Na_2O (3.46~3.96 wt.%) and MgO (1.25~1.68 wt.%) contents, and a high DI differentiation index (70.69~80.45); it is enriched in LREEs and LILEs (e.g., Rb, Ba, and Th) and depleted in HREEs and HFSEs (e.g., Nb, Ta, and Ti). Granites have variable zircon $\varepsilon_{Hf}(t)$ values (−2.5~3.3) with Mesoproterozoic T_{DM2} ages (1310~1013 Ma) and originated from lower crustal melting with mantle inputs and minor upper crustal assimilation. An integrated analysis of magmatic suites in the SBOB, including rock assemblages, geochemical signatures, and zircon $\varepsilon_{Hf}(t)$ values (−2.5 to +3.3), revealed a tectonic transition from advancing to retreating subduction during the Early–Middle Devonian.

Keywords: Early–Middle Devonian; orogeny transition; monzogranite; Sangejing Formation; Beishan orogenic belt

1. Introduction

The Central Asian Orogenic Belt (CAOB), bounded by the Tarim, Siberian, North China, and East European cratons, constitutes the largest Phanerozoic accretionary orogenic belt in the world and was formed through Phanerozoic subduction–accretion processes [1–6]. This orogenic belt recorded the process of the Paleo-Asian Ocean form formation to subduction and extinction. A large number of oceanic crust fragments of the PAO and magmatic–metamorphic–sedimentary activities produced in the process of accretion orogeny in the CAOB [7–11], making the CAOB a pivotal region for elucidating geodynamic transitions during the evolution of the PAO [4,12,13].

Located in the southern margin of the middle segment of the CAOB is the Beishan Orogenic Belt (BOB), which links the Chinese Tianshan Orogenic Belt westward and the Solonker Suture Zone eastward (Figure 1a). The BOB serves as a natural laboratory for unraveling the tectonic evolution of the PAO and is composed of a series of magmatic arcs, microcontinents, ophiolites, accretionary wedges, and others [14]. The tectonic evolution of the BOB is related to the east–west tectonic connection and evolution within the CAOB and is influenced by the interaction between the tectonic domains of the PAO and Tethys. This special tectonic position makes the BOB crucial for studying the mode and time span of accretionary orogeny in the CAOB, as well as the tectonic pattern of NW China [15]. The BOB hosts four north–south-aligned ophiolite belts [Hongshishan (HOB), Jijitaizi-Xiaohuangshan (JXOB), Hongliuhe-Niujuanzi-Xichangjing (HNXOB), Huitongshan-Zhangfangshan (HZOB)] (Figure 1b). Taking the HNXOB as the boundary, the BOB is divided into the Northern BOB (NBOB) and the Southern BOB (SBOB). (Figure 1b). Magmatic rocks formed in different tectonic settings are the key to studying the tectonic evolution of the SBOB and the southern margin of the central PAO [5]. Geochronological data reveal two distinct magmatic events in the SBOB: the early Paleozoic and the Late Paleozoic [16–18].

The Early Paleozoic igneous rocks were formed during the subduction and convergence of the PAO and have geochemical characteristics of arc magmatic rocks [19,20]. However, there are many debates on the formation environment of the late magmatism because of the diversity of magmatic rock types and the complexity of geochemical characteristics of these rocks [21,22], and there is debate as to whether the SBOB was in an extensional setting or convergent setting during the Devonian [8,15]. Some scholars believe that the SBOB was in a post-collisional setting during the Devonian based on the evidence of A-type granites and within-plate basalts (WPBs) outcrop in the western section of the SBOB [5,7,18,23]. However, others also suggest that the SBOB is still dominated by subduction in the Devonian on the basis of arc-like igneous rocks [8,16,24,25]. These controversies hinder the correct understanding of the evolution process of the southern margin of the central PAO.

While Devonian magmatic rocks are widespread across the SBOB, earlier investigations have largely concentrated on its western portion. Conversely, the eastern sector of this region exhibits Devonian igneous outcrops that remain comparatively understudied in the existing literature [5]. A comprehensive analysis of eastern SBOB Devonian magmatism is essential for elucidating the transition from subduction-related to extensional tectonic settings and the oceanic-continental transformation along the central PAO's southern margin.

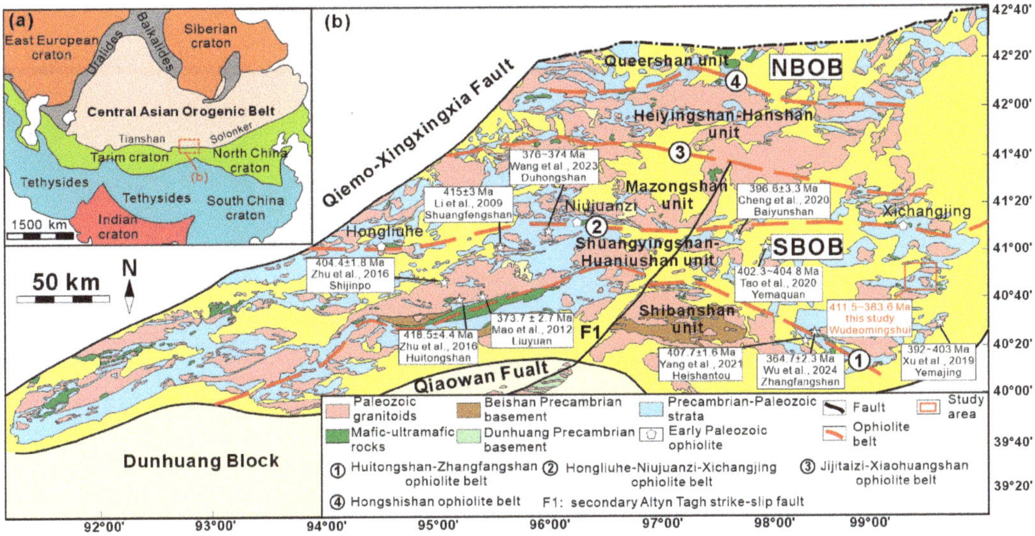

Figure 1. (**a**) A simplified tectonic map of the Central Asian Orogenic Belt (modified after [16]). (**b**) The geological framework of the BOB (modified after [12]). Age data of Devonian magmatic rocks from [5,8,18,19,21,23,24,26,27].

2. Geological Setting

The SBOB is comprises the Shibanshan unit, the Shuangyingshan-Huaniushan (SH) unit, and the HZOB is situated between these two units (Figure 1b).

The study area is located in the SH unit, in which the Proterozoic to Cenozoic strata are distributed. The northern part of the SH unit is predominantly composed of Proterozoic-Ordovician strata, which are composed of metamorphic rocks (including orthometamorphite and parametamorphite) and sedimentary rocks [7,16,25,28,29]. The Ordovician-Devonian strata are mainly outcrops in the southern part of the SH unit [5,17], and the lower Sangejing (SGJ) Formation and upper Dundunshan Formation are composed of the Devonian strata. The SGJ Formation is mainly composed of amygdaloidal basalt, andesite porphyrite, quartz keratophyre, and sandstone with siltstone and conglomerate [22,30]. In addition to the widespread of the Devonian strata, the felsic and mafic intrusive rocks formed in the Devonian are also widely outcrops in the SH unit [18,21,23,31]. These basic-acid intrusive rocks have a lower degree of metamorphic deformation than the Early Paleozoic magmatic rocks with obvious metamorphic deformation in the SH unit.

Volcanic rocks of the SGJ Formation and granites are exposed in the Wudaomingshui (WDMS) area in the east of the SH unit (Figure 1b). The Precambrian Dunhuang Group and Gudongjing Group, the lower Devonian SGJ Formation, the Upper Carboniferous-lower Permian Ganquan Formation, the middle Jurassic toutunhe Formation, the lower Cretaceous Chijinbao Formation and Quaternary outcropping in the WDMS (Figure 2a). The SGJ Formation is composed of amygdaloidal basalt, trachybasalt, andesite, siltstone, sandstone, mudstone, and siliceous rock. This formation forms an unconformable contact with the Dunhuang Group, Ganquan Formation, Chijinbao Formation, and Quaternary, and is invaded by Permian quartz diorite (Figures 2b and 3a,b). Because of strongly weathering, the rocks of the SGJ Formation are generally broken (Figure 3b). The Ordovician, Silurian, Devonian, and Permian quartz diorite, granodiorite, and granite are also outcrop in the study area. The Devonian granite intrudes into the Ordovician quartz diorite and Silurian granodiorite (Figures 2a and 3c,d).

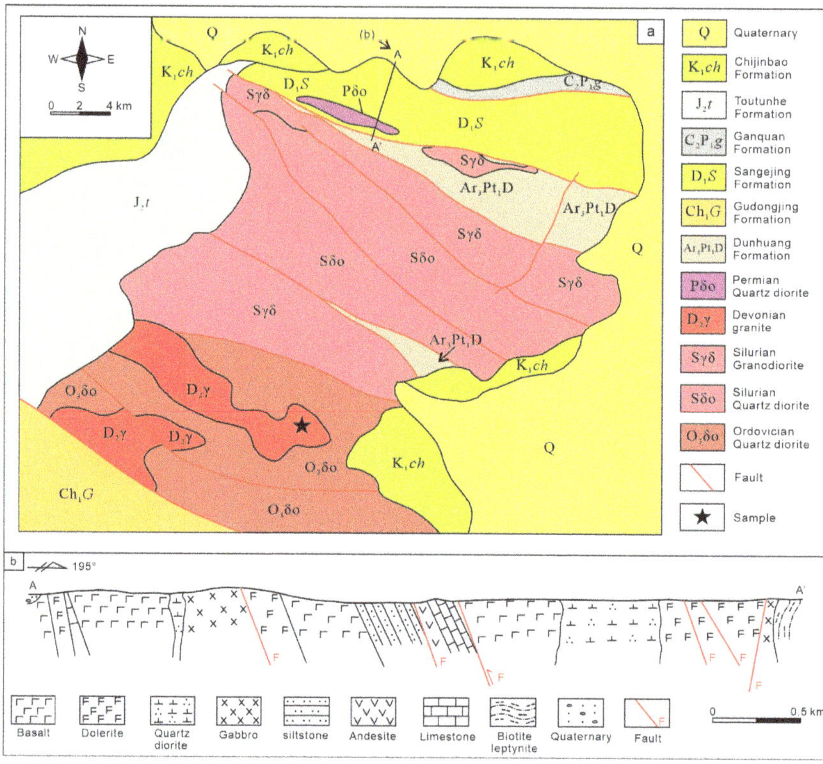

Figure 2. (**a**) Geological map of the Wudaomingshui (WDMS) area; (**b**) profile of the Sangejing Formation.

Figure 3. Field photos and microphotographs showing the geological and petrographical features of the WDMS volcanic rock and granite. (**a**) The Quartz diorite intruded into the volcanic rock; (**b**) volcanic rocks within the SGJ Formation display characteristic amygdaloidal textures; (**c**) granite intruded into the Quartz diorite; (**d**) Field photograph of the granite; (**e**) mineral compositions of the volcanic rock; (**f**) mineral compositions of the granite. Pl: plagioclase; Px: pyroxene; Chl: chlorite; Kfs: potassium feldspar; Bi: biotite; Qtz: quartz.

3. Sample Descriptions

The WDMS grayish-green to dark green volcanic rocks show doleritic texture and vesicular–amygdaloidal structure (Figure 3b). The mineral composition consists of pyroxene (30~40 vol%) and plagioclase (40~60 vol%). Plagioclase occurs as elongate strips with a length of 0.05~1.5 mm and a length–width ratio greater than 3:1. It develops Carlsbad and polysynthetic twinning, and saussuritization is present. Subhedral–anhedral short columnar pyroxene is 0.1~0.5 mm in length and filled between plagioclases. Vesicles and amygdales are well-developed in the rock, with a content of about 10 vol% and a size of 0.1~0.7 mm, and the vesicles are filled chlorite (Figure 3e).

The WDMS grayish-white granite with medium-fine grained granitic texture and massive structure. The mineral composition includes plagioclase (An 25~40) (30~35 vol%), K-feldspar (28~45 vol%), quartz (20~25 vol%), biotite (~5 vol%), and accessory minerals such as apatite, sphene, and metallic minerals. Plagioclase occurs as lath-shaped crystals (~2 mm in length) with polysynthetic twinning and exhibits sericitization. K-feldspar appears as anhedral grains (0.5~2 mm) with grid twinning and have kaolinization alteration. Quartz forms anhedral grains (0.5~1 mm) and the dark mineral biotite occurs as tabular flakes (~0.3 × 0.5 mm) with partial chloritization (Figure 3f).

One volcanic rock was selected for geochronology studies, and one granite sample was used for geochronology and zircon Hf isotope studies. Five volcanic rocks and five granites were used for geochemical studies.

4. Results

4.1. Zircon U-Pb Geochronology Result and Lu-Hf Isotopes

The zircon U-Pb geochronology and in situ Lu-Hf isotope results are listed in Supplementary Tables S1 and S2, respectively.

Twelve zircons from the WDMS volcanic rock are euhedral to subhedral and columnar in shape, with length-to-width ratios of 2:1~3:1. These zircons display inconspicuous concentric oscillation zones (Figure 4a) but have high Th/U ratios (0.38~1.23) (Table S1), indicating a magmatic origin [32]. All twelve analyses yielded similar apparent $^{206}Pb/^{238}U$ ages of 412~403 Ma, with a weighted mean age of 411.5 ± 2.8 Ma (MSWD = 0.43, n = 12) (Figure 4a).

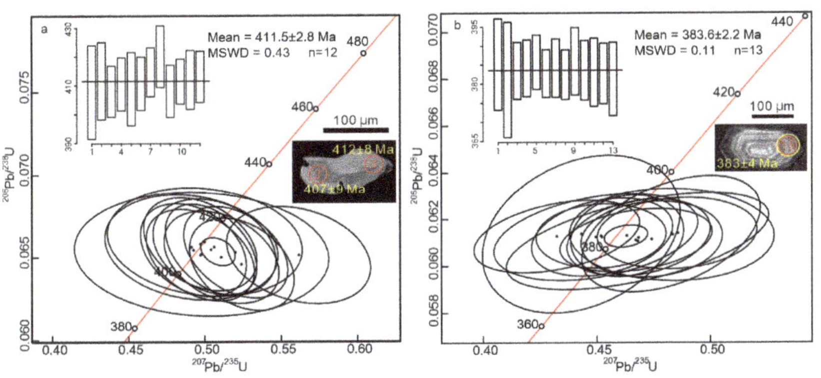

Figure 4. U-Pb concordia diagrams of zircons from Wudaomingshui volcanic rock (**a**) and granite (**b**). Cathodoluminescence images for two zircons from each sample are also shown. Red and yellow circles represent the U-Pb analytical sites and zircon Lu-Hf isotope analysis sites on the Wudaomingshui volcanic rock and granite, respectively.

Thirteen zircon grains from granites were analyzed to determine their U-Pb ages. All these analyzed zircons are euhedral and display granular elongated columnar crystals with length/width ratios of 2:1~3:1 (Figure 4b). Moreover, all these zircons exhibit obvious concentric oscillatory zoning (Figure 4b), and high Th/U ratios (0.25~0.59), indicating a magmatic origin [32]. All these analyses yield close apparent $^{206}Pb/^{238}U$ ages of 382~384 Ma and yield a weighted mean age of 383.6 ± 2.2 Ma (MSWD = 0.11, *n* = 13) (Figure 4b). Thirteen zircon grains were also analyzed for their Lu-Hf isotopic compositions and exhibited low $^{176}Lu/^{177}Hf$ ratios (0.0017467~0.0026975), and the contribution of ^{176}Hf derived from the decay of ^{176}Hf after zircon crystallization can be considered negligible, and the measured $^{176}Hf/^{177}Hf$ ratios can represent the Hf isotopic composition of the magmatic system from which the zircons crystallized [5]. The zircon Hf isotopic composition of WDMS granites is $^{176}Hf/^{177}Hf$ = 0.282478~0.282642, corresponding to $\varepsilon_{Hf}(t)$ = −2.5~3.3, T_{DM2} = 1310~1013 Ma.

4.2. Whole-Rock Major Oxide and Trace Elements

The major oxide and trace element analysis result are presented in Supplementary Table S3.

4.2.1. Major Oxide

WDMS volcanic rocks underwent alteration, as indicated by high loss on ignition (LOI) ratios (1.15~4.23) and petrological characteristics (Figure 3e). Due to the high mobility of elements such as K and Na during alteration, immobile elements are chosen to assess the alkalinity of samples. Volcanic rock samples have low SiO_2 (47.35~55.59 wt.%) and P_2O_5 (0.18~0.30 wt.%) contents and are plotted in the andesite/basalt-andesite fields (Figure 5a). This characteristic indicates that the WDMS volcanic rocks are basalt-basaltic andesites. Samples also have high contents of TiO_2 (1.46~4.16 wt.%), TFe_2O_3 (2.71~6.87 wt.%) and MgO (3.81~8.29 wt.%) with $Mg^{\#}$ values of 43.95~64.77. The Al_2O_3, Na_2O, and K_2O contents are 13.20~16.45 wt.%, 2.81~5.70 wt.%, and 0.24~1.78 wt.%, respectively, and the contents of Na_2O are higher than those of K_2O (Na_2O/K_2O = 1.58~24.24). Samples are plotted in the calc-alkaline fields in the Ce/Yb vs. Ta/Yb diagram (Figure 5b) and indicate these volcanic rocks are in the same calc-alkaline series as volcanic rocks of the SGJ Formation in the SBOB.

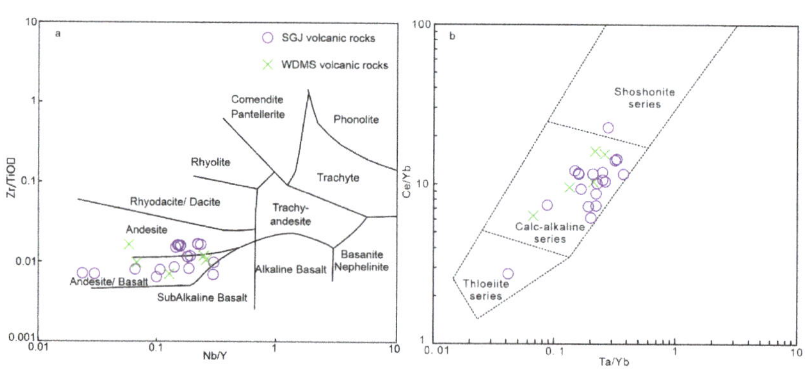

Figure 5. (**a**) Zr/TiO_2 vs. Nb/Y diagram [33]; (**b**) Ce/Yb vs. Ta/Yb diagram [34].

WDMS granite samples have high SiO_2 (65.44~68.86 wt.%) and $Na_2O + K_2O$ (6.64~7.40 wt.%) contents. The Na_2O contents of samples are slightly higher than K_2O contents (Na_2O = 3.46~3.96 wt.%; K_2O = 2.71~3.84 wt.%; Na_2O/K_2O = 0.92~1.45), and samples display high-K calc-alkaline characteristics (Figure 6a). Granites exhibit low TiO_2

(0.32~0.57 wt.%) and P_2O_5 (0.10~0.15 wt.%) contents, high TFe_2O_3 (3.00~4.27 wt.%) and Al_2O_3 (15.51~17.20 wt.%) contents, with A/CNK values of 1.09~1.13, and plotted in the peraluminous field (Figure 6a). Samples also have high MgO contents (1.25~1.68 wt.%) and $Mg^{\#}$ values (34.71~37.75) with low SI consolidation index (9.85~11.80) and high DI differentiation index (70.69~80.45).

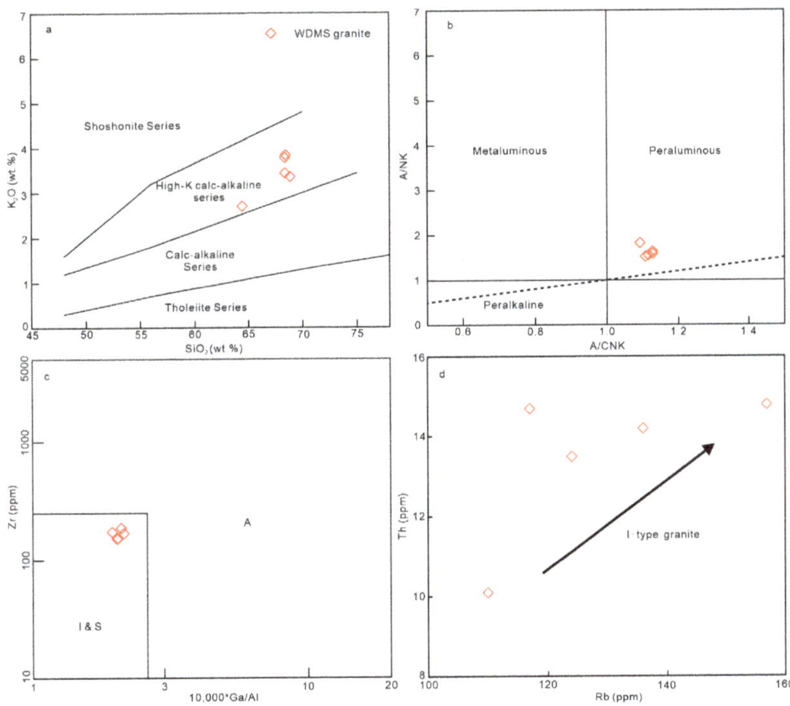

Figure 6. (**a**) K_2O vs. SiO_2 diagram [35]; (**b**) A/NK vs. A/CNK diagram [36]; (**c**) Zr vs. 10,000 *Ga/Al diagram [37]; (**d**) Th vs. Rb diagram.

4.2.2. Trace Elements

The concentration of rare earth elements (REE) in Wudaomingshui volcanic rock samples are 95.10~146.90 ppm. Samples have enrichments of light REEs (LREE) and depletion of heavy REEs (HREE) [$(La/Yb)_N$ = 1.51~5.01; LREE/HREE = 2.55~5.26] and are weakly negative to no Eu anomalies (δEu = 0.84~1.07) (Figure 7a). Samples show positive anomalies in Rb, Ba, Th, and other large ion lithophile elements (LILEs), negative anomalies in Nb, Ta, Ti, and other high-field strength elements (HFSEs) (Figure 7b). In addition, they also have weakly positive anomalies in Zr and Hf (Figure 7b).

The WDMS granites have enrichment of LREEs, depletion of HREEs [$(La/Yb)_N$ = 7.21~9.37, LREE/HREE = 7.41], and with moderate negative Eu anomalies (δEu = 0.52~0.71) (Figure 7a). All granite samples show pronounced positive anomalies in LILEs and negative anomalies in HFSEs (Figure 7b).

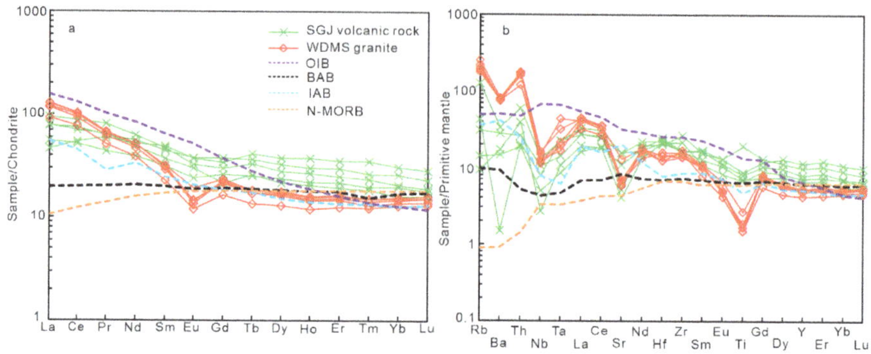

Figure 7. Chondrite-normalized REE diagram (**a**) and primitive-mantle-normalized trace element spider diagram (**b**) for Wudaomingshui volcanic rocks and granites. The values of chondrite and primitive mantle are from Sun and McDonough [38].

5. Discussion

5.1. Petrogenesis

5.1.1. Volcanic Rocks

WDMS volcanic rock samples exhibit elevated Loss on Ignition (LOI) values and pronounced mineral alteration features, indicating substantial post-magmatic alteration. Given the high mobility of incompatible elements such as LILEs, we employed relatively immobile elements, including HSFEs and REEs, to decipher the primary geochemical signatures of the magmatic source.

The Rb/Sr (0.04~0.25) and Lu/Yb (0.13~0.15) ratios of these basalt-basaltic andesites closely overlap with mantle-derived magma ratios (0.06~0.10 and 0.14~0.15, respectively), but distinctly differ from crustal melt signatures (35 and 0.16~0.18) [38,39], supporting a dominant mantle origin. During their ascent through the continental crust, mantle-derived basaltic magmas may undergo variable degrees of fractional crystallization and crustal contamination. Compared to primary basaltic magmas with high $Mg^{\#}$ values (68~75) [40], the WDMS basalt-basaltic andesite samples exhibit moderate MgO contents and Mg# values (except for sample 2013XYQ5-1: MgO = 8.291 wt.%; $Mg^{\#}$ = 64.77), suggesting the influence of fractional crystallization. However, the position correlations of both samples and volcanic rocks of the SGJ Formation in the western segment of the SBOB (Figure 8a) indicate that these rocks have underwent limited fractional crystallization, with partial melting being the dominant process. The WDMS basalt-basaltic andesite samples have low concentrations of Cr (3.7~200.06 ppm) and Ni (1.7~99.3 ppm) compared to these primitive basaltic magmas (Cr > 400 ppm, Ni > 235 ppm) and samples display an olivine fractional crystallization trend in the Ni vs. Nb/Ta diagram (Figure 8b). These characteristics indicate the separation of Mg-Fe-rich minerals (e.g., olivine) during magma evolution.

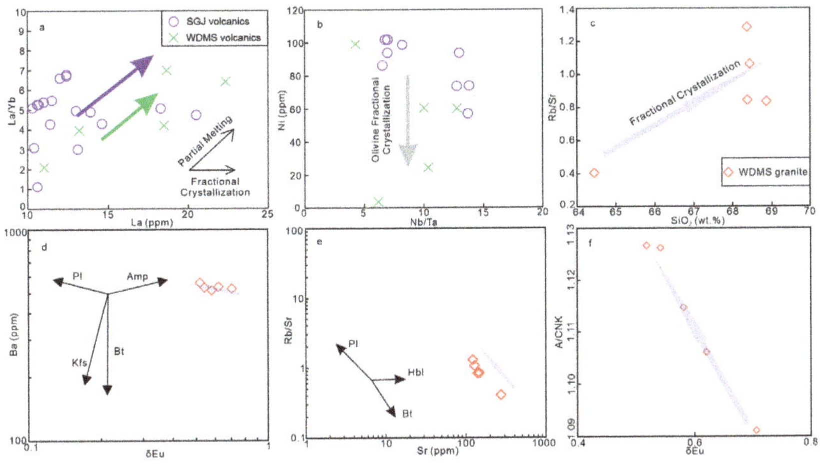

Figure 8. (**a**) La/Yb vs. La diagram [41]; (**b**) Ni vs. Nb/Ta diagram; (**c**) Rb/Sr vs. SiO_2 diagram [42]; (**d**) Ba vs. δEu diagram [5]; (**e**) Ba vs. Rb diagram [5]; (**f**) A/CNK vs. δEu diagram.

Mafic rocks affected by crustal contamination typically exhibit Nb/La < 1 and $(Th/Nb)_{PM}$ > 1 [43]. Basalt-basaltic andesite samples have Nb/La values of 0.14~0.42 and $(Th/Nb)_{PM}$ values of 2.47~14.57. In addition, all samples are plotted within the crustal contamination field (Figure 9a), consistent with the compositional range of the SGJ Formation volcanic rocks in the SBOB, indicating significant crustal input during magma evolution. The $(Th/Yb)_{PM}$ values of samples (1.76~8.32) are between lower crustal (4.6) and upper crustal (28) reference values [44], with closer affinity to the lower crust. Additionally, samples also have Nb/Ta ratios (4.28~12.81) between those indicative of the lower crust (8.3) and mantle (17.39) [38,39]. The above geochemical characteristics suggest that the crustal contamination occurred during the formation of basalt-basaltic andesites (Figure 9b).

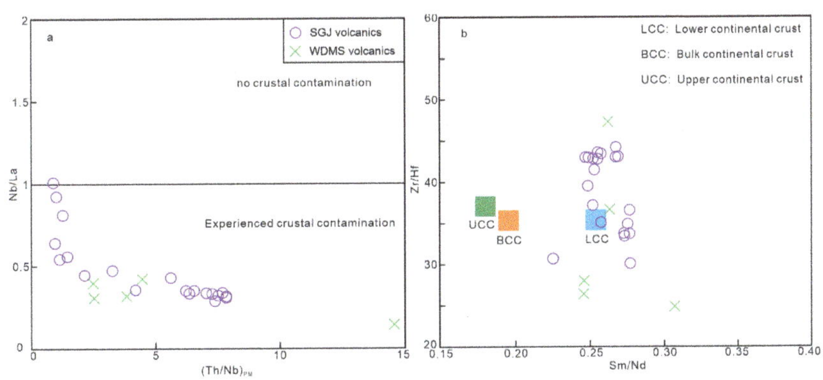

Figure 9. (**a**) Nb/La vs. $(Th/Nb)_{PM}$ diagram [45]; (**b**) Zr/Hf vs. Sm/Nd diagram [46].

The WDMS basalt-basaltic andesite samples have enrichment of LREE and LILE (Figure 7a,b) suggest that they originated from either an enriched asthenosphere or a metasomatized lithosphere modified by subduction-related components [47,48]. Rocks originating from the asthenosphere are typically enriched in LILEs and HFSEs, whereas the samples are enriched in LILEs and depleted in HFSEs (Figure 7b). Additionally, samples are plotted distinctly away from the Ocean Island Basalt (OIB) (Figure 10a), and rocks also exhibit higher La/Nb (2.37~6.95) and La/Ta (20.07~32.20, avg. 27.61) ratios than values of asthenosphere-derived magmas [38]. Therefore, these geochemical signatures indicate that the parent magma did not originate from the asthenosphere but from the lithosphere. This pattern contrasts sharply with typical asthenosphere-derived magmas and instead points to a subduction-modified lithosphere source. The Nb/La ratios of lithosphere and asthenosphere are <0.4 and >1, respectively [49]. The Nb/La values of samples are 0.14~0.42 (avg. 0.31) consistent with the value of lithosphere. Additionally, their negative δNb values (−1.15 to −0.19) indicate a depleted mantle source. At the same time, samples collectively display an enriched evolutionary trend aligned with the SGJ Formation volcanics in the SBOB (Figure 10a). These geochemical features demonstrate that the WDMS basalt-basaltic andesites originated from a metasomatized and depleted lithosphere enriched by subduction-related components. Subduction materials include subduction fluid and subduction melts. Samples have low Th/La (0.09~0.25), Th/Yb (0.30~1.43), and Th/Ta (1.84~7.43, avg. 4.62) ratios with high Ba/Th (6.12~89.00, avg. 48.33) ratios, which are consistent with values of melts generated by the mantle metasomatized by subduction fluids [50,51]. Melts originating from garnet-facies lherzolite typically exhibit Dy/Yb > 2.5, whereas those from spinel-facies lherzolite yield Dy/Yb < 1.5. Melts derived from a garnet-spinel transitional mantle source show intermediate Dy/Yb values (1.5~2.5) [52,53]. The WDMS volcanic rocks display Dy/Yb ratios of 1.81~2.04, plotting near the spinel-garnet lherzolite partial melting trend (Figure 10b), with an estimated partial melting degree of ~10%. In summary, the WDMS volcanic rocks formed through partial melting of the garnet-spinel transitional lherzolite mantle metasomatized by subduction fluid. Subsequent magma evolution involved olivine-dominated fractional crystallization and assimilation-contamination processes with lower crustal materials.

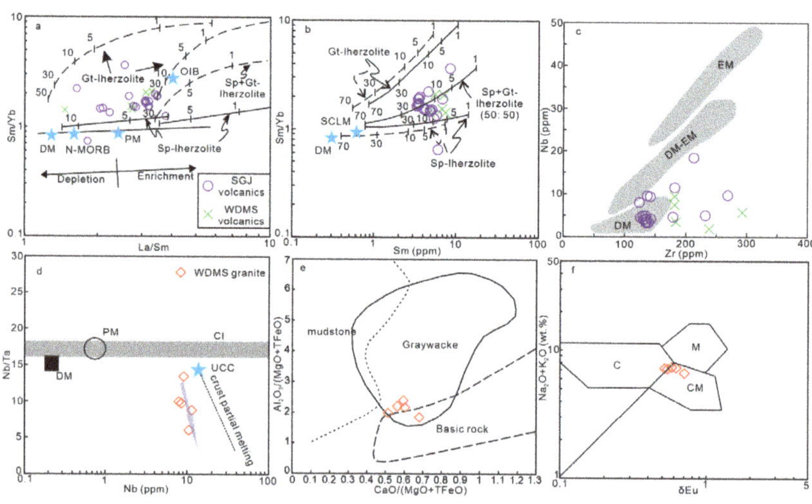

Figure 10. (**a**) Sm/Yb vs. La/Sm diagram [54]; (**b**) Sm/Yb vs. Sm diagram [55]; (**c**) Nb vs. Zr diagram [56]; (**d**) Nb/Ta vs. Nb diagram [56]; (**e**) [Al_2O_3/(MgO + TFeO)] vs. [CaO/(MgO + TFeO)] diagram [57]; (**f**) (Na_2O + K_2O) vs. δEu diagram [58]. PM: primitive mantle; DM: depleted mantle; Sp: spinel; Gt: garnet; EM: enriched mantle; C1: Chondrite; UCC: upper crust; C: crust; M: mantle; CM: crust-mantle mixed source; SCLM: subcontinental lithospheric mantle.

5.1.2. Granites

Granite can be classified as I-, S-, M- and A-type [59–61]. The WDMS granite samples have high K_2O contents (2.71~3.84 wt.%), which is different from the M-type granite with low K_2O contents (<1 wt.%) [62]. Samples exhibit distinct 10,000 × Ga/Al (1.94~2.13), K/Rb (199.87~238.40) and Zr + Nb + Ce + Y (243.90~281.85 ppm) values compared to those of A-type granites [37]. The TFeO/MgO (2.94~3.35) ratios are also different from the iron-enriched signatures (TFeO/MgO > 4.0) of A-type granite [37]. In addition, all samples are plotted within the I- and S-type fields. Therefore, the WDMS granite are I- or S-type. S-type granite is characterized by A/CNK > 1.1, P_2O_5 > 0.2 wt% and enriched in Al-rich phase minerals (e.g., muscovite, garnet) [60,63]. Although samples exhibit weakly peraluminous characteristics (A/CNK = 1.09~1.13), their low P_2O_5 contents (0.10~0.15 wt.%, <0.2 wt.%) and lack of Al-rich minerals in the mineral assemblage distinguish them from S-type granites. In addition, samples have a 10,000 *Ga/Al ratio (2.04) close to that of I-type granite (2.1), and their Th contents increases with the increase in Rb contents (Figure 6c), which indicates that the WDMS granites belong to I-type granite [37].

Notably, samples exhibit low SI consolidation index (9.85~11.80) and high DI differentiation index (70.69~80.45), coupled with a trend in which the Rb/Sr ratios increase with the increasing in SiO_2 contents (Figure 8c). These geochemical trends collectively indicate that the rocks underwent intense fractional crystallization during magmatic evolution. Highly fractionated granites are typically characterized by low ΣREE concentrations, LREE/HREE ratios, K/Rb ratios and TFeO contents (<1 wt%), coupled with strongly negative Eu anomalies [64]. Additionally, they also have elevated Be and significantly reduced Zr concentrations [60]. Granites have high ΣREE concentrations (108.67~145.73 ppm), LREE/HREE ratios (7.46~8.35), K/Rb ratios (199.87~238.40), and TFeO contents (3.68~5.49 wt.%), with weakly negative Eu anomalies, which indicates that samples do not belong to highly fractionated granites. Zr/Hf ratios are critical indicators to determine whether the felsic rocks are highly fractionated granites or not. Highly fractionated granites are characterized by Zr/Hf < 25, while moderately fractionated granites exhibit Zr/Hf values of 25~55 [65,66]. Zr/Hf ratios of samples are 34~49, suggesting the granite underwent moderate fractional crystallization but do not qualify as highly fractionated granitic rocks. Pronounced negative anomalies in Eu, Ba, and Sr (Figure 7) indicate significant plagioclase fractionation (Figure 8d,e). Plagioclase, a major Al-bearing phase, contributes to increasing Al_2O_3 contents and A/CNK values in residual melts during fractionation. Consequently, the strongly peraluminous signatures of the samples (A/CNK = 1.09–1.13) are inferred to have resulted from crystal fractionation-dominated processes rather than pure sedimentary source melting, as demonstrated by the Al_2O_3 vs. A/CNK covariation trend (Figure 8f).

There are three primary genetic mechanisms for I-type granites: (1) crustal anatexis triggered by mantle-derived magma underplating [67]; (2) crust-mantle hybridization [68]; (3) fractional crystallization of mafic magmas [69]. There are limited Silurian-Devonian mafic rocks exposed in the WDMS, and the igneous rocks formed in the Silurian-Devonian are dominated by felsic rocks (Figure 2a). Granites are characterized by high silicon, potassium and total alkali contents, the depletion of Ba, Ti, and Eu, and low Cr-Ni concentrations, indicative of crust-derived magmas [70]. These signatures preclude an origin via fractional crystallization of mafic magmas. The Rb/Sr (0.41~1.29), Ti/Zr (11.13~18.08), and Sm/Nd (0.19~0.20) ratios of the WDMS granites are consistent with crust-derived magma (Rb/Sr > 0.35; Ti/Zr < 20; Sm/Nd < 0.3) [71]. The Nb/Ta ratios (6.02~13.46) and $Mg^{\#}$ values (34.71~37.75) of samples are consistent with melts from the lower crust (Nb/Ta = 8.3~11.4, $Mg^{\#}$ < 45) [71,72], which is supported by samples plotted near the trend in crustal partial melting (Figure 10d). However, the CaO/Na_2O ratios of granite samples are 0.54~0.89, akin to metagreywacke sources (0.3~1.5) [73]. In the A/FM-C/FM diagram (Figure 10e), samples are plotted within the overlapping field of mafic melts and greywacke-derived partial melts, suggesting contributions from upper crustal materials during petrogenesis. In addition to the obvious characteristics of crust-derived magma, the relatively high Zr/Hf ratios (34~49) and high MgO and Na_2O contents of samples suggest the influence of mantle [71]. Magmas derived from or contaminated by the lithosphere typically exhibit La/Ta > 25 [74]. The La/Ta values of samples are 15.80~41.76 (avg. 28.58), which indicate the involvement of mantle-derived components during petrogenesis (Figure 10f). Zircon Hf isotopes serve as a critical tool for investigating continental crustal growth and evolution [75]. The negative zircon $\varepsilon_{Hf}(t)$ values coupled with T_{DM2} ages significantly exceeding their crystallization ages are evidence that the felsic rocks were derived from partial melting from ancient crust; positive $\varepsilon_{Hf}(t)$ values with T_{DM2} ages approximating crystallization ages, suggesting that the felsic magma originated from the partial melting of juvenile crust (derived from depleted mantle sources) or substantial mantle input during magma generation; widely variable $\varepsilon_{Hf}(t)$ values demonstrate contributions from hybrid sources, reflecting the incorporation of both ancient crustal components and juvenile mantle-derived materials during petrogenesis [75]. Samples exhibit high zircon $\varepsilon_{Hf}(t)$ values (−2.51~+3.32) and Mesoproterozoic T_{DM2} ages (1310~1013 Ma), which closely resemble those of the Mesoproterozoic mafic igneous rocks constituting the basement of the SBOB [7]. These features indicate that monzogranites originated from partial melting of Mesoproterozoic mafic magmatic rocks [7]. Additionally, samples display significant variations in $\varepsilon_{Hf}(t)$ values, plotted between the evolution line for depleted mantle and evolution line for crust extracted from mantle at Ca. 1.0 Ga. This geochemical signature further attests to mantle-derived contributions to the crustal magmas, reflecting hybridization between ancient crustal melts and juvenile mantle-derived components during petrogenesis [7,59].

5.2. *Early–Middle Devonian Magmatism in the SBOB*

In the Paleozoic stratigraphic sequence of the SBOB, two major unconformity events occurred during the Silurian-Devonian and Carboniferous-Permian, which indicate tectonic regime transitions in these periods [76]. The Devonian is a critical temporal marker for the Silurian-Devonian unconformity. Investigating the magma source and tectonic settings of Devonian igneous rocks holds significant implications for reconstructing the process of accretionary orogenesis and the ocean–continent transition of the PAO at this tectonic junction. The Devonian magmatic events are subdivided into two stages (Early–Middle Devonian and Late Devonian). The Early–Middle Devonian K-feldspar granite, monzogranite, granodiorite, and volcanics of the SGJ Formation are widely distributed in Huitongshan, Shuangfengshan, Shijinpo, Yemaquan, Yemajing, and Dundunshan [5,8,21,26]. Based on zircon U-Pb ages of volcanic rocks and sedimentary rocks within the stratum, combined with geological evidence of unconformable contacts between this formation and both the underlying strata and overlying Late Devonian Dundunshan Formation, the depositional age of the SGJ Formation is 420~390 Ma (Early Devonian) [26,27]. The SGJ Formation in the WDMS were previously assigned to the Silurian Gongpoquan Formation in the 1:200,000 Wudaoming Geological Map. However, the zircon U-Pb age of basalt-basaltic andesite is 411.5 ± 2.8 Ma (MSWD = 0.43), indicating an Early Devonian rather than the Silurian. Meanwhile, the strata of WDMS and the SGJ Formation in the BOB have similar rock assemblage (such as amygdaloidal basalts, trachybasalts, andesites, siltstones, sandstones, mudstones, and siliceous rocks). In terms of geochronology and rock assemblage characteristics, the two are similar. In addition, both of the WDMS basalt-basaltic rocks and mafic rocks of the SGJ Formation in the BOB have similar geochemical compositions and magma source (Figures 8a–c, 9 and 10a–c). Therefore, we suggest that the strata previously assigned to the Silurian Gongpoquan Formation in the WDMS should be reclassified as the Early–Middle Devonian SGJ Formation. The WDMS granite intrudes into Ordovician quartz diorites and Silurian granodiorites (Figure 2a), indicating a post-Silurian emplacement age. Zircon U-Pb dating of the WDMS granite is 383.6 ± 2.2 Ma (MSWD = 0.11), consistent with field cross-cutting relationships, which constrains its magmatic crystallization to the Middle Devonian. Consequently, the Early Devonian mafic volcanic rocks of the SGJ Formation and Middle Devonian granites in the WDMS collectively represent a tectonic response to the Early–Middle Devonian magmatic events in the SBOB.

5.3. Transition from Advancing to Retreating Accretionary Orogens in the SBOB During the Early–Middle Devonian

Accretionary orogenesis encompasses advancing accretionary orogenesis and retreating accretionary orogenesis, which reflect contrasting plate tectonic dynamics and exhibit markedly divergent geological characteristics [77]. In advancing accretionary orogenesis, sustained compressional advancement of the subducting slab against the overriding plate drives vertical crustal thickening and horizontal shortening. This process is dominated by convergent dynamics, generating igneous rocks exhibiting diagnostic subduction-related geochemical signatures. In contrast, retreating accretionary orogenesis features sustained slab rollback of the descending plate, leading to extensional thinning of the overriding plate. This results in vertical crustal thinning and lateral extension, manifesting as rock assemblages formed in the extensional setting accompanied by extensional structures [77]. The Early–Middle Devonian magmatic rocks in the SBOB comprises not only A-type granites, within-plate basalts (WPB), and highly fractionated I/S-type granites which formed in the extensional setting but also adakites and Nb-enriched basalts were also formed in the SBOB [19–22,30]. This juxtaposition of comparative rock assemblages has sparked ongoing debates regarding the tectonic setting and geodynamic processes governing the SBOB during the Early–Middle Devonian [7,8,21,27]. WDMS mafic rocks have the characteristics of both WPB [high TiO_2 and Zr contents, weakly enriched in Zr and Hf, Zr/Y > 3 (4~7), Th/Nb > 0.11 (0.29~1.74), Th/Ta > 1.6 (1.84~7.43); Figure 11a] and arc basalt [enriched in LILEs, depleted in HFSEs (Nb, Ta), high Al_2O_3 contents (13.20~16.45 wt.%) and low Nb and Ta concentration; Figure 11b] [47,78]. Granites exhibit relatively low Sr concentration (122~271 ppm) and Sr/Y ratios (5.54~11.29), indicating a thinner crustal thickness. However, the Th/Ta and Y/Nb ratios (8.07~18.38 and 1.90~2.79, respectively), which are greater than 6 and 1.2, display characteristics typical of subduction zone granites. Although the contemporaneous formation of these contrasting geochemical signatures and rock assemblages formed in diverse environments poses challenges for determining the Early–Middle Devonian tectonic setting of the SBOB, these features also provide critical constraints for investigating the transition from advancing to retreating accretionary orogenesis in this region. We proposed that the WDMS basalt-basaltic andesites and granites exhibit dual geochemical affinities to both extensional and subduction-related magmatism, and the co-occurrence of A-type granite, WPB, adakite, and Nb-rich basalt formed in Devonian is interpreted to indicate a transition from advancing to retreating accretionary orogenesis at this time. The rationale for this interpretation is as follows: (1) In advancing accretionary orogens, crustal recycling dominates, whereas retreating accretionary orogen are characterized by substantial addition of juvenile material [79]. The WDMS granite originated from partial melting of the basaltic lower crust, and is influenced by mantle-derived magma, which indicates the addition of juvenile material during petrogenesis. Statistical analysis of Hf isotopes for all Devonian felsic rocks in the SBOB reveals that their $\varepsilon_{Hf}(t)$ values generally exhibit a decline evolutionary trend during 420~380 Ma, followed by a subsequent increase during 380~360 Ma (Figure 12). This evolution trend reflects a transition from lateral arc accretion dominated by subduction-driven processes to vertical inputs of mantle-derived materials. Furthermore, the presence of highly variable zircon $\varepsilon_{Hf}(t)$ values in pre-Devonian magmatic rocks, coupled with the predominantly positive $\varepsilon_{Hf}(t)$ values in later magmatic zircons, provides evidence for substantial juvenile material input during Early–Middle Devonian magmatism in the SBOB. (2) Early Paleozoic Basaltic rocks in the

SBOB with tholeiitic and that are Nb-enriched are interpreted to originate from partial melting of the mantle wedge. At same times, the intermediate to felsic rocks dominantly comprise calc-alkaline I-type granitoids, while adakites have high Sr and low Y concentrations identified in the SH unit. These features collectively indicate that the SBOB resided in a subduction-related arc setting during the early Paleozoic [19,20]. However, A-type felsic rocks and mafic rocks have high Zr, TiO_2 contents and Zr/Y ratios, which indicates that an intraplate extensional environment emerged in the SBOB beginning in the Early Devonian [7,8,26], as well as bimodal volcanic rocks also reported in the Yemaquan area of the SBOB [27]. There is variation in the geochemical characteristics and rock assemblages of igneous rocks that formed in early Paleozoic to Devonian reflecting a tectonic transition, that is, the regional transformation from subduction convergence to an extensional tectonic setting. As for basalt-basaltic andesites and granites in this study, although all of them have geochemical characteristics of arc magmatic rocks, volcanic rocks show more obvious geochemical characteristics of WPB [such as Zr/Y > 3 (4~7), Th/Nb > 0.11 (0.29~1.74), Th/Ta > 1.6 (1.84~7.43)] (Figure 11a). Granite samples are plotted in the post-orogenic granite field (Figure 11c) with low Sr/Y (5.54~11.29) and La/Yb (10.05~13.06) ratios, reflecting the shallow melting of magma source. At same time, basalt-basaltic andesites and granites are plotted in the transition zone of arc and extensional magmatic rocks (Figure 11b,d), which also reflects the tectonic transition from advancing to retreating accretionary orogenesis; (3) In addition to the distribution of A-type granites in the Early Devonian, S-type granites are also distributed in this area [80]. These S-type granites have similar characteristics to the Permian S-type granites in New England, and these S-type granites in New England record the transition of the Terra Australis orogenic belt from the convergent to the extensional tectonic setting [81].

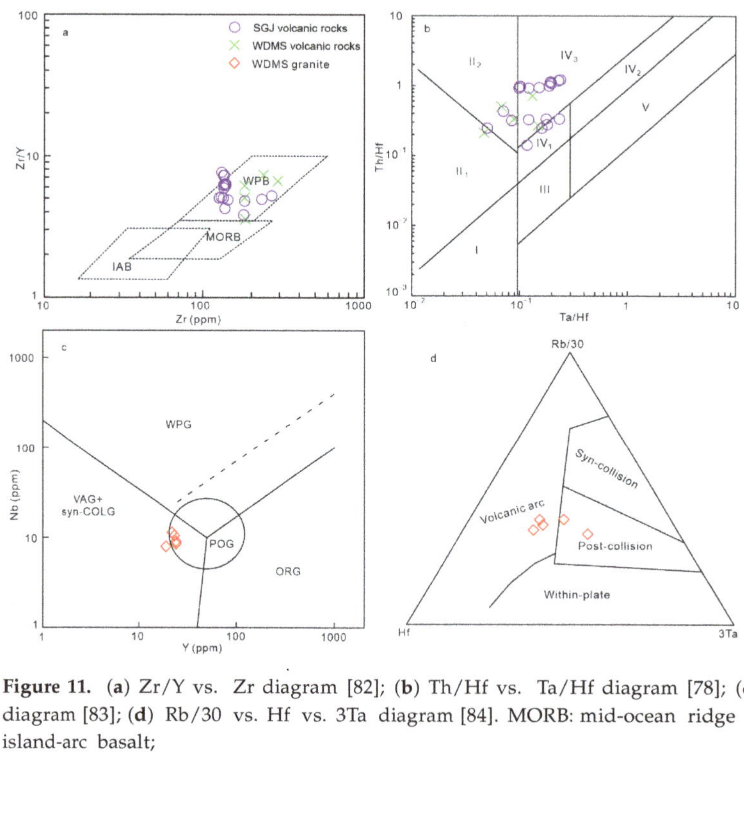

Figure 11. (**a**) Zr/Y vs. Zr diagram [82]; (**b**) Th/Hf vs. Ta/Hf diagram [78]; (**c**) Nb vs. Y diagram [83]; (**d**) Rb/30 vs. Hf vs. 3Ta diagram [84]. MORB: mid-ocean ridge basalt; IAB: island-arc basalt;

WPB: within-plate basalt; VAG: volcanic arc granite; POG: post-orogenic granite; ORG: ocean ridge granite; WPG: within-plate granite; syn-COLG: syn-collisional granite; I: divergent boundary; II: Convergent boundary: II_1: Ocean island arc basalt; II_2: Continental arc basalt; III: Ocean island basalt, Seamount basalt, T-MORB and E- MORB; IV. Continental within-plate basalt: IV_1: Tholeiitic basalts in continental rifts and continental margins; IV_2: Alkali basalts in continental rifts; IV_3: Basalts in continental extensional zones (or incipient rifts); V. Mantle plume-related tholeiitic basalts or N-MORB. The dashed line represents the upper compositional boundary for ORG from anomalous ridge segments.

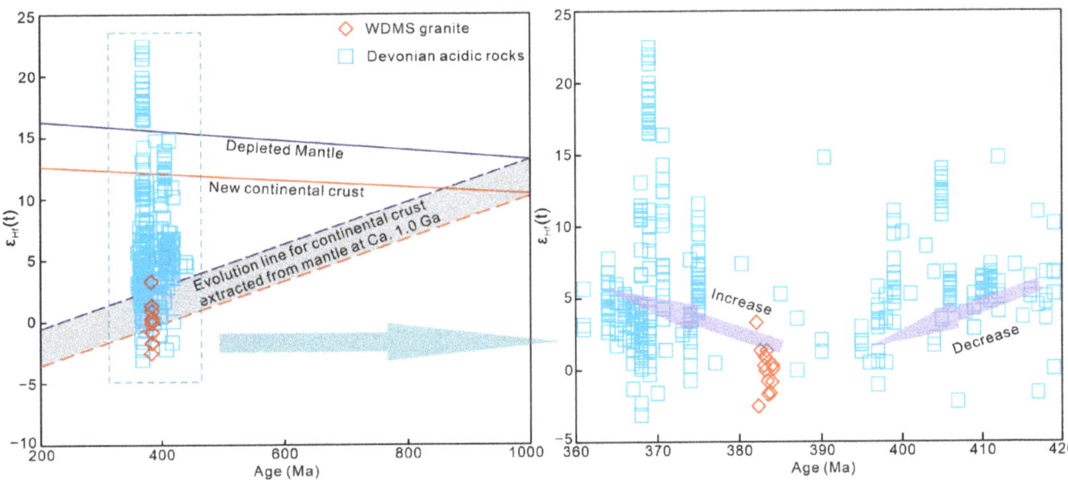

Figure 12. $\varepsilon_{Hf}(t)$-Age diagram of the WDMS granite and Devonian acidic rocks from the SBOB [7]. Ages and $\varepsilon_{Hf}(t)$ values of Devonian acidic rocks from [5,8,18–21,23–27].

From a sedimentological perspective, the Early Paleozoic strata in the SBOB are predominantly composed of pyroclastic rocks, limestones, sandstones, and mudstones, deposited in upper bathyal to neritic environments [85]. A large-scale unconformity separates the Late Paleozoic from the Early Paleozoic successions. The Middle-Lower Devonian SGJ Formation constitutes a molasse-type deposit, showing the characteristics of rapid accumulation [76,86]. Its vertical grain-size profile shows a vertical sequence of coarse–fine–coarse, and the depositional systems transition from subaqueous fan deposits, through fluvial-lacustrine facies to alluvial fan conglomerates. The color variations in sedimentary sequences progressing from reddish-green→green→gray→red collectively record a shallowing–deepening–shallowing cycle of the depositional water column [86]. The late Devonian Dundunshan Formation which in unconformity contact with the SGJ Formation is a continental volcanic sedimentary formation, composed of magmatic rocks (such as rhyolite, rhyolite dacite, andesite) and volcanic clastic rocks (e.g., tuff, volcanic breccia) [18]. The above evidence indicates that the SBOB showed a transition from upper bathyal-neritic facies to littoral-continental facies in the early Paleozoic-Devonian. In addition, magmatic rocks and sedimentary rocks formed in the SBOB during the early Paleozoic have generally suffered deformation and metamorphism, such as the strong mylonitization of granite, indicating that the strong dynamic metamorphism occurred in the region after the formation of these rocks [25]. While the Devonian magmatic rocks and sedimentary rocks did not undergo obvious deformation and metamorphism [87], it is indicated that the strong dynamic metamorphism did not occur in the region after the formation of these rocks, showing that there was a large-scale tectonic event in the SBOB at the turn of the early and late Paleozoic, and that the transformation of the tectonic system occurred in the region.

Characteristics of the sedimentary, deformation, and metamorphism, combined with the diversity of geochemical types and rock assemblages of the Early–Middle Devonian magmatic rocks in the SBOB, support the view that the transition from advancing to retreating accretionary orogenesis in the SBOB during the Early-Mid Devonian.

The core of the transformation of the accretionary orogenic shape is the process that the underlying plate push the overlying plate transform into the process that the underlying plate retreats and leads to the extension and thinning of the overlying plate. The HNXOB and HZOB are located on the north and south sides of the SH unit (Figure 1b). A critical question arises: which of these two ophiolite belts represents the subducting slab that acted as the underlying plate driving the accretionary orogenic regime transition during this tectonic transformation? The zircon U-Pb ages of gabbro and oceanic plagiogranite in the Hongliuhe-Niujuanzi-Xichangjing ophiolite range from 508~536 Ma, indicating an Early–Middle Cambrian formation age [7,88]. It is the product of the north–south bidirectional subduction and closure of the Hongliuhe-Niujuanzi-Xichangjing paleo-ocean basin [18]. The subduction process continued until the Silurian. Affected by this process, a series of arc magmatic rocks were produced on the SH unit [19,20], and the back-arc basin represented by the HZOB was formed by extension on the south side of the tectonic unit at 444.9 $\pm$ 2.1 Ma [18]. The tectonic evolution of the SBOB is affected by the subduction of the Hongliuhe-Niujuanzi-Xichangjing paleo-ocean basin. The Early Devonian A-type granite is the earliest A-type granite in the Beishan orogenic belt. It is widely distributed in the Mazongshan unit and SH unit on both sides of the HBOB, which corresponds to the early–late Paleozoic tectonic regime transformation event in the BOB. At the same time, the Hongliuhe hornblende gabbro and Zhaobishannan A-type granite with the characteristics of within-plate basalt formed at 404 Ma and 408 Ma intruded into the Hongliuhe ophiolite as a nailed rock mass [18], indicating that the Hongliuhe-Niujuanzi-Xichangjing paleo-ocean basin was closed before the Early Devonian. Therefore, the Hongliuhe-Niujuanzi-Xichangjing paleo-ocean basin experienced continuous compressional convergence with the SH unit, without exhibiting extension in the overlying plate caused by retreat of the underlying slab during Early–Middle Cambrian to pre-Early Devonian. Consequently, we propose that the Huitongshan-Zhangfangshan back-arc basin corresponds to the underlying slab during the transition in tectonic system. Zircon U-Pb dating of gabbro from the Huitongshan-Zhangfangshan ophiolite yields crystallization ages of 446~363 Ma [89–91], with the Zhangfangshan ophiolite (363 Ma) exhibiting characteristics of mafic magmatic rocks formed during the back-arc basin spreading stage, indicating that the Huitongshan-Zhangfangshan back-arc basin remained unclosed until the late Devonian [18]. Simultaneously, Devonian arc magmatic rocks (e.g., adakitic rocks) in southern Beishan are predominantly distributed near the Huitongshan-Zhangfangshan ophiolite belt (e.g., Liuyuan and Heishantou areas) [19,24]. In contrast, the SH unit exhibits an increased proportion of magmatic rocks formed in an extensional tectonic setting distal to this HZOB. In summary, we propose that the underlying slab responsible for the Early–Middle Devonian accretionary orogenic transition in the southern Beishan belt corresponds to the Huitongshan-Zhangfangshan back-arc basin. Specifically, the subduction of this back-arc basin beneath the SH unit transitioned from advancing compression to slab retreat during the Early–Middle Devonian, ultimately driving the tectonic shift from progressive advancing to retreating accretionary orogeny in the SBOB.

6. Conclusions

1. The crystallization age of volcanic rock from Wudaomingshui is 411.5 ± 2.8 Ma (MSWD = 0.43, n = 12), and the granite has crystallization age of 383.6 ± 2.2 Ma (MSWD = 0.11, n = 13). The $\varepsilon_{Hf}(t)$ values of granites are −2.5 to 3.3 with T_{DM2} ages from 1310 to 1013 Ma.
2. The Wudaomingshui volcanics are calc-alkaline basalt-basaltic andesite with low SiO_2 contents (47.35~55.59 wt.%), generated from the partial melting of the depleted lithosphere metasomatized by subduction fluid and contaminated by the lower crust. Granites are high-K calc-alkaline I-type felsic rocks, originating from the partial melting of the lower crust, and influenced by the mantle. The upper crust also affected the formation of granite.
3. Wudaomingshui volcanic rocks and granites were formed in the tectonic switching process from advancing to retreating subduction. Combined with the regional geological background, these findings indicate that the transformation from advancing to retreating accretionary orogenesis occurred in the southern Beishan Orogenic Belt during the Early–Middle Devonian.

Supplementary Materials: The following supporting information can be downloaded at: https://www.mdpi.com/article/10.3390/min15060632/s1, Text S1: Analytical methods for zircon U-Pb geochronology, in situ Lu-Hf isotope analysis and whole-rock geochemistry analysis of the Wudaomingshui volcanic rocks and granites; Table S1: Results of U-Pb dating by zircons from the Wudaomingshui volcanic rocks and granites; Table S2: Zircon Lu-Hf isotope analysis of the Wudaomingshui granites; Table S3: Whole-rock major (wt.%) and trace element (ppm) analyses of the Wudaomingshui volcanic rocks and granites [92–104].

Author Contributions: Methodology, T.H., Y.W. and L.W.; Investigation, J.Y., Z.Y., K.L., Z.W. and T.H.; Resources, T.H.; Data curation, T.H., Z.L., Z.W., K.L. and L.W.; Writing—original draft, T.H. and L.W.; Writing—review and editing, T.H., L.W. and Y.W.; Project administration, T.H. and Y.W.; Funding acquisition, T.H. and Y.W. All authors have read and agreed to the published version of the manuscript.

Funding: This research was supported by Gansu Province Key Research and Development Plan—Industrial Projects (No. 24YFGA015, 22YF7GA050), the projects of Ministry of Natural Resources, a new round of Mineral Exploration Breakthrough Strategic Action Science and Technology Support Projects (No. ZKKJ202410, ZKKJ202427) and Gansu Provincial Federation of Trade Unions employee subsidy fund project Study on mineralization and prospecting of porphyry copper and gold deposits in the western part of North Qilian (202411).

Data Availability Statement: The original contributions presented in the study are included in the Supplementary Material.

Acknowledgments: We would like to thank Wei Xu and Xinwei Zhai for their advance with this manuscript. We sincerely thank the editors and all reviewers for their valuable feedback that we have used to improve the quality of our manuscript.

Conflicts of Interest: The authors declare no conflicts of interest.

References

1. Şengör, A.M.C.; Natal'In, B.A.; Burtman, V.S. Evolution of the Altaid Tectonic Collage and Palaeozoic Crustal Growth in Eurasia. *Nature* **1993**, *364*, 299–307. [CrossRef]
2. Şengör, A.M.C.; Sunal, G.; Natal'in, B.A.; Voo, R.V.D. The Altaids: A review of twenty-five years of knowledge accumulation. *Earth-Sci. Rev.* **2022**, *228*, 104013. [CrossRef]
3. Windley, B.F.; Alexeiev, D.; Xiao, W.J.; Kroöner, A.; Badarch, G. Tectonic Models for Accretion of the Central Asian Orogenic Belt. *J. Geol. Soc.* **2007**, *164*, 31–47. [CrossRef]

4. Xiao, W.J.; Windley, B.F.; Han, C.M.; Liu, W.; Wan, B.; Zhang, J.E.; Ao, S.J.; Zhang, Z.Y.; Song, D.F. Late Paleozoic to Early Triassic Multiple Roll-Back and oroclinal Bending of the Mongolia Collage in Central Asia. *Earth-Sci. Rev.* **2018**, *186*, 94–128. [CrossRef]
5. Xu, W.; Xu, X.Y.; Lu, J.C.; Niu, Y.Z.; Chen, G.C.; Shi, J.Z.; Dang, B.; Song, B.; Zhang, Y.X.; Zhang, Q. Geochronology, Petrogenesis and Tectonic Implications of Devonian High-K Acid Magmatic Rocks from Yemajing Area in Beishan Orogen. *Earth Sci.* **2019**, *44*, 2775–2793. (In Chinese with English abstract)
6. Zhao, G.C.; Wang, Y.J.; Huang, B.C.; Dong, Y.P.; Li, S.Z.; Zhang, G.W.; Li, S. Geological reconstructions of the East Asian blocks: From the breakup of Rodinia to the assembly of Pangea. *Earth-Sci. Rev.* **2018**, *186*, 262–286. [CrossRef]
7. Zheng, R.G.; Zheng, J.Y.; Xiao, W.J.; Zhang, J. Long-lived subduction retreating led to continental rifting along the northern Gondwana: Insights from Devonian igneous rocks and ophiolite in the Beishan orogenic collage. *Lithos* **2023**, *454*, 107236. [CrossRef]
8. Zhu, J.; Lv, X.B.; Peng, S.G. U-Pb zircon geochronology, geochemistry and tectonic implications of the early Devonian granitoids in the Liuyuan area, Beishan, NW China. *Geosci. J.* **2016**, *20*, 609–625. [CrossRef]
9. Saktura, W.M.; Buckman, S.; Nutman, A.P.; Belousova, E.A.; Yan, Z.; Aitchison, J.C. Continental origin of the Gubaoquan eclogite and implications for evolution of the Beishan Orogen, Central Asian Orogenic Belt, NW China. *Lithos* **2017**, *294*, 20–38. [CrossRef]
10. He, Z.Y.; Zhang, Z.M.; Zong, K.Q.; Hua, X.; Klemd, R. Metamorphic P–T–t evolution of mafic HP granulites in the northeastern segment of the Tarim Craton (Dunhuang block): Evidence for early Paleozoic continental subduction. *Lithos* **2014**, *196*, 1–13. [CrossRef]
11. Wu, L.; Zhai, X.; Wang, E.; Chen, W.; Song, G.; Zheng, F.; Zhao, J.; Wang, J.; Wang, H. Early Permian Post-Collision Extensional Setting in the Southern Beishan Orogenic Belt: Evidence from the Zhangfangshan Granodiorite and the Baishantang Bimodal Volcanic Rocks. *Minerals* **2023**, *13*, 1468. [CrossRef]
12. Jahn, B.M.; Wu, F.Y.; Chen, B. Granitoids of the Central Asian Orogenic Belt and continental growth in the Phanerozoic. *Trans.—R. Soc. Edinb.* **2000**, *91*, 181–194.
13. Xiao, W.J.; Windley, B.F.; Sun, S.; Li, J.L.; Huang, B.C.; Han, C.M.; Yuan, C.; Sun, M.; Chen, H.L. A Tale of Amalgamation of Three Permo—Triassic Collage Systems in Central Asia: Oroclines, Sutures, and Terminal Accretion. *Annu. Rev. Earth Planet. Sci.* **2015**, *43*, 477–507. [CrossRef]
14. Huang, B.T.; Wang, G.Q.; Li, X.M.; Bu, T.; Dong, Z.C.; Zhu, T. Precambrian tectonic affinity of the Beishan Orogenic Belt: Constraints from Proterozoic metasedimentary rocks. *Precambrian Res.* **2022**, *376*, 106686. [CrossRef]
15. Song, D.F.; Xiao, W.J.; Zeng, H.; Mao, Q.G.; Ao, S.J. Accretionary orogenic processes of the Beishan orogenic belt. *Geol. Bull. China* **2024**, *43*, 2131–2150. (In Chinese with English abstract)
16. Xiao, W.J.; Mao, Q.G.; Windley, B.F.; Han, C.M.; Qu, J.F.; Zhang, J.E.; Ao, S.J.; Guo, Q.Q.; Cleven, N.R.; Lin, S.F.; et al. Paleozoic multiple accretionary and collisional processes of the Beishan orogenic collage. *Am. J. Sci.* **2010**, *310*, 1553–1594. [CrossRef]
17. Mao, Q.G.; Xiao, W.J.; Windley, B.F.; Han, C.M.; Qu, J.F.; Ao, S.J.; Zhang, J.E.; Guo, Q.Q. The Liuyuan complex in the Beishan, NW China: A Carboniferous-Permian ophiolitic fore-arc sliver in the southern Altaids. *Geol. Mag.* **2011**, *149*, 483–506. [CrossRef]
18. Wu, L. Paleo-Oceanic Basin Convergence Process in the Southern Margin of the Middle Central Asian Orogenic Belt: Petrogeochemical and Geochronological Evidence. Doctoral Dissertation, Lanzhou University, Lanzhou, China, 2024. (In Chinese with English abstract).
19. Mao, Q.G.; Xiao, W.J.; Fang, T.H.; Wang, J.B.; Han, C.M.; Sun, M.; Yuan, C. Late Ordovician to early Devonian adakites and Nb-enriched basalts in the Liuyuan area, Beishan, NW China: Implications for early Paleozoic slab–melting and crustal growth in the southern Altaids. *Gondwana Res.* **2012**, *22*, 534–553. [CrossRef]
20. Ao, S.J.; Xiao, W.J.; Han, C.M.; Li, X.H.; Qu, J.F.; Zhang, J.E.; Guo, Q.Q.; Tian, Z.H. Cambrian to early Silurian ophiolite and accretionary processes in the Beishan collage, NW China: Implications for the architecture of the Southern Altaids. *Geol. Mag.* **2012**, *149*, 606–625. [CrossRef]
21. Li, S.; Wang, T.; Tong, Y.; Hong, D.W.; Ouyang, Z.X. Identification of the Early Devonian Shuangfengshan A-type granitesin Liuyuan area of Beishan and its implications to tectonic evolution. *Acta Petrol. Mineral.* **2009**, *28*, 407–422. (In Chinese with English abstract)
22. Guo, Q.Q.; Xiao, W.J.; Hou, Q.L.; Windley, B.F.; Han, C.M.; Yian, Z.H.; Song, D.F. Construction of Late Devonian Dundunshan arc in the Beishan orogen and its implication for tectonics of southern Central Asian Orogenic Belt. *Lithos* **2014**, *184*, 361–378. [CrossRef]
23. Wang, E.; Zhai, X.; Chen, W.; Wu, L.; Song, G.; Wang, Y.; Guo, Z.; Zhao, J.; Wang, J. Late Devonian A-Type Granites from the Beishan, Southern Central Asia Orogenic Belt: Implications for Closure of the Paleo-Asia Ocean. *Minerals* **2023**, *13*, 565. [CrossRef]
24. Yang, Z.X.; Ding, S.H.; Zhang, J.; Fan, X.X.; Kong, W.Q.; Zhao, J.C.; Jing, D.L. The discovery of Early Devonian adakites in Beishan orogenic belt and its geological significance. *Acta Petrol. Mineral.* **2021**, *40*, 185–201. (In Chinese with English abstract)
25. Song, D.F.; Xiao, W.J.; Han, C.M.; Li, J.L.; Qu, J.F.; Guo, Q.Q.; Lin, L.N.; Wang, Z.M. Progressive accretionary tectonics of the Beishan orogenic collage, southern Altaids: Insights from zircon U-Pb and Hf isotopic data of high-grade complexes. *Precambrian Res.* **2013**, *227*, 368–388. [CrossRef]

26. Cheng, X.Y.; Teng, X.J.; Tian, J.; Duan, X.L. Geochemical characteristics, isotopic ages and tectonic environment of Sangejing Formation in Beishan orogenic belt, Inner Mongolia. *Geol. Bull. China* **2020**, *39*, 1461–1473. (In Chinese with English abstract)
27. Tao, G.H.; Li, X.F.; Chen, W.L.; Chen, W.; Lu, G.; Hao, C.; Pan, S.; Wu, K.N. The geochemical characteristics and tectonic significance of Devonian bimodal volcanic rocks in Yemaquan area of Beishan, Inner Mongolia. *Geol. Bull. China* **2022**, *41*, 1783–1797. (In Chinese with English abstract)
28. Ao, S.J.; Xiao, W.J.; Windley, B.F.; Mao, Q.G.; Han, C.M.; Zhang, J.E.; Yang, L.K.; Geng, J.Z. Paleozoic accretionary orogenesis in the eastern Beishan orogen: Constraints from zircon U-Pb and 40Ar/39Ar geochronology. *Gondwana Res.* **2015**, *30*, 224–235. [CrossRef]
29. He, Z.Y.; Klemdb, R.; Yana, L.L.; Zhang, Z.M. The origin and crustal evolution of microcontinents in the Beishan orogen of the southern Central Asian Orogenic Belt. *Earth-Sci. Rev.* **2018**, *185*, 1–14. [CrossRef]
30. Guo, Q.Q.; Chung, S.L.; Xiao, W.J.; Hou, Q.L.; Li, S. Petrogenesis and tectonic implications of late Devonian arc volcanic rocks in southern Beishan orogen, NW China: Geochemical and Nd–Sr–Hf isotopic constraints. *Lithos* **2017**, *278*, 84–96. [CrossRef]
31. Xie, W.; Song, X.Y.; Deng, Y.F.; Wang, Y.S.; Ba, D.H.; Zheng, W.Q.; Li, X.B. Geochemistry and petrogenetic implications of a Late Devonian mafic–ultramafic intrusion at the southern margin of the Central Asian Orogenic Belt. *Lithos* **2012**, *144*, 209–230. [CrossRef]
32. Hoskin, P.W.O.; Black, L.P. Metamorphic zircon formation by solid-state recrystallization of protolith igneous zircon. *J. Metamorph. Geol.* **2000**, *18*, 423–439. [CrossRef]
33. Winchester, J.A.; Floyd, P.A. Geochemical discrimination of different magma series and their differentiation products using immobile elements. *Chem. Geol.* **1977**, *20*, 325–343. [CrossRef]
34. Pearce, J.A. Trace Element Characteristics of Lavas from Destructive Plate Boundaries. In *Andesites: Orogenic Andesites and Related Rocks*; Thorpe, R.S., Ed.; Wiley: Hoboken, NJ, USA, 1982; pp. 528–548.
35. Whiteford, D.G.; Nicholls, I.A.; Taylor, S.R. Spatial variations in the geochemistry of quaternary lavas across the Sunda arc in Java and Bali. *Contrib. Mineral. Petrol.* **1979**, *70*, 341–356. [CrossRef]
36. Maniar, P.D.; Piccoli, P.M. Tectonic discrimination of granitoids. *Geol. Soc. Am. Bull.* **1989**, *101*, 35–643. [CrossRef]
37. Whalen, J.B.; Currie, K.L.; Chappell, B.W. A-type granites: Geochemical characteristics, discrimination and petrogenesis. *Contrib. Mineral. Petrol.* **1987**, *95*, 407–419. [CrossRef]
38. Sun, S.S.; McDonough, W. Chemical and isotopic systematics of oceanic basalts: Implications for mantle composition and processes. *Geol. Soc. Lond. Spec. Publ.* **1989**, *42*, 313–345. [CrossRef]
39. Hofmann, P.F. United Plates of America, the birth of a craton: Early Proterozoic assembly and growth of Laurentia. *Annu. Rev. Earth Planet. Sci.* **1988**, *16*, 543–603. [CrossRef]
40. Langmuir, C.H.; Bender, J.F.; Bence, A.E.; Hanson, G.N.; Taylor, S.R. Petrogenesis of basalts from the famous area: Mid-Atlantic ridge. *Earth Planet. Sci. Lett.* **1977**, *36*, 133–156. [CrossRef]
41. Dostal, J.; Owen, J.V. Cretaceous alkaline lamprophyres from northeastern Czech Republic: Geochemistry and petrogenesis. *Geol. Rundsch.* **1998**, *87*, 67–77. [CrossRef]
42. Raeisi, D.; Gholoizade, K.; Nayebi, N.; Babazadeh, S. Geochemistry and mineral composition of lamprophyre dikes, central Iran: Implications for petrogenesis and mantle evolution. *J. Earth Syst. Sci.* **2019**, *128*, 74. [CrossRef]
43. Qi, L.; Zhou, M.F. Platinum-group elemental and Sr-Nd-Os isotopic geochemistry of Permian Emeishan flood basalts in Guizhou Province, SW China. *Chem. Geol.* **2008**, *248*, 83–103. [CrossRef]
44. Wang, C.Y.; Zhou, M.F.; Keays, R.R. Geochemical constraints on the origin of the Permian Baimazhai mafic–ultramafic intrusion, SW China. *Contrib. Mineral. Petrol.* **2006**, *152*, 309–321. [CrossRef]
45. Li, T.; Liu, L.; Liao, X.Y.; Gai, Y.S.; Ma, T.; Wang, C. Geochemistry, Sr-Nd-Pb Isotopic Compositions and Zircon U-Pb Geochronology of Neoproterozoic Mafic Dyke in the Douling Complex, South Qinling Belt, China. *J. Earth Sci.* **2020**, *31*, 237–248. [CrossRef]
46. Stefan, W.; Carsten, M.; Klaus, M. Nb/Ta, Zr/Hf and REE in the depleted mantle: Implications for the differentiation history of the crust–mantle system. *Earth Planet. Sci. Lett.* **2003**, *205*, 309–324.
47. Pearce, J.A.; Peate, D.W. Tectonic implications of the composition of volcanic arc magmas. *Annu. Rev. Earth Planet. Sci.* **1995**, *23*, 251–285. [CrossRef]
48. Pearce, J.A. Geochemical finger printing of oceanic basalts with applications to ophiolite classification and the search for Archean oceanic crust. *Lithos* **2008**, *100*, 14–48. [CrossRef]
49. Xia, L.Q.; Li, X.M. Basalt geochemistry as a diagnostic indicator of tectonic setting. *Gondwana Res.* **2019**, *64*, 43–67. [CrossRef]
50. Woodhead, J.D.; Hergt, J.M.; Davidson, J.P.; Eggins, S.M. Hafnium isotope evidence for 'conservative' element mobility during subduction zone processes. *Earth Planet. Sci. Lett.* **2001**, *192*, 331–346. [CrossRef]
51. Jeremy, P.; Richard, S.; Robert, K. Special Paper: Adakite-Like Rocks: Their Diverse Origins and Questionable Role in Metallogenesis. *Econ. Geol. Bull. Soc. Econ. Geol.* **2007**, *102*, 537–576.
52. McKenzie, D.; O'Nions, R.K. Partial Melt Distributions from Inversion of Rare Earth Element Concentrations. *J. Petrol.* **1991**, *32*, 1021–1091. [CrossRef]

53. Aldanmaz, E.; Pearce, J.A.; Thirlwall, M.F.; Mitchell, J.G. Petrogenetic evolution of late Cenozoic, post-collision volcanism in western Anatolia, Turkey. *J. Volcanol. Geotherm. Res.* **2000**, *102*, 67–95. [CrossRef]
54. Alvaro, J.J.; Pouclet, A.; Ezzouhairi, H.; Abderrahmane, S. Early Neoproterozoic rift-related magmatism in the Anti-Atlas margin of the West African craton, Morocco. *Precambrian Res.* **2014**, *255*, 433–442. [CrossRef]
55. Zheng, H.; Huang, Q.T.; Kapsiotis, A.; Xia, B. Early cretaceous ophiolites of the Yarlung Zangbo Suture Zone: Insights from dolerites and peridotites from the Baer upper mantle suite, SW Tibet (China). *Int. Geol. Rev.* **2017**, *59*, 1–19. [CrossRef]
56. Yang, G.X.; Li, Y.J.; Xiao, W.J.; Sun, Y. Petrogenesis and tectonic implications of the middle Silurian volcanic rocks in northern West Junggar, NW China. *Int. Geol. Rev.* **2014**, *56*, 869–884. [CrossRef]
57. Kouankap, D.; Pierre, W.; Mufur, A.M.; Ganno, S. Contrasting Ba-Sr Granitoids from Bamenda Area, NW Cameroon: Sources Characteristics and Implications for the Evolution of the Pan African Fold Belt. *J. Geosci. Geomat.* **2018**, *6*, 65–76. [CrossRef]
58. Ye, X.Q.; Xu, Z.T.; Sun, L.Y.; Li, Z.W.; Li, M.M.; Jia, L. Genesis and Tectonic Significance of Miocene Tephrite in Laohushan Volcanic Area, Jilin Province. *Earth Sci.* **2024**, *49*, 1352–1366. (In Chinese with English abstract)
59. Bu, T.; Wang, G.Q.; Huang, B.T.; Guo, L.; Zou, X. Petrogenesis and tectonic significances of Early Devonian granites in Huangqiuquan area from the northern Beishan Orogenic Belt, NW China. *Geol. Rev.* **2024**, *70*, 2193–2211. (In Chinese with English abstract) [CrossRef]
60. Chappell, B.W. Aluminium saturation in I- and S-type granites and the characterization of fractionated haplogranites. *Lithos* **1999**, *46*, 535–551. [CrossRef]
61. Bonin, B. A-Type Granites and Related Rocks: Evolution of a Concept, Problems and Prospects. *Lithos* **2007**, *97*, 129. [CrossRef]
62. Pitcher W, S. Granite type and tectonic environment. *Mt. Build. Process.* **1982**, *4*, 19–40.
63. Chappell, B.W.; White, A.J.R. I-and S-type granites in the Lachlan fold belt. *Earth Environ. Sci. Trans. R. Soc. Edinb.* **1992**, *83*, 1–26. [CrossRef]
64. Wu, F.Y.; Li, X.H.; Yang, J.H.; Zheng, Y.F. Discussions on the petrogenesis of granites. *Acta Petrol. Sin.* **2007**, *23*, 1217–1238. (In Chinese with English abstract)
65. Wu, F.Y.; Liu, X.C.; Ji, W.Q.; Wang, J.M.; Yang, L. Highly fractionated granites: Recognition and research. *Sci. China Earth Sci.* **2017**, *60*, 1201–1219. (In Chinese with English abstract) [CrossRef]
66. Breiter, K.; Lamarão, C.N.; Borges, R.M.K.; Dall'Agnol, R. Chemical characteristics of zircon from A-type granites and comparison to zircon of S-type granites. *Lithos* **2014**, *192*, 208–225. [CrossRef]
67. Wu, F.Y.; Jahn, B.M.; Wilde, S.A.; Lo, C.H.; Yui, T.F.; Lin, Q.; Ge, W.C.; Sun, D.Y. Highly fractionated I-type granites in NE China (I): Geochronology and petrogenesis. *Lithos* **2003**, *66*, 241–273. [CrossRef]
68. Xia, R.; Wang, C.M.; Qing, M.; Li, W.; Carranza, E.J.M.; Guo, X.; Zeng, G.Z. Zircon U-Pb dating, geochemistry and Sr-Nd-Pb-Hf-O isotopes for the Nangetan granodiorites and mafic microgranular enclaves in the East Kunlun Orogen: Record of closure of the Paleo-Tethys. *Lithos* **2015**, *234*, 47–60. [CrossRef]
69. Shao, F.L.; Niu, Y.L.; Liu, Y.; Chen, S.; Kong, J.J.; Duan, M. Petrogenesis of Triassic granitoids in the East Kunlun Orogenic Belt, northern Tibetan Plateau and their tectonic implications. *Lithos* **2017**, *282*, 33–44. [CrossRef]
70. Defant, M.J.; Drummond, M.S. Derivation of some modern arc magmas by melting of young subducted lithosphere. *Nature* **1990**, *347*, 662–665. [CrossRef]
71. Taylor, S.R.; Mclennan, S.M. *The Continental Crust: Its Composition and Evolution: An Examination of the Geochemical Record Preserved in Sedimentary Rocks*; Backwell Scientific Publications: Boston, MA, USA, 1985; pp. 209–230.
72. Rapp, R.P.; Watson, E.B. Dehydration melting of metabasalt at 8–32 kbar: Implications for continental growth and crust-mantle recycling. *J. Petrol.* **1995**, *36*, 891–931. [CrossRef]
73. Jung, S.; Pfänder, J.A. Source composition and melting temperatures of orogenic granitoids: Constraints from CaO/Na_2O, $A1_2O_3/TiO_2$ and accessory mineral saturation thermometry. *Eur. J. Mineral.* **2007**, *19*, 859–870. [CrossRef]
74. Lassiter, J.C.; DePaolo, D.J. Plume/Lithosphere Interaction in the Generation of Continental and Oceanic Flood Basalts: Chemical and Isotopic Constraints. In *Large Igneous Provinces: Continental, Oceanic, and Planetary Flood Volcanism*; Mahoney, J.J., Coffin, M.F., Eds.; American Geophysical Union: Washington, DC, USA, 1997; pp. 335–356. [CrossRef]
75. Cawood, P.A.; Hawkesworth, C.J.; Dhuime, B. The continental record and the generation of continental crust. *Geol. Soc. Am. Bull.* **2013**, *125*, 14–32. [CrossRef]
76. Niu, Y.Z.; Li, C.Y.; Shi, G.R.; Lu, J.C.; Xu, W.; Shi, J.Z. Unconformity-bounded Upper Paleozoic megasequences in the Beishan Region (NW China) and implications for the timing of the Paleo-Asian Ocean closure. *J. Asian Earth Sci.* **2018**, *167*, 11–32. [CrossRef]
77. Cawood, P.A.; Kröner, A.; Collins, W.J.; Kusky, T.M.; Mooney, W.D.; Windley, B.F. Accretionary orogens through Earth history. *Geol. Soc. Lond. Spec. Publ.* **2009**, *318*, 1–36. [CrossRef]
78. Wang, Y.L.; Zhang, C.J.; Xiu, S.Z. Th/Hf-Ta/Hf identification of tectonic setting of basalts. *Acta Petrol. Sin.* **2001**, *17*, 413–421. (In Chinese with English abstract)

79. Collins, W.J.; Belousova, E.A.; Kemp, A.I.; Murphy, J.B. Two contrasting Phanerozoic orogenic systems revealed by hafnium isotope data. *Nat. Geosci.* **2011**, *4*, 333–337. [CrossRef]
80. Ding, J.; Han, C.; Xiao, W.; Wang, Z.; Song, D. Geochronology, geochemistry and Sr-Nd isotopes of the granitic rocks associated with tungsten deposits in Beishan district, NW China, Central Asian Orogenic Belt: Petrogenesis, metallogenic and tectonic implications. *Ore Geol. Rev.* **2017**, *89*, 441–462. [CrossRef]
81. Cawood, P.A.; Leitch, E.C.; Merle, R.E.; Nemchin, A.A. Orogenesis without collision: Stabilizing the Terra Australis accretionary orogen, eastern Australia. *Geol. Soc. Am. Bull.* **2011**, *123*, 2240–2255. [CrossRef]
82. Pearce, J.A.; Norry, M.J. Petrogenetic implications of Ti, Zr, Y, and Nb variations in volcanic rocks. *Contrib. Mineral. Petrol.* **1979**, *69*, 33–47. [CrossRef]
83. Pearce, J.A.; Harris, N.B.W.; Tindle, A.G. Trace element discrimination diagrams for the tectonic interpretation of granitic rocks. *J. Petrol.* **1984**, *25*, 956–983. [CrossRef]
84. Harris, N.B.W.; Pearce, J.A.; Tindle, A.G. Geochemical characteristics of collision-zone magmatim, collision tectonics. *Geol. Soc. Lond. Spec. Publ.* **1986**, *19*, 67–81. [CrossRef]
85. Song, D.; Xiao, W.; Windley, B.F.; Han, C.; Tian, Z. A Paleozoic Japan-type subduction-accretion system in the Beishan orogenic collage, southern Central Asian Orogenic Belt. *Lithos* **2015**, *224*, 195–213. [CrossRef]
86. Liang, J.W.; Chen, Y.L.; Zhang, W.Q.; Wang, G.Q.; Yu, J.Y.; Li, X.M. Detrital Zircon Dating of Devonian and Geological Significance of the Devonian Sangejing Formation in Beishan Area, Gansu Province. *Geol. Sci. Technol. Inf.* **2014**, *33*, 1–9. (In Chinese with English abstract)
87. Song, D.; Xiao, W.; Han, C.; Tian, Z.H. Polyphase deformation of a Paleozoic forearc–arc complex in the Beishan orogen, NW China. *Tectonophysics* **2014**, *632*, 224–243. [CrossRef]
88. Cleven, N.; Lin, S.F.; Guilmette, C.; Xiao, W.J.; Davis, B. Petrogenesis and implications for tectonic setting of Cambrian supra subduction-zone ophiolitic rocks in the central Beishan orogenic collage, Northwest China. *J. Asian Earth Sci.* **2015**, *113 Pt 1*, 369–390. [CrossRef]
89. Mao, Q.G.; Xiao, W.J.; Ao, S.J.; Yang, H. Subduction Initiation of the Southern Branch of the Paleo-Asian Ocean in the Middle Ordovician in the Southern Beishan Orogen. *Earth Space Sci.* **2023**, *10*, e2022EA002350. [CrossRef]
90. Yu, J.Y.; Li, X.G.; Wang, G.Q.; Wu, P.; Yan, Q.J. Zircon U-Pb ages of Huitongshan and Zhangfangshan ophiolite in Beishan of Gansu-Inner Mongolia border area and their significance. *Geol. Bull. China* **2012**, *31*, 2038–2045. (In Chinese with English abstract)
91. Meng, Y.; Zhang, X.; Bai, J.; Wang, K.; Qi, Y.; Zhao, H.; Han, Y. Petrogenesis and Geochronology of the Shazuoquan Ophiolite, Beishan Orogenic Belt: Constraints on the Evolution of the Beishan Ocean. *Minerals* **2023**, *13*, 1067. [CrossRef]
92. Wiedenbeck, M.; Hanchar, J.M.; Peck, W.H.; Sylvester, P.; Vley, J.; Whitehouse, M.; Kronz, A.; Morishita, Y.; Nasdala, L.; Fie-big, J.; et al. Further characterization of the 91500-zircon crystal. *Geostand. Geoanal. Res.* **2004**, *28*, 9–39. [CrossRef]
93. Pearce, N.J.G.; Perkins, W.T.; Westgate, J.A.; Gorton, M.P.; Chenery, S.P. A Compilation of New and Published Major and Trace Element Data for NIST SRM 610 and NIST SRM 612 Glass Reference Materials. *Geostand. Newsl.* **1997**, *21*, 115–144. [CrossRef]
94. Jackson, S.E.; Pearson, N.J.; Griffin, W.L.; Belousova, E.A. The application of laser ablation-inductively coupled plasma-mass spectrometry to in situ U–Pb zircon geochronology. *Chem. Geol.* **2004**, *211*, 47–69. [CrossRef]
95. Vermeesch, P. IsoplotR: A free and open toolbox for geochronology. *Geosci. Front.* **2018**, *9*, 1479–1493. [CrossRef]
96. Yuan, H.L.; Gao, S.; Liu, X.M.; Li, H.M.; Gunther, D.; Wu, F.Y. Accurate U–Pb age and trace element determinations of zircon by laser ablation-inductively coupled plasma-mass spectrometry. *Geostand. Geoanal. Res.* **2004**, *28*, 353–370. [CrossRef]
97. Söderlund, U.; Patchett, P.J.; Vervoort, J.D.; Isachsen, C.E. The 176Lu decay constant determined by Lu–Hf and U–Pb isotope systematics of Precambrian mafic intrusions. *Earth Planet. Sci. Lett.* **2004**, *219*, 311–324. [CrossRef]
98. Bouvier, A.; Vervoort, J.D.; Patchett, P.J. The Lu–Hf and Sm–Nd isotopic composition of CHUR: Constraints from unequili-brated chondrites and implications for the bulk composition of terrestrial planets. *Earth Planet. Sci. Lett.* **2008**, *273*, 48–57. [CrossRef]
99. Nowell, G.M.; Kempton, P.D.; Noble, S.R.; Fitton, J.G.; Saunders, A.D.; Mahoney, J.J.; Taylor, R.N. High precision Hf isotope measurements of MORB and OIB by thermal isnisation mass spectrometry: Insights into the depleted mantle. *Chem. Geol.* **1998**, *149*, 211–233. [CrossRef]
100. Griffin, W.L.; Pearson, N.J.; Belousova, E.; Jackson, S.E.; van Achterbergh, E.; O'Reilly, S.Y.; Shee, S.R. The Hf isotope composition of cratonicmantle: LA-MC-ICPMS analysis of zircon megacrysts in kimberlites. *Geochim. Cosmochim. Acta* **2000**, *64*, 133–147. [CrossRef]
101. Blichert-Toft, J.; Albarede, F. The Lu-Hf geochemistry of chondrites and the evolution of the mantle-crust system. *Earth Planet. Sci. Lett.* **1997**, *148*, 243–258. [CrossRef]
102. Bao, Z.A.; Chen, L.; Zong, C.L.; Yuan, H.L.; Chen, K.Y.; Dai, M.N. Development of pressed sulfide powder tablets for in situ sulfur and lead isotope measurement using LA-MC-ICP-MS. *Int. J. Mass Spectrom.* **2017**, *421*, 255–262. [CrossRef]

103. Li, X.H.; Liu, Y.; Tu, X.L.; Hu, G.Q.; Zeng, W. Precise determination of chemical compositions in silicate rocks using ICP-AES and ICP-MS: A comparative study of sample digestion techniques of alkali fusion and acid dissolution. *Geochemical* **2002**, *31*, 289–294.
104. Gao, J.J.; Liu, J.H.; Li, X.G.; Yan, Q.S.; Wang, X.J.; Wang, H.M. The determination of 52 elements in marine geological samples by an inductively coupled plasma optical emission spectrometry and an inductively coupled plasma mass spectrometry with a high-pressure closed digestion method. *Acta Oceanol. Sin.* **2017**, *36*, 113–121. [CrossRef]

minerals

MDPI

Article

Petrogenesis of Mafic–Ultramafic Cumulates in the Mayudia Ophiolite Complex, NE Himalaya: Evidence of an Island Arc Root in Eastern Neo-Tethys

Sapneswar Sahoo [1], Alik S. Majumdar [1,*], Rajagopal Anand [1], Dwijesh Ray [2] and José M. Fuenlabrada [3]

[1] Department of Applied Geology, Indian Institute of Technology (Indian School of Mines), Dhanbad 826004, Jharkhand, India; sapneswar.17dr000510@agl.ism.ac.in (S.S.); anandr@iitism.ac.in (R.A.)

[2] Planetary Sciences Division, Physical Research Laboratory, Ahmedabad 380009, Gujarat, India; dwijesh@prl.res.in

[3] Geochronology Laboratory, Universidad Complutense de Madrid, Calle José Antonio Novais, 228040 Madrid, Spain; jmfuenla@ucm.es

* Correspondence: asmajumdar@iitism.ac.in; Tel.: +91-(0)-326-223-5450

Abstract: Amphibole-rich cumulates provide crucial information pertaining to the petrogenetic history of suprasubduction zone ophiolites and are, therefore, helpful in constraining the evolution and closure of the Neo-Tethys during the late Cretaceous to the early Tertiary period. Following this, the present contribution examines the meta-hornblendite and meta-hornblende-gabbro lithologies in the Mayudia ophiolite complex (MdOC), NE Himalaya, based on their field and petrographic relations, constituent mineral compositions, whole rock major and trace element chemistry and bulk strontium (Sr)—neodymium (Nd) isotope systematics. MdOC cumulates potentially represent the fossilized record of an island arc root, where amphibole + titanite + magnetite was fractionally crystallized from a super hydrous magma (10.56–13.61 wt.% melt water content) prior to plagioclase in a stable physico-chemical condition (T: 865–940 °C, P: 0.8–1.4 GPa, logfO_2: −8.59–−11.19 unit) at lower crustal depths (30–38 km). Such extreme hydrous nature in the parental magma was generated by the flux melting of the sub-arc mantle wedge with aqueous inputs from the dehydrating slab. A super hydrous magmatic reservoir was, therefore, extant at sub-arc mantle depths in the eastern Neo-Tethys, which has likely modulated the composition of the oceanic crust during intraoceanic subduction.

Keywords: hornblendite; arc cumulate; super hydrous magma; intraoceanic subduction; Neo-Tethys; eastern Himalaya

1. Introduction

Mafic–ultramafic cumulate rocks are common worldwide in various tectonic settings, which provide important insights into the mantle melting process, magmatic differentiation and geodynamic evolution at plate boundaries [1–3]. They are commonly associated with ophiolites formed in mid-ocean ridges, back-arc, island arc and fore-arc basin environments [4], magma chambers produced within subduction-related upper-plate crust [5] and/or xenoliths introduced by alkaline basalt in intracontinental settings [6]. Compared to rift-related cumulates, which are typically linked to asthenopheric upwelling and/or mantle plumes, subduction-related cumulates typically exhibit geochemical signatures indicating hydrous melting, with rare stability of magmatic hydrous amphibole [7,8]. Compositional characteristics of such unique igneous amphiboles (commonly hornblende) are pivotal in understanding various aspects of the physico-chemical conditions (e.g., pressure, temperature, fluid activity, and oxygen fugacity) within sub-arc magma chambers [8,9]. Primary amphibole also imparts a major influence on arc magma compositions, which in turn controls arc crust formation and differentiation processes [10]. Amphibole-rich intrusives may also act as an important sink for subduction-derived fluids, which affects the aqueous budget and fluid-mobile element recycling in the convergent plate boundaries [11]. Studies on amphibole-rich cumulates (e.g., hornblendite, hornblende-gabbro) have, therefore, received significant attention over the years to unravel various aspects of sub-arc processes and geodynamic evolution in orogenic belts [1,2,11–13].

The Indus Tsangpo Suture Zone (ITSZ) in the Himalayan orogenic belt records the subduction of the Neo-Tethyan oceanic and Indian continental lithospheres beneath the Eurasia [14,15]. The Neo-Tethys Ocean opened up to the south of the Paleo-Tethys in the mid-Permian as an ocean or its complexes [16]. A spreading ridge was extant in the Neo-Tethys by the mid-Jurassic [16], which subsequently initiated at least two subduction systems to its north during the Cretaceous to the early Tertiary period, namely—(a) the 'Trans-Tethyan' intraoceanic subduction and (b) the 'Andean-type' subduction of the Neo-Tethyan oceanic slab beneath Eurasia [15]. The collision between India and the Trans-Tethyan subduction zone began in early Palaeocene (ca. 50–55 Ma) and continued until the final continental collision between India and Eurasia in the Middle Eocene (~40 Ma) [14]. This multistage collision consumed the Neo-Tethys, placed the oceanic remnants onto the crystalline rocks of the Greater Himalayan Sequence (GHS) and produced ophiolite complexes and/or mélanges with diversified lithologies, geochemistry, modal mineralogy, and age [17]. Within these ITSZ ophiolites, exposure of amphibole-rich cumulate rocks (e.g., hornblendite, hornblende-gabbro) is extremely rare although they are considered to be an important constituent of suprasubduction zone ophiolites worldwide [18]. In fact, limited occurrences of hornblendite/hornblende-gabbro have only been reported from the Jijal ophiolite complex, western Himalaya and the Mayudia ophiolite complex, eastern Himalaya (Figure 1a), where they are associated with garnetites and peridotite-tectonites, respectively [1,19,20]. Unravelling the petrogenetic history of these unique cumulates is, therefore, crucial to understand the sub-arc processes operating in the eastern and western flank of the ITSZ and the evolution of the Neo-Tethyan oceanic lithosphere during its closure.

To better constrain the sub-arc processes in the eastern Neo-Tethys during Mesozoic subduction, the present contribution focuses on the hornblende-rich cumulate lithologies in the far eastern margin of the ITSZ. These lithologies are exposed within the Mayudia ophiolite complex, NE Himalaya (Figure 1) and include meta-hornblendite and meta-hornblende-gabbro [19,21]. Combining mineral and whole-rock geochemistry with strontium (Sr) and neodymium (Nd) isotope systematics, this study aims to elucidate the magmatic history, geodynamic settings and physico-chemical conditions of the crystallization of the Mayudia meta-hornblendite and meta-hornblende-gabbro protoliths. The unique characteristics of the sub-arc Neo-Tethyan lithosphere have parallelly been examined in relevance to such typical occurrences of hydrous cumulates in the eastern Himalaya, which may have implications to better constrain the evolution in arc magma and crust compositions in the eastern Neo-Tethys during its final closure.

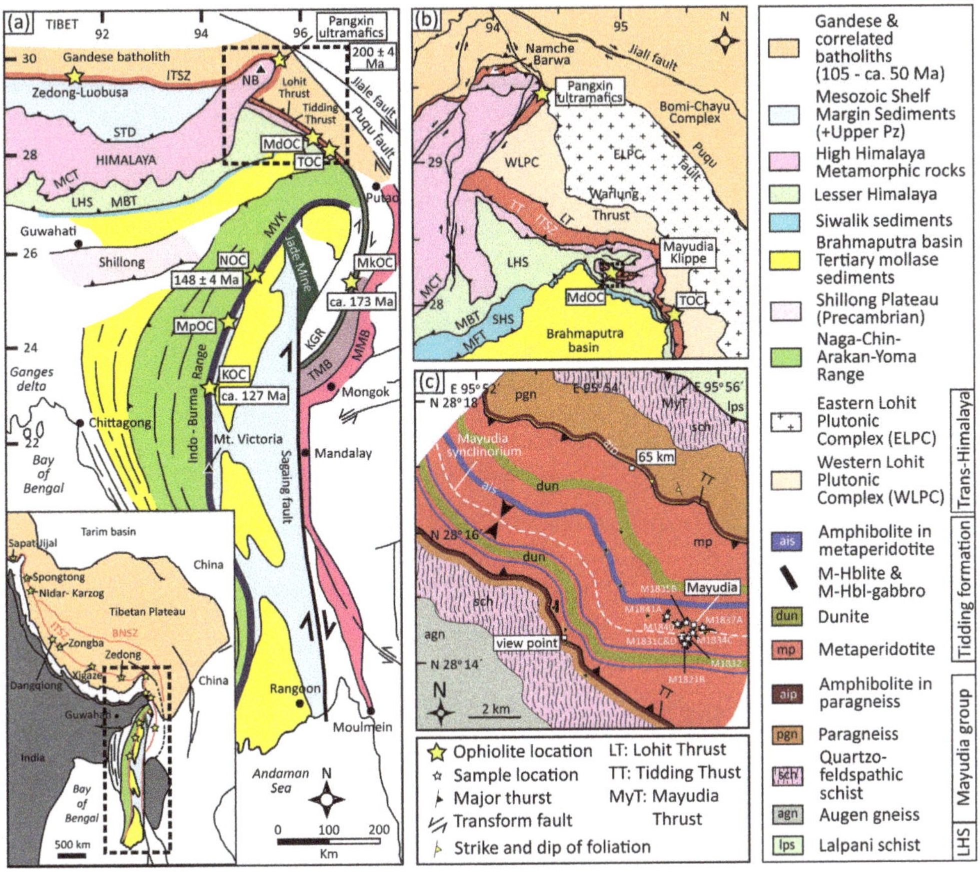

Figure 1. Geological map of (**a**) northeastern Himalaya, modified after [22], (**b**) Eastern Himalayan Syntaxis, following [23], and (**c**) Mayudia Klippe (modified after [24]), demarcating major tectonic and lithological units. The inset map in (**a**) shows the Indus Tsangpo Suture Zone (ITSZ) between India and the Tibetan plateau, modified after [25]. ITSZ ophiolite locations are shown as yellow stars in (**a**,**b**) along with their ages (see text for references). (**a**) elaborates on the area marked by the dotted black box in (**a**) (inset map). Similarly, (**b**,**c**) amplify the area outlined by dotted black boxes in (**a**) and (**b**), respectively. In (**c**), white stars demarcate sample locations, whereas opposing black arrowheads and white dotted lines represent the Mayudia synclinorium and its fold axis, respectively. White rectangular labels as the "view point" and "65 km" are the reference locations along the Roing–Hunli road in the same figure. MdOC: Mayudia ophiolite complex; TOC: Tidding ophiolite complex; NOC: Nagaland ophiolite complex; MpOC: Manipur ophiolite complex; KOC: Kalaymyo ophiolite complex; MkOC: Myitkyina ophiolite complex; STD: South Tibetan Detachment fault; MCT: Main Central Thrust; MBT: Main Boundary Thrust; MFT: Main Frontal Thrust; LHS: Lesser Himalayan Sequence; SHS: Siwalik Himalayan Sequence. M-Hblite: Meta-hornblendite; M-Hbl-gabbro: Meta-hornblende-gabbro. Other abbreviations are according to [22].

2. Geological Background and Field Relationships

In the NE Himalaya, the regional strikes of the major tectonic units bend from ENE-WSW to NW-SE around Namcha Barwa (NB), which is described as the Eastern Himalayan Syntaxis (EHS) (Figure 1a) [26]. Several ophiolite complexes and/or mélange rocks (e.g., Mayudia, Tidding, Nagaland, Manipur, Kalaymyo and Myitkyina) are sporadically exposed on the eastern limb of the EHS (Figure 1a) [21,27–30]. Of these, the Mayudia ophiolite complex (MdOC) provides a near-complete record of an ophiolite succession [19,21,31], bounded by the Lohit and Tidding thrusts on either side (Figure 1b, Table 1) [23]. The occurrence of metaperidotite, dunite, hornblendite, hornblende-zoisite schist, amphibolite, and metabasalt is evident in the MdOC lithospheric sequence (Table 1) [24,31]. These lithologies have originated in different sub-settings (back-arc, fore-arc and/or intra-arc) within a supra-subduction zone environment [21,32]. Lithological comparison of the Mayudia ophiolite rocks and their structural relations with adjacent tectonic units suggests that the Indus-Tsangpo Suture (ITS) is possibly represented by the highly imbricated Tidding Thrust (TT) (Figure 1b) [26]. Multi-generation ductile contractional and lateral-slip deformations modified the geometry of the complex, causing highly tectonized oceanic rocks to occur as slices at different structural levels [23,33]. Three stages of deformation (D_1, D_2, and D_3) were identified in the MdOC rocks [34], where D_2 deformation folded the Tidding thrust and produced the ophiolite-bearing klippe in its frontal portion near Mayudia (i.e., Mayudia Klippe, Figure 1b) [33]. These rocks have also experienced at least three stages of metamorphism under different pressure–temperature conditions (Stage-1: greenschist facies; Stage-2: eclogite facies; Stage-3: amphibolite facies) [26,35].

Table 1. General lithostratigraphy in the study area, as modified after [26,31,33].

Tectonic Domain	Tectonostratigraphic Unit and Subunit	Lithology with Mineralogical Characteristics
Trans-Himalayan Granitoids	Eastern Lohit Plutonic Complex	Leucogranite, quartz diorite, granodiorite, tourmaline-bearing pegmatites Warlung Thrust (WT)
	Western Lohit Plutonic Complex	Gabbro, intrusive leucogranite, aplite dyke, trondhjemite, mylonite gneiss, hornblende granite, tonalite, meta-diorite, dacite, pegmatite Lohit Thrust (LT)
Mayudia Ophiolite Complex (MdOC)	Tidding formation	Metacarbonate (Dolomite + muscovite + quartz) Metapelite grading into metachert (Quartz + biotite + muscovite + plagioclase + epidote ± titanite ± carbonate ± garnet ± chlorite) Metabasalt (Actinolititc amphibole + plagioclase + chlorite) Amphibolite (Hornblende + plagioclase + opaque (mostly magnetite) ± epidote ± titanite) Metaperidotite (olivine + clinopyroxene + serpentine + spinel) with hornblendite–hornblende–zoisite schist (pargasitic amphibole ± plagioclase ± zoisite) intrusives Tidding Thrust (TT)
Mayudia Gneiss	Mayudia Group	Garnet–biotite schist and gneiss, quartzite, and amphibolite Mayudia Thrust (MyT)
Lalpani Group	Lesser Himalayan Sequence (LHS)	Garnet–biotite schist, and paragneiss, micaceous quartzite, marble, carbonaceous schist, carbon-bearing calcareous schist, amphibolite

This study is based on the hornblende-dominated lithologies with conspicuous cumulus textures in the MdOC sequence that are solely restricted within peridotite tectonites as discordant bodies (Figure 2). These hornblende-dominated cumulates can be categorised into two distinct lithotypes based on their modal mineralogy: (1) meta-hornblendite (Hblite)—predominantly (>90 modal%) comprised of very-coarse grained (up to 10 cm) euhedral crystals of cumulus amphiboles (Figure 2a), and (2) meta-hornblende-gabbro (M-Hbl-gabbro)—consisting of large cumulus crystals of melanocratic amphibole and intercumulus grains of leucocratic epidote (Figure 2b). Notably, this meta-hornblende-gabbro lithotype in the MdOC sequence has been described as a hornblende-zoisite schist by previous authors [19,21]. Frequently, meta-hornblendite grades into meta-hornblende-gabbro (Figure 2c), where the leucocratic regions occur as pods (Figure 2d), veins (Figure 2e) and selvages (Figure 2f). Evidence of cross-cutting relations between these cumulate lithologies and adjacent metaperidotite is common, where their lithological contact is sharp without any evidence of chilled margins (Figure 2d,e). Towards the contact, meta-hornblende-gabbro is more common than meta-hornblendite (Figure 2d,e). The interpretation of this study is based on three meta-hornblendite and six meta-hornblende-gabbro samples, which were collected from different structural levels within the Mayudia Klippe (Figure 1c).

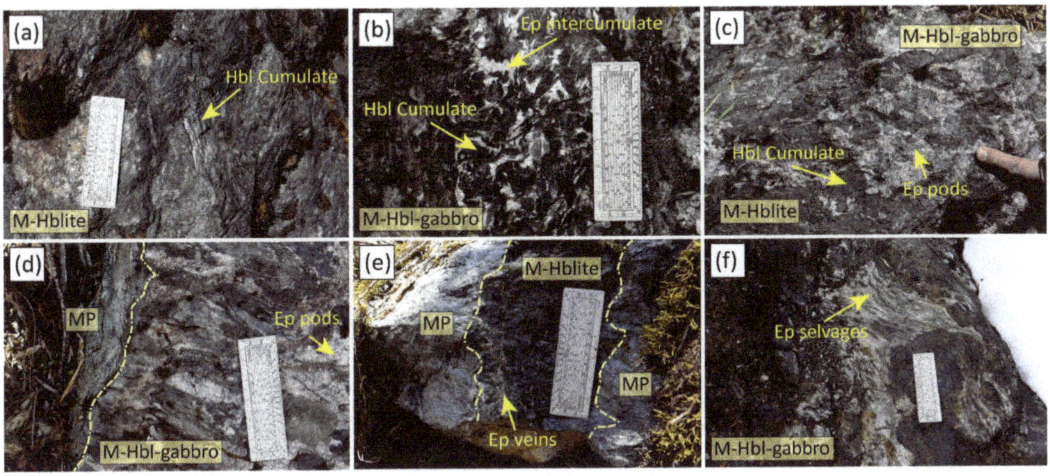

Figure 2. Field photographs of (**a**) meta-hornblendite (M-Hblite), consisting of pegmatitic hornblende and (**b**) meta-hornblende-gabbro (M-Hbl-gabbro), comprised of cumulus hornblende and intercumulus epidote, within the Mayudia ophiolite complex. (**c**) The gradational contact between the two lithotypes. (**d**–**f**) display the occurrence of epidotes as pods, vein and selvages, respectively, with the meta-hornblende-gabbro. The sharp lithological contacts without any evidence of chilled margins are also visible in (**d**,**e**). MP: Metaperidotite.

3. Analytical Techniques

Bulk rock mineral constituents were identified by powdered X-ray diffraction (XRD) analysis, whereas textural characteristics across grain interfaces were examined using visible-light microscopy on polished thin sections. Individual mineral compositions were obtained through Electron Probe Micro Analysis (EPMA), whereas whole-rock major and trace element concentrations were determined by employing Inductively Coupled Plasma–Optical Emission Spectrometry (ICP–OES) and Quadrupole-Inductively Coupled Plasma–Mass Spectrometry (Q-ICP–MS) techniques, respectively. In parallel, bulk $^{143}Nd/^{144}Nd$ and $^{87}Sr/^{86}Sr$ isotope ratios on selected samples were measured using Isotope Dilution–Thermal Ionization Mass Spectrometry (ID–TIMS).

For whole-rock powder preparation, samples were first crushed to <5 mm rock chips, repeatedly washed with ultrapure water and alcohol in an ultrasonic bath and then dried overnight at 60 °C to remove surface moisture. Visibly unweathered rock chips were subsequently powdered using a motorized tungsten carbide mill for 5 min and again left at 60 °C for overnight drying.

For XRD analyses, a monochromatic High-resolution HyPix-3000 semiconductor detector (Manufacturer: Rigaku Corporation, Tokyo, Japan), attached to the Rigaku SmartLab powder X-ray diffractometer (Manufacturer: Rigaku Corporation, Tokyo, Japan) and operated with a Cu anode (source energy 3 KW), was used to collect X-ray peak intensities at the Central Research Facility, IIT (ISM) Dhanbad, India. Here, powdered samples were scanned over a 5–80° 2θ range with a step-size of 0.02° and a counting time of 1s per step. After linear background correction, each XRD spectrum was analyzed in "PANalytical High-score Plus" software (version 5.1) to identify the phases while comparing with a reference database and to perform the Rietveld analysis for modal mineralogy determination.

The EPMA analyses were performed at the Planetary Sciences Division, Physical Research Laboratory, Ahmedabad (India) using a Jeol JXA-8530F Plus Hyper probe (Manufacturer: JEOL Ltd., Tokyo, Japan). Silicates and oxides were analyzed with five wavelength dispersive spectrometers with a pointed probe (diameter zero) at an acceleration voltage of 15 kV and a probe current of 20 nA. Different spectrometers were calibrated using diopside (Si, Mg, Ca), rutile (Ti), corundum (Al), hematite (Fe), synthetic NiO (Ni), albite (Na), orthoclase (K), synthetic Cr_2O_3 (Cr) and rhodonite (Mn). For Na and K, a peak count time of 10 s and a background count time of 5 s were used. For all other elements analyzed, the peak and background intensities were measured for 20 s and 10 s, respectively. The average detection limits were 0.01 wt.% for Si, Al, Fe, Cr, Ca, Mn, and K and 0.02 wt.% for Ti, Mg, Ni and Na. Average constituent mineral compositions and the complete data are shown in Table S1.

The ICP-OES and Q-ICP-MS analyses were carried out at the Australian Laboratory Service (ALS) Geochemistry laboratory, Loughrea, Ireland, following the ME-MS81D method (www.alsglobal.com, accessed on 24 February 2020). For ICP-OES analysis, ~0.2 g of the rock powders was mixed thoroughly with ~0.9 g of lithium metaborate/lithium tetraborate flux and fused at 1050 °C in a muffle furnace. The resulting melt was cooled and dissolved in 100 mL of 4 vol% HNO_3 + 2 vol% HCl. The solutions thus prepared were analyzed and corrected for spectral inter-element interferences to obtain concentrations of SiO_2, TiO_2, Al_2O_3, $Fe_2O_3{}^T$, CaO, MgO, MnO, Na_2O, K_2O and P_2O_5 in wt.% and Cr in ppm. For trace element analysis, ~0.1 g of the rock powders was fused with 0.9 g of lithium metaborate-tetraborate flux at 1025 °C and then dissolving the melt in an acid mixture of HCl, HNO_3 and HF. The resulting solutions were then analyzed using the Q-ICP-MS technique to measure concentrations (in ppm) of Sc, Ti, V, Ni, Co, Ba, Cs, Rb, Sr, U, Th, Pb, high field strength elements (HFSE: Nb, Hf, Zr), rare earth elements (REE: La, Ce, Pr, Nd, Sm, Eu, Gd, Tb, Dy, Ho, Er, Tm, Yb, Lu), and Y. The external precision and accuracy were assessed by repeated analysis of international standards: OREAS 102a, AMIS0304, SY-4, JP-1, UB-N [36]. The average blank level is below detection limits for respective elements. Average analytical uncertainty is commonly better than 5%, 10% and 20% for element concentrations above 1 wt.%, up to 1 ppm and <1 ppm, respectively. In addition, ~1.0 g of the rock powders was heated in the muffle furnace at 1050 °C for one hour to calculate the percent of loss on ignition from the difference in weight. Complete bulk rock data of this study are tabulated in Table S2.

Strontium-neodymium (Sr-Nd) isotope ratios of six selected samples were analyzed at the Geochronology Laboratory, Complutense University of Madrid, Spain following the sample preparation and analytical procedure described in [37]. In short, bulk sample powders were dissolved in ultra-pure reagents, followed by chromatographic separation of Rb, Sr, Sm and Nd isotopes. Isotope ratio analyses were performed using an IsotopX-Phoenix Multicollector Thermal Ionization Mass Spectrometer with data acquired in multidynamic mode. Repeated analyses on NBS-987 standard [36] ($^{87}Sr/^{86}Sr = 0.710248 \pm 0.000003$) yielded $^{87}Sr/^{86}Sr = 0.710242 \pm 0.00014$ (2σ, n = 8) and that for Nd isotopic reference material JNdi-1 ($^{143}Nd/^{144}Nd = 0.512115 \pm 0.000002$ [38]) provided $^{143}Nd/^{144}Nd$ values of 0.512101 ± 0.000008 (2σ, n = 9). The 2σ analytical errors were <0.008% for $^{87}Sr/^{86}Sr$ and

<0.004% for $^{143}Nd/^{144}Nd$. The procedural blanks were <5 ng and <1 ng for Rb-Sr and Sm-Nd, respectively. The analytical results are shown in Table S3.

4. Results

Powder XRD analyses on bulk-rock samples have identified hornblende (hbl) as the dominant (>90 modal%) constituent mineral in meta-hornblendites, whereas meta-hornblende-gabbro is comprised of hornblende (60–70 modal%) and epidote (ep, 20–30 modal%). Chlorite (chl, 5–10 modal%), titanite (ttn, 2–4 modal%) and magnetite (mag, up to 2 modal%) are present in significant proportions in both lithotypes, although their abundances are relatively higher in the meta-hornblendites as compared to the meta-hornblende-gabbros. Any presence of clinopyroxene (cpx), plagioclase (plag) and/or olivine (ol) has not been detected by XRD analysis. In the following sections, the textural–compositional characteristics in bulk-rock samples and individual minerals are presented in detail.

4.1. Petrography and Mineral Chemistry

In meta-hornblendites and meta-hornblende-gabbros, hornblende grains retained adcumulate and orthocumulate textures, respectively, of its precursors (Figure 3a,b) along with the presence of intercumulus epidotes in the latter (Figure 3c). In both lithotypes, hornblende grains preserved a relict interlocking texture, idiomorphic to hypidiomorphic shapes, straight to curved but sharp grain boundaries and evidence of triple junctions (Figure 3a,b). Chlorite always developed after replacing hornblendes either transgressing through the latter and/or at grain boundaries (Figure 3a,c–e). Titanite is detected either as interstitial grains and/or as inclusions (Figure 3d–f). Along with modal proportions, its crystal size is also often higher in the meta-hornblendites than the meta-hornblende-gabbros. Magnetite was identified as inclusion trails within hornblende in both lithologies (Figure 3a,f). Any presence of clinopyroxene, olivine and/or plagioclase was also not identified in the studied samples during the petrographic studies.

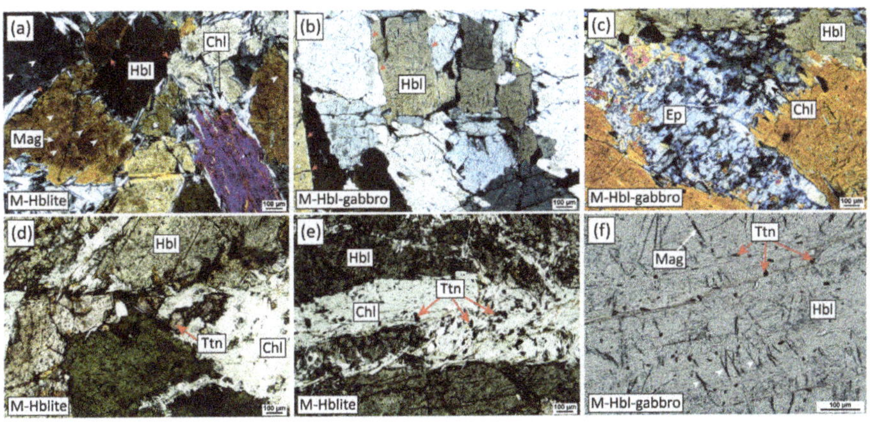

Figure 3. Representative visible-light photomicrographs of (**a**) meta-hornblendite (M-Hblite), exhibiting adcumulate textures and (**b**,**c**) meta-hornblende-gabbro (M-Hbl-gabbro) samples, displaying relict orthocumulate interlocking textures. Red and yellow arrowheads indicate straight to curved but sharp grain boundaries and triple junctions at hornblende (Hbl) interfaces. High dihedral angles at grain interfaces are denoted by yellow dashed lines. White arrowheads point out magnetite inclusion trails within hornblende crystals. (**c**) Occurrence of epidote (Ep) as intercumulus phase within amphibole orthocumulates devoid of any plagioclase feldspar. Titanite (Ttn, indicated by red

arrow) occurs as (**d**) interstitial grains, and/or (**e**) inclusions within meta-hornblendite, whereas both titanite (Ttn) and magnetite (Mag) are present as inclusions within amphibole grains in meta-hornblende-gabbro as demarcated in (**f**) Chlorite (Chl) is evident in both lithotypes replacing amphibole in (**c**–**e**). Mineral abbreviations are after [39].

Compositionally, all hornblendes contain significant proportions of CaO (12.41–13.05 wt.% in M-Hblite and 11.96–12.49 wt.% in M-Hbl-gabbro) and TiO_2 (0.37–0.61 wt.% and 0.25–0.54 wt.%, respectively) with lesser concentrations of Na_2O (1.80–2.66 and 2.22–2.57 wt.%, respectively) and K_2O (0.21–0.33 and 0.27–0.34 wt.%, respectively) (Table S1). Based on stoichiometry, they are classified as pargasite (Figure 4a, [40]), that exhibit strongly linear positive correlations in ^{IV}Al vs. Ti and ^{A}(Na + K) (calculated based on 24 oxygen atoms per formula unit, a.p.f.u.) content (Figure 4b,c). This is opposed to the negative trend identified in their Fe^{Total} vs. Mg (per 24 O a.p.f.u.) and Si vs. (Ca + Na + K) proportions (Figure 4d,e). No correlation is present in their ^{IV}Al vs. ^{VI}Al, Ca, and Mg/(Mg + Fe^{2+}) (per 24 O a.p.f.u.) concentrations (Figure 4f–h). However, the pargasites in the meta-hornblendites contain lower ^{VI}Al but higher Ca in their octahedral sites as compared to that in meta-hornblende-gabbros (Figure 4f,g). Intercumulus epidotes in M-Hbl-gabbro possess significant Al_2O_3 (27.60–32.84 wt.%), low $Fe_2O_3^{calc}$ (up to 7.57 wt.%) and are of zoisitic composition with low pistacite content (Ps = $Fe^{3+}/(Fe^{3+}$ + Al) $\leq$ 0.15) (Table S1). Secondary chlorites, after replacing amphiboles in both lithotypes (Figure 3a–c), exhibit clinochlore compositions with significant SiO_2 (M-Hblite: 28.31–29.29 wt.% and M-Hbl-gabbro: 28.32–30.00 wt.%, respectively) and Al_2O_3 (19.93–22.07 wt.% and 20.03–21.65 wt.%, respectively) contents and Mg# values (80.50 $\pm$ 1.03) (Table S1). Titanite in both lithologies shows narrow-ranging SiO_2 (31.25 $\pm$ 0.18 and 31.32 $\pm$ 0.36, respectively), TiO_2 (37.55 $\pm$ 0.39 and 38.14 $\pm$ 0.62, respectively), CaO (29.74 $\pm$ 0.20 and 29.74 $\pm$ 0.23, respectively) and Al_2O_3 (1.71 $\pm$ 0.09 and 1.39 $\pm$ 0.32, respectively) contents (Table S1).

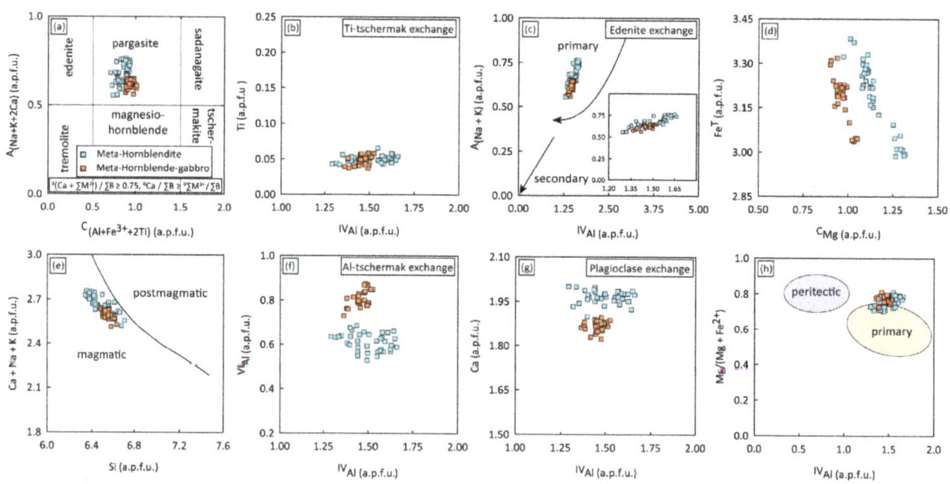

Figure 4. (**a**) Amphibole classification diagram with mineral stoichiometry calculated on 24 (O, OH) basis after [40]. Cation substitution diagrams for amphibole compositions, showing relative variations in (**b**) ^{IV}Al vs. Ti (a.p.f.u.), (**c**) ^{IV}Al vs. ^{A}(Na + K) (a.p.f.u.), (**d**) ^{C}Mg vs. Fe^{T} (a.p.f.u.), (**e**) Si vs. (Ca + Na + K) (a.p.f.u.), (**f**) ^{IV}Al vs. ^{VI}Al (a.p.f.u.), (**g**) ^{IV}Al vs. Ca (a.p.f.u.) and (**h**) ^{IV}Al vs. Mg/(Mg + Fe^{2+}) (a.p.f.u.). Primary and secondary amphibole fields with boundary arrow positions are after [41] in (**c**), the magmatic and post-magmatic amphibole boundary in (**e**) are after [42], whereas primary and peritectic amphibole fields in (**h**) are following [43].

4.2. Whole-Rock Geochemistry

The investigated samples are characterized by variable CaO (6.46–19.85 wt.%), MgO (5.61–14.40 wt.%), $Fe_2O_3^T$ (5.68–12.50 wt.%) contents, and Mg# values (65.03–78.80), with lower SiO_2 (37.20–42.50 wt.%), $Na_2O + K_2O$ (0.78–2.59 wt.%), Co (27–82 ppm), and Sc (22–82 ppm) concentrations (Figures 5 and 6, Table S2). Significant proportions of TiO_2 (0.56–1.21 wt.%) are evident in all the investigated rocks (Table S2). Furthermore, the meta-hornblendites exhibit higher MgO (17.60–20.40 wt.%), $Fe_2O_3^T$ (10.50–12.50 wt.%), Sm (3.00–3.62 ppm), Y (33.20–41.90 ppm), MREE (16.97–21.97 ppm), HREE (8.75–11.21 ppm) and Mg# concentrations (74.39–78.80) than the meta-hornblende-gabbros (5.61–14.40 wt.%, 5.68–10.35 wt.%, 1.31–2.21 ppm, 19.67–34.12 ppm, 8.51–14.84 ppm, 4.38–6.87 ppm and 65.03–74.28, respectively), which is compensated by opposite trends in their SiO_2 (M-Hblite: 37.20–38.30 wt.%; M-Hbl-gabbro: 40.20–42.50 wt.%), CaO (M-Hblite: 6.46–8.03 wt.%; M-Hbl-gabbro: 11.25–19.85 wt.%), and Sr (M-Hblite: 5.20–18.20 ppm; M-Hbl-gabbro: 31.90–501 ppm) contents (Figures 5 and 6, Table S2). High field strength elements (HFSE: Nb, Zr, Hf) and Ti concentrations are comparable in both lithotypes (Nb: 0.30–0.70 ppm and 0.30–0.80 ppm, respectively, Zr: 14.00–17.00 ppm and 11.00–18.00 ppm, respectively, Hf: 0.60–0.80 ppm and 0.60–0.90 ppm, respectively, Ti: 5995–7254 ppm and 3357–6774 ppm, respectively) (Table S2). Strongly negative linear correlations in MgO vs. Al_2O_3 and CaO concentrations are conspicuous across the samples, which contrasts with their positive linear trends in MgO vs. $Fe_2O_3^T$ and TiO_2 variations (Figure 6). In the SiO_2 vs. Mg# diagram, their compositions are akin to the cumulate line of descent (Figure 7a), which has also been observed in other Himalayan hornblende-bearing intrusives [19,20]. Furthermore, all rocks show a clear trend of positive linear correlation in the SiO_2 vs. Sm/Zr plot, where the M-Hblites are characterized by higher Sm/Zr (0.20–0.21) but lower SiO_2 (37.20–38.30 wt.%) as compared to the M-Hbl-gabbros (0.10–0.16; 40.20–42.50 wt.%) (Figure 7b, Table S2). This is opposite to the trend observed in the Eu/Eu* (M-Hblite: 0.61–0.79; M-Hbl-gabbro: 1.15–2.59) ratio against SiO_2 concentrations. (Figure 7c, Table S2). All samples further display low Nb/La ratios (0.10–0.83) and La_N/Sm_N values (0.12–0.62) (Table S2), which has previously been observed in oceanic arc-related rocks [44].

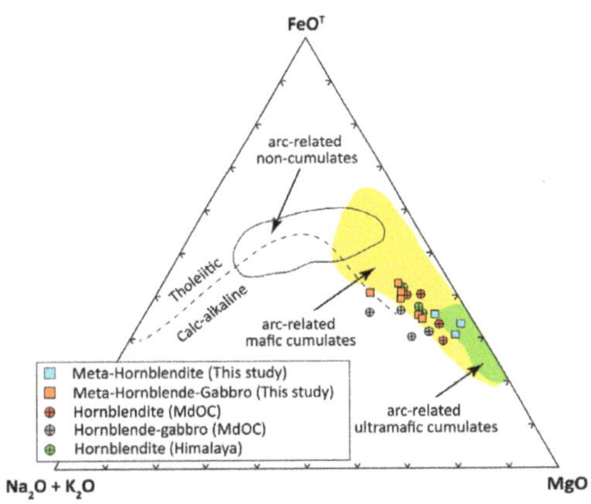

Figure 5. Ternary (Na_2O +K_2O) vs. MgO vs. FeO^T diagram of studied samples, with demarcation of different arc-related and unrelated cumulate fields after [45]. MdOC meta-hornblendite and meta-hornblende-gabbro data are taken from [46] and [19,21], respectively, whereas Himalayan hornblendite compositions are reported from [1,20] as per Table S4.

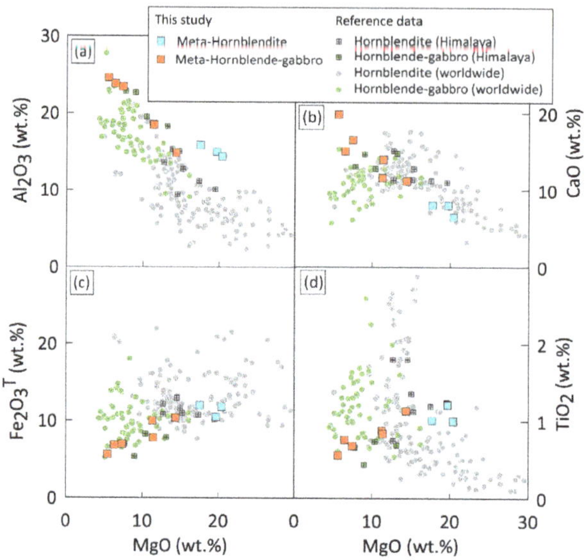

Figure 6. Bulk compositional variations in MgO (wt.%) vs. (**a**) Al_2O_3 (wt.%), (**b**) CaO (wt.%), (**c**) $Fe_2O_3{}^T$ (wt.%) and (**d**) TiO_2 (wt.%) in studied samples. Reference data are from Table S4.

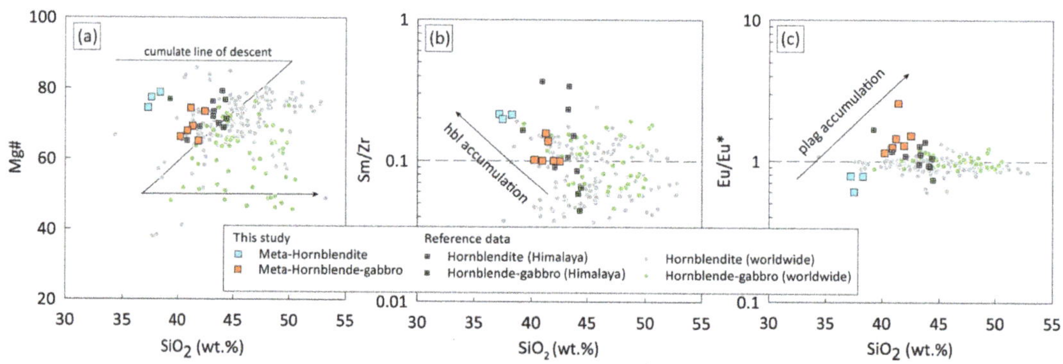

Figure 7. Bulk compositional variations in SiO_2 (wt.%) vs. (**a**) Mg#, (**b**) Sm/Zr, and (**c**) Eu/Eu* in studied samples. The cumulate line of descent is after [20]. Reference data are taken from Table S4.

In the CI chondrite normalized REE diagrams, all samples display a LREE-depleted pattern (Ce_N/Yb_N: 0.20–0.59), where a greater degree of MREE to LREE depletion (Ce_N/Sm_N: 0.26–0.62) is evident as compared to HREE to MREE fractionation (Sm_N/Yb_N: 0.75–1.02) (Figure 8a,b, Table S2). Furthermore, meta-hornblendites exhibit strong negative Eu and Sr anomalies (Eu/Eu* = 0.61–0.79; Sr/Sr* = 0.04–0.13) in contrast to positive Eu-Sr anomalies in the meta-hornblende-gabbros (Eu/Eu* = 1.15–2.59; Sr/Sr* = 2.13–6.62 except M1831D, 0.46) (Figure 8, Table S2). Strong HFSE (Nb, Zr, Hf) and Ti depletion are also evident across the samples in primitive PM-normalized multi-element diagrams, which contrasts with significant enrichment in incompatible elements, such as Ba, Cs, U, and Th (Figure 8c,d).

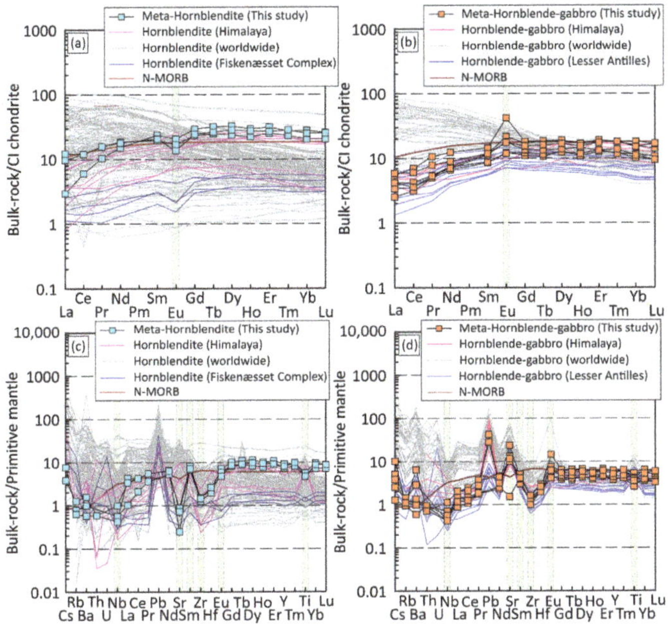

Figure 8. (**a**,**b**) CI chondrite-normalized and (**c**,**d**) PM-normalized concentration diagrams for meta-hornblendite (**a**,**c**) and meta-hornblende-gabbro (**b**,**d**) samples. CI chondrite, primitive mantle and Normal-Mid Ocean Ridge Basalt (N-MORB) values are taken from [47]. Reference data for Himalayan and worldwide hornblendite and hornblende-gabbro are from Table S4. Hornblendite data from Fiskenæsset complex and hornblende-gabbro data from the Lesser Antilles Arc are also plotted for reference after [2,48,49], respectively. Light green vertical stripes point out anomalous behaviours of different elements in these diagrams.

4.3. Sr-Nd Isotope Signatures

Measured $^{87}Sr/^{86}Sr$ and $^{143}Nd/^{144}Nd$ ratios in selected meta-hornblendite and meta-hornblende-gabbro samples vary between 0.703826 ± 3–0.706314 ± 60 and 0.513116 ± 11–0.513303 ± 1, respectively (Table S3). In the $^{87}Sr/^{86}Sr_{(t)}$ and $^{143}Nd/^{144}Nd_{(t)}$ diagram (calculated considering crystallization age (t) of ca. 127 Ma [25]), the samples exhibit compositional consistency with other ITSZ hornblende-bearing cumulate lithologies, while variably diverging from the N-MORB and mantle array compositions (Figure 9a) [1,19]. They further yield $\varepsilon_{Nd(t)}$ (t = 127 Ma) and $f_{Sm/Nd}$ values ranging from +7.71–+12.29–+0.2172–+0.6816, respectively (Figure 9b, Table S3). Interestingly, $\varepsilon_{Nd(t)}$ values of the meta-hornblende-gabbro samples (+7.71–+10.76) are progressively lower as compared to those of the meta-hornblendites (+10.16–+12.29) (Table S3, Figure 9b). In the La/Yb vs. Ba/La diagram, all investigated lithologies further exhibit a greater extent of Ba/La variation and negligible La/Yb change, in a similar manner to that observed in other Himalayan Hbl-cumulates (Figure 10).

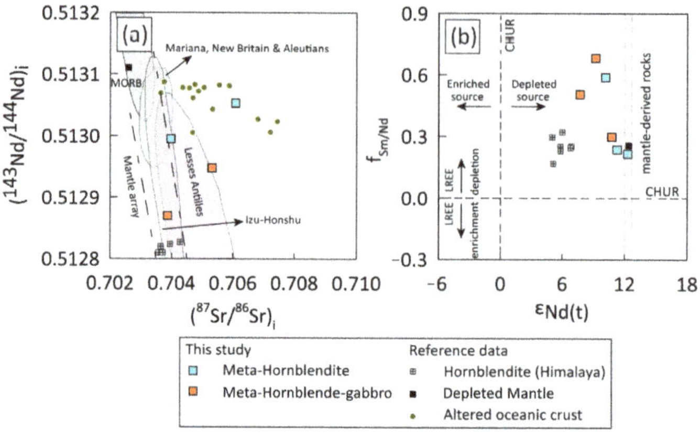

Figure 9. (**a**) Age-corrected $^{87}Sr/^{86}Sr_{(t)}$ vs. $^{143}Nd/^{144}Nd_{(t)}$, and (**b**) $\varepsilon_{Nd(t)}$ vs. $f_{Sm/Nd}$ plots of selected samples from this study. Compositions of depleted mantle (DM) [50], mantle array [51] and MORB plus different worldwide island arcs [52] are also plotted for comparison. Reference Himalayan hornblendite data are from Table S4, and altered oceanic crust data are from [53]. LREE: Light Rare Earth Elements; CHUR: Chondritic Uniform Reservoir.

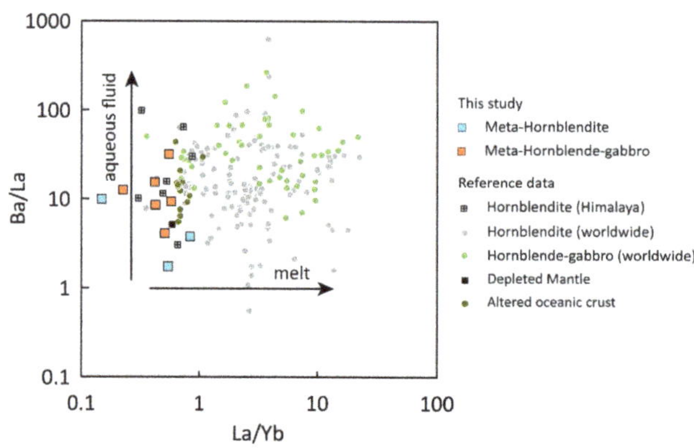

Figure 10. La/Yb vs. Ba/La plot of studied samples.

5. Discussion

5.1. Protolith Signatures

Bulk-rock XRD analyses combined with petrographic studies identify the presence of hornblende, chlorite, titanite, and magnetite ± epidote in the studied lithologies (Figures 2 and 3). The formation of amphibole in mafic–ultramafic rocks has previously been described as the consequence of three distinct processes: (a) direct crystallization from a hydrous silicate melt [48,54], (b) peritectic reaction between clinopyroxene (cpx) ± olivine (ol) and melt [55], and (c) retrograde metamorphism of granulite facies rocks [56]. In the case of amphiboles produced by the melt-induced reaction-replacement process, the presence of cpx ± ol relicts, embedded within an amphibole matrix, and the development of granoblastic and/or poikilitic textures with indented grain boundaries (attesting disequilibrium reactions) are common [43,57]. Major element compositions of peritectic amphiboles also exhibit low ^{IV}Al (a.p.f.u.) and high Mg# values in proximity to clinopyroxene chemistry [43]. A similar compositional distinction for metamorphic amphiboles can be made based on low total alkali ($^{A}(Na + K)$) (a.p.f.u.) but elevated Si (a.p.f.u.) contents [41,58]. Neither of these characteristics is consistent with the textural–compositional observations of the studied pargasites, nor is there any evidence of cpx ± ol in the studied assemblages (Figures 3a–e and 4h) [43]. Rather, the idiomorphic to hypidiomorphic shape of relict pargasitic hornblendes with interlocking textures and sharp straight-to-curved grain boundaries with evidence of triple junctions (Figure 3a–c) clearly indicate an equilibrium condition during amphibole formation, which may only result from magmatic crystallization [59]. Higher proportions of ^{A}(Na + K), ^{IV}Al (a.p.f.u.) and lower concentrations of Si (a.p.f.u.) in pargasites are also more akin to global primary hornblende compositions (Figure 4c,e) [41,43]. An igneous origin of the studied pargasite is, therefore, plausible in both lithotypes, as previously observed in mafic–ultramafic intrusives worldwide [43]. The massive euhedral grain size of amphibole crystals (up to 10 cm) further points to a relatively stable magmatic environment during their crystallization, which might have provided a prolonged residence time for crystal growth [60]. Evidence of titanite and magnetite, either as interstitial phases and/or inclusion trails within the amphibole grains, suggest their comagmatic origin with igneous amphiboles [61]. The epidote grains within the meta-hornblende-gabbro samples, however, exhibit compositional consistency (Ps $\leq$ 0.15, Table S1) with the secondary zoisites identified in other high P/T metamorphic terranes [62] and have previously been interpreted as the consequence of plagioclase breakdown [21]. Likewise, the development of chlorite after replacing pargasites, either along grain boundaries and/or transgressing through the hornblendes, is indicative of a metamorphic origin [63]. Hence, the igneous minerals preserved within the studied assemblages include pargasitic hornblende, titanite and magnetite.

However, the investigated lithologies are variably metamorphosed to chlorite and/or epidote (Figure 3a,c–f). An a-priori assessment of protolith signatures is, therefore, important before further petrogenetic inferences. In this regard, any interpretation (e.g., melt composition, water content) based on the igneous mineral chemistry should be accurate. Given the meta-hornblendites are modally dominated (>92 modal%) by igneous minerals (Hbl + Ttn + Mag), the bulk geochemical characteristics of these ultramafic assemblages should also reflect their magmatic signatures. In the case of the meta-hornblende-gabbros, metamorphic chlorite and epidote constitute up to 38 modal%. Chloritization, however, commonly conserves the bulk geochemical attributes of the protolith assemblages, as revealed in previous studies [64]. Likewise, epidotization tends to retain the bulk elemental characteristics of the parental rock, except for CaO, Sr, and Eu [65]. The possibility of post-magmatic Sr enrichment in meta-hornblende-gabbros cannot be ruled out in the present case. However, the strong linear correlation of CaO (against MgO) and Eu/Eu* (against SiO_2) in the meta-hornblende-gabbros to that in the meta-hornblendites (in contrast to their scattered behaviour) suggests that CaO and Eu possibly have also retained the bulk magmatic signatures. Nevertheless, it is important to emphasize that any interpretation in this study is not based on a particular textural, mineralogical and/or geochemical attribute; rather different observations are combined to support any inference.

Following this, a cumulate igneous origin of these rocks is evident from—(a) the compositional similarities of the meta-hornblendites and the meta-hornblende-gabbros with arc-related ultramafic and mafic cumulates in MgO vs. FeO^T vs. (Na_2O + K_2O) diagram (Figure 5) [45], as observed in previous studies on Himalayan mafic–ultramafic intrusives [1,19–21]; (b) evidence of interlocking cumulate textures, defined by idiomorphic hornblende crystals, in the field and petrographic observations (Figures 2a,b and 3a,b); and (c) concurrent variation in bulk Mg# with SiO_2 (in wt.%) along the cumulate line of descent (Figure 7a) [66,67]. The intercumulus phase in the precursor to the meta-hornblende-gabbros is likely to be plagioclase (before zoisite formation), owing to the covariation in CaO and Al_2O_3 (against MgO) (Figure 6a,b) and positive Eu anomaly in these samples (Table S2, Figures 7c and 8b,d) [21]. The magmatic mineralogy of the meta-hornblendite may, therefore, have been dominated by hornblende (>90 modal %), indicating a hornblendite protolith. In contrast, the precursor of the meta-hornblende-gabbro would have consisted of pargasites (60–70 modal %) + plagioclase (~20–30 modal %) apart from minor titanite and magnetite, which is consistent with hornblende-gabbro mineralogy.

Different elemental behaviours in protolith assemblages are commonly governed by the modal variation in magmatic minerals. For instance, the meta-hornblendites exhibit higher concentrations of MgO, TiO_2, $Fe_2O_3{}^T$, Sc, Ni, Co, MREE (except Eu) and HREE + Y and Mg# values relative to meta-hornblende-gabbro (Figures 6 and 8, Table S2), as these elements are more compatible in amphiboles than other MdOC cumulate minerals [68]. However, greater enrichment in TiO_2 and $Fe_2O_3{}^T$ in the meta-hornblendites than in the meta-hornblende-gabbros (Figure 6c,d, Table S2) might also have been aided by a higher abundance of titanite and magnetite in the former (Figure 3a,d–f) [69]. This contrasts with the higher CaO, Al_2O_3, SiO_2, and Eu contents in the meta-hornblende-gabbros as compared to the meta-hornblendites (Figures 6a,b, 7 and 8). that implies governing role of plagioclase abundance in controlling their behaviours in bulk rock [70]. Comparable concentrations of HFSE (Nb, Zr, Hf) in the studied samples (Table S2), however, suggest that parental source composition rather than modal mineralogy has controlled their abundances in MdOC cumulates [70]. A decreasing trend in MgO, $Fe_2O_3{}^T$, and TiO_2 contents, Sm/Zr ratios, and Mg# values compensated by an opposite trend in CaO, Al_2O_3, and SiO_2 contents, Eu/Eu* ratio in the meta-hornblendites relative to the meta-hornblende-gabbros (Figures 6a,b and 7 and Table S2) further suggest that the protolith of the meta-hornblendite has crystallized earlier from a more primitive magma than that of the meta-hornblende-gabbro [68]. This is also evident from the lack of any correlation between the ^{IV}Al vs. Ca (a.p.f.u.) contents in amphibole, indicating the insignificant role of plagioclase during amphibole formation (Figure 4g) [71,72]. Hence, plagioclase possibly crystallized at a later stage than amphibole, which is supported by textural occurrences of epidote (signifying earlier plagioclase) only within intercumulus spaces (Figure 3c). However, Fe-Ti bearing phases (e.g., titanite, magnetite) might have been crystallized simultaneously along with amphiboles, given the covariation in $Fe_2O_3{}^T$, and TiO_2 along with MgO (Figure 6c,d). Greater modal abundance of titanite and magnetite in the meta-hornblendites (compared to the meta-hornblende-gabbros) also corroborates this interpretation (Figure 3a,d–f). Mineral crystallization sequence in studied cumulates may, therefore, be interpreted as hornblende + titanite + magnetite $\rightarrow$ plagioclase, which has also been previously observed in mafic–ultramafic cumulate rocks worldwide [54,73,74]. Thus, earlier fractionation of pargasites solely constituted the precursor hornblendites, and later fractionation of intercumulus plagioclase within amphibole orthocumulate together formed the hornblende-gabbro protoliths within Mayudia ophiolite complex.

5.2. *Physico-Chemical Conditions*

Amphibole major element compositions are sensitive to various physico-chemical parameters, which include the pressure, temperature, oxygen fugacity and melt water content of the parental magma [8]. Understanding the relative dominance of these intensive variables by using amphibole stoichiometry is, therefore, crucial before any thermodynamic interpretation [71]. Consequently, previous studies have demonstrated that substitution mechanisms are often useful in identifying the controlling parameters during amphibole crystallization [72,75]. In the investigated amphiboles, a simultaneous increase in ^{A}(Na + K) and Ti (a.p.f.u.) with ^{IV}Al (a.p.f.u.) (Figure 4b,c) thus suggest that temperature-sensitive edenite-exchange ($^{IV}Si + {}^A\square = {}^{IV}Al + {}^A(Na + K)$) and Ti–tschermakite substitution ($^{IV}Si + {}^{VI}Mg = {}^{IV}Al + {}^{VI}Ti$) have possibly dictated their composition during crystallization [72,75]. Similar control of oxygen fugacity is evident from a linear increase in pargasite Fe^T with decreasing ^{C}Mg (a.p.f.u.) content (Figure 4d) [71,76]. However, pressure-sensitive Al-tschermakite substitution ($^{IV}Si + ^{VI}Mg = ^{IV}Al + ^{VI}Al$) likely has not played any significant role during the formation of the amphiboles in the MdOC cumulates, as documented by the lack of correlation between ^{IV}Al and ^{VI}Al (Figure 4f). MdOC cumulate pargasites were, therefore, more sensitive to temperature and oxygen fugacity in the parental magma as compared to pressure.

Following this, the application of single-phase geothermobarometers are commonly found to be advantageous to estimate different physico-chemical conditions, owing to better preservation of the subvolcanic records in the minerals produced from hybridized magmas [77,78]. Late-stage formation of plagioclase after amphibole, as proposed earlier, also restricts the suitability of amphibole–plagioclase geothermobarometers to this study [79–82]. Empirical barometers calibrated for amphiboles either forming at shallow crustal depths and/or not genetically linked to calc-alkaline magmas may also not be relevant here [19]. Therefore, this study estimates amphibole crystallization temperatures based on the thermometers calibrated by Putirka [83] and Ridolfi [84], whereas pressure is determined using barometers developed by Krawczynski [8], modified after Larocque and Canil [85] (Table S5). Pressure estimates are converted to depth following Anderson [86], whereas the oxygen fugacity of and H_2O fraction in the MdOC melt (s) are calculated based on Ridolfi [84] (Table S5). Our calculations yield comparable ranges in crystallization temperatures, oxygen fugacities and melt water fractions for the meta-hornblendite (865–940 °C, −8.59 to −11.19 log units, and 10.56–13.58 wt.%, respectively) and meta-hornblende-gabbro amphiboles (902–928 °C, −9.48 to −10.49 log units and 11.14–13.61 wt.%, respectively), but higher-pressure conditions (1.2–1.4 GPa) and depth estimates (~40–48 km) of the magmatic hornblendes in the meta-hornblende-gabbros as compared to the former (0.8–1.1 GPa and ~30–38 km, respectively) (Figure 11). These results are consistent with earlier available data on Himalayan (including MdOC) cumulates (Figure 11) [19,87,88], and suggest a stable physicochemical condition within the magma chamber during meta-hornblendite precursor followed by meta-hornblende-gabbro protolith crystallization. However, the parental rock of the meta-hornblende-gabbro has crystallized at higher pressure conditions than that of the meta-hornblendites (Figure 11a), which is most likely related to a reduction in effective volume within the magma chamber due to the accumulation of amphibole at earlier stages. Thus, evaluation of the depth of precursor hornblende-gabbro formation based on amphibole crystallization pressure data tends to be fallacious. This study, therefore, only relies on the meta-hornblendite amphibole pressure data (0.8–1.1 GPa) for the estimation of cumulate crystallization within the MdOC at a depth of approx. 30–38 km, which implies lower crustal pressures and depths near Mohorovičić discontinuity [89].

Notably, MdOC cumulate amphiboles record a highly oxidizing environment ($logfO_2$: −8.59 to −11.19) above the Fayalite–Magnetite–Quartz (FMQ) buffer at elevated pressure conditions (0.8–1.4 GPa) during crystallization, which corresponds to Ni–NiO (NNO) buffer values of +2.09–+ 3.69 (Figure 11b, Table S5). At such high oxidation and pressure conditions, divergence in Mg# values between bulk rock samples and amphibole grains becomes minimal and magnetite saturates at an early stage of magmatic crystallization [69]. This is corroborated by the similarity in Mg# values in MdOC bulk cumulates (65.03–78.80) and constituent amphiboles (69.39–78.53) and co-magmatic precipitation of magnetite with amphiboles as inclusion trails (Figure 3a,f, Tables S1 and S2). However, such an extremely oxidizing environment at high pressure (0.8–1.4 GPa) can only be sustained when the water content in the parental magma is very high (~10–20 wt.%) (Figure 11a) [8], which correlates well with the inferred H_2O% of the parental melt to the MdOC cumulates (10.56–13.61 wt.% at 0.8–1.4 GPa, Figure 11c, Table S1). Similarly high H_2O contents and oxygen fugacities at high pressures have been reported for the cumulate rocks in the Jijal ophiolite complex (up to 20 wt.% at 1.5 GPa, [87]). Under such circumstances, suppression of plagioclase saturation towards lower temperatures is anticipated [90] because depolymerization in the liquid structure would favour the crystallization of less polymerized phases over highly polymerized structures [91]. The same is reflected by the inferred sequence of plagioclase crystallization post-dating amphibole in MdOC cumulates.

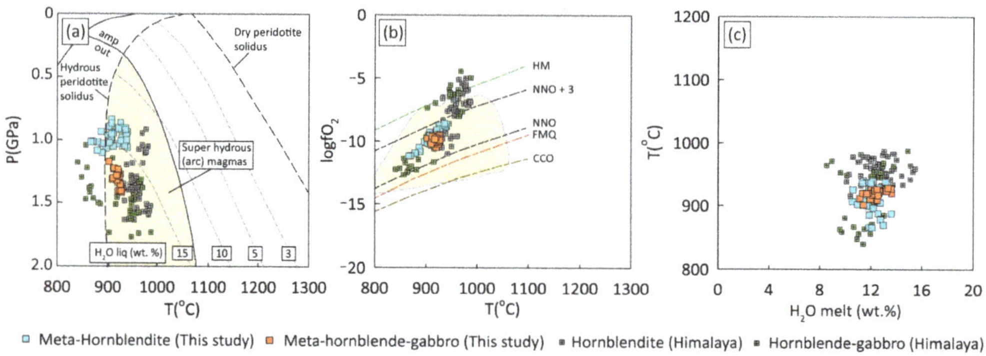

Figure 11. Plot of physico-chemical conditions of amphibole crystallization in (**a**) T (°C) vs. P (GPa), (**b**) T (°C) vs. $logfO_2$, and (**c**) H_2O in melt (wt.%) vs. T (°C) space. Different fields and univariant lines in (**a**) are after [92] and that in (**b**) are after [84]. Reference data for Himalayan hornblendite and hornblende-gabbro are listed in Table S5.

5.3. Parental Magma Characteristics

In MdOC cumulates, linear covariance in CaO, Al_2O_3, $Fe_2O_3{}^T$, and TiO_2 contents, Sm/Zr ratio with MgO and/or SiO_2 contents (Figures 6 and 7b), identical LREE-depleted patterns in the CI-normalized REE diagrams (Figure 8a,b), and comparable HFSE (Nb, Zr, Hf) contents, and εNd(t) values (Tables S2 and S3, and Figure 9b) indicate that their parental magmas were all derived from a single source [68]. This is supported by the gradational contact between hornblendite and meta-hornblende-gabbro in the field (Figure 2c). Positive εNd(t) values in the MdOC cumulates (+7.71–+12.29), coupled with LREE-depleted REE patterns and positive fSm/Nd values (+0.2172–+0.6816), further suggest that they represent the crystallization products of a depleted magma composition (Figures 8a,b and 9b) [93]. However, a greater extent of depletion relative to the mid-oceanic ridge environment is observed from their negative Ti, and HFSE (Nb, Zr, Hf) anomalies and lower LREE concentrations in PM-normalized plots compared to N-MORB (Figure 8c,d). In the Sr–Nd plot, they also exhibit progressive deviation from the N-MORB and mantle array compositions (Figure 9a). A mid-oceanic ridge-related origin of the MdOC parental magma is, therefore, not tenable; rather it suggests an arc-related suprasubduction zone environment for their formation, which is further supported by the following observations: (a) they show compositional resemblance with arc-related cumulates in the triangular MgO vs. FeO^T vs. (Na_2O + K_2O) diagram (Figure 5), as observed earlier in the MdOC cumulates [19], (b) inferred high oxygen fugacities (NNO: +2.09–+3.69) vis-à-vis water contents (10.56–13.61 wt.%) in the parental magma are consistent with an arc-related environment in a subduction zone (Figure 11b), as documented in other Himalayan cumulates [87], (c) all samples are selectively enriched in large ion lithophile elements like Cs and Ba relative to PM and N-MORB values (Figure 8c,d), implying potential subduction input in the parental magma [4]. However, low La_N/Sm_N (0.12–0.62) and Nb/La (0.10–0.83) ratios (Table S2), coupled with positive eNd(t) (+7.71–+12.29) and fSm/Nd values (+0.2172–+0.6816) (Figure 9b), in studied samples reveal that the MdOC cumulates are genetically linked to an oceanic arc setting [44]. The absence of any felsic rock and the presence of metaperidotite in lithological contact with these cumulates also corroborates this interpretation (Figure 2d,e). Highly variable Ba/La ratios and consistent La/Yb values in the studied samples (Figure 10) further imply that a metasomatized aqueous fluid in the parental magma was derived from the dehydrating altered oceanic crust (AOC) in a suprasubduction zone [94].

To reconcile the crystallization process of the MdOC cumulates, a modal fractional crystallization modelling on fore-arc basaltic magma was performed (Figure 12) [70,95]. A fore-arc basalt composition [95] has categorically been considered for parental magma because it represents a relatively primitive melt composition in a suprasubduction zone environment [96]. Whereas the hornblendite composition is modelled assuming 100 modal % amphibole fractionation, the amphibole vs. plagioclase modal proportions is considered at a ratio of 70:30 for the hornblende-gabbro, which are consistent with XRD analysis and petrographic observations. Amphibole/melt and plagioclase/melt partitioning coefficients are taken from Tiepolo [97] and White [70], respectively. Our calculations yield that the meta-hornblendite compositions, particularly their HREE pattern, could be obtained by ~ 30%–60% fractional crystallization from the parental magma (Figure 12a). However, the observed REE pattern of the meta-hornblende-gabbros could not be reproduced by direct crystallization from the residual melt left after the crystallization of the meta-hornblendite protolith. The probable petrogenetic process that led to the formation of the observed REE patterns of the meta-hornblende-gabbros could be modelled by the fractionation of amphibole + plagioclase crystals from a hybridized magma that was generated by the mixing of residual liquid (70%) and a fresh pulse of the parental fore-arc basalt (30%) (Figure 12b). Hybridization of the residual magma composition with fresh parental melt could, therefore, explain the emplacement of the hornblende-gabbros after the hornblendites. Hydrous magma derived by the melting of the sub-arc mantle, metasomatized by slab-derived fluid in a suprasubduction zone and the successive crystallization of the hornblendites and hornblende-gabbros by dynamic fractional crystallization, thus better explains the mineralogical and chemical compositions obtained from the mafic–ultramafic cumulates of the Mayudia ophiolite complex.

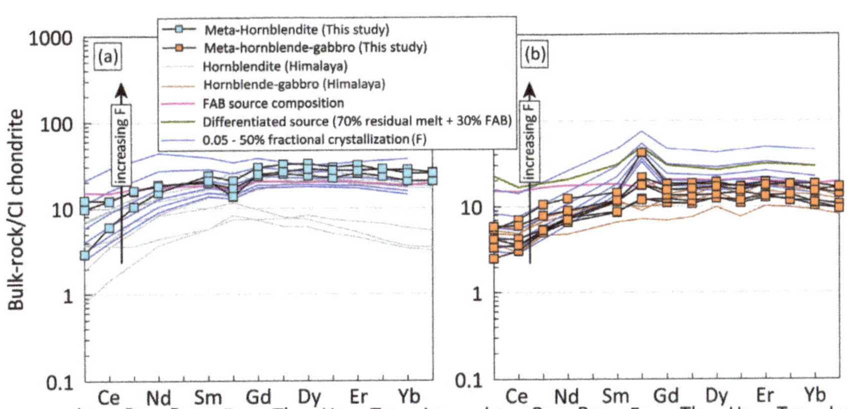

Figure 12. REE modelling of the fractional crystallization process in studied MdOC (**a**) meta-hornblendite protolith, and (**b**) meta-hornblende-gabbro precursors. Source fore-arc basalt composition is after [95], Himalayan samples after Table S4. See text for other details.

5.4. Geodynamic Implications

The closure of the Neo-Tethys preceding the India-Eurasia collision has resulted from two different types of subductions, namely the intraoceanic subduction within the Neo-Tethys and the "Andean type" subduction of the Indian plate beneath Eurasia (Figure 13) [14,66]. Evidence of the "Andean type" subduction of India under Eurasia is well documented by the continental arc-related intermediate to felsic rocks (hornblende- and/or biotite-bearing granites, granodiorites, and diorites of calc-alkaline affinity) in the southern parts of the Karakoram and Lhasa terranes, which are defined as the parts of the Trans-Himalayan batholiths (Gangdese arc) [98–101]. In contrast, fossilized records of intraoceanic Neo-Tethyan subduction are preserved within the ultrabasic to basic cumulate rocks (hornblendite, hornblende-gabbro, pyroxenite, and wehrlite) in the Kohistan, Dras, Spong, Zedong and Pai arcs in the Himalaya [1,102–105]. Whether one or more Trans-Tethyan subduction zones were active during the late Cretaceous to early Tertiary period is, however, debated in the context of the evolution and closure of the Neo-Tethys Ocean. Whereas some authors suggested the presence of only one north-dipping subduction zone within the Neo-Tethys Ocean, others argued for the existence of at least two NNE-dipping intraoceanic subduction systems along the southern margin of the Eurasian plate [102,105,106]. Such ambiguity is even more persistent during the final closure of the eastern Neo-Tethys Ocean, either because of the absence or controversial nature of key evidence from the rock records (arc-type magmatic rocks) in the eastern Himalayas. In fact, the arc-related magmatic rocks bearing signatures of intraoceanic subduction have only been reported to date from the Zedong terrane, and Pai and Mayudia areas in eastern Himalaya [19,103,104]. Such information is further limited from the eastern limb of the EHS [19,107], although geodynamic reconstructions of this region are crucial to understanding the final stage of the evolution and closure of the Neo-Tethys Ocean.

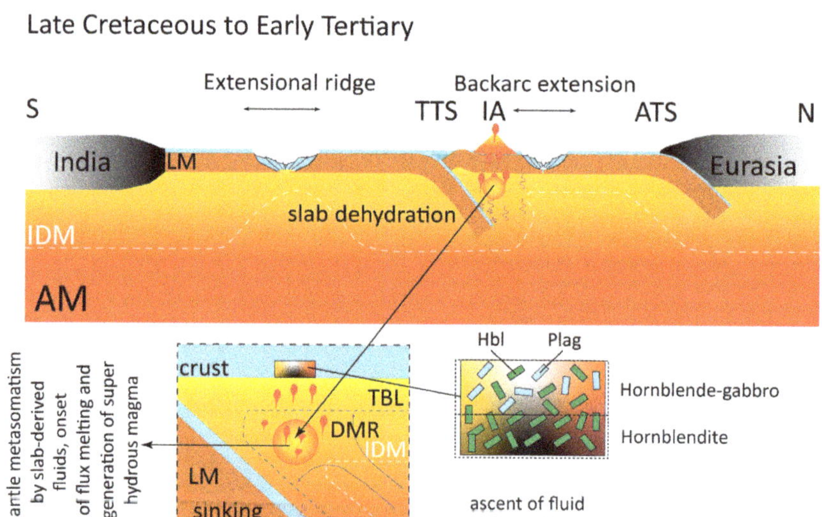

Figure 13. Schematic illustrations of the possible tectonomagmatic history of the eastern Neo-Tethyan oceanic lithosphere, depicting flux melting within the sub-arc mantle and associated formation of hornblende-bearing cumulates (based on the double-subduction model [15]. TTS: Trans-Tethyan Subduction, ATS: Andean Type Subduction, IA: Island Arc, LM: Lithospheric mantle, AM: Asthenospheric mantle, IDM: Initial depth of melting, TBL: Thermal Boundary Layer, Plag: plagioclase feldspar, Hbl: hornblende. Figure is not to scale.

In this context, the geochemical and isotopic affinities of the MdOC cumulates with island arc-related rocks (Figures 8, 9, 10 and 12, Table S2), along with their lithological association with ultrabasic metaperidotites (Figure 2d,e), imply that they formed during intraoceanic subduction and were not influenced by any crustal inputs [89]. Furthermore, the physico-chemical conditions of constituent amphibole formation (Figure 11), coupled with the REE modelling of the crystallization process (Figure 12), suggest the origin of these cumulates through dynamic fractional crystallization of a super hydrous (~10.56–13.61 wt.% H_2O) magma at high pressures (0.8–1.1 GPa) and lower crustal levels (~30–38 km) near the Moho transition zone. The extremely hydrous nature of the sub-arc magma likely resulted in the suppression of clinopyroxene crystallization and the formation of amphibole before plagioclase in this part of the Himalayas. This is common worldwide at deeper crustal levels within a suprasubduction zone [2,18,43,68,87] and is consistent with the root of an island arc complex [1]. It may, therefore, be rational to interpret that the studied pargasitic hornblende-bearing cumulates within the Mayudia ophiolite complex possibly represent the root of a fossilized island arc system at the eastern fringe of the ITSZ, as observed earlier in the Zedong and Pai area in the eastern Himalayas [103,104]. If so, these cumulates potentially have regulated the composition of the ascending magma at sub-arc depths beneath the Mayudia area [68]. For instance, it is expected that amphibole crystallization would cause an increase in SiO_2 and an accompanied decrease in MREE–HREE and water content in the differentiated magma, making it more andesitic and anhydrous [3,108], and contribute to the production of low-K and calc-alkaline granitoid suites [109]. Exposure of diorite-rich granitoid suites has also been previously documented within the eastern Lohit plutonic complex in contact with the MdOC ophiolite complex (Figure 1) [110]. Whether these granitoid suites are genetically linked with the MdOC cumulate is, therefore, a topic of future research.

The MdOC cumulates also signify the presence of a super hydrous magmatic reservoir in the sub-arc mantle, regulating volatile cycling during intraoceanic Neo-Tethyan subduction [11]. Globally, magma with hydrous behaviour (>10 wt.% H_2O) at sub-arc depth is common, as demonstrated by the Gangdese arc [111], Alpine records [92], and global convergence margins [112,113]. However, the occurrence of super hydrous magmas at sub-arc depths during intra-oceanic Neo-Tethyan subduction is relatively rare in the Himalayas and has only been recorded from the Jijal ophiolite complex (up to 20 wt.% at 1.5 GPa) beneath the Kohistan arc, western Himalaya [87]. Given that the different geochemical and isotopic signatures observed in the studied MdOC cumulates are consistent with the cumulate lithologies identified in the Jijal ophiolite complex (Figures 5–11), this may imply that the downgoing sub-arc oceanic lithosphere beneath the MdOC complex may have undergone a similar geodynamic evolution to that of the Kohistan arc during the final closure of the Neo-Tethys. It would, therefore, be rational to conclude that the MdOC cumulates preserve the records of a large-scale intraoceanic subduction system that extended over more than 2000 km from the western fringe of the Nanga Parbat to the eastern extension of the Namcha Barwa.

6. Conclusions

Based on various mineralogical and bulk-rock geochemical and isotopic signatures, in combination with field and petrographic evidence, in the studied MdOC cumulates, the following conclusions can be made:

1. The cumulate lithologies in the MdOC complex varies from hornblendite to hornblende-gabbro, where hornblendite originated prior to hornblende-gabbro due to earlier crystallization of amphibole than plagioclase.
2. Constituent amphiboles have crystallized in a stable physico-chemical condition (T: 865–940 °C, P: 0.8–1.4 GPa, $\log fO_2$: −8.59–−11.19 unit) at a lower crustal depth (30–38 km) from a super hydrous (H_2O in melt: 10.56–13.61 wt.%) parental magma,

which was, in turn, produced due to the flux melting of sub-arc mantle with aqueous inputs from the dehydrating subducting slab.

3. The MdOC cumulates preserve the root of an island arc complex and signify the presence of super hydrous sub-arc mantle reservoirs in the eastern Neo-Tethys Ocean, which have potentially controlled the composition of arc magmas and crust during intraoceanic subduction.

Supplementary Materials: The following supporting information can be downloaded at: https://www.mdpi.com/article/10.3390/min15060572/s1, Table S1: Microprobe analysis data; Table S2: Bulk rock geochemical data; Table S3: Isotope analysis data; Table S4: Reference bulk rock data for hornblende-bearing cumulates from the Himalaya and worldwide locations; Table S5: Estimates of physico-chemical conditions of amphibole crystallization. References [12,13,74,103,114–134] are cited in the Supplementary Materials.

Author Contributions: Conceptualization, S.S. and A.S.M.; methodology, S.S., A.S.M., D.R. and J.M.F.; software, S.S. and A.S.M.; validation, S.S., A.S.M., R.A., D.R. and J.M.F.; formal analysis, S.S., A.S.M., D.R. and J.M.F.; investigation, S.S., A.S.M. and R.A.; resources, S.S. and A.S.M.; data curation, S.S. and A.S.M.; writing—original draft preparation, S.S. and A.S.M.; writing—review and editing, S.S., A.S.M., R.A., D.R. and J.M.F.; visualization, S.S., A.S.M. and R.A.; supervision, A.S.M. and R.A.; project administration, A.S.M.; funding acquisition, A.S.M. All authors have read and agreed to the published version of the manuscript.

Funding: This research was jointly funded by the IIT (ISM) Dhanbad doctoral fellowship, grant number 17DR000510 and the DST-INSPIRE Faculty Award, grant number IFA-17-EAS-056, to S.S. and A.S.M., respectively.

Data Availability Statement: The raw data supporting the findings of this paper will be made available by the corresponding author upon request.

Acknowledgments: This contribution is part of the doctoral thesis of the first author, funded by the IIT (ISM) Dhanbad doctoral fellowship. We express our gratitude to Piyush Sriwastava, IIT Bombay for field assistance. Roing district administration (Arunachal Pradesh), Mustak Ali (transport), and Mayudia guest house caretaker are thanked for their logistic support during the field. Continental Instrument (Lucknow, India) and ALS Geochemistry laboratory (Loughrea, Ireland) are acknowledged for thin section preparation, and bulk-rock analyses, respectively. S.S. and A.S.M. are indebted to the Director, IIT (ISM) Dhanbad for his constant support and encouragement. Constructive comments by four anonymous reviewers have greatly improved the quality of the manuscript.

Conflicts of Interest: The authors declare no conflicts of interest. The funders had no role in the design of the study; in the collection, analyses, or interpretation of data; in the writing of the manuscript; or in the decision to publish the results.

References

1. Dhuime, B.; Bosch, D.; Bodinier, J.-L.; Garrido, C.; Bruguier, O.; Hussain, S.S.; Dawood, H. Multistage Evolution of the Jijal Ultramafic–Mafic Complex (Kohistan, N Pakistan): Implications for Building the Roots of Island Arcs. *Earth Planet. Sci. Lett.* **2007**, *261*, 179–200. [CrossRef]
2. Polat, A.; Fryer, B.J.; Samson, I.M.; Weisener, C.; Appel, P.W.; Frei, R.; Windley, B.F. Geochemistry of Ultramafic Rocks and Hornblendite Veins in the Fiskenæsset Layered Anorthosite Complex, SW Greenland: Evidence for Hydrous Upper Mantle in the Archean. *Precambrian Res.* **2012**, *214*, 124–153. [CrossRef]
3. Davidson, J.; Turner, S.; Handley, H.; Macpherson, C.; Dosseto, A. Amphibole "Sponge" in Arc Crust? *Geology* **2007**, *35*, 787–790. [CrossRef]
4. Dilek, Y.; Furnes, H. Ophiolite Genesis and Global Tectonics: Geochemical and Tectonic Fingerprinting of Ancient Oceanic Lithosphere. *Bulletin* **2011**, *123*, 387–411. [CrossRef]
5. Batanova, V.; Pertsev, A.; Kamenetsky, V.; Ariskin, A.; Mochalov, A.; Sobolev, A. Crustal Evolution of Island-Arc Ultramafic Magma: Galmoenan Pyroxenite–Dunite Plutonic Complex, Koryak Highland (Far East Russia). *J. Petrol.* **2005**, *46*, 1345–1366. [CrossRef]

6. Charlier, B.; Namur, O.; Latypov, R.; Tegner, C. *Layered Intrusions*; Springer: Berlin/Heidelberg, Germany, 2015; ISBN 94-017-9652-1.
7. Hawkesworth, C.; Turner, S.; McDermott, F.; Peate, D.; Van Calsteren, P. U-Th Isotopes in Arc Magmas: Implications for Element Transfer from the Subducted Crust. *Science* **1997**, *276*, 551–555. [CrossRef]
8. Krawczynski, M.J.; Grove, T.L.; Behrens, H. Amphibole Stability in Primitive Arc Magmas: Effects of Temperature, H_2O Content, and Oxygen Fugacity. *Contrib. Mineral. Petrol.* **2012**, *164*, 317–339. [CrossRef]
9. Rooney, T.O.; Nelson, W.R.; Ayalew, D.; Hanan, B.; Yirgu, G.; Kappelman, J. Melting the Lithosphere: Metasomes as a Source for Mantle-Derived Magmas. *Earth Planet. Sci. Lett.* **2017**, *461*, 105–118. [CrossRef]
10. Itano, K.; Morishita, T.; Nishio, I.; Guotana, J.M.; Ogusu, Y.; Ishizuka, O.; Tamura, A. Petrogenesis of Amphibole-Rich Ultramafic Rocks in the Hida Metamorphic Complex, Japan: Its Role in Arc Crust Differentiation. *Lithos* **2021**, *404*, 106440. [CrossRef]
11. Daczko, N.R.; Piazolo, S.; Meek, U.; Stuart, C.A.; Elliott, V. Hornblendite Delineates Zones of Mass Transfer through the Lower Crust. *Sci. Rep.* **2016**, *6*, 31369. [CrossRef]
12. Hamdy, M.M.; Abd El-Wahed, M.A.; El Dien, H.G.; Morishita, T. Garnet Hornblendite in the Meatiq Core Complex, Central Eastern Desert of Egypt: Implications for Crustal Thickening Preceding The ~600 Ma Extensional Regime in the Arabian-Nubian Shield. *Precambrian Res.* **2017**, *298*, 593–614. [CrossRef]
13. García, M.T.; Calderón, M.; de Arellano, C.R.; Hervé, F.; Opitz, J.; Theye, T.; Fanning, C.M.; Pankhurst, R.J.; González-Guillot, M.; Fuentes, F. Trace Element Composition of Amphibole and Petrogenesis of Hornblendites and Plutonic Suites of Cretaceous Magmatic Arcs Developed in the Fuegian Andes, Southernmost South America. *Lithos* **2020**, *372*, 105656. [CrossRef]
14. Martin, C.R.; Jagoutz, O.; Upadhyay, R.; Royden, L.H.; Eddy, M.P.; Bailey, E.; Nichols, C.I.; Weiss, B.P. Paleocene Latitude of the Kohistan–Ladakh Arc Indicates Multistage India–Eurasia Collision. *Proc. Natl. Acad. Sci. USA* **2020**, *117*, 29487–29494. [CrossRef]
15. Jagoutz, O.; Royden, L.; Holt, A.; Becker, T. Anamolously Fast Convergence of India and Eurasia caused by double subduction. *Nat. Geosci.* **2015**, *8*, 475–479. [CrossRef]
16. Metcalfe, I. The Ancient Tethys Oceans of Asia: How Many? How Old? How Deep? How Wide? *UNAEC Asia Pap.* **1999**, *1*, 1–9.
17. Hébert, R.; Bezard, R.; Guilmette, C.; Dostal, J.; Wang, C.; Liu, Z. The Indus–Yarlung Zangbo Ophiolites from Nanga Parbat to Namche Barwa Syntaxes, Southern Tibet: First Synthesis of Petrology, Geochemistry, and Geochronology with Incidences on Geodynamic Reconstructions of Neo-Tethys. *Gondwana Res.* **2012**, *22*, 377–397. [CrossRef]
18. Metcalf, R.V.; Shervais, J.W. Suprasubduction-Zone Ophiolites: Is There Really an Ophiolite Conundrum? *Spec. Pap.-Geol. Soc. Am.* **2008**, *438*, 191.
19. Dutt, A.; Shukla, A.D.; Singh, A.K.; Narayanan, A. A Hydrous Sub-Arc Mantle Domain within the Northeastern Neo-Tethyan Ophiolites: Insights from Cumulate Hornblendites. *Geochemistry* **2024**, *84*, 126122. [CrossRef]
20. Zhang, J.; Wang, R.; Hong, J.; Tang, M.; Zhu, D.-C. Nb-Ta Systematics of Kohistan and Gangdese Arc Lower Crust: Implications for Continental Crust Formation. *Ore Geol. Rev.* **2021**, *133*, 104131. [CrossRef]
21. Roy, S.; Ghosh, B.; Chattopadhaya, S.; Bandyopadhyay, D.; Dhar, A.; Koley, M.; Morishita, T.; Tripathi, S.K. Mayodia Ophiolitic Complex of Arunachal Pradesh, India: A Multistage Evolutionary Record during the Tethyan Closure. *Int. Geol. Rev.* **2024**, *66*, 3017–3049. [CrossRef]
22. Searle, M.; Noble, S.; Cottle, J.; Waters, D.; Mitchell, A.; Hlaing, T.; Horstwood, M. Tectonic Evolution of the Mogok Metamorphic Belt, Burma (Myanmar) Constrained by U-Th-Pb Dating of Metamorphic and Magmatic Rocks. *Tectonics* **2007**, *26*, TC3014. [CrossRef]
23. Haproff, P.J. *Tectonic Evolution of the Easternmost Himalayan Collisional System*; University of California: Los Angeles, CA, USA, 2018.
24. Roy, S.; Bandyopadhyay, D.; Morishita, T.; Dhar, A.; Koley, M.; Chattopadhaya, S.; Karmakar, A.; Ghosh, B. Microtextural Evolution of Chrome Spinels in Dunites from Mayodia Ophiolite Complex, Arunachal Pradesh, India: Implications for a Missing Link in the "Two-Stage" Alteration Mechanism. *Lithos* **2022**, *420*, 106719. [CrossRef]
25. Liu, C.-Z.; Chung, S.-L.; Wu, F.-Y.; Zhang, C.; Xu, Y.; Wang, J.-G.; Chen, Y.; Guo, S. Tethyan Suturing in Southeast Asia: Zircon U-Pb and Hf-O Isotopic Constraints from Myanmar Ophiolites. *Geology* **2016**, *44*, 311–314. [CrossRef]
26. Gururajan, N.; Choudhuri, B. Geology and Tectonic History of the Lohit Valley, Eastern Arunachal Pradesh, India. *J. Asian Earth Sci.* **2003**, *21*, 731–741. [CrossRef]
27. Geng, Q.; Guitang, P.; Zheng, L.; Chen, Z.; Fisher, R.D.; Sun, Z.; Ou, C.; Dong, H.; Wang, X.; Li, S. The Eastern Himalayan Syntaxis: Major Tectonic Domains, Ophiolitic Mélanges and Geologic Evolution. *J. Asian Earth Sci.* **2006**, *27*, 265–285.
28. Liu, C.-Z.; Zhang, C.; Xu, Y.; Wang, J.-G.; Chen, Y.; Guo, S.; Wu, F.-Y.; Sein, K. Petrology and Geochemistry of Mantle Peridotites from the Kalaymyo and Myitkyina Ophiolites (Myanmar): Implications for Tectonic Settings. *Lithos* **2016**, *264*, 495–508. [CrossRef]
29. Singh, A.K.; Chung, S.; Somerville, I.D. Petrogenesis of Mantle Peridotites in Neo-Tethyan Ophiolites from the Eastern Himalaya and Indo-Myanmar Orogenic Belt in the Geo-tectonic Framework of Southeast Asia. *Geol. J.* **2022**, *57*, 4886–4919. [CrossRef]
30. Ovung, N.T.; Ghosh, B.; Ray, J. Petrogenesis of Neo-Tethyan Ophiolites from the Indo-Myanmar Ranges: A Review. *Int. Geol. Rev.* **2020**, *63*, 1437–1449. [CrossRef]

31. Ghosh, B.; Ray, J. Mineral Chemistry of Ophiolitic Rocks of Mayodia-Hunli Area of Dibang Valley District, Arunachal Pradesh, North Eastern India. In *Milestone in Petrology and Future Perspectives*; Geological Society of India: Bangalore, India, 2003; pp. 447–471.
32. Dutt, A.; Singh, A.K.; Srivastava, R.K.; Oinam, G. Evidence of Melt–and Fluid–Rock Interactions in the Refractory Forearc Peridotites and Associated Mafic Intrusives from the Tuting–Tidding Ophiolites, Eastern Himalaya, India: Petrogenetic and Tectonic Implications. *Geol. J.* **2020**, *56*, 2082–2110. [CrossRef]
33. Salvi, D.; Mathew, G.; Kohn, B.; Pande, K.; Borgohain, B. Thermochronological Insights into the Thermotectonic Evolution of Mishmi Hills across the Dibang Valley, NE Himalayan Syntaxis. *J. Asian Earth Sci.* **2020**, *190*, 104158. [CrossRef]
34. Choudhuri, B.; Gururajan, N.; Singh, R. Geology and Sructural Evolution of the Eastern Himalayan Syntaxis. *Himal. Geol.* **2009**, *30*, 17–34.
35. Dutt, A.; Singh, A.K.; Srivastava, R.K.; Oinam, G.; Bikramaditya, R. Geochemical and Metamorphic Record of the Amphibolites from the Tuting–Tidding Suture Zone Ophiolites, Eastern Himalaya, India: Implications for the Presence of a Dismembered Metamorphic Sole. *Geol. Mag.* **2020**, *158*, 787–810. [CrossRef]
36. GeoReM—Database on Geochemical, Environmental and Biological Reference Materials Homepage. Available online: https://georem.mpch-mainz.gwdg.de (accessed on 24 January 2025).
37. Fuenlabrada, J.M. High-precision Sr and Nd Isotope Characterization of BHVO-2 Reference Material by Thermal Ionization Mass Spectrometry. *Rapid Commun. Mass Spectrom.* **2023**, *37*, e9632. [CrossRef]
38. Tanaka, T.; Togashi, S.; Kamioka, H.; Amakawa, H.; Kagami, H.; Hamamoto, T.; Yuhara, M.; Orihashi, Y.; Yoneda, S.; Shimizu, H. JNdi-1: A Neodymium Isotopic Reference in Consistency with LaJolla Neodymium. *Chem. Geol.* **2000**, *168*, 279–281. [CrossRef]
39. Whitney, D.L.; Evans, B.W. Abbreviations for Names of Rock-Forming Minerals. *Am. Mineral.* **2010**, *95*, 185–187. [CrossRef]
40. Hawthorne, F.C.; Oberti, R.; Harlow, G.E.; Maresch, W.V.; Martin, R.F.; Schumacher, J.C.; Welch, M.D. Nomenclature of the Amphibole Supergroup. *Am. Mineral.* **2012**, *97*, 2031–2048. [CrossRef]
41. Keeditse, M.; Rajesh, H.; Belyanin, G.; Fukuyama, M.; Tsunogae, T. Primary Magmatic Amphibole in Archaean Meta-Pyroxenite from the Central Zone of the Limpopo Complex, South Africa. *S. Afr. J. Geol.* **2016**, *119*, 607–622. [CrossRef]
42. Leake, B.E. On Aluminous and Edenitic Hornblendes. *Mineral. Mag.* **1971**, *38*, 389–407. [CrossRef]
43. Zhu, R.-Z.; Smith, D.J.; Wang, F.; Qin, J.-F.; Zhang, C.; Zhao, S.; Liu, M.; Zhang, F.; Zhu, Y.; Lai, S.-C. Hornblendites as a Record of Differentiation, Metasomatism and Magma Fertility in Arc Crust. *Chem. Geol.* **2024**, *650*, 121974. [CrossRef]
44. John, T.; Schenk, V.; Haase, K.; Scherer, E.; Tembo, F. Evidence for a Neoproterozoic Ocean in South-Central Africa from Mid-Oceanic-Ridge–Type Geochemical Signatures and Pressure-Temperature Estimates of Zambian Eclogites. *Geology* **2003**, *31*, 243–246. [CrossRef]
45. Beard, J.S. Characteristic Mineralogy of Arc-Related Cumulate Gabbros: Implications for the Tectonic Setting of Gabbroic Plutons and for Andesite Genesis. *Geology* **1986**, *14*, 848–851. [CrossRef]
46. Ghosh, B.; Mahoney, J.; Ray, J. Mayodia Ophiolites of Arunachal Pradesh, Northeastern Himalaya. *J. Geol. Soc. India* **2007**, *70*, 595–604.
47. Sun, S.-S.; McDonough, W.F. Chemical and Isotopic Systematics of Oceanic Basalts: Implications for Mantle Composition and Processes. *Geol. Soc. Lond. Spec. Publ.* **1989**, *42*, 313–345. [CrossRef]
48. Huang, H.; Polat, A.; Fryer, B.J.; Appel, P.W.; Windley, B.F. Geochemistry of the Mesoarchean Fiskenæsset Complex at Majorqap Qâva, SW Greenland: Evidence for Two Different Magma Compositions. *Chem. Geol.* **2012**, *314*, 66–82. [CrossRef]
49. Cooper, G.F.; Blundy, J.D.; Macpherson, C.G.; Humphreys, M.C.; Davidson, J.P. Evidence from Plutonic Xenoliths for Magma Differentiation, Mixing and Storage in a Volatile-Rich Crystal Mush beneath St. Eustatius, Lesser Antilles. *Contrib. Mineral. Petrol.* **2019**, *174*, 39. [CrossRef]
50. Salters, V.J.; Stracke, A. Composition of the Depleted Mantle. *Geochem. Geophys. Geosyst.* **2004**, *5*, Q05B07. [CrossRef]
51. Zindler, A.; Hart, S. Chemical Geodynamics. *Annu. Rev. Earth Planet. Sci.* **1986**, *14*, 493–571. [CrossRef]
52. Arculus, R.J.; Powell, R. Source Component Mixing in the Regions of Arc Magma Generation. *J. Geophys. Res. Solid Earth* **1986**, *91*, 5913–5926. [CrossRef]
53. Staudigel, H.; Plank, T.; White, B.; Schmincke, H. Geochemical Fluxes during Seafloor Alteration of the Basaltic Upper Oceanic Crust: DSDP Sites 417 and 418. *Subduction Top Bottom* **1996**, *96*, 19–38.
54. Mahdy, N.M.; Abd El-Rahman, Y.; Frische, M.; Ondrejka, M.; Mira, H.I.; Iizuka, T.; Skublov, S.G.; Saleh, G.M.; Ghoniem, M.M.; Mitwally, M. A Remnant Root of a Neoproterozoic Island Arc in the Northern Eastern Desert of Egypt: Evidence from the Whole-Rock and Amphibole Chemistry of the Gattar Gabbro. *Geochemistry* **2024**, *84*, 126113. [CrossRef]
55. Chang, J.; Audétat, A.; Li, J.-W. In Situ Reaction-Replacement Origin of Hornblendites in the Early Cretaceous Laiyuan Complex, North China Craton, and Implications for Its Tectono-Magmatic Evolution. *J. Petrol.* **2021**, *62*, egab030. [CrossRef]
56. Schumacher, J.C. Metamorphic Amphiboles: Composition and Coexistence. *Rev. Mineral. Geochem.* **2007**, *67*, 359–416. [CrossRef]
57. Smith, D.J. Clinopyroxene Precursors to Amphibole Sponge in Arc Crust. *Nat. Commun.* **2014**, *5*, 4329. [CrossRef]

58. Molina, J.F.; Scarrow, J.H.; Montero, P.G.; Bea, F. High-Ti Amphibole as a Petrogenetic Indicator of Magma Chemistry: Evidence for Mildly Alkalic-Hybrid Melts during Evolution of Variscan Basic–Ultrabasic Magmatism of Central Iberia. *Contrib. Mineral. Petrol.* **2009**, *158*, 69–98. [CrossRef]
59. Irvine, T. Terminology for Layered Intrusions. *J. Petrol.* **1982**, *23*, 127–162. [CrossRef]
60. Santana, L.V.; McLeod, C.; Blakemore, D.; Shaulis, B.; Hill, T. Bolivian Hornblendite Cumulates: Insights into the Depths of Central Andean Arc Magmatic Systems. *Lithos* **2020**, *370*, 105618. [CrossRef]
61. Pe-Piper, G. Mineralogy of an Appinitic Hornblende Gabbro and Its Significance for the Evolution of Rising Calc-Alkaline Magmas. *Minerals* **2020**, *10*, 1088. [CrossRef]
62. Enami, M.; Liou, J.; Mattinson, C. Epidote Minerals in High P/T Metamorphic Terranes: Subduction Zone and High-to Ultrahigh-Pressure Metamorphism. *Rev. Mineral. Geochem.* **2004**, *56*, 347–398. [CrossRef]
63. Ferrow, E.; Baginski, B. Chloritisation of Hornblende and Biotite: A HRTEM Study. *Acta Geol. Pol.* **1998**, *48*, 107–113.
64. Deschamps, F.; Godard, M.; Guillot, S.; Hattori, K. Geochemistry of Subduction Zone Serpentinites: A Review. *Lithos* **2013**, *178*, 96–127. [CrossRef]
65. Weber, S.; Diamond, L.W.; Alt-Epping, P.; Brett-Adams, A.C. Reaction Mechanism and Water/Rock Ratios Involved in Epidosite Alteration of the Oceanic Crust. *J. Geophys. Res. Solid Earth* **2021**, *126*, e2020JB021540. [CrossRef]
66. Jagoutz, O.; Müntener, O.; Schmidt, M.W.; Burg, J.-P. The Roles of Flux-and Decompression Melting and Their Respective Fractionation Lines for Continental Crust Formation: Evidence from the Kohistan Arc. *Earth Planet. Sci. Lett.* **2011**, *303*, 25–36. [CrossRef]
67. Müntener, O.; Ulmer, P. Arc Crust Formation and Differentiation Constrained by Experimental Petrology. *Am. J. Sci.* **2018**, *318*, 64–89. [CrossRef]
68. Dessimoz, M.; Müntener, O.; Ulmer, P. A Case for Hornblende Dominated Fractionation of Arc Magmas: The Chelan Complex (Washington Cascades). *Contrib. Mineral. Petrol.* **2012**, *163*, 567–589. [CrossRef]
69. Zhang, J.; Wang, R.; Hong, J. Amphibole Fractionation and Its Potential Redox Effect on Arc Crust: Evidence from the Kohistan Arc Cumulates. *Am. Mineral.* **2022**, *107*, 1779–1788. [CrossRef]
70. White, W.M. *Geochemistry*; John Wiley & Sons: Hoboken, NJ, USA, 2020; ISBN 1-119-43810-1.
71. Shane, P.; Smith, V.C. Using Amphibole Crystals to Reconstruct Magma Storage Temperatures and Pressures for the Post-Caldera Collapse Volcanism at Okataina Volcano. *Lithos* **2013**, *156*, 159–170. [CrossRef]
72. Bachmann, O.; Dungan, M.A. Temperature-Induced Al-Zoning in Hornblendes of the Fish Canyon Magma, Colorado. *Am. Mineral.* **2002**, *87*, 1062–1076. [CrossRef]
73. Sargazi, M.; Zhang, C.-L.; Jing, Y.; Hussain, Z.; Song, Z.-H.; Wang, H.-R.; Liu, X.-Q.; Ye, X.-T. Reconstructing Mantle–Crust Boundary Magmatism through Cimmerian Orogenic Events: Evidence from Deep Crustal Cumulates in Northeastern Pamir. *Contrib. Mineral. Petrol.* **2025**, *180*, 8. [CrossRef]
74. Wang, L.; Zhang, K.; Lin, S.; Bédard, J.H.; Santos, G.S.; He, W.; Yin, C.; Xiao, W. Late Tonian (ca. 785 Ma) Subduction-Related Mafic-Ultramafic Cumulates in the West Cathaysia Terrane, South China. *Precambrian Res.* **2023**, *387*, 106980. [CrossRef]
75. Kiss, B.; Harangi, S.; Ntaflos, T.; Mason, P.R.; Pál-Molnár, E. Amphibole Perspective to Unravel Pre-Eruptive Processes and Conditions in Volcanic Plumbing Systems beneath Intermediate Arc Volcanoes: A Case Study from Ciomadul Volcano (SE Carpathians). *Contrib. Mineral. Petrol.* **2014**, *167*, 986. [CrossRef]
76. Ridolfi, F.; Renzulli, A. Calcic Amphiboles in Calc-Alkaline and Alkaline Magmas: Thermobarometric and Chemometric Empirical Equations Valid up to 1130 °C and 2.2 GPa. *Contrib. Mineral. Petrol.* **2012**, *163*, 877–895. [CrossRef]
77. Ridolfi, F.; Renzulli, A.; Perugini, D.; Cesare, B.; Braga, R.; Del Moro, S. Unravelling the Complex Interaction between Mantle and Crustal Magmas Encoded in the Lavas of San Vincenzo (Tuscany, Italy). Part II: Geochemical Overview and Modelling. *Lithos* **2016**, *244*, 233–249. [CrossRef]
78. Gorini, A.; Ridolfi, F.; Piscaglia, F.; Taussi, M.; Renzulli, A. Application and Reliability of Calcic Amphibole Thermobarometry as Inferred from Calc-Alkaline Products of Active Geothermal Areas in the Andes. *J. Volcanol. Geotherm. Res.* **2018**, *358*, 58–76. [CrossRef]
79. Holland, T.; Blundy, J. Non-Ideal Interactions in Calcic Amphiboles and Their Bearing on Amphibole-Plagioclase Thermometry. *Contrib. Mineral. Petrol.* **1994**, *116*, 433–447. [CrossRef]
80. Hammarstrom, J.M.; Zen, E. Aluminum in Hornblende: An Empirical Igneous Geobarometer. *Am. Mineral.* **1986**, *71*, 1297–1313.
81. Johnson, M.C.; Rutherford, M.J. Experimental Calibration of the Aluminum-in-Hornblende Geobarometer with Application to Long Valley Caldera (California) Volcanic Rocks. *Geology* **1989**, *17*, 837–841. [CrossRef]
82. Molina, J.; Moreno, J.; Castro, A.; Rodríguez, C.; Fershtater, G. Calcic Amphibole Thermobarometry in Metamorphic and Igneous Rocks: New Calibrations Based on Plagioclase/Amphibole Al-Si Partitioning and Amphibole/Liquid Mg Partitioning. *Lithos* **2015**, *232*, 286–305. [CrossRef]
83. Putirka, K. Amphibole Thermometers and Barometers for Igneous Systems and Some Implications for Eruption Mechanisms of Felsic Magmas at Arc Volcanoes. *Am. Mineral.* **2016**, *101*, 841–858. [CrossRef]

84. Ridolfi, F. Amp-TB2: An Updated Model for Calcic Amphibole Thermobarometry. *Minerals* **2021**, *11*, 324. [CrossRef]
85. Larocque, J.; Canil, D. The Role of Amphibole in the Evolution of Arc Magmas and Crust: The Case from the Jurassic Bonanza Arc Section, Vancouver Island, Canada. *Contrib. Mineral. Petrol.* **2010**, *159*, 475–492. [CrossRef]
86. Anderson, D.L. *Theory of the Earth*; Cambridge University Press: Cambridge, UK, 1989; ISBN 0-86542-335-0.
87. Urann, B.; Le Roux, V.; Jagoutz, O.; Müntener, O.; Behn, M.; Chin, E. High Water Content of Arc Magmas Recorded in Cumulates from Subduction Zone Lower Crust. *Nat. Geosci.* **2022**, *15*, 501–508. [CrossRef]
88. Jagoutz, O. Arc Crustal Differentiation Mechanisms. *Earth Planet. Sci. Lett.* **2014**, *396*, 267–277. [CrossRef]
89. Zheng, Y.-F.; Chen, Y.-X. Continental versus Oceanic Subduction Zones. *Natl. Sci. Rev.* **2016**, *3*, 495–519. [CrossRef]
90. Müntener, O.; Kelemen, P.B.; Grove, T.L. The Role of H 2 O during Crystallization of Primitive Arc Magmas under Uppermost Mantle Conditions and Genesis of Igneous Pyroxenites: An Experimental Study. *Contrib. Mineral. Petrol.* **2001**, *141*, 643–658. [CrossRef]
91. Gaetani, G.A.; Grove, T.L.; Bryan, W.B. The Influence of Water on the Petrogenesis of Subduction related Igneous Rocks. *Nature* **1993**, *365*, 332–334. [CrossRef]
92. Müntener, O.; Ulmer, P.; Blundy, J.D. Super hydrous Arc Magmas in the Alpine Context. *Elements* **2021**, *17*, 35–40. [CrossRef]
93. Hao, H.; Campbell, I.H.; Park, J.-W. Nd-Hf Isotopic Systematics of the Arc Mantle and Their Implication for Continental Crust Growth. *Chem. Geol.* **2022**, *602*, 120897. [CrossRef]
94. Tian, L.; Castillo, P.R.; Hawkins, J.W.; Hilton, D.R.; Hanan, B.B.; Pietruszka, A.J. Major and Trace Element and Sr–Nd Isotope Signatures of Lavas from the Central Lau Basin: Implications for the Nature and Influence of Subduction Components in the Back-Arc Mantle. *J. Volcanol. Geotherm. Res.* **2008**, *178*, 657–670. [CrossRef]
95. Reagan, M.K.; Ishizuka, O.; Stern, R.J.; Kelley, K.A.; Ohara, Y.; Blichert-Toft, J.; Bloomer, S.H.; Cash, J.; Fryer, P.; Hanan, B.B. Fore-arc Basalts and Subduction Initiation in the Izu-Bonin-Mariana System. *Geochem. Geophys. Geosyst.* **2010**, *11*, Q03X12. [CrossRef]
96. Schmidt, M.W.; Jagoutz, O. The Global Systematics of Primitive Arc Melts. *Geochem. Geophys. Geosyst.* **2017**, *18*, 2817–2854. [CrossRef]
97. Tiepolo, M.; Oberti, R.; Zanetti, A.; Vannucci, R.; Foley, S.F. Trace-Element Partitioning between Amphibole and Silicate Melt. *Rev. Mineral. Geochem.* **2007**, *67*, 417–452. [CrossRef]
98. Chapman, J.B.; Scoggin, S.H.; Kapp, P.; Carrapa, B.; Ducea, M.N.; Worthington, J.; Oimahmadov, I.; Gadoev, M. Mesozoic to Cenozoic Magmatic History of the Pamir. *Earth Planet. Sci. Lett.* **2018**, *482*, 181–192. [CrossRef]
99. Searle, M.; Hacker, B. Structural and Metamorphic Evolution of the Karakoram and Pamir Following India–Kohistan–Asia Collision. *Geol. Soc. Lond. Spec. Publ.* **2019**, *483*, 555–582. [CrossRef]
100. Zhu, D.-C.; Wang, Q.; Chung, S.-L.; Cawood, P.A.; Zhao, Z.-D. Gangdese Magmatism in Southern Tibet and India–Asia Convergence since 120 Ma. *Geol. Soc. Lond. Spec. Publ.* **2019**, *483*, 583–604. [CrossRef]
101. Zhu, D.-C.; Wang, Q.; Weinberg, R.F.; Cawood, P.A.; Zhao, Z.; Hou, Z.-Q.; Mo, X.-X. Continental Crustal Growth Processes Recorded in the Gangdese Batholith, Southern Tibet. *Annu. Rev. Earth Planet. Sci.* **2023**, *51*, 155–188. [CrossRef]
102. Buckman, S.; Aitchison, J.C.; Nutman, A.P.; Bennett, V.C.; Saktura, W.M.; Walsh, J.M.; Kachovich, S.; Hidaka, H. The Spongtang Massif in Ladakh, NW Himalaya: An Early Cretaceous Record of Spontaneous, Intra-Oceanic Subduction Initiation in the Neotethys. *Gondwana Res.* **2018**, *63*, 226–249. [CrossRef]
103. Zhang, L.-L.; Liu, C.-Z.; Wu, F.-Y.; Ji, W.-Q.; Wang, J.-G. Zedong Terrane Revisited: An Intra-Oceanic Arc within Neo-Tethys or a Part of the Asian Active Continental Margin? *J. Asian Earth Sci.* **2014**, *80*, 34–55. [CrossRef]
104. Zhang, Z.; An, W.; Palin, R.M.; Ding, H.; Dong, X.; Tian, Z. Intraoceanic Subduction System Within the Neo-Tethys: Evidence From Late Cretaceous Arc Magmatic Rocks of the Eastern Himalaya. *Geochem. Geophys. Geosyst.* **2023**, *24*, e2023GC011214. [CrossRef]
105. Walsh, J.M.; Buckman, S.; Nutman, A.P.; Zhou, R. Age and Provenance of the Nindam Formation, Ladakh, NW Himalaya: Evolution of the Intraoceanic Dras Arc before Collision with India. *Tectonics* **2019**, *38*, 3070–3096. [CrossRef]
106. Searle, M.P. Timing of Subduction Initiation, Arc Formation, Ophiolite Obduction and India–Asia Collision in the Himalaya. *Geol. Soc. Lond. Spec. Publ.* **2019**, *483*, 19–37. [CrossRef]
107. Abdullah, S.; Misra, S.; Ghosh, B. Melt-Rock Interaction and Fractional Crystallization in the Moho Transition Zone: Evidence from the Cretaceous Naga Hills Ophiolite, North-East India. *Lithos* **2018**, *322*, 197–211. [CrossRef]
108. Barclay, J.; Carmichael, I. A Hornblende Basalt from Western Mexico: Water-Saturated Phase Relations Constrain a Pressure–Temperature Window of Eruptibility. *J. Petrol.* **2004**, *45*, 485–506. [CrossRef]
109. Collins, W.J.; Murphy, J.B.; Johnson, T.E.; Huang, H.-Q. Critical Role of Water in the Formation of Continental Crust. *Nat. Geosci.* **2020**, *13*, 331–338. [CrossRef]
110. Bikramaditya, R.; Chung, S.-L.; Singh, A.K.; Lee, H.-Y.; Lin, T.-H.; Iizuka, Y. Age and Isotope Geochemistry of Magmatic Rocks of the Lohit Plutonic Complex, Eastern Himalaya: Implications for the Evolution of Transhimalayan Arc Magmatism. *J. Geol. Soc.* **2020**, *177*, 379–394. [CrossRef]

111. Lu, Y.-J.; Loucks, R.R.; Fiorentini, M.L.; Yang, Z.-M.; Hou, Z.-Q. Fluid Flux Melting Generated Postcollisional High Sr/Y Copper Ore–Forming Water-Rich Magmas in Tibet. *Geology* **2015**, *43*, 583–586. [CrossRef]
112. Prouteau, G.; Scaillet, B. Experimental Constraints on the Origin of the 1991 Pinatubo Dacite. *J. Petrol.* **2003**, *44*, 2203–2241. [CrossRef]
113. Carmichael, I.S. The Andesite Aqueduct: Perspectives on the Evolution of Intermediate Magmatism in West-Central (105–99 W) Mexico. *Contrib. Mineral. Petrol.* **2002**, *143*, 641–663. [CrossRef]
114. Caro, G.; Bourdon, B. Non-Chondritic Sm/Nd Ratio in the Terrestrial Planets: Consequences for the Geochemical Evolution of the Mantle–Crust System. *Geochim. Cosmochim. Acta* **2010**, *74*, 3333–3349. [CrossRef]
115. Jacobsen, S.; Wasserburg, G. Sm-Nd Isotopic Evolution of Chondrites and Achondrites, II. *Earth Planet. Sci. Lett.* **1984**, *67*, 137–150. [CrossRef]
116. Lugmair, G.; Shimamura, T.; Lewis, R.; Anders, E. Samarium-146 in the Early Solar System: Evidence from Neodymium in the Allende Meteorite. *Science* **1983**, *222*, 1015–1018. [CrossRef]
117. Steiger, R.H.; Jäger, E. Subcommission on Geochronology: Convention on the Use of Decay Constants in Geo-and Cosmochronology. *Earth Planet. Sci. Lett.* **1977**, *36*, 359–362. [CrossRef]
118. Villares, F.; Blanco-Quintero, I.F.; Reyes, P.S.; Proenza, J.A.; Cartagena, R.; Lázaro, C.; Garcia-Casco, A. Petrogenesis of the Tampanchi Ultramafic–Mafic Complex (Ecuador): Geodynamic Implications for the Northwestern Margin of South America during the Late Cretaceous. *Gondwana Res.* **2022**, *105*, 514–534. [CrossRef]
119. Fazlnia, A.; Alizade, A. Petrology and Geochemistry of the Mamakan Gabbroic Intrusions, Urumieh (Urmia), Iran: Magmatic Development of an Intra-Oceanic Arc. *Period. Mineral.* **2013**, *82*, 263–290.
120. Kettanah, Y.A.; Ismail, S.A.; Karo, N.M. Mineralogy, Geochemistry, and 40Ar/39Ar Dating of Hornblendite/Amphibolite Pods in Penjween Ophiolite of Northeastern Iraq: Implications for Petrogenesis and Tectonics. *J. Afr. Earth Sc.* **2022**, *193*, 104591. [CrossRef]
121. Guo, L.; Jagoutz, O.; Shinevar, W.J.; Zhang, H.-F. Formation and Composition of the Late Cretaceous Gangdese Arc Lower Crust in Southern Tibet. *Contrib. Mineral. Petrol.* **2020**, *175*, 1–26. [CrossRef]
122. Triantafyllou, A.; Berger, J.; Baele, J.; Bruguier, O.; Diot, H.; Ennih, N.; Monnier, C.; Plissart, G.; Vandycke, S.; Watlet, A. Intra-Oceanic Arc Growth Driven by Magmatic and Tectonic Processes Recorded in the Neoproterozoic Bougmane Arc Complex (Anti-Atlas, Morocco). *Precambrian Res.* **2018**, *304*, 39–63. [CrossRef]
123. Xu, W.; Zhu, D.-C.; Wang, Q.; Weinberg, R.F.; Wang, R.; Li, S.-M.; Zhang, L.-L.; Zhao, Z.-D. Constructing the Early Mesozoic Gangdese Crust in Southern Tibet by Hornblende-Dominated Magmatic Differentiation. *J. Petrol.* **2019**, *60*, 515–552. [CrossRef]
124. Chen, B.; Suzuki, K.; Tian, W.; Jahn, B.; Ireland, T. Geochemistry and Os–Nd–Sr Isotopes of the Gaositai Alaskan-Type Ultramafic Complex from the Northern North China Craton: Implications for Mantle–Crust Interaction. *Contrib. Mineral. Petrol.* **2009**, *158*, 683–702. [CrossRef]
125. Gao, Y.; Santosh, M.; Hou, Z.; Wei, R.; Ma, G.; Chen, Z.; Wu, J. High Sr/Y Magmas Generated through Crystal Fractionation: Evidence from Mesozoic Volcanic Rocks in the Northern Taihang Orogen, North China Craton. *Gondwana Res.* **2012**, *22*, 152–168. [CrossRef]
126. Hou, Z.; Li, Q.; Gao, Y.; Lu, Y.; Yang, Z.; Wang, R.; Shen, Z. Lower-Crustal Magmatic Hornblendite in North China Craton: Insight into the Genesis of Porphyry Cu Deposits. *Econ. Geol.* **2015**, *110*, 1879–1904. [CrossRef]
127. Koepke, J.; Seidel, E.; Kreuzer, H. Ophiolites on the Southern Aegean Islands Crete, Karpathos and Rhodes: Composition, Geochronology and Position within the Ophiolite Belts of the Eastern Mediterranean. *Lithos* **2002**, *65*, 183–203. [CrossRef]
128. Pettigrew, N.T.; Hattori, K.H. The Quetico Intrusions of Western Superior Province: Neo-Archean Examples of Alaskan/Ural-Type Mafic–Ultramafic Intrusions. *Precambrian Res.* **2006**, *149*, 21–42. [CrossRef]
129. Daczko, N.; Emami, S.; Allibone, A.; Turnbull, I. Petrogenesis and Geochemical Characterisation of Ultramafic Cumulate Rocks from Hawes Head, Fiordland, New Zealand. *N. Z. J. Geol. Geophys.* **2012**, *55*, 361–374. [CrossRef]
130. Ma, L.; Wang, Q.; Li, Z.-X.; Wyman, D.A.; Jiang, Z.-Q.; Yang, J.-H.; Gou, G.-N.; Guo, H.-F. Early Late Cretaceous (ca. 93 Ma) Norites and Hornblendites in the Milin Area, Eastern Gangdese: Lithosphere–Asthenosphere Interaction during Slab Roll-Back and an Insight into Early Late Cretaceous (ca. 100–80 Ma) Magmatic "Flare-up" in Southern Lhasa (Tibet). *Lithos* **2013**, *172*, 17–30.
131. Ma, X.; Fan, H.-R.; Santosh, M.; Guo, J. Petrology and Geochemistry of the Guyang Hornblendite Complex in the Yinshan Block, North China Craton: Implications for the Melting of Subduction-Modified Mantle. *Precambrian Res.* **2016**, *273*, 38–52. [CrossRef]
132. Santosh, M.; Teng, X.-M.; He, X.-F.; Tang, L.; Yang, Q.-Y. Discovery of Neoarchean Suprasubduction Zone Ophiolite Suite from Yishui Complex in the North China Craton. *Gondwana Res.* **2016**, *38*, 1–27. [CrossRef]

133. Peltonen, P.; Kontinen, A.; Huhma, H. Petrogenesis of the Mantle Sequence of the Jormua Ophiolite (Finland): Melt Migration in the Upper Mantle during Palaeoproterozoic Continental Break-Up. *J. Petrol.* **1998**, *39*, 297–329. [CrossRef]
134. Liu, A.-K.; Chen, B.; Suzuki, K.; Liu, L. Petrogenesis of the Mesozoic Zijinguan Mafic Pluton from the Taihang Mountains, North China Craton: Petrological and Os–Nd–Sr Isotopic Constraints. *J. Asian Earth Sci.* **2010**, *39*, 294–308. [CrossRef]

minerals

MDPI

Article

Magma Mixing Origin for the Menyuan Granodioritic Pluton in the North Qilian Orogenic Belt, China

Shugang Xia [1], Yu Qi [1], Shengyao Yu [1,2,*], Xiaocong Jiang [1], Xiangyu Gao [3], Yue Wang [1], Chuanzhi Li [1], Qian Wang [4,*], Lintao Wang [1] and Yinbiao Peng [1,5]

1 Frontiers Science Center for Deep Ocean Multispheres and Earth System, Key Lab of Submarine Geosciences and Prospecting Techniques, Ministry of Education, College of Marine Geosciences, Ocean University of China, Qingdao 266100, China; xiashugang7@163.com (S.X.); qi15204273588@126.com (Y.Q.); jxc2363964495@163.com (X.J.); wy0610@stu.ouc.edu.cn (Y.W.); lichuanzhi@stu.ouc.edu.cn (C.L.); wlt@stu.ouc.edu.cn (L.W.); ybpeng@uw.edu (Y.P.)

2 Laboratory for Marine Mineral Resources, Qingdao National Laboratory for Marine Science and Technology, Qingdao 266037, China

3 School of Resources and Environmental Engineering, Shandong University of Technology, Zibo 255000, China; gaoxy@sdut.edu.cn

4 College of Earth Science and Engineering, Shandong University of Science and Technology, Qingdao 266590, China

5 Isotope Laboratory, Department of Earth and Space Sciences, University of Washington, Seattle, WA 98195-1310, USA

* Correspondence: yushengyao@ouc.edu.cn (S.Y.); wangqian2021@sdust.edu.cn (Q.W.)

Abstract: Magma mixing or mingling is not just a geological phenomenon that widely occurs in granitoid magmatism, but a complex dynamic process that influences the formation of mafic microgranular enclaves (MMEs) and the diversity of granitic rocks. Herein, we carried out a comprehensive study that encompassed the petrology, mineral chemistry, zircon U-Pb ages, Lu-Hf isotopes, whole-rock elements, and Sr-Nd isotope compositions of the Menyuan Granodioritic Pluton in the northern margin of the Qilian Block, to elucidate the petrogenesis and physical and chemical processes occurring during magma mixing. The Menyuan Granodioritic Pluton is mainly composed of granodiorites accompanied by numerous mafic microgranular enclaves (MMEs) and is intruded by minor gabbro dikes. LA-ICP-MS zircon U-Pb dating reveals that these rocks possess a similar crystallization age of ca. 456 Ma. The Menyuan host granodiorites, characterized as metaluminous to weakly peraluminous, belong to subduction-related I-type calc-alkaline granites. The MMEs and gabbroic dikes have relatively low SiO_2 contents and high $Mg^{\#}$ values, probably reflecting a mantle-derived origin. They are enriched in large ion lithophile elements (LILEs) and light, rare earth elements (LREEs) but are depleted in high field strength elements (HFSEs), indicating continental arc-like geochemical affinities. The host granodiorites yield relatively enriched whole-rock Sr-Nd and zircon Hf isotopic compositions ($^{87}Sr/^{86}Sr_i$ = 0.7072–0.7158; $\varepsilon_{Nd}(t) = -9.21$ to -4.23; $\varepsilon_{Hf}(t) = -8.8$ to -1.2), implying a derivation from the anatexis of the ancient mafic lower continental crust beneath the Qilian Block. The MMEs have similar initial Sr isotopes but distinct whole-rock Nd and zircon Hf isotopic compositions compared with the host granodiorites ($^{87}Sr/^{86}Sr_i$ = 0.7078–0.7089; $\varepsilon_{Nd}(t) = -3.88$ to -1.68; $\varepsilon_{Hf}(t) = -0.1$ to $+4.1$). Field observation, microtextural and mineral chemical evidence, geochemical characteristics, and whole-rock Nd and zircon Hf isotopic differences between the host granodiorites and MMEs suggest insufficient magma mixing of lithospheric mantle mafic magma and lower continental crust felsic melt. In combination with evidence from regional geology, we propose that the anatexis of the ancient mafic lower continental crust and subsequent magma mixing formed in an active continental arc setting, which was triggered by the subducted slab rollback and mantle upwelling during the southward subduction of the Qilian Proto-Tethys Ocean during the Middle-Late Ordovician.

Keywords: magma mixing; granodiorite and mafic microgranular enclaves; North Qilian Orogenic Belt; Middle-Late Ordovician; Sr-Nd-Hf isotopes

1. Introduction

Magma mixing, or the mingling of distinct compositions, is a common phenomenon in the process of magmatic evolution and has a significant impact on the genesis and compositional diversity of granitoids [1–3]. Compared with magma mingling, which refers to physical dispersion without chemical exchanges between compositionally distinct magmas, magma mixing emphasizes magma interactions accompanied by mechanical and chemical exchanges [4–6]. The temperature, viscosity, crystallinity, and rheology of both interacting magmas affect the degree of magma homogenization of the magmatic system [7]. Calc-alkaline granitic plutons commonly contain abundant mafic microgranular enclaves (MMEs), and the MMEs and their host granitoids are considered to be the crucial objects to study the genesis and diversity of granites and the interaction between crust and mantle [7–13]. Several potential models have been put forward to account for the genesis of MMEs, including (1) wall-rock xenoliths [14,15]; (2) refractory remnants from the magma source [16–18]; (3) autoliths associated with early-formed cumulate minerals within the cognate magmatic system [19–24]; and (4) hybrids generated by the mixing/mingling between mafic and felsic magmas [25–27]. Magmatic enclaves occur as a representation of hybrid magma blobs with obvious chemical–physical disequilibrium features, which are considered as one of the most clear pieces of evidence for insufficient magma mixing [5,28–30]. To date, there exist two major mechanisms for the formation of magmatic enclaves: heterogeneous magma mixing and homologous magma mixing [19,21,25,31,32]. Verifying the relationship between the MMEs and host granitoids could provide additional insights into the petrogenesis and source of granitoids, the interaction between both magmatic systems, and deep geodynamic processes.

The Qilian Orogen lies in the northeastern margin of the Qinghai-Tibet Platea, stretching in a NW-SE direction. It is bounded by the Alxa Block on the north, the Tarim Block on the west, the Qaidam Block on the south, and the West Qinling Orogenic Belt on the southeast (Figure 1a,b). The North Qilian Belt is considered a typical oceanic suture zone located between the Qilian Block and Alxa Block (Figure 1b [33,34]). The extensive Early Paleozoic granitic magmatism is widespread in the North Qilian Belt, and the majority of these instances of granitic magmatism are suggested to be associated with oceanic–continental subduction and subsequent continental collision [33,35–41]. In the past decades, an ever-growing quantity of geochemical and isotopic chronological investigations has been conducted in the North Qilian Belt to ascertain the origin of early Paleozoic granitoid plutrons and their MMEs, such as the Chaidanuo, Jinzichuan, Beilinggou, Shidonggou, Niuxinshan, Tuole, Xiagucheng, Yeniutan, Leigongshan, and other plutons [33,37,42–49]. Therefore, the North Qilian Belt is an ideal study area for studying the petrogenesis of the MMEs and their host rocks and the geochemical behavior of elements and isotopes in magmatic systems during mixing. This study is concentrated on the Menyuan Granodioritic Pluton, which is situated on the northern margin of the Qilian Block, which contains abundant enclaves and synplutonic dikes. The petrology, whole-rock geochemistry, Sr-Nd-Hf isotopes, zircon U-Pb ages, and mineral chemistry of the MMEs and their host granodiorites are discussed in this paper. By integrating these new results and prior geological data, our objective is to shed light on the relationship between the MMEs and host granitoids, as well as the possibility of efficient magma mixing during the crust–mantle interaction process during the subduction stage of the Qilian Proto-Tethys Ocean.

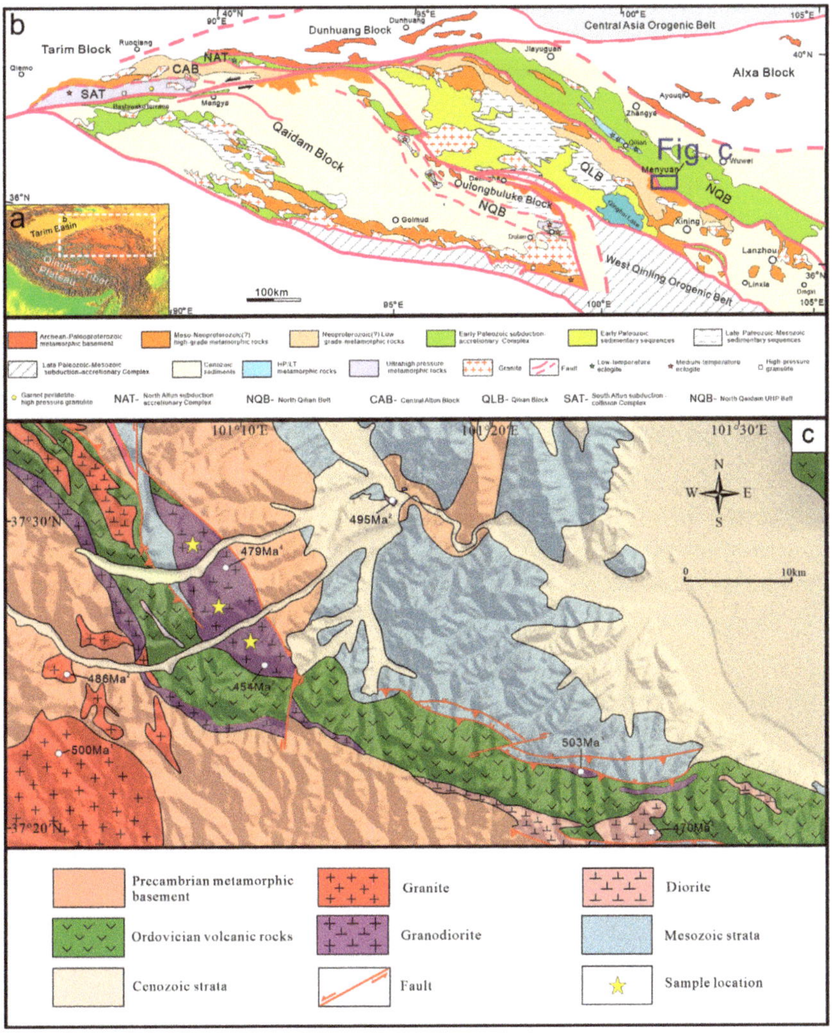

Figure 1. (**a**) Topographic map of Tibetan Plateau. (**b**) Geological sketch map of the Qilian Orogenic Belt (modified from [50]). (**c**) Simplified geological map of the Menyuan area showing the distribution of the Menyuan Granodioritic Pluton and sample locations (modified from [37]). The data were compiled from Data sources: 1. [38]; 2. [37]; 3. [51]; 4. [52].

2. Geological Background

From north to south, the Qilian orogen can be partitioned into three distinct tectonic units: the North Qilian Belt, the Qilian Block, and the South Qilian Belt (Figure 1b [41,53–55]). The overall Qilian orogen records the complete tectonic evolution of oceanic spreading (the Proto-Tethys Ocean), oceanic plate subduction and accretion, and continent–continent (arc–continent) collision [56–58].

The North Qilian Belt stretches in a NW-SE direction for over 1000 km, lying between the Qilian Block to the southwest and the Alxa Block to the northeast (Figure 1b [34]), and is confirmed to be a classic subduction–accretion orogenic belt featuring a complex trench-arc-basin system [33]. The North Qilian Belt predominantly consists of early Paleozoic ophiolite suites, high-pressure/low-temperature (HP/LT) metamorphic rocks, granitoid plutons related to subduction and collision, and arc-related volcanic rocks, overlain by Silurian flysch, Devonian molasse, and Carboniferous–Triassic sedimentary cover sequences [59]. There are two ophiolite belts in the North Qilian Belt: Neoproterozoic–Cambrian ophiolite belts (550–496 Ma) of oceanic ridge origin in the south (including the Aoyougou, Yushigou, and Dongcaohe ophiolites) and Cambrian–Ordovician Ophiolite Belts (490–448 Ma) of back-arc oceanic basin origin in the north (including Jiugequan, Biandukou, and Laohushan ophiolites) [33,54,60]. The widespread granitoids with zircon U-Pb ages ranging from 520 Ma to 440 Ma are predominantly found as elongated bodies within island-arc volcanic rocks in the North Qilian Belt [33,35,39–41,46,61]. These granitoids are distinguished by a peraluminous granitic batholith, as well as numerous medium/high-K calc-alkaline I-type diorite-granodiorite-granite intrusions, along with a small number of minor tonalite-trondhjemite associations [35,40,62–65].

The Qilian Block consists primarily of a substantial Precambrian basement sequence, a thin Phanerozoic sedimentary overlying layer, and early Paleozoic igneous rocks [33,66–71]. A few studies have suggested that the lower Precambrian crystallized basement was partitioned into the western Tuolai Group and the Beidahe Group [72,73], the Huangyuan Group in the center [69], and the eastern Maxianshan Group [71,74], and recorded metamorphism from greenschist facies to granulite facies [75–77]. Voluminous Early Paleozoic basic to felsic magmatic rocks related to subduction and collision intrude into the Precambrian basement, and they are distinguished by calc-alkaline to high-K calcalkaline and shoshonitic [37,38,51,66,78–83].

An Early Paleozoic continental arc, characterized by arc magmatic rocks and associated high-temperature metamorphic rocks, has been discovered in the northern margin of the Qilian Block. The high-temperature metamorphic rocks, traditionally believed to be the part of the lower continental crust from the Qilian Block, consist mainly of granitic gneiss, metasedimentary rocks, and amphibolite (with mafic granulite occurring locally), which experienced different degrees of anatexis during the Cambrian–Ordovician period [38]. Our study area is located in the northern margin of the Qilian Block, with extensively exposed Early Paleozoic mafic to felsic plutonic intrusions, comprising gabbro, diorite, granodiorite, and granite with zircon U-Pb ages of ca. 500–446 Ma (Figure 1c). Geochemical and zircon Hf isotopic analyses show that the Early Paleozoic intrusive rocks have arc-related affinity [37,51]. The Menyuan Granodioritic Pluton, as a large and complex intrusion, provides a unique opportunity and new perspective for researching the petrogenesis and the crust–mantle interaction during the oceanic subduction process.

3. Petrology

A typical granitoid exposure with abundant mafic enclaves was identified in the Menyuan area approximately 100 km NE of Xi'ning City, adjacent to Menyuan County, Qinghai Province, China (Figures 1c and 2a). The Menyuan Granodioritic Pluton in this study intruded into the Precambrian basement including Meso-Neoproterozoic gneiss, schist, and local amphibolite. The Menyuan Granodioritic Pluton is mainly composed of granodiorite intruded by a 1–3-m wide dike of gabbro (Figure 2b). The granodiorite contains a large number of mafic microgranular enclaves (MMEs) throughout the whole pluton (Figure 2c,d).

Figure 2. (**a–d**) Field photographs of granodiorite, microgranular mafic enclave, and gabbroic dike from the Menyuan Granodioritic Pluton. Abbreviations: MME, microgranular mafic enclave.

Granodiorites of the Menyuan Granodioritic Pluton show classic massive textures, and the dominant minerals are of medium- to coarse-grain size (Figure 3a,b). Petrographically, the thin sections reveal that the main minerals contain plagioclase (40–50 vol.%), alkali-feldspar (20–30 vol.%), quartz (10–20 vol.%), and amphibole (15 vol.%) with minor biotite and accessory minerals (e.g., zircon, apatite, sphene, and titanite) (total 5 vol.%). The secondary minerals mainly include sericite, epidote, and opaque phases. Plagioclase typically presents as euhedral to subhedral in shape, displaying clear polysynthetic twinning partly altered to sericite. Almost all alkali feldspars are turbid with the characteristics of plagioclase alteration. These alkali feldspars are microcline featuring typical tartan twinning, and within the alkali feldspar crystals, fine-grained quartz grains can be observed (Figure 3a,b). The amphibole occurs as subhedral tabular crystals and displays pleochroism. Some amphiboles are replaced by alkali feldspar and biotite. The biotite is mainly associated with amphiboles and occurs as subhedral morphologies. The MMEs are widespread in the Menyuan granodiorites and have irregular and rounded morphologies, ranging in size from ten centimeters to thirty centimeters in length, with a chilled margin; they are distributed irregularly in the host rocks (Figure 2c,d). Thin sections indicate that the MMEs have a clear boundary with the host granodiorites (Figure 3c). The MMEs are fine-to medium-fine-grained and dark-colored and show typical magmatic textures (Figure 3d–g). They mainly consist of subhedral plagioclase (~60 vol.%), amphibole (~30 vol.%), minor biotite (~5 vol.%), quartz (~3 vol.%), along with accessory minerals, such as apatite and zircon. The MME contains more mafic minerals than the host granodiorites. In addition, the characteristic microtextures observed within the MMEs encompass euhedral-subhedral plagioclase phenocryst, hornblende featuring zonal and poikilitic structures, skeletal biotite, acicular apatite, and slender hornblende (Figure 3d–g). The gabbro dike intruding into the host granodiorite displays a dark gray color, featuring a medium- to medium-fine-grained texture. It mainly consists of plagioclase (50–60 vol.%), amphibole (30–35 vol.%), and pyroxene (5–10 vol.%), with trace amounts of accessory minerals such as Fe–Ti oxides (Figure 3h).

Figure 3. Microphotographs of granodiorite, microgranular mafic enclave, and gabbroic dike from the Menyuan Granodioritic Pluton: (**a**,**b**) medium to coarse-grained granodiorite; (**c**) contact between medium- to coarse-grained host granodirite and fine-grained in MME; (**d**) poikilitic-textured plagioclase and hornblende in MME; (**e**) plagioclase xenocrysts surrounded by groundmass minerals within MME; (**f**) fine-grained hornblende cluster and skeletal biotite in MME; (**g**) acicular apatite and slender hornblende in MME; (**h**) medium-grained gabbro dike. Mineral abbreviations: Qtz = quartz; Pl = plagioclase; Kfs = potassic feldspar; Bt = biotite; Hbl = hornblende; Ap = apatite; Cpx = Clinopyroxene.

4. Analytical Methods

In total, more than 30 representative samples were employed for zircon U-Pb dating, thin-section analyses, as well as zircon Hf and whole-rock Sr-Nd isotopes analyses. The detailed analytical methods are described in Appendix A.

5. Results

5.1. Mineral Composition

For the electron probe microanalysis, representative zonal plagioclase and amphibole grains were chosen from both the host granodiorites and MMEs from the Menyuan Granodioritic Pluton. All analyses were conducted on fresh mineral domains to ascertain their compositions. The results are presented in Table S1 and Table S2, respectively.

5.1.1. Plagioclase

Plagioclase grains in the host granodiorite mainly plot in the bytownite and andesine fields in the Ab-Or-An diagram (Figure 4a). Some of the plagioclases in the host granodiorite recorded reversed compositional zonation with variation in the anorthite (An) content from the core (An_{42}) to the rim (An_{74}) (Figure 4b). In addition, plagioclase phenocryst in the MMEs mainly plot in the oligoclase and andesine fields in the Ab-Or-An diagram (Figure 4a). The anorthite (An) content of plagioclase in the MMEs is relatively lower compared to that in the host granodiorite. The plagioclase in the MMEs exhibits complex oscillatory zoning structure from core to rim, and the range of anorthite (An) content varies from 24 to 44 (Figure 4c).

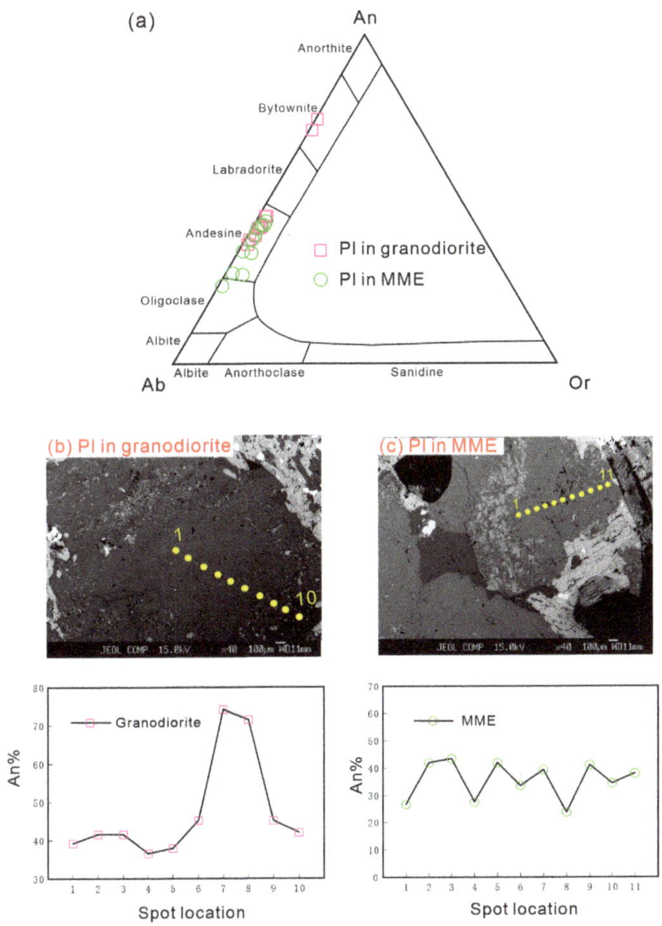

Figure 4. (**a**) Ab-Or-An diagrams for the plagioclase [84]; (**b**,**c**) anorthite contents of plagioclase crystals illustrating oscillatory zoning. Ab—albite; Or—potassium feldspar; An—anorthite.

5.1.2. Amphibole

Based on the classification proposed by Leake et al. (1997) [85] all amphiboles from the tested samples are calcic hornblende ($Ca_B \geq 1.5$) and can be divided into two groups: $(Na + K)_A \geq 0.5$ and <0.5, respectively. Most amphibole analyses fall within the group of $(Na + K)_A < 0.5$, located in the magnesiohornblende field (Figure 5a); the remaining amphiboles analysis spots classified under the group of $(Na + K)_A \geq 0.5$ mainly lie in the edenite field (Figure 5b). Based on the Ca-(Fe^{2+} + Fe^{3+})-Mg ternary diagram, the amphiboles in both the host granodiorite and the MMEs are plotted into the region of crust-mantle mixed-source, implying that the amphiboles may be the products of magmatic mixing(Figure 5c; [86]).

We applied the Al-in-amphibole barometry from Mutch et al. (2016) [87] and the Ti-in-amphibole thermometer from Liao et al. (2021) [88] to estimate the crystallization pressures and temperatures of the host granodiorites. The calculated pressures and temperatures for the host granodiorites range from 1.99 to 3.09 kbar and 714 to 774 °C, respectively. In addition, the Amp-Pl barometry from Molina et al. (2015) [89] and the Ti-in-amphibole thermometer from Liao et al. (2021) [88] were used to calculate the crystallization conditions of the MMEs, yielding pressures of 2.01–3.69 kbar and temperatures of 723–795 °C. Based on the continental crustal pressure gradient of 3.7 km per 1 kbar [90], the Menyuan Granodioritic Pluton might have been emplaced at an intermediate-crustal or shallower depth (7.4–13.7 km).

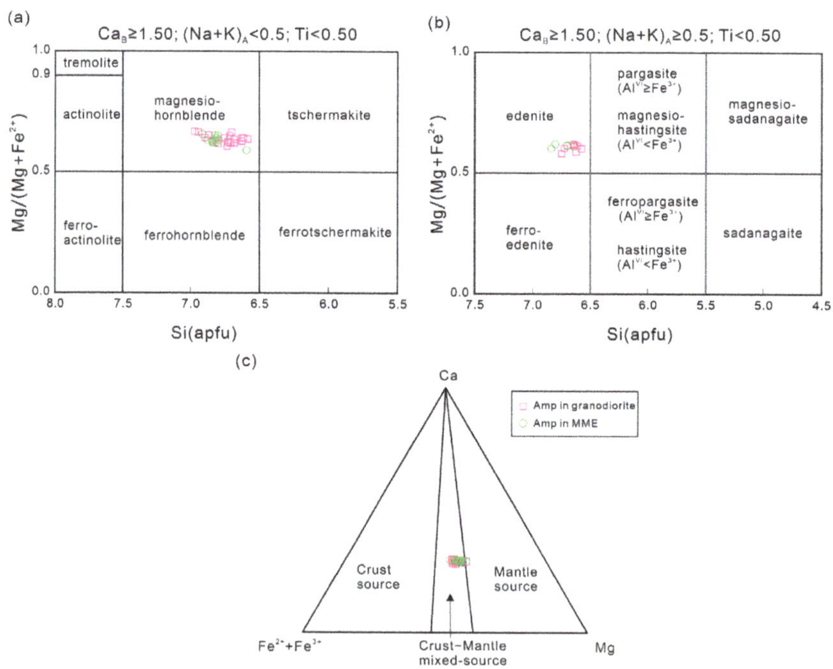

Figure 5. (**a**,**b**) Amphibole classification diagram [91] for the host granodiorite and MME. (**c**) The Ca-(Fe^{2+} + Fe^{3+})-Mg diagram of amphibole for discriminating the magma source [92].

5.2. Zircon U-Pb Geochronology

The zircon cathodoluminescence (CL) images and Concordia diagrams of three representative samples are listed and presented in Table S3 and Figure 6. The majority of the studied zircons exhibit subhedral to euhedral prismatic, and most of them measure 100 μm to 200 μm in length and 50 μm to 100 μm in width. All the analyzed zircons exhibit clear oscillatory zoning with high Th/U ratios ranging from 0.22 to 0.9, which suggest a magmatic origin [93].

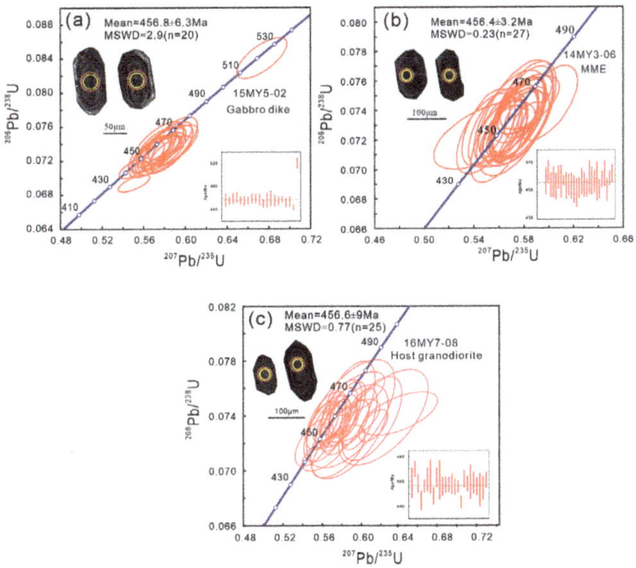

Figure 6. Cathodoluminescence images of representative zircons and U-Pb Concordia diagrams of magmatic rocks from the Menyuan Granodioritic Pluton. (**a**) Gabbro dike; (**b**) MME; (**c**) host granodiorite. Yellow circles and red circles represent the location of zircon U-Pb dating and Hf isotopic analysis, respectively.

Twenty zircon spot analyses from the gabbro dike 15MY5-02 were obtained. The $^{206}Pb/^{238}U$ ages range from 520 Ma to 443 Ma, yielding a weighted average of 456.8 ± 6.3 Ma (MSWD = 2.9; Figure 6a), which can be interpreted as the emplacement age of the gabbro dike. The Th/U ratio of zircons varied between 0.57 and 0.90.

The $^{206}Pb/^{238}U$ ages given by twenty-seven analytical spots on the zircons from the MME 14MY3-06 range from 464 Ma to 448 Ma, with a weighted mean age of 456.4 ± 3.2 Ma (MSWD = 0.23; Figure 6b), which represents the crystallization age of the mafic microgranular enclaves in the Menyuan Pluton. The Th/U ratios of all zircons are greater than 0.4 (except for one value of 0.22), ranging from 0.22 to 0.79.

A total of twenty-five dating spots were performed on twenty-five zircons from the host granodiorite 16MY7-08. The $^{206}Pb/^{238}U$ ages given by 25 analytical spots on the zircons range from 470 Ma to 443 Ma, with a weighted mean age of 456.6 ± 9 Ma (MSWD = 0.77; Figure 6c), which indicates the emplacement age of the Menyuan host granodiorite. All the measured zircons showed Th/U ratios that varied between 0.36 and 0.83.

5.3. Whole-Rock Geochemistry

Field investigations and thin-section microstructures reveal that the studied samples are essentially fresh with minimal chemical alterations. The whole-rock major and trace element data are presented in Table S4.

The chemical classification reveals a broad lithological variety for the host rocks, enclaves, and gabbro dikes (Figure 7a), from relatively mafic to felsic composition (gabbro-diorite to granite). The host rocks have relatively high and variable SiO_2 contents (60.03–73.34 wt.%), plotting within the diorite, granodiorite, and granite fields, exhibiting sub-alkaline features in the TAS diagram (Figure 7a). The host rocks have high contents of Al_2O_3 (12.68–17.44 wt.%) and K_2O (1.94–3.58 wt.%) and low contents of TiO_2 (0.08–0.73 wt.%), FeO^T (0.73–5.82 wt.%), MgO (0.26–2.64 wt.%), and CaO (2.07–3.47 wt.%) (with corresponding $Mg^{\#}$ values of 35–59). The total alkali content ($Na_2O + K_2O$) ranges from 5.94 wt.% to 7.45 wt.%, with Na_2O/K_2O ratios of 1.02–2.36. The magma was calc-alkaline to high-K calc-alkaline series (Figure 7b,c) and metaluminous to weakly peraluminous with A/CNK ratios of 0.80–1.15 (Figure 7d).

In contrast, the MMEs are characterized by lower SiO_2 (49.77–54.26 wt.%), TiO_2 (0.59–0.94 wt.%), and Al_2O_3 (13.47–15.69 wt.%), but higher MgO (6.93–8.62 wt.%), FeO^T (6.68–9.12 wt.%), CaO (6.80–10.24 wt.%) (with corresponding $Mg^{\#}$ values of 58–68), and total alkali ($Na_2O + K_2O$ = 4.18–6.82 wt.%) contents with variable Na_2O/K_2O ratios of 0.64–2.95. The magma was calc-alkaline to shoshonitic (Figure 7b,c) and metaluminous with A/CNK ratios of 0.56–0.70 (Figure 7d). The gabbro dikes plot within the monzogabbro and monzodiorite fields, and most of the samples belong to alkaline series in the TAS diagram (Figure 7a). In addition, the gabbro dikes are also characterized by low SiO_2 (48.05–54.15 wt.%) and high Al_2O_3 (16.23–18.16 wt.%), FeO^T (7.45–10.84 wt.%), MgO (3.52–5.20 wt.%), and CaO (7.00–9.03 wt.%) (with corresponding $Mg^{\#}$ values of 44–49), combined with low TiO_2 (0.86–1.38 wt.%) contents. The total alkali content ($Na_2O + K_2O$) ranges from 5.26 wt.% to 6.39 wt.%, with Na_2O/K_2O ratios of 1.04–2.51. The magma was high-K calc-alkaline to shoshonite series (Figure 7b,c) and metaluminous with A/CNK ratios of 0.71–0.82 (Figure 7d).

These studied samples all exhibit similar chondrite-normalized REE distribution patterns, characterized by an enrichment of light rare earth elements (LREEs) in contrast to heavy rare earth elements (HREEs), with no significant Eu anomalies (Figure 8a–c). The primitive mantle-normalized, multi-element diagrams indicate comparable chemical features with enrichment in Ba, Th, Nd, U, and Pb, and depletion in Nb, Ta, Zr, Hf, P, and Ti (Figure 8d–f).

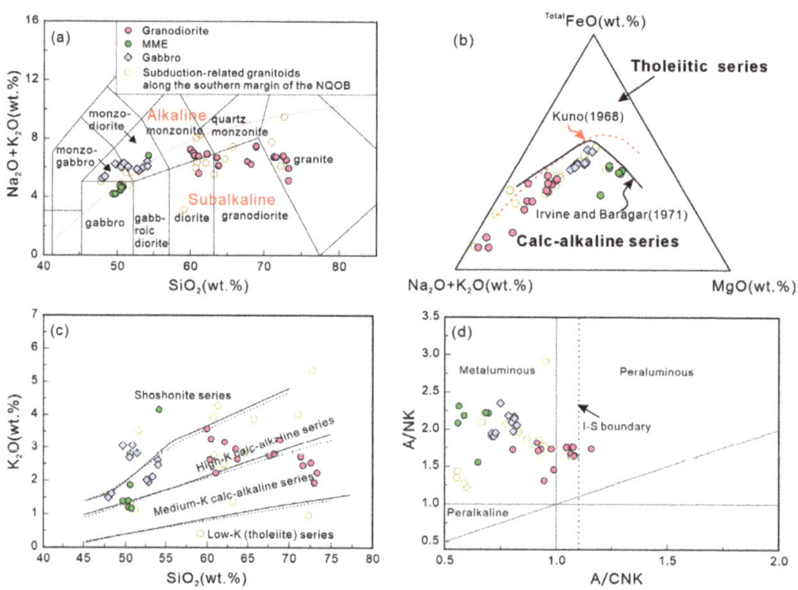

Figure 7. Geochemical discrimination diagrams for the magmatic rocks from the Menyuan Granodioritic Pluton. (**a**) Total alkalis versus silica diagram [94]; (**b**) triangle plot of $^{Total}FeO$ versus ($Na_2O + K_2O$ wt.%) versus MgO (wt.%) (after [95]); (**c**) SiO_2 (wt.%) versus K_2O (wt.%) diagram (after [96]); (**d**) molar $Al_2O_3/(CaO + Na_2O + K_2O)$ versus molar $Al_2O_3/(Na_2O + K_2O)$ (after [97]). The data for the subduction-related granitoids along the southern margin of the North Qilian Orogenic Belt (NQOB) are from [38,40,51,64].

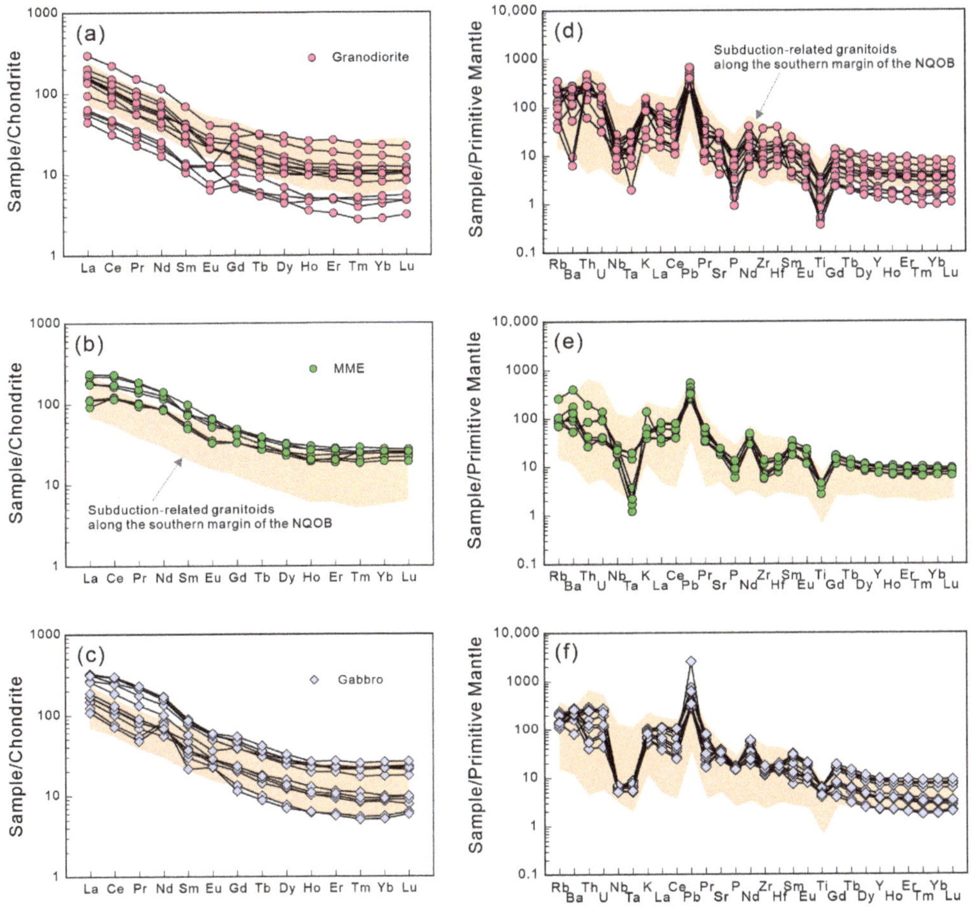

Figure 8. Chondrite-normalized REE distribution pattern diagrams (**a**–**c**) and primitive mantle-normalized trace element distribution pattern diagrams (**d**–**f**) for the magmatic rocks from the Menyuan Granodioritic Pluton. The chondrite normalization values are after [98], and primitive mantle normalization values are after [99]. The compositions for the subduction-related granitoids along the southern margin of the North Qilian Orogenic Belt (NQOB) are from [38,40,51,64].

5.4. Zircon Hf and Whole-Rock Sr-Nd Isotopes

Zircon Hf and whole-rock Sr-Nd isotopic compositions for the studied samples are presented in Tables S5 and S6. The Lu-Hf isotopic analysis points are located in the same zircon grains that were used for the U-Pb dating. The initial $^{176}Hf/^{177}Hf$ ratios and $\varepsilon_{Hf}(t)$ values of the zircons were calculated based on their crystallization ages. The whole-rock initial Sr-Nd isotopic compositions and $\varepsilon_{Nd}(t)$ values were determined using a weighted mean age of 456 Ma.

The host granodiorite shows negative $\varepsilon_{Hf}(t)$ values of −8.8 to −1.2 (average −4, 24 zircons; Figure 9a) with corresponding two-stage Hf model ages of 1764–1356 Ma. Zircon grains from the MME show variable $\varepsilon_{Hf}(t)$ values of −0.1 to +4.1 (average +2.5, 27 zircons; Figure 9a) with corresponding depleted mantle model ages of 1075–905 Ma. The gabbro samples have positive $\varepsilon_{Hf}(t)$ values of +4.0 to +8.2 (average +6.13, 20 zircons; Figure 9a) with corresponding depleted mantle model ages of 906–743 Ma.

The host granodiorites have variable initial $^{87}Sr/^{86}Sr$ ratios of 0.7072–0.7158 and $\varepsilon_{Nd}(t)$ values of −9.21 to −4.23 (Figure 9b,c), with two-stage Nd model ages (T_{DM2}) of 1939–1535 Ma, which are closed to the Nd model ages of the Precambrian basement metamorphic rocks of the Qilian Block (ca. 1.8–2.6 Ga, [100,101]). In comparison, the MMEs have slightly lower initial $^{87}Sr/^{86}Sr$ ratios of 0.7078–0.7089 and higher $\varepsilon_{Nd}(t)$ values of −3.88 to −1.68 (Figure 9b,c), with corresponding mantle-depleted Nd model ages (T_{DM1}) of 1711–1306 Ma. The gabbro samples have homogeneous and depleted initial $^{87}Sr/^{86}Sr$ ratios of 0.7049–0.7054 and $\varepsilon_{Nd}(t)$ values of +1.23 to +2.56 with T_{DM1} model ages of 1187–903 Ma (Figure 9b,c), and plot in the field of the subcontinental lithospheric mantle in the North Qilian Belt.

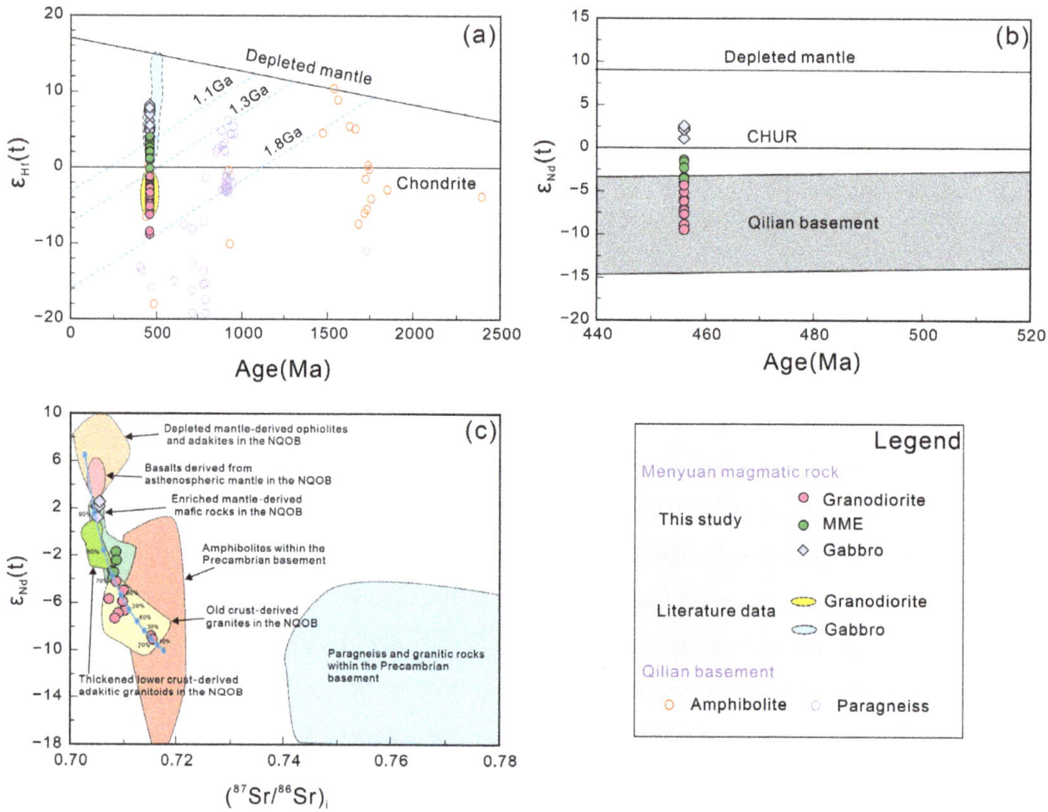

Figure 9. Plots of (**a**) $\varepsilon_{Hf}(t)$ and (**b**) $\varepsilon_{Nd}(t)$ values versus ages and (**c**) $(^{87}Sr/^{86}Sr)_i$ versus $\varepsilon_{Nd}(t)$ diagram for samples from the NQOB. In (**c**), a binary mixing curve between the depleted mantle-derived Paleozoic basalt (Sample LH57, Sr = 201, $(^{87}Sr/^{86}Sr)_i$ = 0.7038, Nd = 12.2, $\varepsilon_{Nd}(t)$ = 6.57; [102]) and amphibolite within the Precambrian Qilian basement (Average: Sr = 250, $(^{87}Sr/^{86}Sr)_i$ = 0.7187, Nd = 46.5, $\varepsilon_{Nd}(t)$ = −10.01; [103]) are also shown. Blue dots on the binary mixing curve represent the mixing ratios. Data sources: Menyuan magmatic rocks [38]; the Precambrian basement rocks in the Qilian Block [101,103–105]; enriched mantle-derived mafic rocks in the NQOB [106–108]; depleted mantle-derived ophiolites and adakites in the NQOB [102,109–111]; coeval arc-type mafic rocks derived from metasomatized asthenospheric mantle in the NQOB [61]; thickened lower crust-derived adakitic rocks and old crust-derived granites in the NQOB [43,46,81,101,112].

6. Discussion

6.1. Petrogenesis and Source Nature of the Menyuan Granodioritic Pluton

6.1.1. Petrogenesis of the Menyuan Granodioritic Pluton

The principal rock-forming minerals of the Menyuan Granodioritic Pluton are plagioclase and hornblende, lacking typical peraluminous minerals like muscovite, garnet, and cordierite. The host granodiorites are metaluminous to weakly peraluminous with A/CNK ratios ranging from 0.80 to 1.16 (with an average of 0.85), low K_2O/Na_2O ratios (0.42–0.98), and low Zr saturation temperatures (with an average of 746 °C, see Table S4), indicating affinity with I-type granites [113]. Furthermore, the studied samples of the Menyuan Granodioritic Pluton have a significant difference from those of sanukitoid magma featuring depleted CaO and enriched MgO elements [114] (Figure S1). These rock samples show arc-type magmatic features with enrichment of large ion lithophile elements (LILEs, e.g., Rb and Pb) and LREEs but depletion of HFSEs (e.g., Nb, Ta, Zr, Hf and Ti). Integrated with the geochemical characteristics of arc magma and low $(Na_2O + K_2O)/CaO$ and TFeO/MgO ratios (Figure 10), the Menyuan granodiorites are classified as unfractionated calc-alkaline I-type granites. The fractional crystallization from mantle-derived basaltic magma, partial melting of a single heterogeneous source, such as pure crust or lithospheric mantle, as well as magma mixing between mantle-derived mafic melts and crust-derived felsic melts are the three primary mechanisms for the petrogenesis of calc-alkaline I-type granites [115–117].

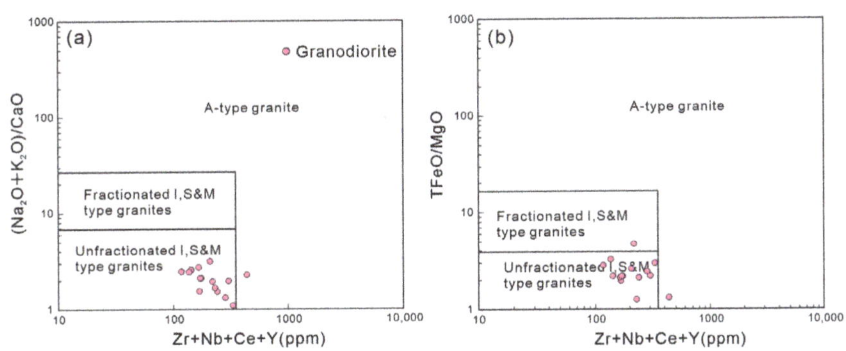

Figure 10. (**a**) Zr + Nb + Ce + Y (ppm) versus $(Na_2O + K_2O)/CaO$ diagram; (**b**) Zr + Nb + Ce + Y (ppm) versus TFeO/MgO diagram (after [118]).

Before discussing the petrogenesis and characteristics of the source nature, it is essential to consider the potential of crustal contamination during the ascent and emplacement of magma [119,120]. Crustal contamination is expected to result in positive Zr-Hf anomalies and positive correlation between $(^{87}Sr/^{86}Sr)_i$ ratios and SiO_2 contents or negative correlation between $\varepsilon_{Nd}(t)$ values and SiO_2 contents [121]. However, our data exhibited significant negative Zr-Hf anomalies (Figure 8), with no obvious correlation between SiO_2 contents and both $(^{87}Sr/^{86}Sr)_i$ ratios and $\varepsilon_{Nd}(t)$ values for the MMEs and gabbro dikes (Figure 11a,b), and the $\varepsilon_{Nd}(t)$ values of host granodiorites do not decrease significantly with the increase of SiO_2 contents (Figure 11b). Meanwhile, ancient inherited zircon grains are not present in the Menyuan-dated samples, indicating that contamination by ancient crustal materials is insignificant or negligible for the Menyuan Granodioritic Pluton. Thus, the geochemical characteristics of the Menyuan Granodioritic Pluton can effectively reflect their petrogenesis and magma sources.

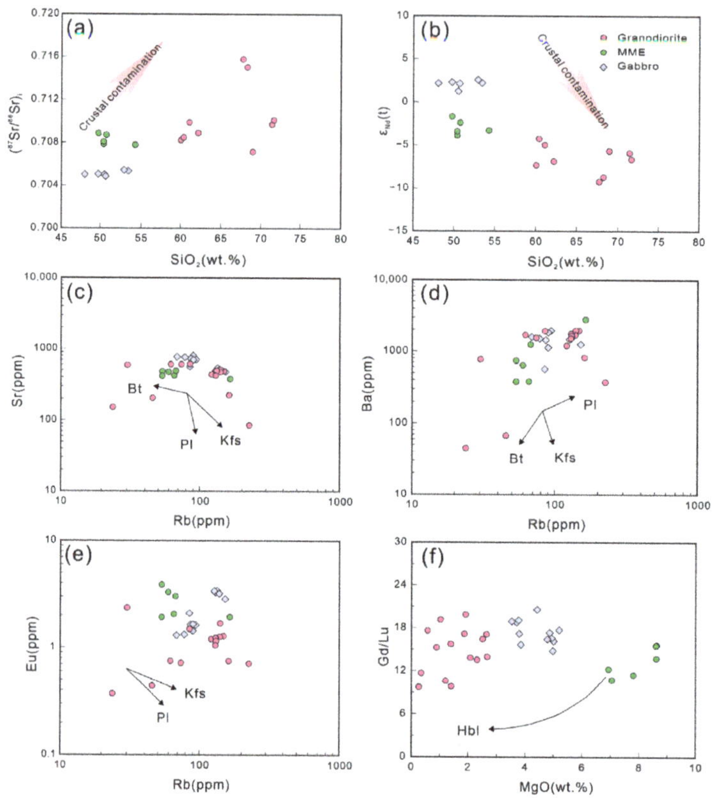

Figure 11. (**a**) SiO_2 (wt.%) versus $(^{87}Sr/^{86}Sr)_i$ diagram; (**b**) SiO_2 (wt.%) versus $\varepsilon_{Nd}(t)$ diagram; (**c**) Rb (ppm) versus Sr (ppm) diagram; (**d**) Rb (ppm) versus Ba (ppm) diagram; (**e**) Rb (ppm) versus Eu (ppm) diagram; (**f**) MgO (wt.%) versus Gd/Lu diagram. Abbreviations. Bt = biotite; Hbl = hornblende; Kfs = potassic feldspar; Pl = plagioclase.

Several genetic mechanisms have accounted for the formation of MMEs in the granitoids and their connection to the host granites: (1) xenolith: fragments from wall rocks that are incorporated during the ascent of the host magmas (e.g., [120,122]); (2) refractory remnants from the magma source region (e.g., [123,124]); (3) autoliths associated with early-formed cumulate minerals originating from a cogenetic magma (e.g., [20]); and (4) hybrid magma blobs resulting from mafic and felsic magma mixing (e.g., [7,125]). The question widely applied to their formation are whether they are of co-genetic origin or that of heterogeneous magma mixing [21,25,31,126]. The MMEs hosted in the Menyuan Granodioritic Pluton are ellipsoidal or irregular in shape with clear edges and they display apparent magmatic textures, and the MMEs are distributed throughout the Menyuan Granodioritic Pluton (Figure 2c,d). The fact that the MMEs and the host granodiorites have indistinguishable crystallization ages and a similar mineral assemblage eliminates the possibility that the MMEs are restites or xenoliths from magma sources or country rocks.

The Menyuan host granodiorites have variable SiO_2 (60.03–73.34 wt.%), high total FeO (0.73–5.82 wt.%) and MgO (0.26–2.64 wt.%) contents, with relatively high $Mg^{\#}$ (35.28–58.66, mean value = 43.76) values, which belong to calc-alkaline and high-K calc-alkaline rocks, and have metaluminous-to-weakly-peraluminous features (Figure 7d). Compared to the host granodiorites, MMEs have relatively lower SiO_2 (49.77–54.26 wt.%) and higher MgO (6.93–8.62 wt.%) contents and $Mg^{\#}$ (57.73–67.54) concentrations, reflecting a significant mantle contribution [121,127]. The Nb/Ta ratios of our studied samples are well above the average range of the continental crust and are inconsistent with typical crustal origin. Experimental studies have revealed that melts that originated from the partial melting of pure lower crust with no mantle material input generally correspond to low $Mg^{\#}$ melts (<42), but high $Mg^{\#}$ melts (>60) are derived from mantle-derived basic rocks [128,129]. Moreover, the host granodiorites have significantly enriched Sr-Nd isotope compositions with initial $^{87}Sr/^{86}Sr$ ratios (0.7072–0.7158) and whole-rock $\varepsilon_{Nd}(t)$ values (−9.21 to −4.23), which are obviously distinct from the lithospheric mantle-derived melts in the North Qilian Belt (Figure 9c). Therefore, they are not controlled by partial melting of the mafic lower crust or lithospheric mantle.

Previous studies have shown that the mafic minerals crystallized at an early stage in the magma chamber are characterized by coarse-grained and cumulate textures [130]. However, the predominantly rock-forming minerals of the MMEs are obvious smaller compared to their host granites, and the MMEs are randomly distributed without mafic minerals' cumulate texture. Likewise, given that the REEs are strongly incompatible elements, if the MMEs and host granites have a cognate origin and host granodiorites are derived from fractional crystallization of the basaltic magma, the contents of REEs of the host granodiorites were higher than the MMEs (e.g., [96]), but the MMEs exhibit REE contents comparable to those of the host granodiorites (Figure 8a). In addition, the fractional crystallization of the amphibole, biotite, or plagioclase was not detected in the diagrams of fractionation trends for the Menyuan Granodioritic Pluton (Figure 11c–f). The fractionation crystallization from mantle-derived basaltic magma demands contemporaneous and larger amounts of mafic rocks than granites; however, the Menyuan granodiorites cropped out as large-scale plutons and gabbro dikes have an intrusive contact with the granodiorite. These observations are inconsistent with cognate or autolith origin whereby the MMEs were produced during early-formed phases within the same magma chamber as the host granodiorites. Combined with a clear whole-rock Sr-Nd and zircon Hf isotopic contrast between the host granodiorites and MMEs (Figure 9), we suggest that they are unlikely to have been formed by fractional crystallization of a single heterogeneous source. The wide ranges of major element contents and trace elemental ratios of the host granites and MMEs reflect contrasting mantle and crustal sources. Thus, the heterogeneous magma mixing from different magma sources is the most possible petrogenesis for the Menyuan Granodioritic Pluton.

There are several pieces of evidence suggesting that the petrogenesis of the Menyuan Granodioritic Pluton may be related to incomplete magmatic mixing. The MMEs are irregular ellipsoidal in shape, and have sharp and occasionally diffusive contact with the Menyuan granodiorites, which indicate that parental mafic magma was still partially ductile while being injected into the host granitic magma and during their mixing [1,131]. Plagioclase xenocrysts captured from the host granodiorite and surrounded by fine-grained mafic minerals can be identified in MME (Figure 3e). The megacrysts of plagioclase in MMEs can be considered as the consequence of mechanical transfer occurring between felsic and mafic magmas in the course of the magmatic mingling process [115,132]. Felsic back-veins with a mineral assemblage resembling that of the host granodiorites are detected in the MMEs (Figure 2d), and they are caused by the formation of fractures due to magma-quenching crystallization when hotter mafic magma was injected into relatively cooler felsic host magma [7,116]. In addition, disequilibrium textures are common in the MMEs and host granodiorites, which are considered to be evidence of mechanical stirring of minerals between felsic and mafic magmas during magma mingling. Mostly, the plagioclase grains are rounded and embayed with dissolution textures in the MMEs, which indicates that the plagioclases crystallized in the felsic magma chamber and experienced remelting in relatively high-temperature mafic magma after mechanical transfer. The occurrences of skeletal biotite and sieve-textured hornblende in the MMEs and some euhedral-subhedral plagioclase grains rimmed with fine-grained mafic minerals in the host granodiorites probably reflect that partial dissolution and overgrowth of principal rock-forming minerals have occurred in the hybrid magma [117]. The occurrence of acicular apatite and slender hornblende (Figure 3g) can be considered to be a result of undercooling and rapid crystallization under water saturation conditions when the hotter basic magma was injected into relatively cool felsic magma [131,133]. Furthermore, plagioclase in the host granodiorites exhibit reversed compositional zonation with the rapid increase in anorthite (An) content (Figure 4b), indicating the chemical and thermal changes during plagioclase crystallization, which may be related to magmatic mixing [134–136]. Plagioclase in MMEs exhibits complex oscillatory zoning structure with oscillating orthite (An) content (Figure 4c). This may be attributed to the injection of high-temperature basaltic magma that caused the dissolution of the previous low-An plagioclase in the shallow magma chamber, leading to the formation of sieve textures [137]. Subsequently, the sieve-textured plagioclase continued to grow in the hybrid magma, and during the magma mixing process, these plagioclase macrocrysts were successively captured by the MMEs. These features suggest that the Menyuan Granodioritic Pluton was most likely derived from a mixed source of mantle and crustal origin, while the MMEs represent mafic-hybrid magma globules scattered throughout the host granodiorites by convection.

The elemental geochemical and isotopic characteristics of the whole rocks further suggest that the magma mixing was insufficient during the formation of the Menyuan Granodioritic Pluton. The MgO versus TFeO distribution shown on the whole-rock discrimination diagram conforms to the magma mixing trend (Figure 12a). In the Al_2O_3/CaO versus Na_2O/CaO and CaO/Al_2O_3 versus K_2O/MgO diagrams, there exist clear linear and curved correlations which are also in accordance with the magma mixing trend (Figure 12b,c). The behavioral differences between compatible and incompatible elements are conducive to the identification of magmatic mixing [138,139]. The Rb/V versus 1/V diagram plots along the mixing arrays of the melts, which can indicate magma mixing (Figure 12d). Additionally, similar normalized elemental patterns (Figure 8) for the studied samples of the Menyuan Granodioritic Pluton reflect that elemental diffusion and re-equilibration of trace elements that occurred during the magma mixing process. Zircon Hf isotopes characterized by high closure temperatures, high stability, and early crystallization in magmatic systems can effectively resist later isotopic exchange and re-equilibration between the interacting magmas [140]. As illustrated in Figure 9a, diverse Hf isotopic compositions recorded the magma source and mixing process. For the Sr-Nd isotopic system, Sr exhibits much higher mobility than Nd in both fluids and melts [141,142], which results in comparable initial $^{87}Sr/^{86}Sr$ ratios (ranging from to 0.7072 to 0.7101) but somewhat distinct $\varepsilon_{Nd}(t)$ values for the Menyuan MMEs and host granodiorites during the magma mixing process (Figure 9c). Nevertheless, the wide range of whole-rock Sr-Nd and zircon Hf isotopic compositions observed in the MMEs and host granodiorites (Figure 9a–c) indicates restricted chemical exchange between the two magma end-members from distinct sources during magma mixing, and isotopic homogenization was not achieved prior to the complete solidification of the hybrid magma [140]. Consequently, we deduce that the most likely explanation for the formation of the Menyuan Granodioritic Pluton involves the incomplete mixing of magma derived from mantle and crustal sources, and the MMEs are the relics of local heterogeneity within the mixed melt serving as direct evidence of magma mixing. Effective magma mixing requires the onset of local superheating of the felsic magma [143]. The occurrence of contemporaneous Menyuan gabbroic dikes indicates a regional mafic magma injection event, and the mafic mantle-derived magma underplating provides adequate energy for the partial melting of the lower crust to generate crustal melts.

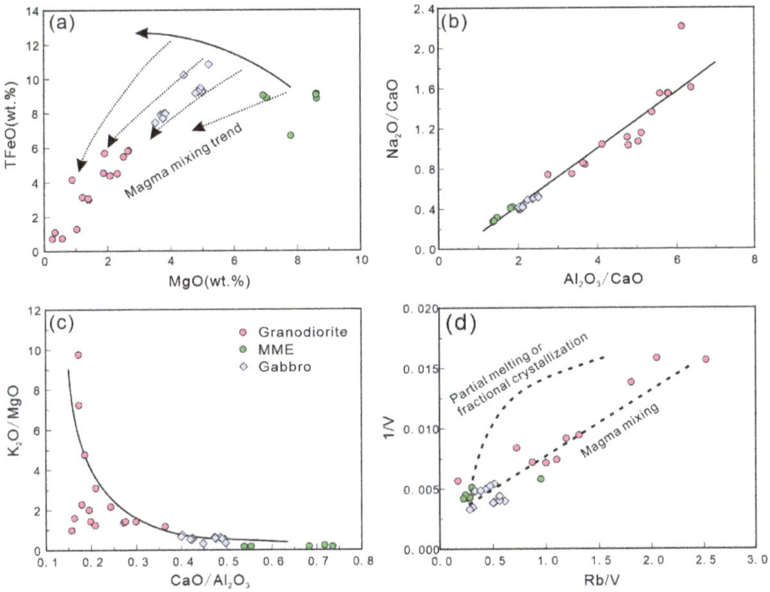

Figure 12. Major and trace element covariation diagrams indicating magma mixing for the studied samples from the Menyuan Granodioritic Pluton ((**a**)—after [144]; (**b**,**c**)—after [138]; (**d**)—after [139]).

6.1.2. Source Nature of the Menyuan Granodioritic Pluton

The key to clarifying the crust–mantle magma mixing process lies in revealing the components of crust–mantle magmatic end-members. The host granodiorites, MMEs, and gabbro dikes in this study share similar diagenetic ages, normalized elements distribution patterns, as well as spatiotemporal correlation, suggesting that they may have a genetic connection. Notably, similar geochemical and isotopic compositions between the MMEs and coeval gabbro dikes indicate that they probably originated from a similar mantle source. The finding of magma mixing origin illustrates that the geochemical characteristics of the MMEs and coeval gabbro dikes cannot accurately represent the original compositions of the magma derived from the mantle. However, we can deduce some crucial information about the mantle source from the geochemical and isotopic features of these studied rocks. The gabbro dikes are characterized by much more positive and homogeneous $\varepsilon_{Nd}(t)$ (+1.23 to +2.56) and $\varepsilon_{Hf}(t)$ (+4.0 to +8.2) values compared to other Menyuan samples in this study with depleted mantle Hf model ages of 906–743 Ma. Additionally, the MMEs have moderately enriched $\varepsilon_{Nd}(t)$ (−3.88 to −1.68) and $\varepsilon_{Hf}(t)$ (−0.1~+4.1) values with depleted mantle Hf model ages of 1075–905 Ma, which indicate an early Neoproterozoic-enriched mantle source. Notably, the Nd isotope data of the MMEs and the gabbro dikes are lower than those of the coeval depleted mantle-derived basalts-basaltic andesites ($\varepsilon_{Nd}(t)$ = +3.0 to +5.9) in the North Qilian Belt (Figure 9c [61]), but they overlap with those of enriched mantle-derived mafic rocks in the North Qilian Belt. The MMEs and the gabbro dikes in the Menyuan Granodioritic Pluton are distinctly enriched in LILEs (Ba, K, and Sr) and LREEs, and are depleted in HFSEs (Nb, Ta, Zr, Hf, and Ti; Figure 8e,f), which is consistent with arc-like trace elements distribution patterns. In the La/Yb versus Nb/La diagram, most samples of the MMEs and the gabbro dikes fall in the source region of the lithosphere mantle (Figure 13a). Combined with the relatively low La/Ba and high Th/Yb ratios (Figure 13b,c), these geochemical characteristics indicate that the mafic magma was produced by partial melting of a subduction-modified continental lithospheric mantle source [145–147]. Combined with the Nb/Y versus Ba, Nb/Y versus Rb/Y, and Sr/Nd versus Th/Yb diagrams (Figure 13d–f), these mafic samples display a fluid-related enrichment trend, indicating that the lithospheric mantle source was probably metasomatized by subducted slab-derived fluids prior to melting [148]. Moreover, the presence of hornblende in the studied samples from the Menyuan Granodioritic Pluton also reflects the participation of fluids. Thus, we suggest that the mafic magma end-member in this study was derived from partial melting of the enriched lithospheric mantle metasomatized by slab-derived fluids in the North Qilian Belt.

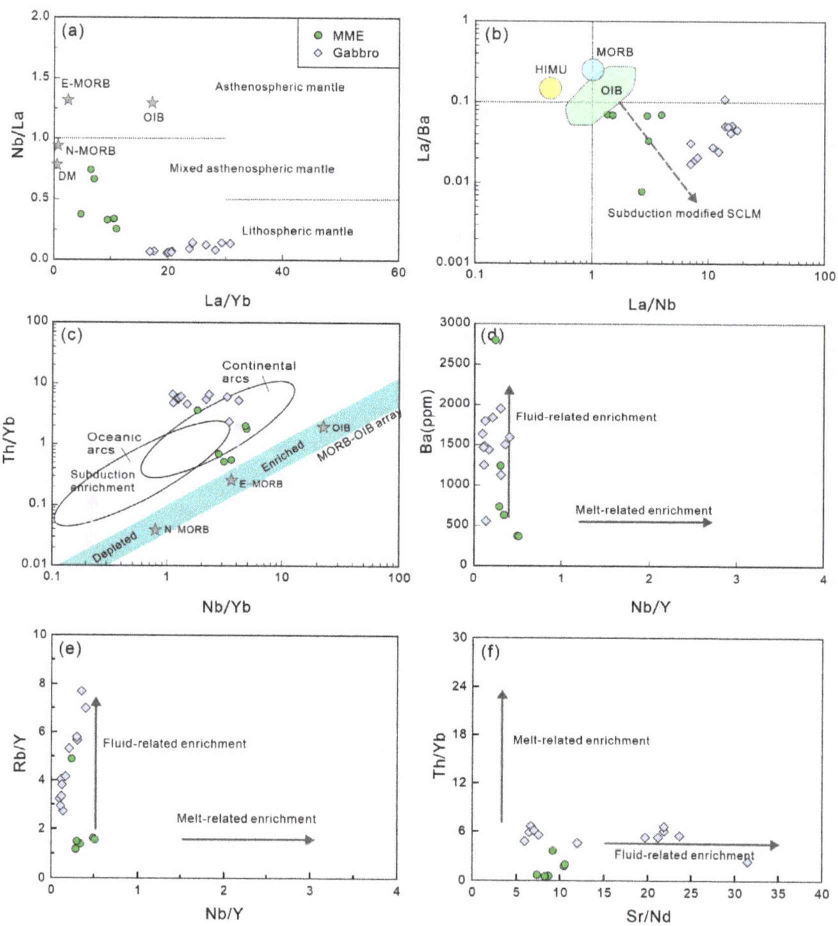

Figure 13. Diagrams of (**a**) La/Yb versus Nb/La (after [149]); (**b**) La/Nb versus La/Ba (after [69]); (**c**) Nb/Yb versus Th/Yb (after [150]); (**d**) Nb/Y versus Ba; (**e**) Nb/Y versus Rb/Y; (**f**) Sr/Nd versus Th/Yb ((**d**–**f**) after [148]) for the MMEs and gabbro dikes from the Menyuan Granodioritic Pluton. MORB—mid-oceanic-ridge basalt; N-MORB—normal MORB; E-MORB—enriched MORB; OIB—oceanic-island basalt; DM—depleted mantle; HIMU—high U/Pb mantle source; SCLM—subcontinental lithospheric mantle. The fields for OIB, MORB, and HIMU in (**b**) are from [147].

The Menyuan host granodiorites are K-rich, calc-alkaline, and high-K calc-alkaline rocks in geochemistry and they have negative $\varepsilon_{Hf}(t)$ values of −8.8 to −1.2, corresponding with two-stage Hf model ages of 1764–1356 Ma, which indicates that they might be predominantly associated with the recycling of Paleoproterozoic-to-early-Mesoproterozoic crustal materials within the North Qilian Orogenic Belt. Their relatively low $\varepsilon_{Nd}(t)$ values range from −9.21 to −4.23, with two-stage Nd model ages of 1939–1535 Ma, which are close to the isotopic composition range of the meta-basaltic rocks from the Qilian basement, but deviate from data related to the paragneiss and Neoproterozoic granitoids from the Precambrian basement (Figure 9c). It can be inferred that the crustal-derived source of the Menyuan Granodioritic Pluton was produced by the partial melting of meta-basaltic rocks from the Qilian basement. Abundant meta-basaltic rocks have been identified in the Qilian Precambrian basement, and their mantle-depleted Nd model ages range from 1.8 to 2.6 Ga [100,101]. Considering all the mentioned facts, it is reasonable to confirm that the ancient mafic lower continental crustal rocks from the Qilian Block were the crustal source of the Menyuan Granodioritic Pluton, which is reinforced by the presence of amphibolite enclaves within the Menyuan Granodioritic Pluton [38]. In this scenario, we applied a two-component mixing model of a whole rock Sr-Nd isotope to roughly simulate the interaction between the ancient mafic lower continental crust and depleted mantle-derived melts. The simple binary mixing simulation indicates that the proportions of mantle components involved in the formation of the Menyuan Granodioritic Pluton varied from 10% to 30%, while the proportions of ancient crustal materials varied from 70% to 90% (Figure 9c).

As discussed above, the magma originating from the lithospheric mantle and the melts of the ancient mafic lower continental crust beneath the Qilian Block made an substantial contribution to the magmatic sources of the Middle-Late Ordovician Menyuan Granodioritic Pluton.

6.2. Geodynamic Implication for the Menyuan Granodioritic Pluton

The LA-ICP-MS zircon U-Pb dating data obtained from the granodiorites, MMEs, and gabbroic dikes suggest that the magma emplacement time for the Menyuan Granodioritic Pluton can be constrained to the Middle-Late Ordovician period (470–443 Ma). These ages are consistent with the previous magmatic ages (ranging from 520 to 440 Ma) associated with the subduction of the North Qilian Ocean in the northern margin of the Qilian Block and adjacent regions [35–38,40,51,55,151]. For example, the ca. 466 Ma felsic volcanic rocks in the Qingshuigou were attributed to an arc setting [152].The quartz diorites (ca. 476 Ma) from the Niuxinshan and granodiorites (ca. 463 Ma) from the Minyueyaogou were formed in an active continental margin setting [153]. The 503–450 Ma I-type granite and mafic rocks, 461–443 Ma arc volcanic rocks, and 508–450 Ma metamorphic rocks at Menyuan Lenglongling-Dabanshan in the northern margin of the Qilian Block were formed during the south-dipping subduction of the North Qilian Ocean in the Early Paleozoic [37,38,51,154,155]. In this regard, the Menyuan Granodioritic Pluton in this study was likely produced in an active continental arc setting related to the southward subduction of the North Qilian Oceanic slab. All studied samples from the Menyuan Granodioritic Pluton are calc-alkaline in composition and have similar normalized elemental patterns with enrichment in LILEs (e.g., Ba, Rb, and Pb) and depletion in HFSEs (e.g., Nb, Ta, and Ti; Figure 8e,f), typically characterizing subduction-related magmatism [145,156–158]. In addition, the MMEs and gabbro dike samples are mainly distributed in the island-arc basalt field in the Y/15-La/10-Nb/8 and Ti/100-Zr-Y*3 ternary diagrams (Figure 14a,b), and show affinities close to the continental arcs in the Nb/Yb versus Th/Yb diagram (Figure 13c), which support an active continental arc setting. Therefore, the U-Pb zircon ages from the Menyuan Granodioritic Pluton (ca. 456 Ma), in combination with the chemical characteristics described above, suggest subduction-related magmatism. The Menyuan Granodioritic Pluton can be regarded as continental arc rocks formed in an active continental arc setting, which was associated with the southward subduction of the North Qilian Ocean beneath the northern margin of the Qilian Block in the Middle-Late Ordovician period.

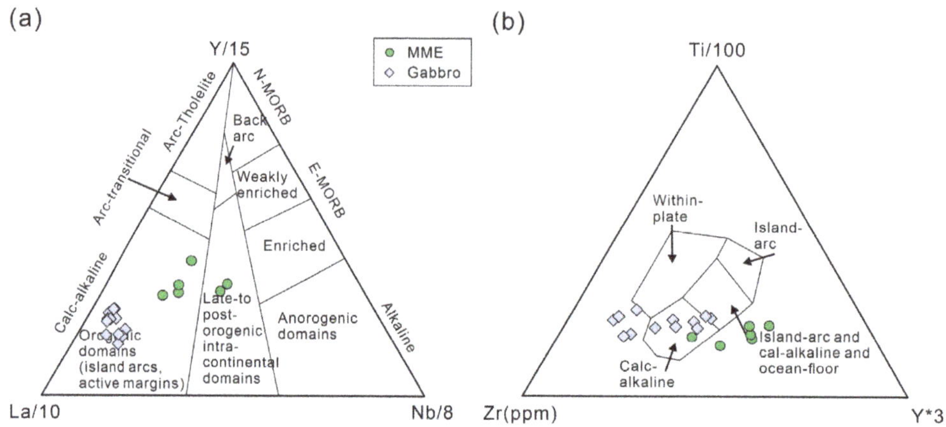

Figure 14. Plots of (**a**) Y/15-La/10-Nb/8 (after [159]); (**b**) Ti/100-Zr-Y*3 (after [160]).

The triggering mechanism of Middle-Late Ordovician arc magmatism along the northern margin of the Qilian Block during the subduction-dominated stage constitutes a pivotal scientific question that needs to be discussed. The subducted slab rollback and breakoff can induce asthenospheric upwelling coupled with partial melting of the overriding lithospheric mantle. Then, the thermal perturbation migrates upward, initiating crustal anatexis within the continental crust, which establishes favorable conditions for extensive crust–mantle interaction [161]. The slab breakoff model is commonly invoked to explain syn- to post-collisional magmatism in orogenic systems, wherein detachment initiates after the complete consumption of the oceanic lithosphere, which seems inadequate to interpret the arc magmatism of the Menyuan Granodioritic Pluton [162,163]. Therefore, we consider that the formation of Menyuan Granodioritic Pluton in this study was most likely initiated by the slab rollback under the force of gravity during the Middle-Late Ordovician period.

Clarifying the petrogenetic affiliation of the Menyuan Granodioritic Pluton can provide important insights to the emplacement process of orogenic magmas and extensive crust–mantle interactions. The Menyuan Granodioritic Plutonic rocks were related to mixing between the lithospheric mantle-derived magma and ancient continental crustal melts in the active continental arc setting. During subduction, the subducted slab rollback caused partial melting of the overlying lithospheric mantle metasomatized by slab-derived fluids and the thermal upwelling of high-temperature, mantle-derived basaltic magmas. Meanwhile, the high-temperature basaltic magmatic underplating could provide sufficient energy for the anatexis of the ancient mafic continental lower crust beneath the Qilian Block to form primitive felsic melts. These crust-derived felsic melts subsequently experienced variable mixing with mantle-derived basaltic magma during ascent, ultimately emplacing as the Menyuan Granodioritic Pluton in the upper crust along the northern margin of the Qilian Block (Figure 15a,b). Consequently, our results advocate for a magmatic underplating mechanism in which the upwelling of high-temperature mantle-derived magma induced extensive crust–mantle interaction.

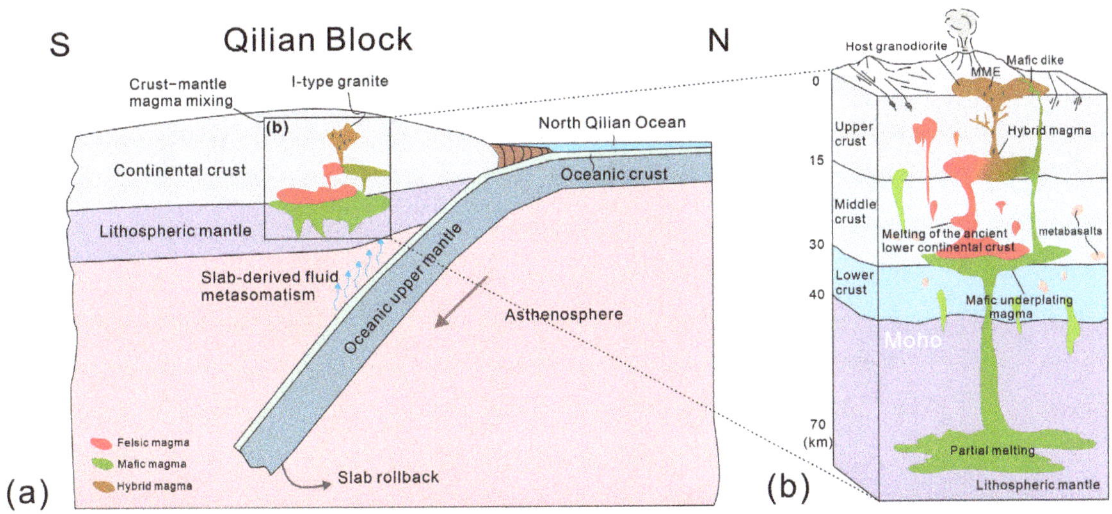

Figure 15. (**a**,**b**) A schematic geodynamic model illustrating the Middle-Late Ordovician magmatism in this study in the northern margin of the Qilian Block. The crust section through the Early Paleozoic continental arc in the northern margin of the Qilian Block is from [38,164].

7. Conclusions

(1) The Menyuan Granodioritic Pluton primarily consists of enclave-bearing granodiorites with minor gabbroic dike. These studied rocks samples have a consistent LA-ICP-MS zircon U-Pb age of ca. 456 Ma, suggesting a Middle-Late Ordovician mafic to felsic magmatism within the North Qilian orogen;

(2) The host granodiorites and the MMEs exhibit similar crystallization ages, trace elemental geochemistry, and Sr isotopic compositions, yet display heterogeneous whole-rock Nd and zircon Hf isotopic characteristics. This can be interpreted as a result of insufficient chemical and isotopic equilibration during the magma mixing process;

(3) The Menyuan host granodiorites and the MMEs were likely derived from a hybrid magma, which was composed of components derived from the lithospheric mantle and ancient mafic lower continental crust. The felsic magma was generated from the anatexis of the meta-basalt from the Precambrian Qilian basement, and the mafic magma originated from the melting of the enriched lithospheric mantle that intruded into the lower continental crust through underplating. Then, the mafic magmas mixed with felsic melts to form arc magmatic rocks ranging from gabbroic to granodioritic in the shallow crust;

(4) The geochemical and isotopic characteristics of the gabbro dikes indicate that they originated from the lithospheric mantle metasomatized by slab-derived fluids. Their presence reflects the existence of evolved mafic magmas in the region that did not directly participate in mixing with the granodiorites and provides evidence for mantle-derived magmatism;

(5) Detailed field observations, petrographic, and geochemical data suggest that the Menyuan Granodioritic Pluton formed in an active continental-arc setting induced by the northward rollback of the Qilian Proto-Tethys Ocean in the Middle-Late Ordovician period.

Supplementary Materials: The following supporting information can be downloaded at: https://www.mdpi.com/article/10.3390/min15040391/s1: Figure S1: MgO (wt.%) versus CaO (wt.%) diagram (after [114]). Table S1: Plagioclase compositions of the host granodiorite and MME from the Menyuan Granodioritic Pluton; Table S2: Amphibole compositions of the host granodiorite and MME from the Menyuan Granodioritic Pluton; Table S3: LA-ICP-MS zircon U-Pb isotope data for the MME, host granodiorite, and gabbro dike from the Menyuan Granodioritic Pluton; Table S4:

Whole-rock major and trace element compositions for the MME, host granodiorite, and gabbro dike from the Menyuan Granodioritic Pluton [165]; Table S5: LA-MC-ICP-MS zircon Lu-Hf isotope compositions for the MME, host granodiorite, and gabbro dike from the Menyuan Granodioritic Pluton; Table S6: Whole-rock Sr-Nd isotope compositions for the MME, host granodiorite, and gabbro dike from the Menyuan Granodioritic Pluton.

Author Contributions: Conceptualization, methodology, and writing—original draft preparation, S.X.; supervision, investigation, S.Y. and Q.W.; writing—review and editing, Y.Q., X.J. and X.G.; data curation, Y.W., C.L., L.W. and Y.P.; funding acquisition, S.Y. All authors have read and agreed to the published version of the manuscript.

Funding: This research was funded by the Taishan Scholars (grant number ts20231214), National Natural Science Foundation of China (grant number 42372247, U2344214 and 42402218), and the Natural Science Youth Foundation of Shandong Province (grant number ZR2022QD055).

Data Availability Statement: Data presented in the study are included in the article and Supplementary Materials. Further inquiries can be directed to the corresponding authors.

Conflicts of Interest: The authors declare no conflicts of interest.

Appendix A

Appendix A.1. LA-ICP-MS Zircon U-Pb Dating

The zircons were isolated by using heavy liquid and standard magnetic technology. The zircons were placed in epoxy resin and ground down to approximately half of their original thickness to obtain Cathodoluminescence (CL) and perform subsequent U-Pb analysis operations. By using laser ablation multicollector inductively coupled plasma mass spectrometry (LA-MC-ICP-MS), zircon U-Pb dating analysis of representative fresh samples was carried out in the Institute of Mineral Resources, CAGS (Chinese Academy of Geological Sciences), Beijing, China. In the process of laser ablation, the laser spot size was 30 μm, the carrier gas uesd helium, and the make-up gas was argon. Hou et al. (2009) have described detailed operating instructions and simplified data processing for the laser ablation system [166]. Zircon GJ-1 served as an external standard for U-Pb dating, and was analyzed twice every 5–10 measurements. To correct for the time-dependent drift of U-Th-Pb isotopic ratios, linear interpolation was applied every 5–10 analyses using GJ-1 as a reference [167]. The Isoplot 3.0 program was used to generate zircon U-Pb Concordia diagrams and calculate the weighted average value.

Appendix A.2. Zircon Lu-Hf Isotope Analysis

The Lu-Hf isotope analysis of zircon was performed in the Institute of Mineral Resources, CAGS, Beijing, China and by using the Neptune multi-receiver ICP-MS and New wave UP 213 laser ablative microprobe. The relevant description of the instrument and the specific method of data collection can be referred to in the literature published by Hou et al. (2007) [168]. In this experiment, the carrier gas uesd helium, and the ablation diameter was determined to be 42 μm based on zircon size. The international zircon standard GJ-1 was used as the standard material. For mass identification calibration, the index normalization of Yb isotope ratio was carried out using $^{173}Yb/^{172}Yb = 1.35274$, and the Hf isotope ratios were calibrated using $^{179}Hf/^{177}Hf = 0.7325$. In the calculation of $\varepsilon_{Hf}(t)$ values, the values of chondrites used were $^{176}Lu/^{177}Hf = 0.0336$ and $^{176}Hf/^{177}Hf = 0.282785$. The specific analysis process is consistent with the content elaborated by Griffin et al. (2000) [169].

Appendix A.3. Whole-Rock Geochemical

The analysis of rare earth elements, major elements, and trace elements in the whole rock was performed by using X-ray Fluorescence (XRF) and inductively coupled plasma-mass spectrometry (ICP-MS) at the CAGS, Beijing, China. The analysis error was controlled within 5%. The determination of rare earth elements and trace elements in whole rock was completed by using the ICP-MS technique. At the same time, the whole-rock major elements were completed by using the Primus II XRF technique. The separation of rare earth elements was realized by cation exchange technology. The measurement precision of elements was 10% if the abundances were below 10 ppm, whereas the measurement precision of elements was approximately 5% if the abundances were above 10 ppm. Details of the analytical procedures of major and trace elements can be referred to in the study by Liu et al. (2008) [170].

Appendix A.4. Whole-Rock Sr-Nd Isotopes

The analysis of Sr and Nd isotopes was carried out at Beijing Createch Testing Technology. The isotope dilution method was employed to determine the concentrations of Sr and Nd. After Sr and Nd were separated and purified using a double-packed column, they were analyzed by a Neptune Plus MC-ICP–MS instrument. The ratios of $^{87}Sr/^{86}Sr$ and $^{143}Nd/^{144}Nd$ were corrected by the exponential fractionation law and the assumption that $^{88}Sr/^{86}Sr = 8.375209$ and $^{146}Nd/^{144}Nd = 0.7219$, respectively. The $^{143}Nd/^{144}Nd$ ratios of GSP-2 and BHVO-2 international Certified Reference Materials were 0.511374 ± 5 and 0.512984 ± 5, respectively, which were identified with standard values within analytical errors. The detailed analyses on the Sr-Nd isotope can be referred to in the study by Li et al. (2012) [171].

Appendix A.5. Mineral Analysis

Before analysis, the sample probe sections were carefully examined under a microscope to select target minerals for imaging. The geochemical analysis of major elements in minerals was conducted by the CAGS, Beijing, China, using a JEOL JXA 8900 electron probe microanalyzer (EPMA). The measurement accuracy of major oxides was superior to 1%. Quantitative analysis was conducted under experimental conditions of an acceleration voltage of 15 kV, a probe current of 20 nA, as well as a beam spot size of 5 μm. In addition, secondary electron (SE) and backscattered electron (BSE) images were utilized for image analysis. The results were corrected using the ZAF method. Detailed procedures and parameters are as those described in the study by Lai et al. (2016) [172].

References

1. Kumar, S.; Rino, V.; Pal, A.B. Field Evidence of Magma Mixing from Microgranular Enclaves Hosted in Palaeoproterozoic Malanjkhand Granitoids, Central India. *Gondwana Res.* **2004**, *7*, 539–548. [CrossRef]
2. Turnbull, R.; Weaver, S.; Tulloch, A.; Cole, J.; Handler, M.; Ireland, T. Field and geochemical constraints on mafic–felsic interactions, and processes in high-level arc magma chambers: An example from the Halfmoon Pluton, New Zealand. *J. Petrol.* **2010**, *51*, 1477–1505. [CrossRef]
3. Wiebe, R.A.; Hawkins, D.P. Growth and Impact of a Mafic-Silicic Layered Intrusion in the Vinalhaven Intrusive Complex, Maine. *J. Petrol.* **2015**, *56*, 273–298. [CrossRef]
4. Flinders, J.; Clemens, J.D. Non-linear dynamics, chaos, complexity and enclaves in granitoid magmas. *Earth Environ. Sci. Trans. R. Soc. Edinb.* **1996**, *87*, 217–223.
5. Ortiz, A.; Becchio, R.; Juárez, S.; Suzaño, N.; Hauser, N.; Sola, A.; Bardelli, L.; Simón, V.; Quiroga, M.; Bordese, S. Magma mixing, zircon U–Pb ages and Hf isotopes: Insights for the Miocene magmatic plumbing system in the Soroche Porphyry, Puna Argentina, Central Andes. *J. S. Am. Earth Sci.* **2021**, *109*, 103291. [CrossRef]
6. Wilcox, R.E. The Idea of Magma Mixing: History of a Struggle for Acceptance. *J. Geol.* **1999**, *107*, 421–432. [CrossRef]
7. Barbarin, B. Mafic magmatic enclaves and mafic rocks associated with some granitoids of the central Sierra Nevada batholith, California: Nature, origin, and relations with the hosts. *Lithos* **2005**, *80*, 155–177. [CrossRef]
8. Kumar, S.; Bora, S.; Sharma, U.K.; Yi, K.; Kim, N. Early Cretaceous subvolcanic calc-alkaline granitoid magmatism in the Nubra-Shyok valley of the Shyok Suture Zone, Ladakh Himalaya, India: Evidence from geochemistry and U–Pb SHRIMP zircon geochronology. *Lithos* **2017**, *277*, 33–50. [CrossRef]

9. Renjith, M.L.; Charan, S.N.; Subbarao, D.V.; Babu, E.; Rajashekhar, V.B. Grain to outcrop-scale frozen moments of dynamic magma mixing in the syenite magma chamber, Yelagiri Alkaline Complex, South India. *Geosci. Front.* **2014**, *5*, 801–820. [CrossRef]
10. Paterson, S.R.; Pignotta, G.S.; Vernon, R.H. The significance of microgranitoid enclave shapes and orientations. *J. Struct. Geol.* **2004**, *26*, 1465–1481. [CrossRef]
11. Castro, A.; Moreno-Ventas, I.; de la Rosa, J.D. H-type (hybrid) granitoids: A proposed revision of the granite-type classification and nomenclature. *Earth-Sci. Rev.* **1991**, *31*, 237–253. [CrossRef]
12. Didier, J.; Barbarin, B. *Enclaves and Granite Petrology, Developments in Petrology*; Elsevier Science: Amsterdam, The Netherlands, 1991.
13. Fernández, C.; Castro, A. Mechanical and structural consequences of magma differentiation at ascent conduits: A possible origin for some mafic microgranular enclaves in granites. *Lithos* **2018**, *320–321*, 49–61. [CrossRef]
14. Chappell, B.W.; Wyborn, D. Origin of enclaves in S-type granites of the Lachlan Fold Belt. *Lithos* **2012**, *154*, 235–247.
15. Vernon, R.H. Restite, xenoliths and microgranitoid enclaves in granites. *J. Proc. R. Soc. N. S. W.* **1983**, *116*, 77–103.
16. Acosta-Vigil, A.; Buick, I.; Hermann, J.; Cesare, B.; Rubatto, D.; London, D.; Morgan, G.B.V.I. Mechanisms of Crustal Anatexis: A Geochemical Study of Partially Melted Metapelitic Enclaves and Host Dacite, SE Spain. *J. Petrol.* **2010**, *51*, 785–821. [CrossRef]
17. Chappell, B.W.; White, A.J.R.; Wyborn, D. The Importance of Residual Source Material (Restite) in Granite Petrogenesis. *J. Petrol.* **1987**, *28*, 1111–1138. [CrossRef]
18. Collins, W.J. Evaluation of petrogenetic models for Lachlan Fold Belt granitoids: Implications for crustal architecture and tectonic models. *Aust. J. Earth Sci.* **1998**, *45*, 483–500.
19. Dahlquist, J.A. Mafic microgranular enclaves: Early segregation from metaluminous magma (Sierra de Chepes), Pampean Ranges, NW Argentina. *J. S. Am. Earth Sci.* **2002**, *15*, 643–655.
20. Donaire, T.; Pascual, E.; Pin, C.; Duthou, J.-L. Microgranular enclaves as evidence of rapid cooling in granitoid rocks: The case of the Los Pedroches granodiorite, Iberian Massif, Spain. *Contrib. Mineral. Petrol.* **2005**, *149*, 247–265.
21. Huang, H.; Niu, Y.L.; Nowell, G.; Zhao, Z.D.; Yu, X.H.; Zhu, D.C.; Mo, X.X.; Ding, S. Geochemical constraints on the petrogenesis of granitoids in the East Kunlun Orogenic belt, northern Tibetan Plateau: Implications for continental crust growth through syn-collisional felsic magmatism. *Chem. Geol.* **2014**, *370*, 1–18. [CrossRef]
22. Niu, Y.L.; Zhao, Z.D.; Zhu, D.C.; Mo, X.X. Continental collision zones are primary sites for net continental crust growth–A testable hypothesis. *Earth-Sci. Rev.* **2013**, *127*, 96–110. [CrossRef]
23. Rodríguez, C.; Pereira, M.F.; Martí, J.; Bach, N.; Scahmuells, N. Ultra-high temperatures recorded in parental magmas from hybrid zones. *GSA Bull.* **2025**. [CrossRef]
24. Rodríguez, C.; Aragón, E.; Díaz-Alvarado, J.; Castro, A.; Pedreira, R.; Sánchez-Navas, A.; Martí, J. Fractionation by compositional magma splitting: An example from Cerro Munro, Argentina. *Lithos* **2021**, *400–401*, 106396. [CrossRef]
25. Chen, B.; Chen, Z.C.; Jahn, B.M. Origin of mafic enclaves from the Taihang Mesozoic orogen, north China craton. *Lithos* **2009**, *110*, 343–358. [CrossRef]
26. Guo, F.; Nakamuru, E.; Fan, W.; Kobayoshi, K.; Li, C. Generation of Palaeocene Adakitic Andesites by Magma Mixing; Yanji Area, NE China. *J. Petrol.* **2007**, *48*, 661–692. [CrossRef]
27. Plail, M.; Edmonds, M.; Woods, A.W.; Barclay, J.; Humphreys, M.C.S.; Herd, R.A.; Christopher, T. Mafic enclaves record syn-eruptive basalt intrusion and mixing. *Earth Planet. Sci. Lett.* **2018**, *484*, 30–40.
28. Abdel-Karim, A.-A.M.; Azer, M.K.; Mogahed, M.M. Neoproterozoic concentric intrusive complex of gabbro-diorite-tonalite-granodiorite association, Rahaba area, southern Eastern Desert of Egypt: Implications for magma mixing of arc intrusive rocks. *Lithos* **2021**, *404–405*, 106423. [CrossRef]
29. Mandon, C.L.; Christenson, B.W.; Seward, T.M.; Schipper, C.I. Magma mixing, degassing and late sulfide saturation: Insights into the 1976–2000 eruptive sequence at White Island, New Zealand. *J. Volcanol. Geotherm. Res.* **2021**, *417*, 107299. [CrossRef]
30. Perugini, D. *What is Magma Mixing? The Mixing of Magmas: Field Evidence, Numerical Models, Experiments*; Springer: Cham, Switzerland, 2021; pp. 3–12.
31. Chen, S.; Niu, Y.; Li, J.; Sun, W.; Zhang, Y.; Hu, Y.; Shao, F. Syn-collisional adakitic granodiorites formed by fractional crystallization: Insights from their enclosed mafic magmatic enclaves (MMEs) in the Qumushan pluton, North Qilian Orogen at the northern margin of the Tibetan Plateau. *Lithos* **2016**, *248–251*, 455–468. [CrossRef]
32. Holden, P.; Halliday, A.N.; Stephens, W.E. Neodymium and strontium isotope content of microdiorite enclaves points to mantle input to granitoid production. *Nature* **1987**, *330*, 53–56.
33. Song, S.G.; Niu, Y.L.; Su, L.; Xia, X.H. Tectonics of the North Qilian orogen, NW China. *Gondwana Res.* **2013**, *23*, 1378–1401. [CrossRef]
34. Tao, L.; Zhang, H.; Wu, J.; Zhou, X.; Zhang, L. Magma Generation of Magnetite-Rich Intermediate-Mafic Rocks and Its Mantle Processes in the Southwestern Alxa Block, NW China. *J. Earth Sci.* **2022**, *33*, 161–176. [CrossRef]
35. Chen, Y.X.; Song, S.G.; Niu, Y.l.; Wei, C.J. Melting of continental crust during subduction initiation: A case study from the Chaidanuo peraluminous granite in the North Qilian suture zone. *Geochim. Cosmochim. Acta* **2014**, *132*, 311–336. [CrossRef]

36. Fu, C.; Yan, Z.; Aitchison, J.C.; Xiao, W.; Buckman, S.; Wang, B.; Li, W.; Li, Y.; Ren, H. Multiple subduction processes of the Proto-Tethyan Ocean: Implication from Cambrian intrusions along the North Qilian suture zone. *Gondwana Res.* **2020**, *87*, 207–223. [CrossRef]
37. Peng, Y.B.; Yu, S.Y.; Zhang, J.X.; Li, S.Z.; Tong, L.X.; Sun, D.Y. Early Paleozoic arc magmatism and metamorphism in the northern Qilian Block, western China: Petrological and geochronological constraints. *Geol. J.* **2017**, *52*, 339–364. [CrossRef]
38. Peng, Y.B.; Yu, S.Y.; Zhang, J.X.; Li, Y.S.; Li, S.Z.; Lv, P. Building a continental arc section: Constraints from Paleozoic granulite-facies metamorphism, anatexis, and magmatism in the northern margin of the Qilian Block, northern Tibet Plateau. *GSA Bull.* **2022**, *134*, 1301–1318. [CrossRef]
39. Qiao, J.R.; Dong, J.; Song, S.G.; Allen, M.B.; Wang, C.; Xia, X.H.; Su, L. Development of a Modern-Style Trench-Arc-Backarc System in the Proto-Tethys Ocean (Qilian Orogenic Belt, NW China). *J. Petrol.* **2023**, *64*, egad029. [CrossRef]
40. Wu, C.L.; Xu, X.Y.; Gao, Q.M.; Li, X.M.; Lei, M.; Gao, Y.H.; Frost, R.B.; Wooden, J.L. Early Palaezoic grranitoid magmatism and tectonic evolution in North Qilian, NW China. *Acta Petrol. Sin.* **2010**, *26*, 1027–1044.
41. Xia, L.Q.; Li, X.M.; Yu, J.Y.; Wang, G.Q. Mid-late neoproterozoic to early paleozoic volcanism and tectonic evolution of the Qilianshan, NW China. *GeoResJ* **2016**, *9–12*, 1–41. [CrossRef]
42. Chen, S.; Niu, Y.; Sun, W.; Zhang, Y.; Li, J.; Guo, P.; Sun, P. On the origin of mafic magmatic enclaves (MMEs) in syn-collisional granitoids: Evidence from the Baojishan pluton in the North Qilian Orogen, China. *Mineral. Petrol.* **2015**, *109*, 577–596. [CrossRef]
43. Tseng, C.Y.; Yang, H.J.; Yang, H.Y.; Liu, D.Y.; Wu, C.L.; Cheng, C.K.; Chen, C.H.; Ker, C.M. Continuity of the North Qilian and North Qinling orogenic belts, Central Orogenic System of China: Evidence from newly discovered Paleozoic adakitic rocks. *Gondwana Res.* **2009**, *16*, 285–293. [CrossRef]
44. Yan, M.Q.; Li, G.M.; Shi, W.J.; Huizenga, J.M.; Turlin, F.; Xu, C.W.; Moritz, R.; Wei, J.H. Recognizing crystal accumulation in plutonic system: Evidence from mafic microgranular enclaves in Early Silurian Qingshan granodiorite, North Qilian Orogenic Belt, NW China. *Lithos* **2024**, *468*, 107518.
45. Yang, H.; Zhang, H.F.; Xiao, W.J.; Luo, B.J.; Gao, Z.; Tao, L.; Zhang, L.Q.; Guo, L. Petrogenesis of Early Paleozoic high Sr/Y intrusive rocks from the North Qilian orogen: Implication for diachronous continental collision. *Lithosphere* **2020**, *12*, 53–73.
46. Yu, S.Y.; Zhang, J.X.; Qin, H.P.; Sun, D.Y.; Zhao, X.L.; Cong, F.; Li, Y.S. Petrogenesis of the early Paleozoic low-Mg and high-Mg adakitic rocks in the North Qilian orogenic belt, NW China: Implications for transition from crustal thickening to extension thinning. *J. Asian Earth Sci.* **2015**, *107*, 122–139.
47. Zhu, X.H.; Chen, D.L.; Feng, Y.M.; Yunfei, R.E.N.; Zhang, X. Granitic magmatism and tectonic evolution in the Qilian Mountain Range in NW China: A review. *Earth Sci. Front.* **2022**, *29*, 241.
48. Zhang, L.Q.; Zhang, H.F.; Zhang, S.S.; Xiong, Z.L.; Luo, B.J.; Yang, H.; Pan, F.B.; Zhou, X.C.; Xu, W.C.; Guo, L. Lithospheric delamination in post-collisional setting: Evidence from intrusive magmatism from the North Qilian orogen to southern margin of the Alxa block, NW China. *Lithos* **2017**, *288*, 20–34.
49. Wang, N.; Wu, C.L.; Lei, M.; Chen, H.J. Petrogenesis and tectonic implications of the Early Paleozoic granites in the western segment of the North Qilian orogenic belt, China. *Lithos* **2018**, *312–313*, 89–107. [CrossRef]
50. Zhang, J.X.; Yu, S.Y.; Li, Y.S.; Yu, X.X.; Lin, Y.H.; Mao, X.H. Subduction, accretion and closure of Proto-Tethyan Ocean: Early Paleozoic accretion/collision orogeny in the Altun-Qilian-North Qaidam orogenic system. *Acta Petrol. Sin.* **2015**, *31*, 3531–3554.
51. Huang, H.; Niu, Y.L.; Nowell, G.; Zhao, Z.D.; Yu, X.H.; Mo, X.X. The nature and history of the Qilian Block in the context of the development of the Greater Tibetan Plateau. *Gondwana Res.* **2015**, *28*, 209–224. [CrossRef]
52. Peng, Y.B.; Yu, S.Y.; Zhang, J.X.; Li, S.Z.; Sun, D.Y.; Tong, L.X. Early Paleozoic arc magmatism and metamorphism in the northern Qilian Block, western China: A case study of Menyuan-Kekeli. *Acta Petrol. Sin.* **2017**, *33*, 3925–3941.
53. Yan, Z.; Fu, C.L.; Aitchison, J.C.; Niu, M.L.; Buckman, S.; Xiao, W.J.; Zhou, R.J.; Chen, L.; Li, J.H. Arc-continent collision during culmination of Proto-Tethyan Ocean closure in the Central Qilian belt, NE Tibetan Plateau. *Geol. Soc. Am. Bull.* **2022**, *134*, 3079–3098. [CrossRef]
54. Yu, S.Y.; Peng, Y.B.; Zhang, J.X.; Li, S.Z.; Santosh, M.; Li, Y.S.; Liu, Y.J.; Gao, X.Y.; Ji, W.T.; Lv, P. Tectono-thermal evolution of the Qilian orogenic system: Tracing the subduction, accretion and closure of the Proto-Tethys Ocean. *Earth-Sci. Rev.* **2021**, *215*, 103547. [CrossRef]
55. Zhao, L.M.; Li, Y.L.; Cheng, S.; Li, Z.Y.; Zheng, J.P.; Qiu, H.N.; Bai, X.J.; Xiao, W.J.; Brouwer, F.M. Geochronology and geochemistry of early Paleozoic magmatism in the Qilian orogen: Constraints on closure of the Proto-Tethys Ocean. *Gondwana Res.* **2024**, *126*, 223–242. [CrossRef]
56. Zhang, J.X.; Yu, S.Y.; Mattinson, C.G. Early Paleozoic polyphase metamorphism in northern Tibet, China. *Gondwana Res.* **2017**, *41*, 267–289. [CrossRef]
57. Dong, Y.; Sun, S.; Santosh, M.; Zhao, J.; Sun, J.; He, D.; Shi, X.; Hui, B.; Cheng, C.; Zhang, G. Central China orogenic belt and amalgamation of East Asian continents. *Gondwana Res.* **2021**, *100*, 131–194. [CrossRef]

58. Zhang, J.X.; Mattinson, C.; Yu, S.Y.; Li, Y.S.; Yu, X.X.; Mao, X.H.; Lu, Z.L.; Peng, Y.B. Two contrasting accretion v. collision orogenies: Insights from Early Paleozoic polyphase metamorphism in the Altun–Qilian–North Qaidam orogenic system, NW China. *Geol. Soc. Lond. Spec. Publ.* **2019**, *474*, 153–181. [CrossRef]
59. Wang, C.; Song, S.G.; Zhao, G.C.; Allen, M.B.; Su, L.; Gao, T.Y.; Wen, T.; Feng, D. Calc-Alkaline Plutons in a Proto-Tethyan Intra-Oceanic Arc (Qilian Orogen, NW China): Implications for the Construction of Arc Crust. *J. Petrol.* **2023**, *64*, egac131. [CrossRef]
60. Fu, D.; Huang, B.; Johnson, T.E.; Wilde, S.A.; Jourdan, F.; Polat, A.; Windley, B.F.; Hu, Z.; Kusky, T. Boninitic blueschists record subduction initiation and subsequent accretion of an arc–forearc in the northeast Proto-Tethys Ocean. *Geology* **2022**, *50*, 10–15. [CrossRef]
61. Li, X.Y.; Li, S.Z.; Yu, S.Y.; Santosh, M.; Zhao, S.J.; Guo, X.Y.; Cao, H.H.; Wang, Y.M.; Huang, Z.B. Early Paleozoic arc–back-arc system in the southeastern margin of the North Qilian Orogen, China: Constraints from geochronology, and whole-rock elemental and Sr-Nd-Pb-Hf isotopic geochemistry of volcanic suites. *Gondwana Res.* **2018**, *59*, 9–26. [CrossRef]
62. Chen, Y.X.; Cui, Y.; Song, S.G.; Fu, S.M.; Zhang, L.p.; Sun, W.D.; Xiao, T.F. Late Cambrian tonalite-trondhjemite association in the eastern segment of North Qilian suture zone: Petrogenesis and geodynamic implications. *Int. Geol. Rev.* **2022**, *64*, 1431–1449. [CrossRef]
63. Chen, Y.X.; Cui, Y.; Song, S.G.; Wu, K.; Sun, W.D.; Xiao, T.F. Petrogenesis and tectonic implications of cambrian Nb-enriched I-and aluminous A-type granites in the North Qilian suture zone. *Int. Geol. Rev.* **2021**, *63*, 1090–1109.
64. Wu, C.L.; Gao, Y.H.; Frost, B.R.; Robinson, P.T.; Wooden, J.L.; Wu, S.P.; Chen, Q.L.; Lei, M. An early Palaeozoic double-subduction model for the North Qilian oceanic plate: Evidence from zircon SHRIMP dating of granites. *Int. Geol. Rev.* **2011**, *53*, 157–181.
65. Wu, C.L.; Yang, J.S.; Yang, H.Y.; Wooden, J.; Shi, R.D.; Chen, S.Y.; Zheng, Q.G. Dating of two types of granite from north Qilian, China. *Acta Petrol. Sin.* **2004**, *20*, 425–432.
66. Tung, K.A.; Yang, H.Y.; Yang, H.J.; Smith, A.; Liu, D.Y.; Zhang, J.X.; Wu, C.L.; Shau, Y.H.; Wen, D.J.; Tseng, C.Y. Magma sources and petrogenesis of the early–middle Paleozoic backarc granitoids from the central part of the Qilian block, NW China. *Gondwana Res.* **2016**, *38*, 197–219.
67. Xiao, W.J.; Windley, B.F.; Yong, Y.; Yan, Z.; Yuan, C.; Liu, C.Z.; Li, J.L. Early Paleozoic to Devonian multiple-accretionary model for the Qilian Shan, NW China. *J. Asian Earth Sci.* **2009**, *35*, 323–333. [CrossRef]
68. Yan, Z.; Fu, C.L.; Aitchison, J.C.; Buckman, S.; Niu, M.L.; Cao, B.; Sun, Y.; Guo, X.Q.; Wang, Z.Q.; Zhou, R.J. Retro-foreland basin development in response to Proto-Tethyan Ocean closure, NE Tibet Plateau. *Tectonics* **2019**, *38*, 4229–4248.
69. Li, Y.L.; Xiao, W.J.; Li, Z.Y.; Wang, K.; Zheng, J.P.; Brouwer, F.M. Early Neoproterozoic magmatism in the Central Qilian block, NW China: Geochronological and petrogenetic constraints for Rodinia assembly. *Bulletin* **2020**, *132*, 2415–2431.
70. Li, Y.L.; Xiao, W.J.; Zheng, J.P.; Brouwer, F.M. Northward subduction of the South Qilian ocean: Insights from early Paleozoic magmatism in the South-Central Qilian belts. *Geosyst. Geoenviron.* **2022**, *1*, 100013. [CrossRef]
71. Li, Z.Y.; Li, Y.L.; Xiao, W.J.; Zheng, J.P.; Brouwer, F.M. Geochemical and zircon U-Pb-Hf isotopic study of metasedimentary rocks from the Huangyuan Group of the Central Qilian block (NW China): Implications for paleogeographic reconstruction of Rodinia. *Precambrian Res.* **2020**, *351*, 105947. [CrossRef]
72. Fu, G.M.; Su, J.P.; Hu, N.G.; Wang, Y.B. Geochemistry and tectonic setting of amphibolites in the Paleoproterozoic Beidahe Group-complex in the western segment of the Qilian Mountains. *Geol. China* **2005**, *32*, 546–547.
73. Wang, L.; Yu, F.; Liu, Z.X.; Xu, X.F.; Jia, Y.Y. Reconstruction of protoliths of metamorphic rocks of the Tuolai Group and its tectonic setting. *Geol. Bull. China* **2016**, *35*, 1448–1455.
74. Tung, K.A.; Yang, H.Y.; Liu, D.Y.; Zhang, J.X.; Yang, H.J.; Shau, Y.H.; Tseng, C.Y. The amphibolite-facies metamorphosed mafic rocks from the Maxianshan area, Qilian block, NW China: A record of early Neoproterozoic arc magmatism. *J. Asian Earth Sci.* **2012**, *46*, 177–189.
75. Li, Y.L.; Xiang, H.; Zheng, J.P.; Qiu, H.N.; Bai, X.J.; Brouwer, F.M. Petrology and P-T-t Path of Huangyuan Group and Maxianshan Group in the Central Qilian Block, NW China: Implications for Tectonic Evolution of the Proto-Tethys Ocean. *J. Petrol.* **2022**, *63*, egac077. [CrossRef]
76. Tung, K.A.; Yang, H.Y.; Liu, D.Y.; Zhang, J.X.; Yang, H.J.; Shau, Y.H.; Tseng, C.Y. The Neoproterozoic granitoids from the Qilian block, NW China: Evidence for a link between the Qilian and South China blocks. *Precambrian Res.* **2013**, *235*, 163–189.
77. Li, Z.Y.; Li, Y.L.; Zhao, L.M.; Zheng, J.P.; Brouwer, F.M. Petrology and metamorphic PT paths of metamorphic zones in the Huangyuan Group, Central Qilian block, NW China. *J. Earth Sci.* **2019**, *30*, 1280–1292.
78. Li, X.; Wang, C.; Wu, M.; Zhang, J.; Zhu, X.; Meert, J.G.; Sun, X.; Li, H.; Hao, J.; Yu, Z. Crystal accumulation induced Sr/Y variation: Insights from a rear-arc pluton in the western Central Qilian Belt, NW China. *J. Asian Earth Sci.* **2023**, *256*, 105806.
79. Liu, C.F.; Wu, C.; Song, Z.J.; Liu, W.C.; Zhang, H.Y. Petrogenesis and tectonic significance of Early Paleozoic magmatism in the northern margin of the Qilian block, northeastern Tibetan Plateau. *Lithosphere* **2019**, *11*, 365–385.

80. Ma, D.Q.; Wang, C.; Li, H.; Zhu, X.H.; Hao, J.B.; Yu, Z.P.; Sun, X.K.; Zhang, S.; Li, X. Petrogenesis of mafic-intermediate dykes from the Central Qilian belt, NW China: Significance of the role of subducted compositions in sub-arc mantle. *Geol. J.* **2023**, *58*, 2790–2804.
81. Yang, H.; Zhang, H.F.; Luo, B.J.; Gao, Z.; Guo, L.; Xu, W.C. Generation of peraluminous granitic magma in a post-collisional setting: A case study from the eastern Qilian orogen, NE Tibetan Plateau. *Gondwana Res.* **2016**, *36*, 28–45.
82. Yang, H.; Zhang, H.F.; Xiao, W.J.; Tao, L.; Gao, Z.; Luo, B.J.; Zhang, L.Q. Multiple Early Paleozoic granitoids from the southeastern Qilian orogen, NW China: Magma responses to slab roll-back and break-off. *Lithos* **2021**, *380*, 105910.
83. Yong, Y.; Xiao, W.J.; Yuan, C.; Yan, Z.; Li, J.L. Geochronology and geochernistry of Paleozoic granitic plutons from the eastern Central Qilian and their tectonic implications. *Acta Petrol. Sin.* **2008**, *24*, 855–866.
84. Deer, W.A.; Howie, R.A.; Zussman, J. *An introduction to the rock-forming minerals*; Longman Scientific & Technical: Essex, UK, 1992.
85. Leake, B.; Woolley, A.; Birch, W.; Grice, J.; Hawthorne, F.; Kato, A.; Kisch, H.; Vladimir, K.; Laird, J.; Mandarino, J.; et al. Nomenclature of amphiboles: Report of the subcommittee on amphiboles of the International. *Can. Mineral.* **1997**, *35*, 219–246.
86. Jiang, C.Y.; An, S.Y. On chemical characteristics of calcific amphiboles from igneous rocks and their petrogenesis significance. *J. Mineral. Petrol.* **1984**, *3*, 1–9.
87. Mutch, E.J.F.; Blundy, J.D.; Tattitch, B.C.; Cooper, F.J.; Brooker, R.A. An experimental study of amphibole stability in low-pressure granitic magmas and a revised Al-in-hornblende geobarometer. *Contrib. Mineral. Petrol.* **2016**, *171*, 1–27. [CrossRef]
88. Liao, Y.; Wei, C.J.; Rehman, H.U. Titanium in calcium amphibole: Behavior and thermometry. *Am. Mineral.* **2021**, *106*, 180–191. [CrossRef]
89. Molina, J.F.; Moreno, J.A.; Castro, A.; Rodríguez, C.; Fershtater, G.B. Calcic amphibole thermobarometry in metamorphic and igneous rocks: New calibrations based on plagioclase/amphibole Al-Si partitioning and amphibole-liquid Mg partitioning. *Lithos* **2015**, *232*, 286–305.
90. Tulloch, A.J.; Challis, G.A. Emplacement depths of Paleozoic-Mesozoic plutons from western New Zealand estimated by hornblende-AI geobarometry. *N. Z. J. Geol. Geophys.* **2000**, *43*, 555–567. [CrossRef]
91. Leake, B.E.; Wooley, A.R.; Arps, C.E.S.; Birch, W.D.; Gilbert, M.C.; Grice, J.D.; Hawthorne, F.C.; Kato, A.; Kisch, H.J.; Krivovichec, V.G.; et al. Nomenclature of amphiboles; report of the Subcommittee on Amphiboles of the International Mineralogical Association Commission on New Minerals and Mineral Names. *Eur. J. Mineral.* **1997**, *9*, 623–651.
92. Xie, Y.W.; Zhang, Y.Q. Peculiarities and genetic significance of hornblende from granite in the Hengduansan region. *Acta Mineral. Sin.* **1990**, *10*, 35–45.
93. Hoskin, P.W.O.; Schaltegger, U. The composition of zircon and igneous and metamorphic petrogenesis. *Rev. Mineral. Geochem.* **2003**, *53*, 27–62.
94. Middlemost, E.A.K. Naming materials in the magma/igneous rock system. *Earth-Sci. Rev.* **1994**, *37*, 215–224. [CrossRef]
95. Wager, L.R.; Deer, W.A. Geological investigations in East Greenland. Part III. The petrology of the Skaergaard intrusion, Kangerdlugssuag, East Geenland. *Meddelser Gronl.* **1939**, *105*, 1–352.
96. Rollinson, H.; Pease, V. *Using Geochemical Data: To Understand Geological Processes*, 2nd ed.; Cambridge University Press: Cambridge, UK, 2021.
97. Maniar, P.D.; Piccoli, P.M. Tectonic discrimination of granitoids. *Geol. Soc. Am. Bull.* **1989**, *101*, 635–643.
98. Boynton, W.V. Cosmochemistry of the rare earth elements: Meteorite studies. In *Developments in Geochemistry*; Elsevier: Amsterdam, The Netherlands, 1984; Volume 2, pp. 63–114.
99. Sun, S.S.; McDonough, W.F. Chemical and isotopic systematics of oceanic basalts: Implications for mantle composition and processes. *Geol. Soc. Lond. Spec. Publ.* **1989**, *42*, 313–345.
100. Wan, Y.S.; Zhang, J.X.; Yang, J.S.; Xu, Z.Q. Geochemistry of high-grade metamorphic rocks of the North Qaidam mountains and their geological significance. *J. Asian Earth Sci.* **2006**, *28*, 174–184.
101. Zhang, H.F.; Zhang, B.R.; Harris, N.; Zhang, L.; Chen, Y.L.; Chen, N.S.; Zhao, Z.D. U–Pb zircon SHRIMP ages, geochemical and Sr–Nd–Pb isotopic compositions of intrusive rocks from the Longshan–Tianshui area in the southeast corner of the Qilian orogenic belt, China: Constraints on petrogenesis and tectonic affinity. *J. Asian Earth Sci.* **2006**, *27*, 751–764.
102. Qian, Q.; Zhang, Q. Geochemical Characteristics and Tectonic Setting of the Laohushan Basalts, North Qilian Mountains. *Acta Geol. Sin.* **2001**, *75*, 452–457. [CrossRef]
103. Xu, W.C. U–Pb Zircon Geochronology, Geochemical and Sr–Nd–Pb Isotopic Compositions of the Hualong Group in Qilian Mountains: Constraints on Tectonic Evolution of Crustal Basement. Ph.D. Thesis, China University of Geosciences, Wuhan, China, 2007.
104. Yan, Z.; Aitchison, J.; Fu, C.L.; Guo, X.Q.; Niu, M.L.; Xia, W.J.; Li, J.L. Hualong Complex, South Qilian terrane: U–Pb and Lu–Hf constraints on Neoproterozoic micro-continental fragments accreted to the northern Proto-Tethyan margin. *Precambrian Res.* **2015**, *266*, 65–85.

105. Li, Y.L.; Tong, X.; Zhu, Y.H.; Lin, J.W.; Zheng, J.P.; Brouwer, F.M. Tectonic affinity and evolution of the Precambrian Qilian Block: Insights from petrology, geochemistry and geochronology of the Hualong Group in the Qilian Orogen, NW China. *Precambrian Res.* **2018**, *315*, 179–200.
106. Tao, L.; Zhang, H.; Gao, Z.; Yang, H.; Zhang, L.; Guo, L.; Pan, F. Across-arc geochemical and Sr–Nd–Hf isotopic variations of mafic intrusive rocks at the southern Central Qilian block, China. *Gondwana Res.* **2018**, *59*, 108–125.
107. Yang, H.; Zhang, H.F.; Luo, B.J.; Zhang, J.; Xiong, Z.L.; Guo, L.; Pan, F.B. Early Paleozoic intrusive rocks from the eastern Qilian orogen, NE Tibetan Plateau: Petrogenesis and tectonic significance. *Lithos* **2015**, *224*, 13–31.
108. Wang, T.; Wang, Z.Q.; Yan, Z.; Ma, Z.H.; He, S.F.; Fu, C.L.; Wang, D.S. Geochronological and geochemical evidence of amphibolite from the Hualong Group, northwest China: Implication for the early Paleozoic accretionary tectonics of the Central Qilian belt. *Lithos* **2016**, *248*, 12–21.
109. Chen, Y.X.; Xia, X.H.; Song, S.G. Petrogenesis of Aoyougou high-silica adakite in the North Qilian orogen, NW China: Evidence for decompression melting of oceanic slab. *Chin. Sci. Bull.* **2012**, *57*, 2289–2301.
110. Hou, Q.; Zhao, Z.; Zhang, H.; Zhang, B.; Chen, Y. Indian Ocean-MORB-type isotopic signature of Yushigou ophiolite in North Qilian Mountains and its implications. *Sci. China Ser. D Earth Sci.* **2006**, *49*, 561–572. [CrossRef]
111. Hou, Q.Y.; Zhao, Z.; Zhang, B.R.; Zhang, H.; Zhang, L.; Chen, Y. On the boundary of Tethyan tectonic domain on northeastern margin of the Tibetan Plateau. *Acta Petrol. Sin.* **2006**, *22*, 567–577.
112. Wang, K.X.; Yu, C.D.; Yan, J.; Liu, X.D.; Liu, W.H.; Pan, J.Y. Petrogenesis of Early Silurian granitoids in the Longshoushan area and their implications for the extensional environment of the North Qilian Orogenic Belt, China. *Lithos* **2019**, *342*, 152–174.
113. Chappell, B.W.; Stephens, W.E. Origin of infracrustal (I-type) granite magmas. *Earth Environ. Sci. Trans. R. Soc. Edinb.* **1988**, *79*, 71–86.
114. Castro, A.; Rodríguez, C.; Gutiérrez-Alonso, G.; de la Rosa, J.D. A post-collisional batholith from Southern Iberia rooted in the Earth's mantle: Los Pedroches batholith. *Lithos* **2023**, *454–455*, 107245. [CrossRef]
115. Troll, V.R.; Schmincke, H.-U. Magma Mixing and Crustal Recycling Recorded in Ternary Feldspar from Compositionally Zoned Peralkaline Ignimbrite 'A', Gran Canaria, Canary Islands. *J. Petrol.* **2002**, *43*, 243–270. [CrossRef]
116. Sun, Y.; Ma, C.; Liu, B. Important role of magma mixing in generating the Late Cretaceous Shima intrusion along the Middle-Lower Yangtze River belt: Evidence from petrology, geochemistry, and zircon U-Pb-Hf isotopes. *Lithos* **2021**, *390–391*, 106143. [CrossRef]
117. Wiebe, R.A.; Smith, D.; Sturm, M.; King, E.M.; Seckler, M.S. Enclaves in the Cadillac Mountain Granite (Coastal Maine): Samples of Hybrid Magma from the Base of the Chamber. *J. Petrol.* **1997**, *38*, 393–423. [CrossRef]
118. Whalen, J.; Currie, K.; Chappell, B. A-type granites: Geochemical characteristics, discrimination and petrogenesis. *Contrib. Mineral. Petrol.* **1987**, *95*, 407–419. [CrossRef]
119. DePaolo, D.J. Trace element and isotopic effects of combined wallrock assimilation and fractional crystallization. *Earth Planet. Sci. Lett.* **1981**, *53*, 189–202. [CrossRef]
120. Maas, R.; Nicholls, I.A.; Legg, C. Igneous and Metamorphic Enclaves in the S-type Deddick Granodiorite, Lachlan Fold Belt, SE Australia: Petrographic, Geochemical and Nd-Sr Isotopic Evidence for Crustal Melting and Magma Mixing. *J. Petrol.* **1997**, *38*, 815–841. [CrossRef]
121. Rudnick, R.; Gao, S. Composition of the Continental Crust. Treatise Geochem 3:1-64. *Treatise Geochem.* **2003**, *3*, 1–64. [CrossRef]
122. Meng, Y.K.; Xiong, F.H.; Xu, Z.Q.; Ma, X.X. Petrogenesis of Late Cretaceous mafic enclaves and their host granites in the Nyemo region of southern Tibet: Implications for the tectonic-magmatic evolution of the Central Gangdese Belt. *J. Asian Earth Sci.* **2019**, *176*, 27–41. [CrossRef]
123. Chappell, B.W.; White, A.J.R.; Williams, I.S.; Wyborn, D.; Hergt, J.M.; Woodhead, J.D. Discussion and Reply: Evaluation of petrogenetic models for Lachlan Fold Belt granitoids: Implications for crustal architecture and tectonic models. *Aust. J. Earth Sci.* **1999**, *46*, 827–836. [CrossRef]
124. White, A.J.R.; Chappell, B.W.; Wyborn, D. Application of the Restite Model to the Deddick Granodiorite and its Enclaves: A Reinterpretation of the Observations and Data of Maas et al. (1997). *J. Petrol.* **1999**, *40*, 413–421. [CrossRef]
125. Clemens, J.D.; Stevens, G. What controls chemical variation in granitic magmas? *Lithos* **2012**, *134*, 317–329. [CrossRef]
126. Feeley, T.C.; Wilson, L.F.; Underwood, S.J. Distribution and compositions of magmatic inclusions in the Mount Helen dome, Lassen Volcanic Center, California: Insights into magma chamber processes. *Lithos* **2008**, *106*, 173–189. [CrossRef]
127. Gao, P.; Zheng, Y.F.; Zhao, Z.F. Experimental melts from crustal rocks: A lithochemical constraint on granite petrogenesis. *Lithos* **2016**, *266–267*, 133–157. [CrossRef]
128. Martin, H.; Smithies, R.H.; Rapp, R.P.; Moyen, J.-F.; Champion, D. An overview of adakite, tonalite-trondhjemiten-granodiorite (TTG), and sanukitoid: Relationships and some implications for crustal evolution. *Lithos* **2005**, *79*, 1–24. [CrossRef]
129. Rapp, R.P.; Watson, E.B. Dehydration Melting of Metabasalt at 8–32 kbar: Implications for Continental Growth and Crust-Mantle Recycling. *J. Petrol.* **1995**, *36*, 891–931. [CrossRef]

130. Barbarin, B.; Didier, J. Conclusions: Enclaves and Granite Petrology. In *Enclaves and Granite Petrology*; Didier, J., Barbarin, B., Eds.; Elsevier: Amsterdam, The Netherlands, 1991; pp. 545–549.
131. Vernon, R.H. Microgranitoid enclaves in granites—Globules of hybrid magma quenched in a plutonic environment. *Nature* **1984**, *309*, 438–439.
132. Blundy, J.D.; Sparks, R.S.J. Petrogenesis of Mafic Inclusions in Granitoids of the Adamello Massif, Italy. *J. Petrol.* **1992**, *33*, 1039–1104. [CrossRef]
133. Piccoli, P.M.; Candela, P.A. Apatite in Igneous Systems. *Rev. Mineral. Geochem.* **2002**, *48*, 255–292. [CrossRef]
134. Pietranik, A.; Koepke, J. Interactions between dioritic and granodioritic magmas in mingling zones: Plagioclase record of mixing, mingling and subsolidus interactions in the Gęsiniec Intrusion, NE Bohemian Massif, SW Poland. *Contrib. Mineral. Petrol.* **2009**, *158*, 17–36.
135. Zhang, J.Y.; Ma, C.Q.; Zhang, C.; Li, J.W. Fractional crystallization and magma mixing: Evidence from porphyritic diorite-granodiorite dykes and mafic microgranular enclaves within the Zhoukoudian pluton, Beijing. *Mineral. Petrol.* **2014**, *108*, 777–800.
136. Wang, R.Q.; Qiu, J.S.; Yu, S.B.; Lin, L.; Xu, H. Magma mixing origin for the Quxu intrusive complex in southern Tibet: Insights into the early Eocene magmatism and geodynamics of the southern Lhasa subterrane. *Lithos* **2019**, *328–329*, 14–32. [CrossRef]
137. Tsuchiyama, A. Dissolution kinetics of plagioclase in the melt of the system diopside-albite-anorthite, and origin of dusty plagioclase in andesites. *Contrib. Mineral. Petrol.* **1985**, *89*, 1–16.
138. Langmuir, C.H.; Vocke, R.D.; Hanson, G.N.; Hart, S.R. A general mixing equation with applications to Icelandic basalts. *Earth Planet. Sci. Lett.* **1978**, *37*, 380–392. [CrossRef]
139. Schiano, P.; Monzier, M.; Eissen, J.P.; Martin, H.; Koga, K.T. Simple mixing as the major control of the evolution of volcanic suites in the Ecuadorian Andes. *Contrib. Mineral. Petrol.* **2010**, *160*, 297–312. [CrossRef]
140. Griffin, W.L.; Wang, X.; Jackson, S.E.; Pearson, N.J.; O'Reilly, S.Y.; Xu, X.; Zhou, X. Zircon chemistry and magma mixing, SE China: In-situ analysis of Hf isotopes, Tonglu and Pingtan igneous complexes. *Lithos* **2002**, *61*, 237–269. [CrossRef]
141. Defant, M.J.; Drummond, M.S. Derivation of some modern arc magmas by melting of young subducted lithosphere. *Nature* **1990**, *347*, 662–665. [CrossRef]
142. Hermann, J.; Spandler, C.; Hack, A.; Korsakov, A.V. Aqueous fluids and hydrous melts in high-pressure and ultra-high pressure rocks: Implications for element transfer in subduction zones. *Lithos* **2006**, *92*, 399–417. [CrossRef]
143. Weidendorfer, D.; Mattsson, H.B.; Ulmer, P. Dynamics of magma mixing in partially crystallized magma chambers: Textural and petrological constraints from the basal complex of the Austurhorn intrusion (SE Iceland). *J. Petrol.* **2014**, *55*, 1865–1903. [CrossRef]
144. Zorpi, M.J.; Coulon, C.; Orsini, J.B.; Cocirta, C. Magma mingling, zoning and emplacement in calc-alkaline granitoid plutons. *Tectonophysics* **1989**, *157*, 315–329. [CrossRef]
145. Elliott, T. Tracers of the slab. *Geophys. Monogr.-Am. Geophys. Union* **2003**, *138*, 23–46.
146. McCulloch, M.T.; Gamble, J.A. Geochemical and geodynamical constraints on subduction zone magmatism. *Earth Planet. Sci. Lett.* **1991**, *102*, 358–374. [CrossRef]
147. Saunders, A.D.; Storey, M.; Kent, R.W.; Norry, M.J. Consequences of plume-lithosphere interactions. *Geol. Soc. Lond. Spec. Publ.* **1992**, *68*, 41–60. [CrossRef]
148. Kepezhinskas, P.; McDermott, F.; Defant, M.J.; Hochstaedter, A.; Drummond, M.S.; Hawkesworth, C.J.; Koloskov, A.; Maury, R.C.; Bellon, H. Trace element and Sr-Nd-Pb isotopic constraints on a three-component model of Kamchatka arc petrogenesis. *Geochim. Cosmochim. Acta* **1997**, *61*, 577–600.
149. Wang, W.; Pandit, M.K.; Zhao, J.H.; Chen, W.T.; Zheng, J.P. Slab break-off triggered lithosphere-asthenosphere interaction at a convergent margin: The Neoproterozoic bimodal magmatism in NW India. *Lithos* **2018**, *296*, 281–296. [CrossRef]
150. Pearce, J.A. Geochemical fingerprinting of oceanic basalts with applications to ophiolite classification and the search for Archean oceanic crust. *Lithos* **2008**, *100*, 14–48. [CrossRef]
151. Ma, Z.; Liu, Z.; Liu, Y.X.; Wang, J.R. Geochemical Characteristics and Tectonic Significance of Shibandun Mafic-ultramafic Cumulates in Western Part of the Central Qilian. *Geoscience* **2018**, *32*, 667.
152. Zhang, J.X.; Xu, Z.Q.; Chen, W.; Xv, H.F. Discussion on the age of the subduction—Accretion complex/volcanic arc in the middle section of the North Qilian Mountains. *Acta Petrol. Mineral.* **1997**, *2*, 17–24.
153. Wu, C.L.; Yao, S.Z.; Yang, J.S.; Zeng, L.S.; Chen, S.Y.; Li, H.B.; Qi, X.X.; Wooden, J.L.; Mazdab, F.K. Double subduction of the Early Paleozoic North Qilian oceanic plate: Evidence from granites in the central segment of North Qilian, NW China. *Geol. China* **2006**, *33*, 1197–1208.
154. Wang, G.Q.; Li, X.M.; Xu, X.Y.; Yu, J.Y.; Sun, J.M.; Wu, P. LA-ICP-MS U-Pb dating of zircons from basic volcanic rocks in the Honggou copper polymetallic deposit, Menyuan area, Qinghai. *Geol. Bull. China* **2011**, *30*, 1060–1065.
155. Guo, Z.P.; Zhao, X.M.; Bai, Y.; Zhang, J.W.; Kong, H.l. Zircon U-Pb and molybdenite Re-Os dating of the Langlike copper deposit in North Qilian Mountain and its geological implications. *Geol. China* **2015**, *42*, 691–701.
156. Schmidt, M.W.; Jagoutz, O. The global systematics of primitive arc melts. *Geochem. Geophys. Geosystems* **2017**, *18*, 2817–2854.

157. Stern, C.R.; Kilian, R. Role of the subducted slab, mantle wedge and continental crust in the generation of adakites from the Andean Austral Volcanic Zone. *Contrib. Mineral. Petrol.* **1996**, *123*, 263–281.
158. Wade, B.P.; Barovich, K.M.; Hand, M.; Scrimgeour, I.R.; Close, D.F. Evidence for early Mesoproterozoic arc magmatism in the Musgrave Block, central Australia: Implications for Proterozoic crustal growth and tectonic reconstructions of Australia. *J. Geol.* **2006**, *114*, 43–63.
159. Cabanis, B. Le diagramme La/10-Y/15-Nb/8: Un outil pour la discrimination des séries volcaniques et la mise en évidence des processus de mélange et/ou de contamination crustale. *Comptes Rendus l'Academie Sci. Ser. 2* **1989**, *309*, 2023–2029.
160. Pearce, J.A.; Cann, J.R. Tectonic setting of basic volcanic rocks determined using trace element analyses. *Earth Planet. Sci. Lett.* **1973**, *19*, 290–300. [CrossRef]
161. Zhang, X.F.; Pang, Z.S.; Chen, L.X.; Liu, J.L.; Zhang, D.H.; Zhou, Y.; Wang, B.R.; Teng, C.; Cao, J. Petrogenesis and tectonic setting of the Early Permian gabbro–granite complex in the southeastern Central Asian Orogenic Belt, central Inner Mongolia, China. *J. Asian Earth Sci. X* **2021**, *5*, 100059.
162. Hu, X.M.; Garzanti, E.; Moore, T.; Raffi, I. Direct stratigraphic dating of India-Asia collision onset at the Selandian (middle Paleocene, 59 ± 1 Ma). *Geology* **2015**, *43*, 859–862. [CrossRef]
163. Fang, W.; Dai, L.Q.; Zheng, Y.F.; Zhao, Z.F.; Ma, L.T. Tectonic transition from oceanic subduction to continental collision: New geochemical evidence from Early-Middle Triassic mafic igneous rocks in southern Liaodong Peninsula, east-central China. *Bulletin* **2020**, *132*, 1469–1488.
164. Zhang, J.X.; Yu, S.Y.; Li, Y.S.; Mao, X.H.; Teng, X.; Wu, Y.W.; Guo, Q. Proto-Tethys orogenic system in the northern Qinghai-Tibet Plateau: Resubdivision of lithotectonic units in Qilian-North Qaidam and discussion on several controversial scientific issues. *Acta Petrol. Sin.* **2023**, *39*, 3507–3532.
165. Watson, E.B.; Harrison, T.M. Zircon saturation revisited, temperature and composition effects in a variety of crustal magma types. *Earth Planet. Sci. Lett.* **1983**, *64*, 295–304. [CrossRef]
166. Hou, K.J.; Li, Y.H.; Tian, Y.Y. In situ U-Pb zircon dating using laser ablation-multi ion counting-ICP-MS. *Miner. Depos.* **2009**, *28*, 481–492.
167. Liu, Y.S.; Gao, S.; Hu, Z.C.; Gao, C.G.; Zong, K.Q.; Wang, D.B. Continental and oceanic crust recycling-induced melt–peridotite interactions in the Trans-North China Orogen: U–Pb dating, Hf isotopes and trace elements in zircons from mantle xenoliths. *J. Petrol.* **2010**, *51*, 537–571.
168. Hou, K.J.; Li, Y.H.; Zou, T.R.; Qu, X.M.; Shi, Y.R.; Xie, G.Q. Laser ablation-MC-ICP-MS technique for Hf isotope microanalysis of zircon and its geological applications. *Acta Petrol. Sin.* **2007**, *23*, 2595–2604.
169. Griffin, W.L.; Pearson, N.J.; Belousova, E.; Jackson, S.E.; Achterbergh, E.V.; O'Reilly, S.Y.; Shee, S.R. The Hf isotope composition of Cratonic mantle: LA-MC-ICP-MS analysis of zircon megacrysts in kimberlites. *Geochim. Cosmochim. Acta* **2000**, *64*, 133–147.
170. Liu, Y.S.; Hu, Z.C.; Gao, S.; Günther, D.; Xu, J.; Gao, C.G.; Chen, H.H. In situ analysis of major and trace elements of anhydrous minerals by LA-ICP-MS without applying an internal standard. *Chem. Geol.* **2008**, *257*, 34–43.
171. Li, C.F.; Li, X.H.; Li, Q.L.; Guo, J.H.; Li, X.H.; Yang, Y.H. Rapid and precise determination of Sr and Nd isotopic ratios in geological samples from the same filament loading by thermal ionization mass spectrometry employing a single-step separation scheme. *Anal. Chim. Acta* **2012**, *727*, 54–60. [CrossRef]
172. Lai, Z.Q.; Zhao, G.T.; Han, Z.Z.; Liu, B.; Bu, X.J.; Leng, C.X. Back-arc magma processes in the Okinawa Trough: New insights from textural and compositional variations of plagioclase in basalts. *Geol. J.* **2016**, *51*, 346–356.

minerals

Article

A Geochemical and Sr–Nd–Hf–O Isotopic Study of the Early Silurian Shandan Adakites in the Longshoushan Area: Implications for the Collisional Setting of the Proto–Tethyan North Qilian Orogen, Northwest China

Zhihan Bai [1], Yang Yang [1,2,*], Xijun Liu [1,2,3], Pengde Liu [3], Gang Chen [1], Xiao Liu [1,2], Rongguo Hu [1,2], Hao Tian [1], Yande Liu [1], Wenmin Huang [1] and Yao Xiao [1]

1 Guangxi Key Laboratory of Hidden Metallic Ore Deposits Exploration, College of Earth Sciences, Guilin University of Technology, Guilin 541004, China; zhihan2525@163.com (Z.B.); xijunliu@glut.edu.cn (X.L.); herochenchenchen@gmail.com (G.C.); liuxiao@glut.edu.cn (X.L.); hurongguo@glut.edu.cn (R.H.); nmth521@163.com (H.T.); liuyd@glut.edu.cn (Y.L.); huangwenmin@glut.edu.cn (W.H.); xiaoyao@glut.edu.cn (Y.X.)

2 Collaborative Innovation Center for Exploration of Nonferrous Metal Deposits and Efficient Utilization of Resources in Guangxi, Guilin University of Technology, Guilin 541004, China

3 National Key Laboratory of Arid Area Ecological Security and Sustainable Development, Xinjiang Institute of Ecology and Geography, Chinese Academy of Sciences, Urumqi 830011, China; liupengde23@mails.ucas.ac.cn

* Correspondence: yangyang007@glut.edu.cn

Abstract: The North Qilian Orogen experienced a series of late Neoproterozoic to early Paleozoic tectonic events, including the opening and closure of the Proto-Tethyan Qilian Ocean, as well as post-subduction processes. This study investigated the Shandan adakites in the Longshoushan area of the North Qilian Orogen, focusing on zircon U–Pb geochronology, whole-rock geochemistry, and Sr–Nd–Hf–O isotopic compositions. The Shandan adakites yield ages of ca. 446–440 Ma, suggesting they crystallized during the collision between the Alxa and Qilian blocks following the closure of the Proto-Tethyan North Qilian Ocean. High Sr/Y (40.9–117) ratios and enrichments in light rare earth elements indicate that the Shandan adakites were formed by partial melting of thickened magnesian lower crust. They have relatively rich ε_{Nd} (t) (−7.66 to −6.32), ε_{Hf}(t) (3.30 to −12.4), and $\delta^{18}O$ (5.34‰–7.52‰). Zircon Hf–O and whole-rock Sr–Nd isotopes confirm significant contributions from the ancient crust and mantle-derived melts, suggesting complex crust–mantle interactions in their magma sources. We propose that the Shandan adakites formed during the (early) post-collisional stage of orogenesis. Based on regional geological evidence and previous studies, we suggest the Alxa and Central Qilian blocks collided during ca. 446–440 Ma, leading to the thickening of the lower crust. After ca. 440 Ma, the tectonic setting of the Northern Qilian Orogen transitioned from a collisional to a post-collisional stage.

Keywords: Proto-Tethys; North Qilian Orogenic Belt; adakite; Sr–Nd–Hf–O isotopes

1. Introduction

Adakites are characterized by high Sr contents, elevated Sr/Y ratios (>40), and relatively low Y and Yb contents. Adakites have long been debated in igneous petrology, with ongoing discussions about their formation processes under different environmental conditions [1,2]. Recent studies indicate that adakites can form due to (1) interactions between melts derived from subducting oceanic crust and mantle peridotites [3–7]; (2) partial melting of thickened lower crust [8–12]; (3) partial melting of delaminated lower crust [13–15];

 https://doi.org/10.3390/min15040352

(4) partial melting of subducted continental crust [16]; and (5) fractional crystallization of tholeiitic magmas [2]. In general, adakites record information on processes such as oceanic subduction, continental collision, and lithospheric extension. Studies of their petrogenesis are important for understanding crust–mantle interactions during subduction and continental crustal growth.

The Qilian Orogenic Belt is a key part of the Central Orogenic Belt in China, which was formed during the early Paleozoic [17–20]. From north to south, the Qilian Orogenic Belt is divided into the North Qilian Belt, Central Qilian Block, and Southern Qilian Belt, forming a composite orogenic system that has experienced a long and complex history [21] (Figure 1b). The Northern Qilian Belt, along the southwestern margin of the Alxa Block, consists of a series of intrusive plutons along the Longshoushan Fault that comprise the Longshoushan structural belt [22].

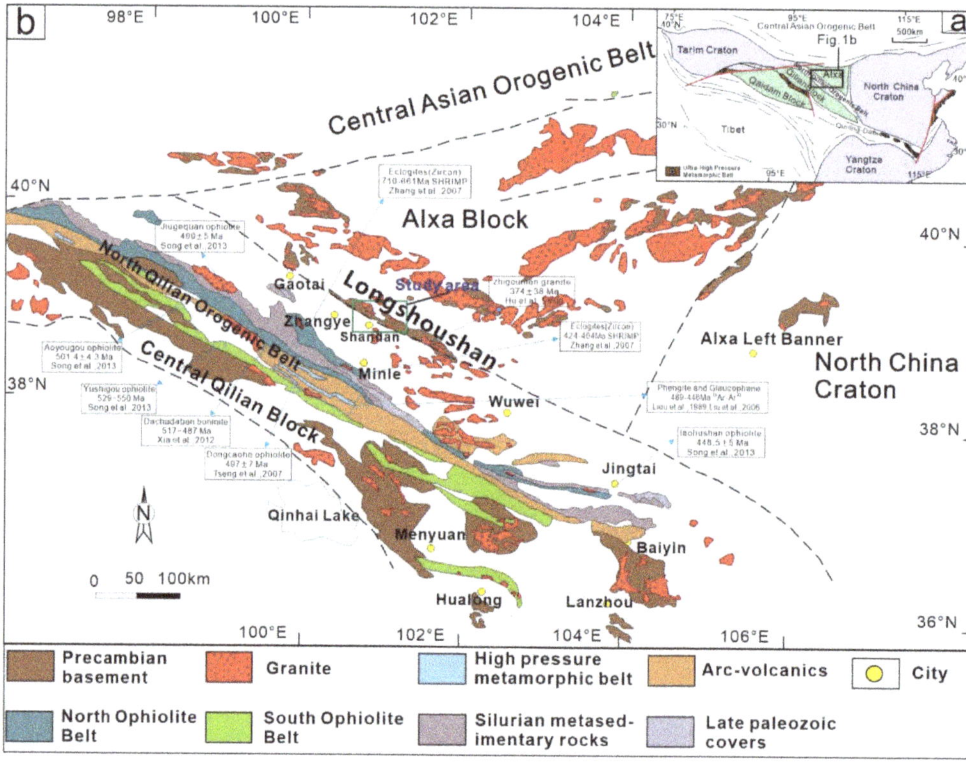

Figure 1. (**a**) Location of the Qilian Mountains. (**b**) Schematic geologic map of the North Qilian Block [18,21,23]. The zircon U–Pb ages in Fig.1b are from Hu et al. [24], Zhang et al. [25], Liou et al. [26], Liu et al. [27], Xia et al. [28], Song et al. [18], and Tseng et al. [29].

The Longshoushan area is located at the southern margin of the Alxa Block and the northern margin of the Northern Qilian Belt. It contains extensive exposures of early Paleozoic intermediate-acid magmatic rocks [23], which are considered to be related to the northward subduction of the North Qilian oceanic lithosphere and subsequent collisional processes [18,23,30]. Previous studies of the ca. 445–403 Ma intermediate-acid magmatic rocks in the Northern Qilian region suggest that magmatism after the Paleozoic closure of the North Qilian Ocean occurred in various tectonic settings, including a compressional environment associated with a collisional orogeny [31], a post-collisional extensional environment [22,31–33], and a transitional setting from compression to extension [18,34–36].

Moreover, the origins and source regions of the ca. 445–403 Ma intermediate-acid magmatic rocks in the Longshoushan area are also debated. Several differing views have been proposed regarding the origins of the granitic rocks in this region: (1) formation by partial melting of ancient Precambrian crust of the Longshoushan Group, with a minimal contribution from mantle material [30,37]; (2) mixing between adakitic magmas generated by crustal melting and mantle-derived magmas [38,39]; and (3) the silicic magmas being produced by fractional crystallization of intermediate to mafic magmas [40,41]. As such, research on the genesis petrogenesis of the adakites in the Longshoushan area is critical for resolving conflicting models of crust–mantle interaction during the Proto-Tethyan closure.

In this paper, we present zircon U–Pb geochronological, whole-rock geochemical, and Sr–Nd isotope and zircon Lu–Hf–O isotope data for the Shandan adakites from the Longshoushan area. Our aim was to investigate the age, petrogenesis, source, and tectonic significance of these rocks, thereby constraining the tectonic evolution of the Northern Qilian Orogenic Belt.

2. Geological Setting

The Longshoushan area is located at the southwestern margin of the Alxa Block in the western North China Craton [42]. It is 30 km wide and 300 km long, with its main part occurring in a NW–SE-trending belt that gradually transitions to a nearly E–W orientation in the east (Figure 2). It extends eastward from Hongyashan, and to the west is separated from the Tarim Plate by the Altin Fault. To the south, it is adjacent to the Minle Basin of the Hexi Corridor, from which it is separated by a deep fault at the southern margin of the Longshoushan. To the north, it is bounded by a deep fault between the northern margin of the Longshoushan and the Chaoshui Basin. The Longshoushan area has experienced multiple episodes of superimposed tectonic deformation, with an overall structural trend that trends approximately WNW–ESE. Its main structural features are faults and folds, followed by ductile shear zones. The Longshoushan Fault was active during the early Paleozoic Qilian Orogeny, with its southern segment exhibiting right-lateral motion and trending WNW–ESE.

The Longshoushan area consists primarily of the Late Mesoproterozoic Dunzigou Formation and Neoproterozoic Hanmushan Formation. The Dunzigou Formation consists of an upper unit of calc–silicate marble with meta chert bands, intercalated with quartzite, quartz conglomerate, and hematite lenses, and local occurrences of marble; the middle unit consists of purplish phyllite, quartzite, calc–silicate marble, and tuffaceous phyllite; and the lower unit consists of quartzite interbedded with calc–silicate marble, phyllite, and sericite–quartz schist [43]. The upper Hanmushan Formation, which is Neoproterozoic in age and exposed mainly in the northern part of Shandan County, consists of quartz sandstone, slate, and volcanic rocks that are intruded by early Paleozoic granitoids [44,45]. The Longshou Mountains consist mainly of metasedimentary rocks and granitic gneiss, The magmatic and metamorphic zircons from the granitic gneiss yield ages of 2.04–2.07 Ga and 1.89–1.93 Ga, respectively, confirming that the Longshou Mountains formed mainly in the Paleoproterozoic [46,47]. Voluminous igneous rocks crop out in the Longshoushan area, ranging from ultramafic to mafic compositions, and the granitoid intrusions have multiple phases, with magmatic events recorded from the Paleoproterozoic to Paleozoic. Numerous outcrops of Paleozoic intrusive and volcanic rocks occur in this region. For example, in the central Jiling area, exposures include gabbro, diorite, porphyritic granite, quartz monzonite, and granodiorite in the eastern Hexibao area; peridotite in the Xiaokouzi area; dioritic granite in the western Zhigoumen area; and monzogranite in the Hongshiquan area.

The study area is located ~5 km west of Shandan County in the Hexi Corridor Belt in Gansu Province and 6–10 km north of the Longshoushan Fault (Figure 2).

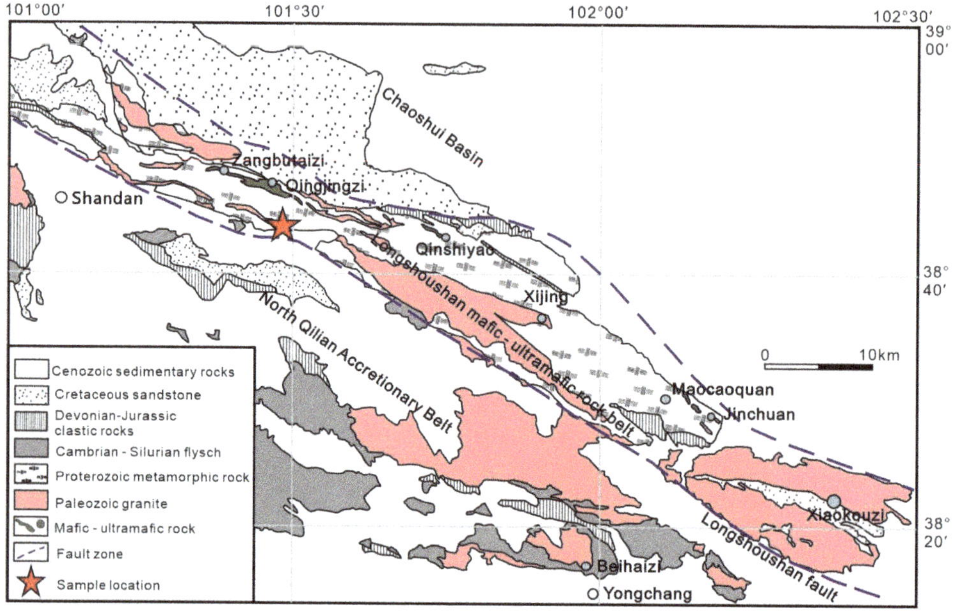

Figure 2. Simplified geological map of the Longshoushan–Shandan area (modified after Duan et al. [48]).

3. Analytical Methods

All analyses were carried out at the Guangxi Key Laboratory of Hidden Metallic Ore Deposit Exploration, Guilin University of Technology, Guangxi, China.

3.1. Zircon U–Pb Dating and Hf–O Isotopes

We separated zircon grains from three adakites (samples 21SD-280, 21SD-281, and 21SD-285) using conventional heavy liquid and magnetic methods. The zircons were mounted in epoxy resin and polished with 0.25 μm diamond paste. The internal textures of the grains were carefully documented with transmitted and reflected light photomicrographs and cathodoluminescence (CL) images. The CL imaging was performed using a JXA-8230R electron microprobe (JEOL Ltd., Tokyo, Japan).

Zircon U–Pb dating and trace element analyses were carried out simultaneously by laser ablation–inductively coupled plasma–mass spectrometry (LA–ICP–MS) comprising a quadrupole ICP–MS instrument (Agilent 7900; Agilent Technologies, Santa Clara, CA, USA) coupled to a GeoLas HD 193 nm ArF excimer laser (Lambda Physik, Göttingen, Germany) with an automatic positioning system. The laser spots were 24 μm in diameter. Each analysis included 20–30 s of background acquisition (i.e., a gas blank) followed by 50 s of data acquisition (U, Th, ^{204}Pb, ^{206}Pb, ^{207}Pb, and ^{208}Pb) from the sample. Trace element contents were calibrated using the NIST 610 glass and Si as external and internal standards, respectively. Isotopic fractionation was corrected using the Plešovice zircon as an external standard. ICPMSDataCal [49] was used to calculate trace element contents and U–Pb isotopic ratios. Concordia ages and diagrams were obtained with Isoplot/Ex [50].

Hafnium isotopic analyses of the same zircon grains were conducted with an ArF excimer LA system coupled to a Neptune Plus multi-collector (MC)–ICP–MS (Thermo Scientific, Waltham, MA, USA) with a laser beam diameter of 44 μm, repetition rate of 6 Hz, and beam energy density of 6 J/cm^2. During data acquisition, the GJ-1 zircon standard was analyzed as an external standard to check the data quality.

Zircon O isotopic compositions were measured with a Cameca IMS-1280-HR secondary ion mass spectrometer (CAMECA, Gennevilliers, France) at the SKLaBIG, GIG-CAS, Guangzhou, China. The detailed analytical procedures were similar to those described by Li et al. [51] and Yang et al. [52]. The $^{133}Cs^{+}$ primary beam was accelerated at 10 kV with an intensity of ~2 nA and a beam size of ~20 μm. A normal incidence electron gun was activated to compensate for sample charging during the analysis. A mass resolution of 2500 was obtained in multi-collection mode. A nuclear magnetic resonance (NMR) probe was used for magnetic field control. ^{18}O and ^{16}O were measured in two off-axis Faraday cups (H1 and L2), respectively. Each analysis took <3 min, consisting of pre-sputtering (30 s), automatic secondary beam centering (60 s), and integration of O isotopes (4 s/cycle × 16 cycles).

Measured $^{18}O/^{16}O$ ratios are reported as $\delta^{18}O$ in per mil (‰) relative to the O isotopic composition of Vienna Standard Mean Ocean Water with $(^{18}O/^{16}O)_{VSMOW}$ = 0.0020052 [53]. The instrumental mass fractionation factor (IMF) for zircon was obtained using the zircon standard Penglai [54] ($\delta^{18}O$ = 5.31‰). Fourteen measurements of the Qinghu zircon standard yielded a weighted-mean $\delta^{18}O$ value of 5.6 ± 0.1‰ (2σ), which is within error of its reported value of 5.4 ± 0.2‰ [55].

3.2. Whole-Rock Data

Four unaltered samples were selected and crushed into small chips. Any chips that were altered or included secondary veins were removed and the remaining chips were soaked in 4 N HCl for 30 min to leach any alteration minerals. The rock chips were then powdered to <200 mesh using an alumina ceramic shatter box.

Major element compositions were determined with a ZSX Primus II X-ray fluorescence spectrometer. The loss-on-ignition (LOI) value of each sample was measured before the major element analyses.

Trace element analyses were undertaken with an Agilent 7500CX ICP–MS instrument (Agilent Technologies, Santa Clara, CA, USA) using the acid dissolution method of Liu et al. [56]. First, ~50 mg of sample powder was placed in a bomb and 0.5 mL of purified HNO_3 and 1.0 mL of HF were added. The bomb was then placed in a high-pressure tube at 190 °C for 48 h. After evaporation, 0.5 mL of purified HNO_3 was added to the bomb, and this step was repeated. Subsequently, 4 mL of purified 4 N HNO_3 was added to the bomb, which was then placed in the high-pressure tube at 170 °C for 4 h. Finally, the solution was diluted 1000 times, and 10 ppb of Rh was added as an internal standard to correct for instrument drift.

The Sr–Nd isotopic compositions were determined using the method of Zhang et al. [57]. The sample powder was dissolved in a Savillex Teflon beaker, Eden Prairie, MN, USA, by adding 2 mL of 22 N HF and 1 mL of 8 N HNO_3 and then heated to 120 °C on a hot plate for 5–7 d. After drying, 3 mL of HNO_3 was added to dissolve the sample again. The Sr was separated using SR-B50-A (100–150 μm) resin. Rare earth elements (REEs) were separated with AG 50-X8 cation exchange resin, and Nd was purified with HDEHP resin. The blanks contained <300 pg Rb and Sr, and <100 pg Sm and Nd. Mass fractionation was corrected by assuming an $^{88}Sr/^{86}Sr$ ratio of 8.375209 and a $^{146}Nd/^{144}Nd$ ratio of 0.7219. Instrument stability was monitored using NBS-987 and JNdi-1 standard solutions for the Sr and Nd isotopic compositions, respectively. Repeated analyses of NBS-987 and JNdi-1 yielded a mean $^{87}Sr/^{86}Sr$ ratio of 0.710294 ± 0.000016 (n = 40; 2 SD) and mean $^{143}Nd/^{144}Nd$ ratio of 0.512081 ± 0.000008 (n = 40; 2 SD), respectively.

4. Results

4.1. Petrography

In this study, four representative adakite samples were collected from the Shandan area in the central–eastern Longshoushan Belt. The adakite samples include two granodiorites and two potassic granites. The granodiorite samples are gray–white or pink, with a medium–coarse-grained porphyritic texture and massive structure (Figures 3 and 4). The phenocrysts consist of euhedral plagioclase (50–60 vol.%) that is ~3 mm in size and has poorly developed multiple twinning. The groundmass is mainly subhedral–anhedral plagioclase (5 vol.%; 1–2 mm), K-feldspar (10 vol.%; 2–3 mm), quartz (20 vol.%; 1–2 mm), and biotite (5 vol.%; 0.5–1.0 mm). The biotite is brown, flaky, and shows pronounced pleochroism. The potassic granite is pinkish red and contains phenocrysts of euhedral K-feldspar (50–60 vol.%; 2–3 mm) in a groundmass of subhedral–anhedral plagioclase (5 vol.%; ~1 mm), K-feldspar (5 vol.%; 1–2 mm), quartz (25 vol.%; 1–2 mm), and biotite (5 vol.%; 0.5–1.0 mm).

Figure 3. (**a–d**) Field photographs of adakites in the Shandan area.

Figure 4. (**a**,**c**) Plane-polarized and (**b**,**d**) cross-polarized light photomicrographs of adakites in the Shandan area. Bi = biotite, Kfs = K-feldspar, Pl = plagioclase, and Qtz = quartz.

4.2. Zircon U–Pb Geochronology and Hf–O Isotopic Compositions

The zircon U–Pb age data for the Shandan adakites are presented in Table S1. In these samples, the zircon grains are predominantly euhedral, with elongated columnar to short and stubby shapes. The grains are relatively large, with sizes of 100–200 μm, and exhibit clearly defined oscillatory zoning. All 50 analyses showed Th/U = 0.07–1.32, which is typical of magmatic zircons [58]. The contents of Th (144.12–15702.57 ppm) and U (1917.97–16,843.40 ppm) in SD-285 were larger than those of Th (318.89–4286.21 ppm) and U (327.12–4966.33 ppm) in SD-280 and SD-281. They have negative Eu anomaly (Eu/Eu* = 0.04–0.90). Zircon U–Pb dating of the three adakite samples (SD-280, SD-281, and SD-285) yielded weighted-mean ages of 444.3 ± 2.3 Ma (2σ; n = 14; MSWD = 0.18), 446.2 ± 2.8 Ma (2σ; n = 11; MSWD = 0.34), and 440.2 ± 1.9 Ma (2σ; n = 25; MSWD = 0.18), respectively. These represent the crystallization ages of the representative magmas (Figure 5).

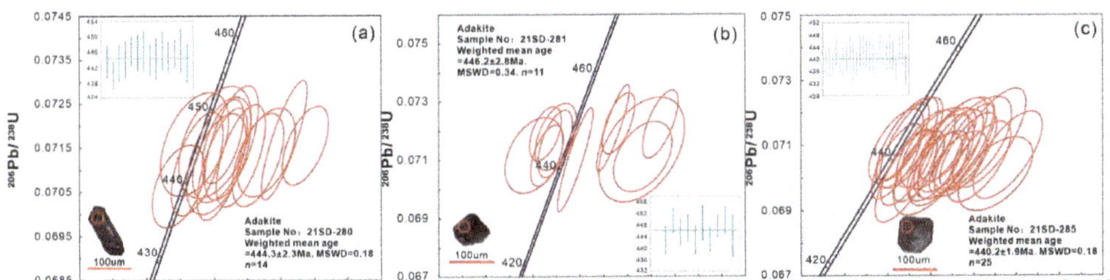

Figure 5. Zircon U–Pb concordia diagrams and weighted average diagram for adakites sample ((**a**) SD-280, (**b**) SD-281, (**c**) SD-2855) from the Shandan area, along with representative cathodoluminescence images of zircons. MSWD—mean square of weighted deviates. Red circles (32 μm) represent the locations of zircon LA–MC–ICPMS U–Pb dating.

In situ zircon Lu–Hf isotopic analyses were undertaken on adakite samples SD-280, SD-281, and SD-285 (Table S2). The initial $^{176}Hf/^{177}Hf$ ratios are 0.282145–0.282591. The corresponding $\varepsilon_{Hf}(t)$ values vary from 3.30 to −9.57, with an average of −2.01, and the calculated crustal model ages (T_{DM2}) are 2.04–1.22 Ga, indicating a magma source dominated by ancient crustal materials (Figure 6a). In situ zircon O isotope data for the Shandan adakite samples SD-280 and SD-285 are listed in Table S2. The adakite samples have a range of $\delta^{18}O$ values (5.34 ± 0.15‰ to 7.52 ± 0.51‰) with corresponding variations in zircon $\varepsilon_{Hf}(t)$ values (+2.48 to −12.4) (Figure 6b). The zircon crustal Hf model ages are 2214–1271 Ma.

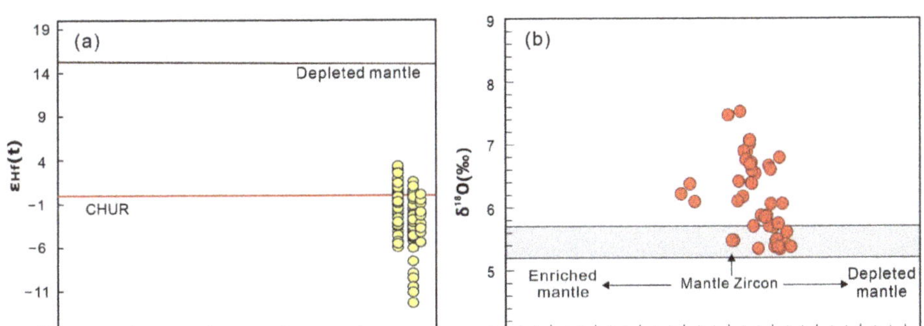

Figure 6. (**a**) Plot of zircon $\varepsilon_{Hf}(t)$ values versus age (modified after Zhang et al. [23]). (**b**) Plot of zircon $\delta^{18}O$ versus $\varepsilon_{Hf}(t)$ values (the average $\delta^{18}O$ value of mantle-derived magmatic zircon is 5.3 ± 0.3‰; Valley et al. [59]).

4.3. Whole-Rock Major and Trace Elements

The SiO_2 contents of the Shandan adakites are 66.8–73.2 wt.%. In a $Na_2O + K_2O$–SiO_2 diagram, the samples lie in the granodiorite and granite fields (Figure 7a). The Na_2O contents are 1.91–4.24 wt.% and K_2O contents are 2.29–7.72 wt.%, with K_2O/Na_2O = 0.54–4.05. In a K_2O–SiO_2 diagram, the samples lie mainly in the medium- to high-K series fields (Figure 7b). The MgO contents of the adakites are 0.26–1.57 wt.%, with Mg# = 41–49. The A/CNK values are 0.92–1.09. In an A/NK–A/CNK diagram, the samples lie in the nearly peraluminous to metaluminous fields (Figure 7c). In a ($Na_2O + K_2O - CaO$)–SiO_2 diagram, the samples exhibit both alkaline–calcic and calc–alkalic affinities (Figure 7d).

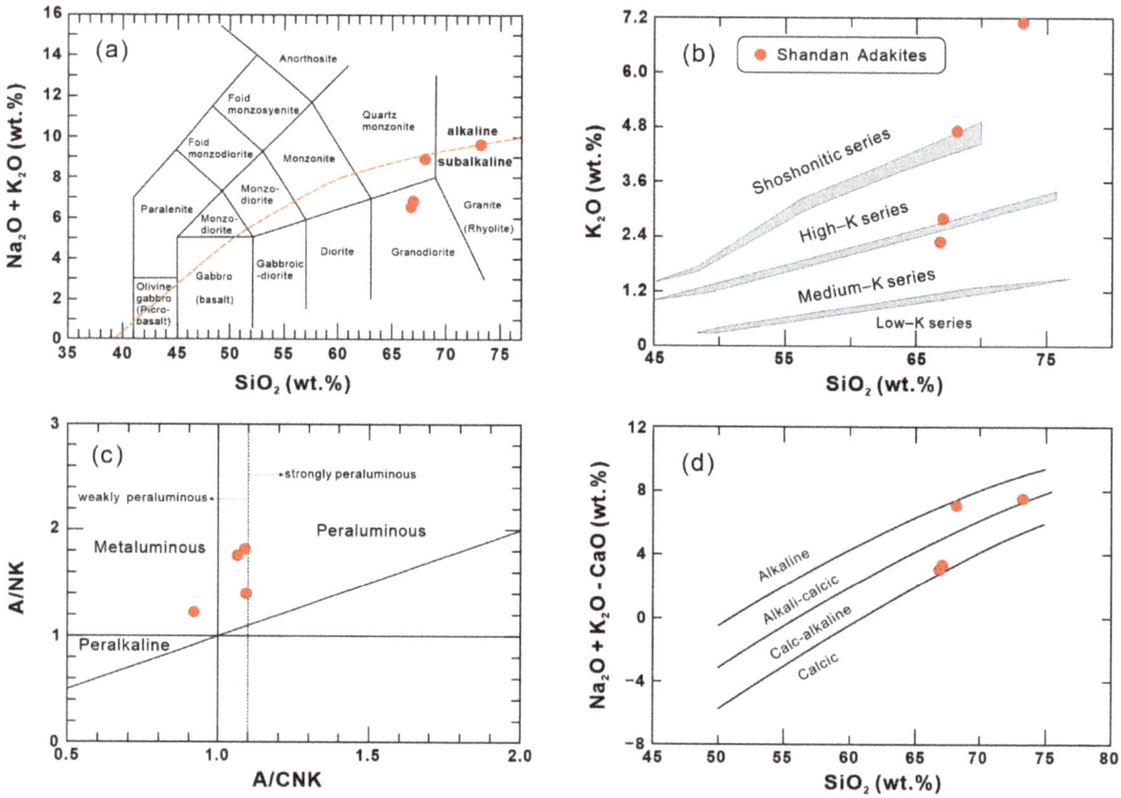

Figure 7. (**a**) $Na_2O + K_2O$ versus SiO_2; (**b**) K_2O versus SiO_2; (**c**) A/NK versus A/CNK; and (**d**) $Na_2O + K_2O - CaO$ versus SiO_2 diagrams.

The total REE (∑REE) contents of the adakites are 28.9–218 ppm. On a chondrite-normalized REE diagram, the samples are LREE enriched and heavy REE depleted, with $(La/Yb)_n$ values of 4.05–37.8, similar to continental crust (Figure 8a). In addition, the adakites exhibit positive Eu anomalies (Eu/Eu* = 1.07–5.01). On a primitive-mantle-normalized trace element diagram, the adakites are enriched in large-ion lithophile elements (e.g., Rb, Ba, Th, U, and Pb), indicating the magma source was rich in minerals such as K-feldspar and mica, and are depleted in high-field-strength elements (e.g., Nb, Ta, and Ti), similar to island arc rocks associated with subduction zones and the continental crust (Figure 8b). The major and trace element data are listed in Table S3.

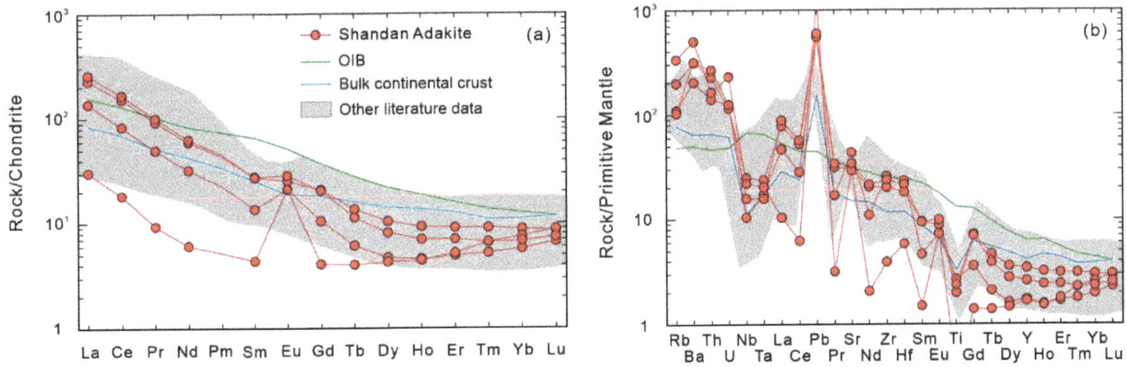

Figure 8. (**a**) Chondrite-normalized REE patterns and (**b**) primitive-mantle-normalized trace element diagrams for the Shandan adakites. Ocean island basalt (OIB), continental crust, and normalizing values are from Sun et al. [60]. The grey background represents the geochemical data of Maozangsi High-Mg adakitic rocks, quartz diorites from Mengjiadawan (MJDW), and granodiorites from Mengjiadawan (MJDW), Lianhuashan (LHS), and Yangqiandashan (YQDS). These data are from Yu et al. [61] and Zhang et al. [23]. The study area is located in the North Qilian Orogenic Belt and the southern margin of the Alxa Block.

4.4. Whole-Rock Sr–Nd Isotopic Compositions

Strontium–Nd isotopic analyses were conducted on four Shandan adakite samples. The samples have $^{87}Sr/^{86}Sr$ = 0.707867–0.710006, and age-corrected $(^{87}Sr/^{86}Sr)_t$ ratios of 0.706468–0.706535 (Figure 9). The adakites have low $\varepsilon_{Nd}(t)$ values, ranging from −7.66 to −6.32. The Sr–Nd isotopic characteristics are typical of crust-derived granites, and the granites have two-stage model ages (T_{DM2}) of 1.81–1.70 Ga, consistent with Paleoproterozoic crustal reservoirs in the North China Craton [62]. The whole-rock initial Sr–Nd isotopic compositions of the Shandan adakites are listed in Table S4.

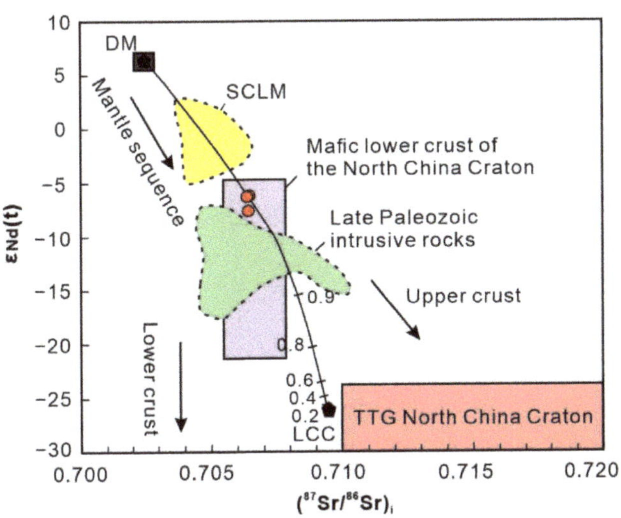

Figure 9. Plot of $\varepsilon_{Nd}(t)$ versus $(^{87}Sr/^{86}Sr)_i$ for whole-rock samples of the Shandan adakites. The Sr–Nd isotope data for late Paleozoic intrusive rocks on the northern margin of the North China Craton are from Zhang et al. [23]; Sr–Nd isotope data for the continental lithospheric mantle of the North China Craton are from Yang et al. [63]; Sr–Nd isotope data for Archean mafic lower crust and TTG gneisses from the North China Craton are from Chen et al. [62].

5. Discussion

5.1. Petrogenesis

The Shandan adakites have relatively high SiO_2 (66.8–73.2 wt.%) and CaO (1.85–3.49 wt.%) contents, and differentiation index values (DI = 90.57–96.96), indicating the magmas experienced fractional crystallization. In Harker diagrams, all samples display well-defined linear relationships between SiO_2 and MgO, Fe_2O_3T, TiO_2, MnO, and P_2O_5. The negative correlations amongst these elements suggest the adakites experienced fractional crystallization of biotite, apatite, and Fe–Ti oxides, while the negative correlation between MnO and SiO_2 indicates that Mn-bearing minerals also fractionated. In the process of crystallization differentiation, hornblende is preferentially combined with MREE [64]. When magmatic composition evolves to feldspathic composition, hornblende differentiation produces a stable or reduced $(Dy/Yb)_N$ ratio (Figure 10f), which indicates that Shandan adakites are separated and crystallized by biotite and hornblende at the same time, which is similar to the characteristics of high Mg Adakite. According to the classification criteria of adakites, Sr/Y > 40, Yb < 1.9, and La/Yb > 20 were used as diagnostic parameters [65]. Shandan adakites have high Sr/Y ratios (40.9–117) and Y contents (7.67–15.8). They are characterized by low Yb (0.98–1.50) and high La/Yb ratios (5.65–52.75, mean 31.81), and lie in the adakite field on the Sr/Y-Y diagram. (Figure 11a). Moreover, the adakites have high SiO_2 contents and moderate Mg# values and lie in the adakite field on a SiO_2–Mg# diagram (Figure 11b).

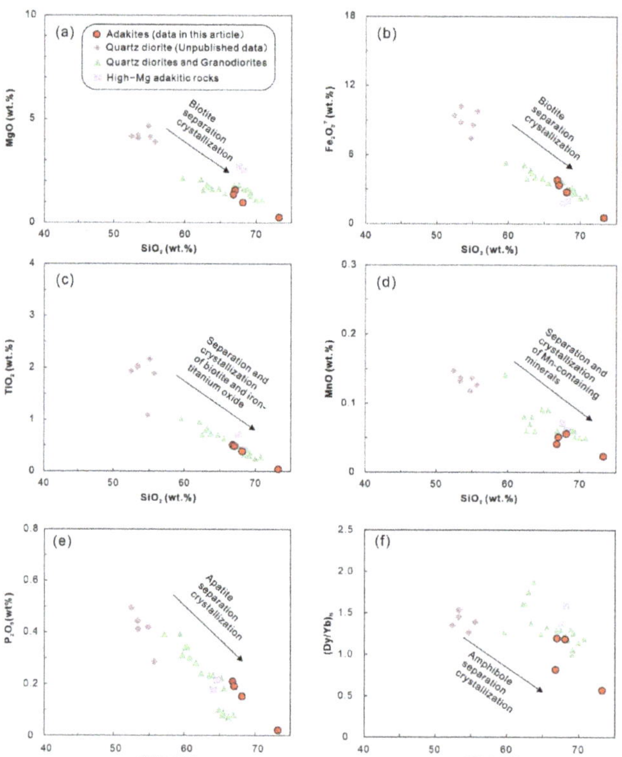

Figure 10. (**a**–**f**) Plots of MgO, Fe_2O_3T, TiO_2, MnO, and P_2O_5, $(Dy/Yb)_N$ versus SiO_2 for the adakites and granodiorites from the Shandan area (unpublished data). The geochemical data of Maozangsi High-Mg adakitic rocks are from Yu et al. [61], while those of quartz diorites from Mengjiadawan (MJDW) and granodiorites from Mengjiadawan (MJDW), Lianhuashan (LHS), and Yangqiandashan (YQDS) in the North Qilian Orogenic Belt and the southern margin of the Alxa Block are from Zhang et al. [23].

Existing petrogenetic models for adakites include (1) interactions between melts derived from subducting oceanic crust and mantle peridotites [3–6]; (2) partial melting of thickened lower crust [9–12,15,66]; (3) partial melting of delaminated lower crust [13–15]; (4) partial melting of subducted continental crust [16]; and (5) fractional crystallization of tholeiitic magma [2]. We now evaluate these potential adakite formation mechanisms in terms of the studied samples.

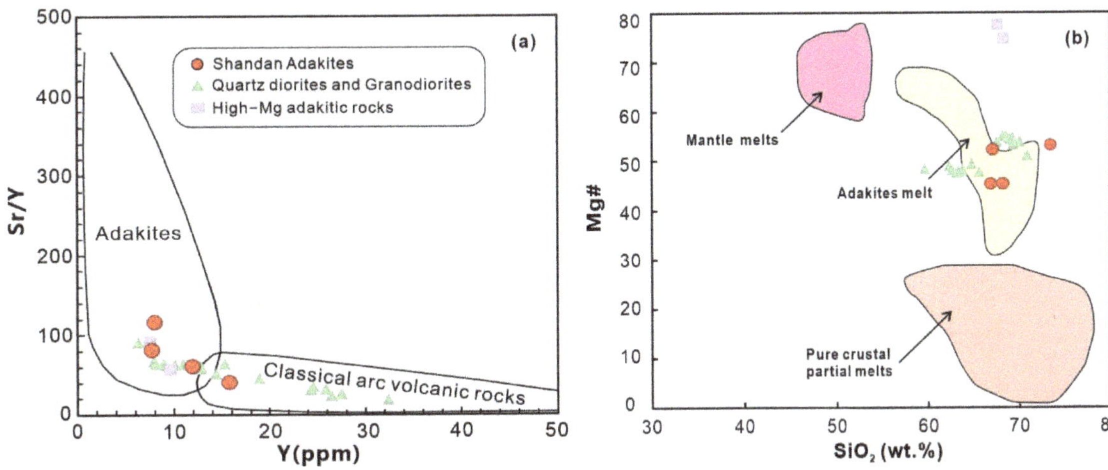

Figure 11. Plots of (**a**) Sr/Y versus Y (modified from Defant et al. [65]) and (**b**) Mg# versus SiO_2 (modified from Zhao et al. [67]). The geochemical data of Maozangsi High-Mg adakitic rocks are from Yu et al. [61], while those of quartz diorites from Mengjiadawan (MJDW) and granodiorites from Mengjiadawan (MJDW), Lianhuashan (LHS), and Yangqiandashan (YQDS) in the North Qilian Orogenic Belt and the southern margin of the Alxa Block are from Zhang et al. [23].

(1) Adakites formed by partial melting of subducted oceanic crust typically have relatively unradiogenic Sr isotopic compositions and high $\varepsilon_{Nd}(t)$ values. However, the adakites in the Shandan area have enriched Sr–Nd isotopic compositions of $(^{87}Sr/^{86}Sr)_i = 0.706468–0.706535$ and $\varepsilon_{Nd}(t) = -7.66$ to -6.32, which are inconsistent with derivation by melting of oceanic crust. Moreover, adakitic melts produced by partial melting of subducting oceanic crust tend to react with mantle peridotite during ascent, resulting in elevated MgO contents. In contrast, the Shandan adakites have low MgO contents, further excluding their formation by the melting of the oceanic crust. Furthermore, compared with adakitic granite–diorite complexes from Niuxinshan in northern Qilian [68], which were formed by the melting of the oceanic crust, the Shandan adakites have lower TiO_2 and MgO contents (Figure 12). (2) Magmas generated by partial melting of delaminated lower crust inevitably interact with the mantle during ascent, and the resulting adakites typically have relatively high contents of compatible elements (e.g., Cr and Ni) and MgO. In contrast, the Shandan adakites have relatively low Cr, Ni, and MgO contents, unlike the late Silurian (423 Ma) high-Mg adakites produced by partial melting of the delaminated lower crust in the eastern Qilian Orogen [69]. Furthermore, compared with adakites formed by the melting of the delaminated lower crust, the Shandan adakites have relatively low TiO_2 contents (Figure 12). (3) Both fractional crystallization of mantle-derived tholeiitic magma and fractional crystallization and coupled assimilation of ancient crust can produce adakites, typically as a product of high-pressure crystallization of island arc magmas at the top of the upper mantle [3]. However, these processes generally form intermediate dioritic rocks [70] and are associated with coeval, regionally extensive mafic–ultramafic magmatism. In contrast, the adakites in the Shandan area have high SiO_2 contents (Figure 12), and there

are no evidence of widespread coeval mafic–ultramafic rocks. Moreover, there is no continuous differentiation trend between the Shandan quartz diorites and adakites, and thus the adakites were not derived by fractional crystallization of quartz diorite (Figure 10). Therefore, the formation by fractional crystallization of tholeiitic magma and assimilation of ancient crust can be ruled out. (4) Adakites derived by partial melting of subducted continental crust are characterized by $K_2O > 3$ wt.% [16], whereas two of the Shandan adakite samples have $K_2O < 3$ wt.% (2.29–2.80 wt.%), thereby excluding the partial melting of subducted continental crust as a viable mechanism for their generation.

In TiO_2–SiO_2 and MgO–SiO_2 diagrams (Figure 12), the Shandan adakites lie in the field for partial melting of the thickened lower crust. Furthermore, on a Sr–Nd isotope diagram, the Shandan adakites lie in the field corresponding to the melting of the magnesian lower crust of the North China Craton (Figure 9). In addition, the Shandan adakites have similar major element compositions as the adakitic granites from Dachaidan, which were formed by partial melting of the thickened lower crust (Figure 12). The zircon saturation temperatures of the Shandan adakites are 674–835 °C, indicating they were the products of high-temperature partial melting, similar to the formation of A-type granites under high-temperature conditions (>800 °C) [71]. Based on the above considerations, we propose that the Shandan adakites were generated by high-temperature partial melting of thickened, magnesian lower crust as a result of underplating by mantle-derived magmas.

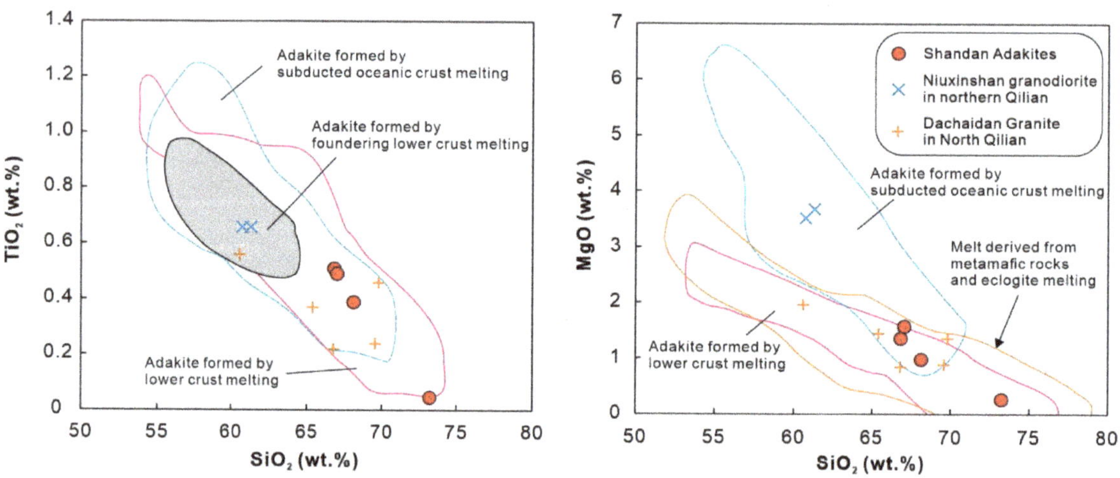

Figure 12. Plots of TiO_2 versus SiO_2 and MgO versus SiO_2 (modified from Wang et al. [72]). Data for the Niuxinshan granodiorite in northern Qilian are from Wu et al. [68]; data for the Dachaidan adakitic granite in northern Qilian are our unpublished data.

5.2. Mantle Source

Having established a crustal origin, we now evaluate mantle contributions to the adakitic magma. The Shandan adakites have negative $\varepsilon_{Nd}(t)$ values (−7.66 to −6.32) and negative Nb and Ta anomalies, with trace element patterns similar to those of continental crust, indicative of a magnesian lower crustal source. Igneous rocks with similar U–Pb ages, but different $\varepsilon_{Hf}(t)$ values, are generally considered to have formed from magmas generated by mixing from different source regions [73,74]. The negative zircon $\varepsilon_{Hf}(t)$ values suggest that ancient continental crust had a significant role in the magma formation. Positive $\varepsilon_{Hf}(t)$ values for granites can be interpreted as representing mantle-derived magmas or newly formed crust [70]. The $\varepsilon_{Hf}(t)$ values of the Shandan adakites range from +3.30 to −12.4, indicative of contributions from older crustal material and also enriched mantle components, suggesting the Shandan adakites formed by crust–mantle interactions. Continental crust generally has high La/Nb (~2.5) and low Ce/Pb (~4) ratios [75], and there is no significant correlation with SiO_2. The Ce/Pb and SiO_2 diagrams (Figure 13a) show that the samples are mainly influenced by mantle sources, and the influence of crustal mixing is limited. Further constraints on the mantle source can be obtained from the La/Ba versus La/Nb diagram, in which the Shandan adakites have high La/Nb (0.46–5.41) and low La/Ba (0.00–0.03) ratios, which is similar to the characteristics of high-Mg adakite and subcontinental lithospheric mantle melts affected by subduction (Figure 13b [76]). Therefore, the adakites were likely derived from a relatively enriched continental lithospheric mantle source region. Zircon $\delta^{18}O$ values for mantle-derived magmas are 5.3 ± 0.3‰ [51]. The Shandan adakites have zircon $\delta^{18}O$ = 5.34‰–7.52‰, which is similar to or higher than typical mantle-derived zircon values, indicating the parental magma of the Shandan adakites formed by partial melting due to the interactions between thickened magnesian lower crust and continental lithospheric mantle.

Trace element and isotopic data suggest that the Shandan adakites formed by partial melting due to interactions between enriched continental lithospheric mantle and magnesian lower crust. Using a two-end-member $(^{87}Sr/^{86}Sr)_t$ versus $\varepsilon_{Nd}(t)$ mixing model, we quantitatively estimated the contribution of crustal material to the Shandan adakites (Figure 9), using lower continental crust (LCC [77]) as the crustal end-member and depleted mantle (DM) as the mantle end-member [78,79]. The results indicate the Shandan adakites formed by partial melting of 5%–6% LCC. The lower crust in the North China Craton is mainly composed of Archaean gneiss (TTG). The TTG gneiss in the Beishan complex in the west of the Alashan Block has undergone magmatic crystallization in the late NeoArchaean period and high-pressure metamorphism in the early Paleoproterozoic period, which proves that there is a neoarchaean metamorphic basement in the block [80]. Therefore, TTG gneiss in the western Alxa Block is selected as the crustal end member (~2.5Ga gneiss), and depleted mantle is selected as the mantle end member. Based on zircon Hf isotopes, further estimates of the crustal contribution to the magma suggest the Shandan adakites formed by partial melting and mixing of 5%–15% crustal material with 95%–85% mantle-derived magma (Figure 14).

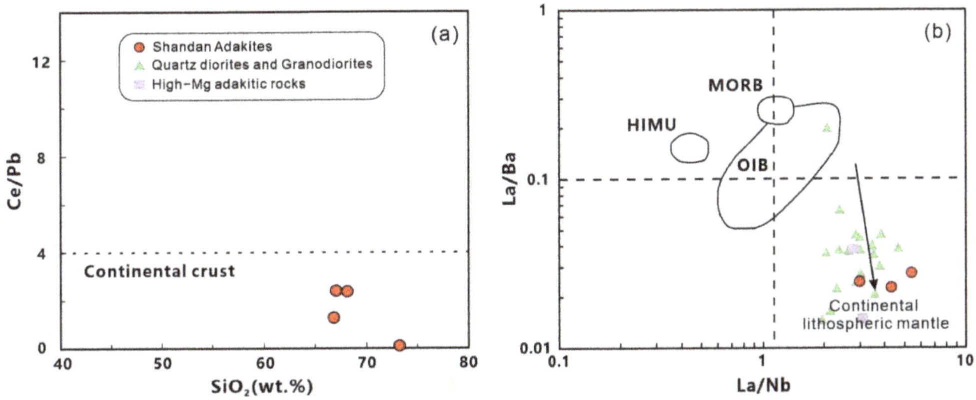

Figure 13. (**a**) Ce/Pb versus SiO_2 diagram (continental crust values are from Rudnick et al. [75]), (**b**) La/Ba versus La/Nb diagram (modified from Saunders et al. [81]). The geochemical data of Maozangsi High-Mg adakitic rocks are from Yu et al. [79], while those of quartz diorites from Mengjiadawan (MJDW) and granodiorites from Mengjiadawan (MJDW), Lianhuashan (LHS), and Yangqiandashan (YQDS) in the North Qilian Orogenic Belt and the southern margin of the Alxa Block are from Zhang et al. [23].

The Shandan adakites have low $\varepsilon_{Nd}(t)$ values (−7.66 to −6.32) and high $\varepsilon_{Hf}(t)$ values (3.30 to −12.4), indicating that the Nd–Hf isotope system of the Shandan adakite may show different evolution trends under certain circumstances. Typically, the Lu–Hf and Sm–Nd isotopic systems exhibit a similar behavior during magma evolution and will show a decoupling phenomenon. However, in some samples, the zircon Hf isotope and whole-rock Nd isotope ratios show the same behavior. The Shandan adakites primarily formed by partial melting of the thickened magnesian lower crust, and the Lu–Hf isotopic system may have been modified by variable degrees of lower crustal alteration. The thickened magnesian lower crust was garnet-rich, and thus had a high Lu/Hf ratio, leading to the formation of melts with high $^{176}Hf/^{177}Hf$ ratios [82]. This process leads to the magmas having higher $\varepsilon_{Hf}(t)$ values at a given $\varepsilon_{Nd}(t)$ value, resulting in differences between the Nd–Hf isotopes of the Shandan adakites.

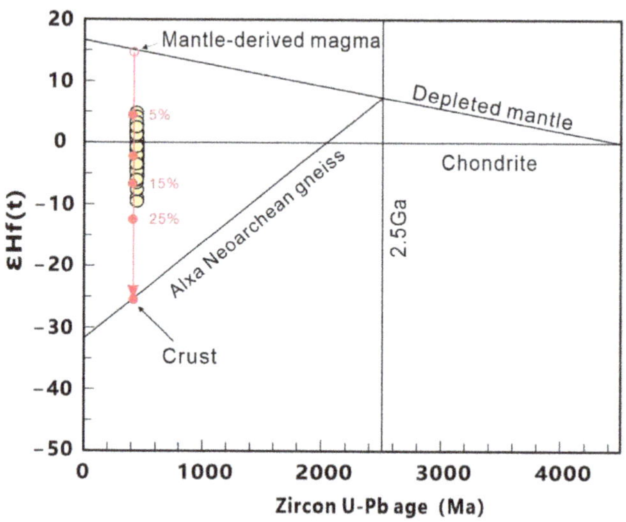

Figure 14. Simulated crustal contamination (the base map is modified from Duan et al. [48]; The Hf isotope evolution of the Alxa basement (2.5 Ga gneiss [80]) was determined using the parameters for the upper crust given by Amelin et al. [83]).

5.3. Geodynamic Processes

The tectonic evolution of the North Qilian Orogenic Belt and Alxa Block involved multiple stages, including oceanic subduction, arc/continent–continent collision, post-collision, and continental subduction, which resulted in the formation of numerous ophiolites and intermediate to silicic igneous rocks. These igneous rocks provide important insights into geodynamic processes and the tectonic evolution, and also document interactions between the lithospheric mantle and continental crust [84]. In the Longshoushan area, located at the northern margin of the North Qilian Orogenic Belt and the southern margin of the Alxa Block, there are extensive exposures of early Paleozoic intermediate to silicic igneous rocks [23,85]. Previous studies have shown that geochemical characteristics can be used to identify the tectonic and geodynamic settings of granites.

In the early Paleozoic, the North Qilian oceanic lithosphere was subducted and the ocean eventually closed, leading to the collision between the Qilian and Alxa blocks [86]. The porphyritic biotite granite, granodiorite, monzogranite, and K-feldspar granite in the Qingshanbao area have a Late Ordovician age (441 Ma), suggesting derivation by melting of the ancient crust of the Paleoproterozoic Longshoushan Formation, with some addition of mantle material. These rocks formed during the collisional orogenic phase in a compressional environment [31]. Wei et al. [87] investigated Late Ordovician (444 Ma) granites from the Hexibao area, which are syn-collision granites, and proposed that northward subduction of the North Qilian oceanic lithosphere occurred at this time, marking the initial collisional phase of the North Qilian Ocean. Chen et al. [88] identified Late Ordovician (447 Ma) quartz monzogranite in the Longshoushan area, and suggested it was the product of magmatism during the subduction of the North Qilian oceanic plate beneath the Alxa Block. Niu et al. [34] identified early Silurian (438 Ma) diorite in the Qingshanbao area, and suggested it formed during the transition from a compressional to an extensional setting. In addition, studies of the 430 Ma Huoshan granite, 427 Ma diabase, 424 Ma Jinfusi S-type granite, and 427–414 Ma Wuwei–Jinchang intrusions [23,89–91] suggest that between 435 and 414 Ma, the Alxa Block and Central Qilian–Qaidam blocks transitioned from the syn- to post-collisional stages. Other intermediate to silicic igneous rocks in the North Qilian area include the 516–505 Ma Qaidanu peraluminous granites, 501 Ma Kekeli quartz diorite, 477 Ma Niuxinshan granite, and 457–441 Ma adakitic granites [23,61,90]. Previous studies of the 457–441 Ma adakitic granites in the northern North Qilian Orogenic Belt [23,61,90] indicate that these formed during continent–continent collision [23,61]. The formation of these igneous rocks was closely related to oceanic subduction and subsequent collision. Previous studies have suggested that early Paleozoic magmatism in the Longshoushan area occurred mainly in two phases: Early to Late Ordovician and Late Ordovician to Early Devonian. The Early to Late Ordovician magmatism was primarily associated with the post-collisional stage after the closure of the North Qilian Ocean, while the Late Ordovician to Early Devonian magmatism occurred in a post-collisional setting between the Alxa and Qilian blocks.

Based on previous studies of silicic igneous rocks in the Longshoushan area and northern Qilian region, as well as the 446–440 Ma age of the Shandan adakites obtained in this study (Figure 5), it is likely these rocks formed during the closure of the North Qilian Ocean and collision between the Alxa and Qilian blocks. In Rb versus Y + Nb and Nb versus Y tectonic discrimination diagrams the Shandan adakites lie in the transitional zone between syn-collisional and post-collisional granites, which is similar to that of high-Mg adakite. (Figure 15a,b [92,93]). Therefore, we suggest that the Shandan adakites formed during the collisional to early post-collisional stages between the Alxa and Qilian blocks.

Based on our trace element and Sr–Nd–Hf–O isotope data for the studied adakites, The Shandan adakites have relatively enriched $\varepsilon_{Nd}(t)$, $\varepsilon_{Hf}(t)$ values (3.30 to −12.4) and zircon oxygen isotopes higher than the mantle ($\delta^{18}O$ = 5.34‰–7.52‰), and similar to the subcontinental lithospheric mantle magma in the La/Ba La/Nb diagram. So, we suggest they formed by partial melting resulting from the interaction between the magnesian lower crust and subcontinental lithospheric mantle. The formation of these rocks is attributed primarily to the high-temperature conditions of the thickened magnesian lower crust caused by the underplating of mantle-derived magma. As such, the Shandan adakites likely formed by partial melting induced by heat supplied by asthenospheric mantle upwelling, which triggered interactions between the subcontinental lithospheric mantle and thickened magnesian lower crust (Figure 16). The primary magmas underwent fractional crystallization of biotite, hornblende, apatite, Fe–Ti oxides, and Mn-bearing minerals to form the adakites. The continued subduction of the North Qilian oceanic lithosphere led to ocean closure and collision between the Alxa and Central Qilian blocks at ca. 446 Ma [18]. During 446–440 Ma, the Alxa and Central Qilian blocks collided and the lower crust was thickened. After 440 Ma, the tectonic environment in the Longshoushan area transitioned from the collision to the post-collisional stage.

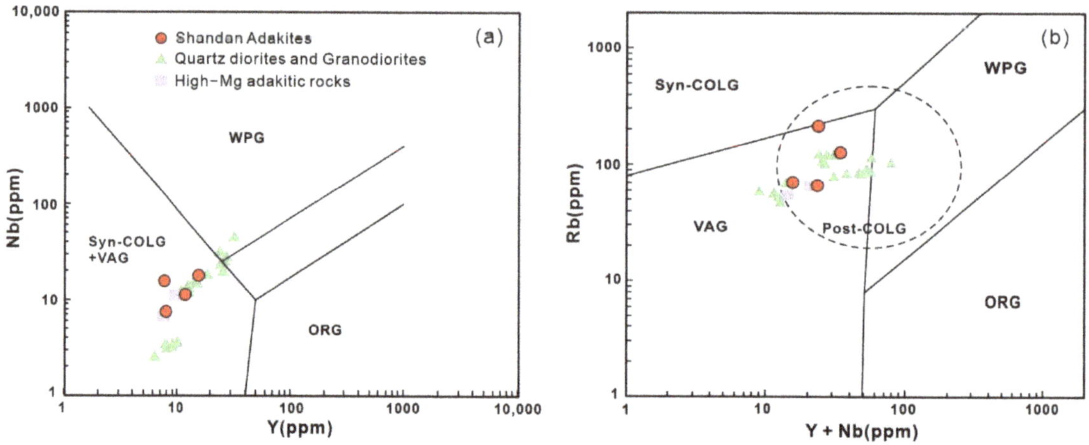

Figure 15. Plots of (**a**) Nb versus Y (modified after Pearce et al. [92]) and (**b**) Rb versus Y + Nb (modified after Eby et al. [93]). ORG = plagiogranite; Syn-COLG = syn-collision granite; Post-COLG = post-collision granite; VAG = volcanic arc granite; WPG = within-plate granite. The geochemical data of Maozangsi High-Mg adakitic rocks are from Yu et al. [79], while those of quartz diorites from Mengjiadawan (MJDW) and granodiorites from Mengjiadawan (MJDW), Lianhuashan (LHS), and Yangqiandashan (YQDS) in the North Qilian Orogenic Belt and the southern margin of the Alxa Block are from Zhang et al. [23].

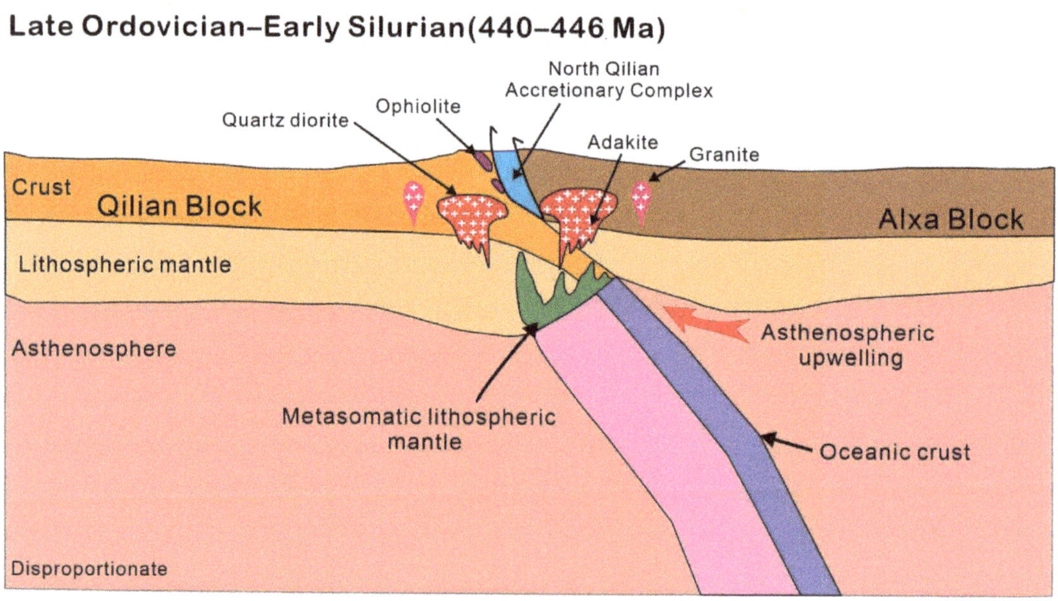

Figure 16. Early Paleozoic tectonic evolution model of Shandan area (modified after Xia et al. [28] and Qiao et al. [94]).

6. Conclusions

1. The Shandan adakites were formed during the collisional to early post-collisional stages between the Alxa and Qilian blocks following the closure of the Proto-Tethyan North Qilian Ocean at ca. 446–440 Ma. The adakites have high Sr/Y (40.9–117) ratios

and specific geochemical characteristics, revealing that their genesis is related to the partial melting of the mafic lower crust.

2. The relatively enriched $\varepsilon_{Hf}(t)$ (3.30 to −12.4) and $\delta^{18}O$ (5.34‰–7.52‰) of the Shandan adakites indicate that they were formed by the melting of the thickened lower crust and facilitated by mantle-derived heat input, which was accompanied by significant crust–mantle interaction.
3. The Alxa and Central Qilian blocks then collided, which caused lower-crustal thickening and a transition to post-collisional tectonics after ca. 440 Ma.

Supplementary Materials: The following supporting information can be downloaded at: https://www.mdpi.com/article/10.3390/min15040352/s1, Table S1. LA–ICP–MS zircon U–Pb isotope data for Shandan adakites from the Longshoushan area. Table S2. Zircon Hf–O isotope ages of the Shandan adakites in the Longshoushan area. Table S3. Whole-rock major and trace element data of Shandan adakites in the Longshoushan area. Table S4. Whole-rock Sr–Nd isotope and content data of representative samples of Shandan adakites in Longshoushan area.

Author Contributions: Conceptualization, Y.Y. and X.L. (Xijun Liu); methodology, Z.B. and Y.L.; validation, Z.B., Y.Y. and X.L. (Xijun Liu); formal analysis, Z.B.; investigation, resources, Y.Y.; data curation, Z.B. and H.T.; writing-original draft preparation, Z.B. and P.L.; writing-review and editing, W.H. and Y.X.; visualization, G.C., X.L. (Xiao Liu) and R.H.; supervision, Y.Y.; project administration and funding acquisition, X.L. (Xijun Liu). All authors have read and agreed to the published version of the manuscript.

Funding: This study was supported financially by funds from the Deep Earth probe and Mineral Resources Exploration-National Science and Technology Major Project: 2024ZD1001503, Guangxi Science and Technology Project: No. GuikeAD24010023, National Natural Science Foundation of China: No. 42473063, 42203051, and the Natural Science Foundation of Guangxi Province for Young Scholars: No. GuikeAD23026175. This research is a contribution to "Xinjiang Tianchi Distinguished Expert" by Xi-Jun Liu.

Data Availability Statement: The data in this paper are reliable and have not been published elsewhere.

Acknowledgments: We would like to express our sincere gratitude to the editors and reviewers for their careful review and constructive comments, which have significantly improved the quality of this manuscript.

Conflicts of Interest: The authors declare that they have no known competing financial interests or personal relationships that could have appeared to influence the work reported in this paper.

References

1. Castillo, P.R. An overview of adakite petrogenesis. *Chin. Sci. Bull.* **2006**, *51*, 257–268. [CrossRef]
2. Castillo, P.R. Adakite petrogenesis. *Lithos* **2012**, *134*, 304–316. [CrossRef]
3. Rapp, R.; Shimizu, N.; Norman, M.; Applegate, G. Reaction between slab-derived melts and peridotite in the mantle wedge: Experimental constraints at 3.8 GPa. *Chem. Geol.* **1999**, *160*, 335–356. [CrossRef]
4. Carmichael, I.S. The andesite aqueduct: Perspectives on the evolution of intermediate magmatism in west-central (105–199 W) Mexico. *Contrib. Mineral. Petrol.* **2002**, *143*, 641–663. [CrossRef]
5. Martin, H.; Smithies, R.; Rapp, R.; Moyen, J.-F.; Champion, D. An overview of adakite, tonalite–trondhjemite–granodiorite (TTG), and sanukitoid: Relationships and some implications for crustal evolution. *Lithos* **2005**, *79*, 1–24. [CrossRef]
6. Macpherson, C.G.; Dreher, S.T.; Thirlwall, M.F. Adakites without slab melting: High pressure differentiation of island arc magma, Mindanao, the Philippines. *Earth Planet. Sci. Lett.* **2006**, *243*, 581–593. [CrossRef]
7. Tang, G.; Wang, Q.; Wyman, D.A.; Li, Z.-X.; Zhao, Z.-H.; Jia, X.-H.; Jiang, Z.-Q. Ridge subduction and crustal growth in the Central Asian Orogenic Belt: Evidence from Late Carboniferous adakites and high-Mg diorites in the western Junggar region, northern Xinjiang (west China). *Chem. Geol.* **2010**, *277*, 281–300. [CrossRef]
8. Chung, S.-L.; Chu, M.-F.; Zhang, Y.; Xie, Y.; Lo, C.-H.; Lee, T.-Y.; Lan, C.-Y.; Li, X.; Zhang, Q.; Wang, Y. Tibetan tectonic evolution inferred from spatial and temporal variations in post-collisional magmatism. *Earth Sci. Rev.* **2005**, *68*, 173–196. [CrossRef]

9. Wang, Q.; McDermott, F.; Xu, J.-f.; Bellon, H.; Zhu, Y.-t. Cenozoic K-rich adakitic volcanic rocks in the Hohxil area, northern Tibet: Lower-crustal melting in an intracontinental setting. *Geology* **2005**, *33*, 465–468. [CrossRef]
10. Atherton, M.P.; Petford, N. Generation of sodium-rich magmas from newly underplated basaltic crust. *Nature* **1993**, *362*, 144–146. [CrossRef]
11. Petford, N.; Atherton, M. Na-rich partial melts from newly underplated basaltic crust: The Cordillera Blanca Batholith, Peru. *J. Petrol.* **1996**, *37*, 1491–1521. [CrossRef]
12. Condie, K.C. TTGs and adakites: Are they both slab melts? *Lithos* **2005**, *80*, 33–44. [CrossRef]
13. Xu, J.-F.; Shinjo, R.; Defant, M.J.; Wang, Q.; Rapp, R.P. Origin of Mesozoic adakitic intrusive rocks in the Ningzhen area of east China: Partial melting of delaminated lower continental crust? *Geology* **2002**, *30*, 1111–1114.
14. Kay, R.W.; Kay, S.M. Delamination and delamination magmatism. *Tectonophysics* **1993**, *219*, 177–189. [CrossRef]
15. Rudnick, R.L. Making continental crust. *Nature* **1995**, *378*, 571–578. [CrossRef]
16. Wang, Q.; Wyman, D.A.; Xu, J.; Dong, Y.; Vasconcelos, P.M.; Pearson, N.; Wan, Y.; Dong, H.; Li, C.; Yu, Y. Eocene melting of subducting continental crust and early uplifting of central Tibet: Evidence from central-western Qiangtang high-K calc-alkaline andesites, dacites and rhyolites. *Earth Planet. Sci. Lett.* **2008**, *272*, 158–171. [CrossRef]
17. Xu, Z.-Q.; Yang, J.-S.; Li, H.-B.; Yao, J.-X. The Early Palaeozoic Terrene Framework and the Formation of the High-Pressure (HP) and Ultra-High Pressure (UHP) Metamorphic Belts at the Central Orogenic Belt (COB). *Acta Geol. Sin.* **2006**, *80*, 1793–1806.
18. Song, S.; Niu, Y.; Su, L.; Xia, X. Tectonics of the north Qilian orogen, NW China. *Gondwana Res.* **2013**, *23*, 1378–1401. [CrossRef]
19. Xia, L.-Q.; Li, X.-M.; Yu, J.-Y.; Wang, G.-Q. Mid−Late Neoproterozoic to Early Paleozoic volcanism and tectonic evolution of the Qilian Mountain. *Geol. China* **2016**, *43*, 1087–1138. [CrossRef]
20. Dong, Y.; Sun, S.; Santosh, M.; Zhao, J.; Sun, J.; He, D.; Shi, X.; Hui, B.; Cheng, C.; Zhang, G. Central China orogenic belt and amalgamation of East Asian continents. *Gondwana Res.* **2021**, *100*, 131–194. [CrossRef]
21. Xia, L.-Q.; Li, X.-M.; Yu, J.-Y.; Wang, G.-Q. Mid-late neoproterozoic to early paleozoic volcanism and tectonic evolution of the Qilianshan, NW China. *GeoResJ* **2016**, *9–12*, 1–41. [CrossRef]
22. Wang, Z.-Z.; Chen, X.-H.; Shao, Z.-G.; Li, B.; Ding, W.-C.; Zhang, Y.-P.; Wang, Y.-C.; Zhang, Y.-Y.; Xu, S.-L.; Qin, X. Petrogenesis of the Late Silurian-Early Devonian granites in the Longshoushan-Helishan area, Gansu Province, and its tectonic implications for the Early Paleozoic evolution of the southwestern Alxa Block. *Acta Geol. Sin.* **2020**, *94*, 2243–2261. [CrossRef]
23. Zhang, L.; Zhang, H.; Zhang, S.; Xiong, Z.; Luo, B.; Yang, H.; Pan, F.; Zhou, X.; Xu, W.; Guo, L. Lithospheric delamination in post-collisional setting: Evidence from intrusive magmatism from the North Qilian orogen to southern margin of the Alxa block, NW China. *Lithos* **2017**, *288–289*, 20–34. [CrossRef]
24. Hu, N.; Xu, A.; Yang, J. Characteristics and tectonic environment of Zhigoumen pluton in Longshoushan area. *J. Earth Sci. Environ.* **2005**, *27*, 5–11.
25. Zhang, J.; Meng, F.; Wan, Y. A cold Early Palaeozoic subduction zone in the North Qilian Mountains, NW China: Petrological and U-Pb geochronological constraints. *J. Metamorph. Geol.* **2007**, *25*, 285–304. [CrossRef]
26. Liou, J.; Wang, X.; Coleman, R.G.; Zhang, Z.M.; Maruyama, S. Blueschists in major suture zones of China. *Tectonics* **1989**, *8*, 609–619. [CrossRef]
27. Liu, Y.J.; Neubauer, F.; Genser, J.; Takasu, A.; Ge, X.H.; Handler, R. 40Ar/39Ar ages of blueschist facies pelitic schists from Qingshuigou in the Northern Qilian Mountains, western China. *Isl. Arc* **2006**, *15*, 187–198. [CrossRef]
28. Xia, X.; Song, S.; Niu, Y. Tholeiite–Boninite terrane in the North Qilian suture zone: Implications for subduction initiation and back-arc basin development. *Chem. Geol.* **2012**, *328*, 259–277. [CrossRef]
29. Tseng, C.-Y.; Yang, H.-J.; Yang, H.-Y.; Liu, D.; Tsai, C.-L.; Wu, H.; Zuo, G. The Dongcaohe ophiolite from the North Qilian Mountains: A fossil oceanic crust of the Paleo-Qilian ocean. *Chin. Sci. Bull.* **2007**, *52*, 2390–2401. [CrossRef]
30. Qian, W.; Xilin, Z.; Yuanku, M.; Shengyao, Y.; Yanan, L. Early Devonian Post-collisional Granitic Magmatism in the North Qilian Orogenic Belt, Western China: Insights into Lithospheric Delamination and Orogenic Collapse. *Acta Geol. Sin.* **2024**, *98*, 352–367. [CrossRef]
31. Liu, W.-H.; Pan, J.-Y.; Liu, X.-D.; Wang, K.-X.; Wang, G.; Xue, P. Petrogenesis and tectonic implication of Qingshanbao pluton in longshou moutains, gansu: Constraints from elemental geochemistry, zircon U-Pb age and Sr-Nd isotopes. *Miner. Pet.* **2019**, *39*, 26–40.
32. Sun, B.-L.; Qian, Q.; Zhang, J.-X. Zircon U-Pb geochronology, Hf-O isotopes, whole-rock geochemistry of the Dafosi and Jinfosi granite plutons, Gansu Province and geological implications. *Acta Petrol. Sin.* **2017**, *33*, 3091–3108. (In Chinese with English Abstract)
33. Leng, T.-C.; Liu, W.-H.; Yu, C.-D.; Liu, X.-D.; Wang, W.; Li, Y.-X. Discussion on Geochemical Characteristics and Genesis of Hongshiquan Syenite in Longshoushan Area, Gansu Province. *Mineral. Petrol.* **2021**, *41*, 72–83. [CrossRef]
34. Niu, Y.-B.; Liu, W.-H.; Liu, X.-D.; Wang, K.-X.; Wang, G. Geochemistry and petrogenesis of Qingshanbao pluton of Longshoushan metallogenic belt in Gansu. *Sci. Technol. Eng.* **2018**, *18*, 11–21. (In Chinese with English Abstract)

35. Yi, C.; ShuGuang, S. Deformation and metamorphism of HP belt in the North Qilian Mountains and their implications for exhumation. *Acta Petrol. Sin.* **2009**, *25*, 2235–2246.
36. Xu, Y.-J.; Du, Y.-S.; Cawood, P.A.; Guo, H.; Huang, H.; An, Z.-H. Detrital zircon record of continental collision: Assembly of the Qilian Orogen, China. *Sediment. Geol.* **2010**, *230*, 35–45. [CrossRef]
37. Zhao, X.; Liu, C.; Wang, J.; Zhao, Y.; Wang, L.; Zhang, Q. Detrital zircon U–Pb ages of Paleozoic sedimentary rocks from the eastern Hexi Corridor Belt (NW China): Provenance and geodynamic implications. *Sediment. Geol.* **2016**, *339*, 32–45. [CrossRef]
38. Wang, K.-X.; Yu, C.-D.; Yan, J.; Liu, X.-D.; Liu, W.-H.; Pan, J.-Y. Petrogenesis of Early Silurian granitoids in the Longshoushan area and their implications for the extensional environment of the North Qilian Orogenic Belt, China. *Lithos* **2019**, *342–343*, 152–174. [CrossRef]
39. Xue, F.; Yang, F.; Ren, W.; Santosh, M.; Qian, Z.; Huang, Y.; Tan, Z. Petrogenesis of the Early Paleozoic Dioritic–Granitic Magmatism in the Eastern North Qilian Orogen, NW China: Implications for Tethyan Tectonic Evolution. *Lithosphere* **2024**, *2024*, lithosphere_2023_2297. [CrossRef]
40. Cox, K.G. Continental magmatic underplating. Philosophical Transactions of the Royal Society of London. *Ser. A Phys. Eng. Sci.* **1993**, *342*, 155–166. [CrossRef]
41. Bryan, S.E.; Ferrari, L.; Reiners, P.W.; Allen, C.M.; Petrone, C.M.; Ramos-Rosique, A.; Campbell, I.H. New insights into crustal contributions to large-volume rhyolite generation in the mid-Tertiary Sierra Madre Occidental province, Mexico, revealed by U–Pb geochronology. *J. Petrol.* **2008**, *49*, 47–77. [CrossRef]
42. Tang, Z.-L.; Bai, Y.-L. Geotectonic framework and metallogenic system in the southwest margin of North China paleocontinent. *Earth Sci. Front.* **1999**, 78–90.
43. Gong, J.; Zhang, J.; Wang, Z.; Yu, S.; Li, H.; Li, Y. Origin of the Alxa Block, western China: New evidence from zircon U–Pb geochronology and Hf isotopes of the Longshoushan Complex. *Gondwana Res.* **2016**, *36*, 359–375. [CrossRef]
44. Xue, S.; Ling, M.-X.; Liu, Y.-L.; Zhang, H.; Sun, W. The genesis of early Carboniferous adakitic rocks at the southern margin of the Alxa Block, North China. *Lithos* **2017**, *278–281*, 181–194. [CrossRef]
45. Liu, J.; Yin, C.; Zhang, J.; Qian, J.; Li, S.; Xu, K.; Wu, S.; Xia, Y. Tectonic evolution of the Alxa Block and its affinity: Evidence from the U-Pb geochronology and Lu-Hf isotopes of detrital zircons from the Longshoushan Belt. *Precambr. Res.* **2020**, *344*, 105733. [CrossRef]
46. Gong, J.-H.; Zhang, J.-X.; Yu, S.-Y. The origin of Longshoushan Group and associated rocks in the southern part of the Alxa block: Constraint from LA-ICP-MS U-Pb zircon dating. *Acta Petrol. Mineral.* **2011**, *30*, 795–818.
47. Tung, K.; Yang, H.; Liu, D.; Zhang, J.; Tseng, C.; Wan, Y. SHRIMP U-Pb geochronology of the detrital zircons from the Longshoushan Group and its tectonic significance. *Chin. Sci. Bull.* **2007**, *52*, 1414–1425. [CrossRef]
48. Duan, J.; Li, C.; Qian, Z.; Jiao, J. Geochronological and geochemical constraints on the petrogenesis and tectonic significance of Paleozoic dolerite dykes in the southern margin of Alxa Block, North China Craton. *J. Asian Earth Sci.* **2015**, *111*, 244–253. [CrossRef]
49. Liu, Y.; Hu, Z.; Gao, S.; Günther, D.; Xu, J.; Gao, C.; Chen, H. In situ analysis of major and trace elements of anhydrous minerals by LA-ICP-MS without applying an internal standard. *Chem. Geol.* **2008**, *257*, 34–43. [CrossRef]
50. Ludwig, K. A geochronological toolkit for Microsoft Excel. *Isoplot* **2003**, *3*, 1–70.
51. Li, X.; Li, W.; Wang, X.; Li, Q.; Liu, Y.; Tang, G. Role of mantle-derived magma in genesis of early Yanshanian granites in the Nanling Range, South China: In situ zircon Hf-O isotopic constraints. *Sci. China Ser. D Earth Sci.* **2009**, *52*, 1262–1278. [CrossRef]
52. Yang, Q.; Xia, X.; Zhang, W.; Zhang, Y.; Xiong, B.; Xu, Y.; Wang, Q.; Wei, G. An evaluation of precision and accuracy of SIMS oxygen isotope analysis. *Solid Earth Sci.* **2018**, *3*, 81–86. [CrossRef]
53. Baertschi, P. Absolute18O content of standard mean ocean water. *Earth Planet. Sci. Lett.* **1976**, *31*, 341–344. [CrossRef]
54. Li, X.H.; Long, W.G.; Li, Q.L.; Liu, Y.; Zheng, Y.F.; Yang, Y.H.; Chamberlain, K.R.; Wan, D.F.; Guo, C.H.; Wang, X.C. Penglai zircon megacrysts: A potential new working reference material for microbeam determination of Hf–O isotopes and U–Pb age. *Geostand. Geoanal. Res.* **2010**, *34*, 117–134. [CrossRef]
55. Li, X.; Tang, G.; Gong, B.; Yang, Y.; Hou, K.; Hu, Z.; Li, Q.; Liu, Y.; Li, W. Qinghu zircon: A working reference for microbeam analysis of U-Pb age and Hf and O isotopes. *Chin. Sci. Bull.* **2013**, *58*, 4647–4654. [CrossRef]
56. Xijun, L.; Ji-Feng, X.; Castillo, P.R.; Wenjiao, X.; Yu, S.; Zhiguo, Z.; Xuan-Ce, W.; Songjian, A.; Baohua, W.; Rongguo, H.; et al. Long-lived low Th/U Pacific-type isotopic mantle domain: Constraints from Nd and Pb isotopes of the Paleo-Asian Ocean mantle. *Earth Planet. Sci. Lett.* **2021**, *567*, 117006. [CrossRef]
57. Zhang, Z.; Liu, X.; Xiao, W.; Xu, J.-F.; Shi, Y.; Gong, X.; Hu, R.; Liu, P.; Song, Y.; Xiao, Y. Geochemistry and Sr–Nd–Hf–Pb isotope systematics of late Carboniferous sanukitoids in northern West Junggar, NW China: Implications for initiation of ridge-subduction. *Gondwana Res.* **2021**, *99*, 204–218. [CrossRef]
58. Hoskin, P.W.O.; Black, L.P. Metamorphic zircon formation by solid-state recrystallization of protolith igneous zircon. *J. Metamorph. Geol.* **2000**, *18*, 423–439. [CrossRef]

59. Valley, J.W.; Kinny, P.D.; Schulze, D.J.; Spicuzza, M.J. Zircon megacrysts from kimberlite: Oxygen isotope variability among mantle melts. *Contrib. Mineral. Petrol.* **1998**, *133*, 1–11. [CrossRef]
60. Sun, S.S.; McDonogh, W.F. Chemicalandisotopicsystematics ofoceanicbasalts: Implicationsfor mantlecompositionand processes. *Geol. Lond. Spec.* **1989**, *42*, 313–345.
61. Yu, S.; Zhang, J.; Qin, H.; Sun, D.; Zhao, X.; Cong, F.; Li, Y. Petrogenesis of the early Paleozoic low-Mg and high-Mg adakitic rocks in the North Qilian orogenic belt, NW China: Implications for transition from crustal thickening to extension thinning. *J. Asian Earth Sci.* **2015**, *107*, 122–139. [CrossRef]
62. Chen, Y.; Song, S.; Niu, Y.; Wei, C. Melting of continental crust during subduction initiation: A case study from the Chaidanuo peraluminous granite in the North Qilian suture zone. *Geochim. Cosmochim. Acta* **2014**, *132*, 311–336. [CrossRef]
63. Yang, Y.-H.; Wu, F.-Y.; Wilde, S.A.; Liu, X.-M.; Zhang, Y.-B.; Xie, L.-W.; Yang, J.-H. In situ perovskite Sr–Nd isotopic constraints on the petrogenesis of the Ordovician Mengyin kimberlites in the North China Craton. *Chem. Geol.* **2009**, *264*, 24–42. [CrossRef]
64. Johnson, K. Experimental cpx/and garnet/melt partitioning of REE and other trace elements at high pressures: Petrogenetic implications. *Mineral. Mag. A* **1994**, *58*, 454–455.
65. Defant, M.J.; Drummond, M.S. Derivation of some modern arc magmas by melting of young subducted lithosphere. *Nature* **1990**, *347*, 662–665. [CrossRef]
66. Chung, S.-L.; Liu, D.; Ji, J.; Chu, M.-F.; Lee, H.-Y.; Wen, D.-J.; Lo, C.-H.; Lee, T.-Y.; Qian, Q.; Zhang, Q. Adakites from continental collision zones: Melting of thickened lower crust beneath southern Tibet. *Geology* **2003**, *31*, 1021–1024. [CrossRef]
67. Zhao, T.; Cawood, P.A.; Wang, K.; Zi, J.-W.; Feng, Q.; Nguyen, Q.M.; Tran, D.M. Neoarchean and Paleoproterozoic K-rich granites in the Phan Si Pan Complex, north Vietnam: Constraints on the early crustal evolution of the Yangtze Block. *Precambrian Res.* **2019**, *332*, 105395. [CrossRef]
68. Wu, C.-L.; Yao, S.-Z.; Yang, J.-S.; Zeng, L.-S.; Chen, S.-Y.; Li, H.-B.; Qi, X.-X.; Wooden, J.L.; Mazdab, F.K. Double subduction of the Early Paleozoic North Qilian oceanic plate: Evidence from granites in the central segment of North Qilian, NW China. *Geol. China* **2006**, *33*, 1197–1208.
69. Zhao, J.-L.; Huang, X.-J.; Hu, P.-Q.; Yang, Z.-X.; Fan, Y.; Wang, E.-T.; Yang, F.-B.; Zhang, J.-Y. Petrogenesis and tectonic implications of the Silurian adakitic granitoids in the eastern segment of the Qilian Orogenic Belt, Northwest China. *Acta Geochim.* **2024**, *43*, 72–86. [CrossRef]
70. Zheng, Y.; Chen, Y.; Dai, L.; Zhao, Z. Developing plate tectonics theory from oceanic subduction zones to collisional orogens. *Sci. China Earth Sci.* **2015**, *58*, 1045–1069. [CrossRef]
71. King, P.L.; White, A.; Chappell, B.; Allen, C. Characterization and origin of aluminous A-type granites from the Lachlan Fold Belt, southeastern Australia. *J. Petrol.* **1997**, *38*, 371–391. [CrossRef]
72. Wang, Q.; Xu, J.-F.; Jian, P.; Bao, Z.-W.; Zhao, Z.-H.; Li, C.-F.; Xiong, X.-L.; Ma, J.-L. Petrogenesis of adakitic porphyries in an extensional tectonic setting, Dexing, South China: Implications for the genesis of porphyry copper mineralization. *J. Petrol.* **2006**, *47*, 119–144. [CrossRef]
73. Belousova, E.; Griffin, W.; O'Reilly, S.Y. Zircon crystal morphology, trace element signatures and Hf isotope composition as a tool for petrogenetic modelling: Examples from Eastern Australian granitoids. *J. Petrol.* **2006**, *47*, 329–353. [CrossRef]
74. Yang, J.-H.; Wu, F.-Y.; Wilde, S.A.; Xie, L.-W.; Yang, Y.-H.; Liu, X.-M. Tracing magma mixing in granite genesis: In situ U–Pb dating and Hf-isotope analysis of zircons. *Contrib. Mineral. Petrol.* **2007**, *153*, 177–190. [CrossRef]
75. Rudnick, R.L.; Gao, S. Composition of the Continental Crust. *Treatise Geochem.* **2014**, *4*, 1–51.
76. Thompson, P.M.; Kempton, P.D.; Kerr, A.C. Evaluation of the effects of alteration and leaching on Sm–Nd and Lu–Hf systematics in submarine mafic rocks. *Lithos* **2008**, *104*, 164–176. [CrossRef]
77. Jahn, B.-m.; Wu, F.; Lo, C.-H.; Tsai, C.-H. Crust–mantle interaction induced by deep subduction of the continental crust: Geochemical and Sr–Nd isotopic evidence from post-collisional mafic–ultramafic intrusions of the northern Dabie complex, central China. *Chem. Geol.* **1999**, *157*, 119–146. [CrossRef]
78. Workman, R.K.; Hart, S.R. Major and trace element composition of the depleted MORB mantle (DMM). *Earth Planet. Sci. Lett.* **2005**, *231*, 53–72. [CrossRef]
79. Zindler, A.; Hart, S. Chemical geodynamics. *Annu. Rev. Earth Planet. Sci.* **1986**, *14*, 493–571.
80. Gong, J.; Zhang, J.; Yu, S.; Li, H.; Hou, K. Ca. 2.5 Ga TTG rocks in the western Alxa Block and their implications. *Chin. Sci. Bull.* **2012**, *57*, 4064–4076. [CrossRef]
81. Saunders, A.D.; Storey, M.; Kent, R.; Norry, M. Consequences of plume-lithosphere interactions. *Geol. Soc. Lond. Spec. Publ.* **1992**, *68*, 41–60. [CrossRef]
82. Schmitz, M.D.; Vervoort, J.D.; Bowring, S.A.; Patchett, P.J. Decoupling of the Lu-Hf and Sm-Nd isotope systems during the evolution of granulitic lower crust beneath southern Africa. *Geology* **2004**, *32*, 405–408. [CrossRef]
83. Amelin, Y.; Lee, D.-C.; Halliday, A.N.; Pidgeon, R.T. Nature of the Earth's earliest crust from hafnium isotopes in single detrital zircons. *Nature* **1999**, *399*, 252–255. [CrossRef]
84. Dewey, J.F. Extensional collapse of orogens. *Tectonics* **1988**, *7*, 1123–1139. [CrossRef]

85. Zhang, J.; Yu, S.; Mattinson, C.G. Early Paleozoic polyphase metamorphism in northern Tibet, China. *Gondwana Res.* **2017**, *41*, 267–289. [CrossRef]
86. Song, S.G.; Zhang, G.B.; Zhang, C.; Zhang, L.-F.; Wei, C.-J. Dynamic process of oceanic subduction and continental collision: Petrological constraints of HP-UHP belts in Qilian-Qaidam, the northern Tibetan Plateau. *Chin. Sci. Bull.* **2013**, *58*, 2240–2245. [CrossRef]
87. Wei, Q.-Q.; Hao, L.-B.; Lu, J.-L.; Zhao, Y.-Y.; Zhao, X.-Y.; Shi, H.-L. LA-MC-ICP-MS Zircon U-Pb Dating of Hexipu granite and Its Geological Implications. *Bull. Mineral. Petrol. Geochem.* **2013**, *32*, 729–735, (In Chinese with English Abstract).
88. Chen, Y.-J.; Wang, G.; Waang, W.; He, J.-J.; Li, X.-C.; Zhang, C.; Guo, B.-J. Geological Characteristics and Significance of Middle-coarse Grain Granites in Jiling Area of Longshoushan. *Uranium Geol.* **2022**, *38*, 1111–1121. [CrossRef]
89. Chen, S.; Niu, Y.; Li, J.; Sun, W.; Zhang, Y.; Hu, Y.; Shao, F. Syn-collisional adakitic granodiorites formed by fractional crystallization: Insights from their enclosed mafic magmatic enclaves (MMEs) in the Qumushan pluton, North Qilian Orogen at the northern margin of the Tibetan Plateau. *Lithos* **2016**, *248–251*, 455–468. [CrossRef]
90. Tseng, C.-Y.; Yang, H.-J.; Yang, H.-Y.; Liu, D.; Wu, C.; Cheng, C.-K.; Chen, C.-H.; Ker, C.-M. Continuity of the North Qilian and North Qinling orogenic belts, Central Orogenic System of China: Evidence from newly discovered Paleozoic adakitic rocks. *Gondwana Res.* **2009**, *16*, 285–293. [CrossRef]
91. Tseng, C.-Y.; Zuo, G.-C.; Yang, H.-J.; Yang, H.-Y.; Tung, K.-A.; Liu, D.-Y.; Wu, H.-Q. Occurrence of Alaskan-type mafic-ultramafic intrusions in the North Qilian Mountains, northwest China: Evidence of Cambrian arc magmatism on the Qilian Block. *Isl. Arc* **2009**, *18*, 526–549. [CrossRef]
92. Pearce, J. Sources and settings of granitic rocks. *Epis. J. Int. Geosci.* **1996**, *19*, 120–125. [CrossRef]
93. Eby, G.N. Chemical subdivision of the A-type granitoids: Petrogenetic and tectonic implications. *Geology* **1992**, *20*, 641–644.
94. Qiao, J.; Dong, J.; Song, S.; Allen, M.B.; Wang, C.; Xia, X.; Su, L. Development of a Modern-Style Trench-Arc-Backarc System in the Proto-Tethys Ocean (Qilian Orogenic Belt, NW China). *J. Petrol.* **2023**, *64*, egad029. [CrossRef]

minerals

MDPI

Article

Zircon LA-ICP-MS Dating and Geochemical Characteristics of Rhyolites from the Qushi Area, Tengchong Terrane, Yunnan Province

Xiong Mo [1,2], Chen Gong [1], Yan Shang [1], Jinglong Wu [1], Jialin Wu [2], Ronghui Qi [2], Xiaofeng Wang [3], Qi Guan [2] and Xu Kong [1,*]

1 Ordos Institute of Technology, Ordos 017010, China
2 Yunnan Institute of Geology Survey, Kunming 650216, China
3 Yunnan Planning and Design Institute of Land Resources, Kunming 650216, China
* Correspondence: konx163@163.com

Abstract: The Qushi rhyolites, situated in the eastern sector of the Tengchong terrane, are critical to understanding the Early Cretaceous tectono-magmatic evolution of the Eastern Tethyan Tectonic Domain. Zircon LA-ICP-MS U-Pb geochronology indicates crystallization ages of 118.3–120.5 Ma, with Ti-in-zircon temperatures of 641–816 °C (mean = 716 °C), representing the Early Cretaceous magmatic activity in the Tengchong terrane. Inherited zircons within the rhyolites yield a zircon age of ca. 198.5 Ma, with corresponding Ti-in-zircon temperatures of 615–699 °C (mean = 657 °C), implying the potential presence of an Early Jurassic igneous basement beneath the Qushi region. Geochemically, the rhyolites are classified as calc-alkaline and weakly to moderately peraluminous (A/CNK = 1.07–2.86). These rocks display signatures typical of acidic magmas, marked by significant enrichments in light rare earth elements (LREE: La and Ce) and large ion lithophile elements (LILE: Rb, K, Th and U) while simultaneously exhibiting depletions in high-field-strength elements (HFSE: Nb, Ta, Ti, and P) and heavy rare earth elements (HREE). Trace element signatures further reveal marked depletions in Sr (12.4–244.7 ppm) and Ba while displaying enrichments in Zr and Hf. These geochemical features, including the huge range of the Sr content and A/CNK ratios, suggest both I-type and S-type granite affinities. The Early Cretaceous volcanism of the Qushi rhyolites is likely attributed to the combined effects of subduction and the closure of the Meso-Tethyan Ocean (MTO). This volcanic activity is interpreted to result from subduction-related processes associated with the MTO, potentially involving slab rollback, slab break-off, and subsequent asthenospheric upwelling. The formation of these rhyolites may also be linked to the final closure of the MTO, characterized by the Late Cretaceous collision and amalgamation of the Burma and Tengchong terranes.

Keywords: zircon U–Pb ages; rhyolites; Qushi area; Tengchong terrane; Yunnan province

1. Introduction

The Eastern Tethyan Tectonic Domain (ETTD), situated between the Eurasian Plate and the Indian Plate, records a prolonged geodynamic evolution spanning approximately 400 million years [1]. This domain preserves critical evidence of multi-stage tectonic processes, including the rifting of Gondwana-derived terranes, the opening and closure of the Neo-Tethyan Ocean (NTO), and subsequent continental collision and accretionary orogenesis; thus, it has attracted substantial attention from tectonic and igneous geologists [2–9].

The ETTD comprises a series of accreted crustal blocks (including the Himalayan, Songpan–Ganzi, Lhasa, and Qiangtang terranes) separated by the Indus–Yarlung, Bangong–Nujiang, and Jinsha suture zones (Figure 1a) [6,10,11]. Although paleomagnetic evidence indicates that the Lhasa block underwent a ca. 87° clockwise rotation during the Eocene (ca. 40 Ma), tectonic correlation analyses confirm its structural kinship with the Tengchong terrane in the southeastern ETTD [12,13].

Robust evidence from the Lhasa terrane confirms voluminous Mesozoic magmatic activity, supported by comprehensive geochronological and petrogenetic analyses [2,14–16]. The magmatism in the Lhasa terrane can be categorized into three distinct stages. The first stage, occurring between 140 and 120 Ma, is characterized by calc-alkaline granites [15,16], potentially related to the southward subduction of the Bangong–Nujiang oceanic crust beneath the Lhasa terrane [15,17]. The second stage, spanning 118 to 110 Ma, includes granitoids and volcanic rocks, with granitoids being widely distributed within the Lhasa terrane. This phase may be associated with slab break-off of the Bangong–Nujiang oceanic crust [2,14,15,18,19]. The third stage, during the Late Cretaceous (94 to 72 Ma), is marked by granitic rocks, which may be the result of the delamination of the lower crust of the central–northern Lhasa terrane after the collision of the Lhasa and Qiangtang terranes [20–24].

Compared with the Lhasa terrane, the Jurassic–Cretaceous magmatic activity in the Tengchong terrane remains controversial. The igneous activities spanning the Triassic to Jurassic periods in the Tengchong terrane are scattered across the region (Figure 1b). The Early–Middle Triassic igneous events (250–236 Ma) might have resulted from partial melting of the thickened lower crust after the Tengchong and Baoshan terranes collided and merged. The Late Triassic igneous events (219–213 Ma) could be related to the collision with the Indochina terrane [25]. Additionally, the Early Jurassic magmatic events (198–180 Ma) in the Tengchong terrane could be linked to the subduction of the Meso-Tethys Ocean and the collision between the Tengchong terrane and Eurasian continent [25], or they could have formed in the post-collisional tectonic setting response to the Paleo-Tethyan regime [26]. The Early Cretaceous magmatic events have been ascribed to two primary mechanisms: (1) the development of post-collisional arc systems linked to the subduction of the Meso-Tethyan oceanic crust [27–29]; and (2) flat subduction processes of the far-field Neo-Tethyan oceanic crust. The geological environment of the Late Cretaceous magmatic events have been variably attributed to (1) syn-collisional [30,31] or (2) extensional settings [32,33]. The Tengchong terrane may exhibit a spatially westward migration pattern with three high-magma-addition-rate events during the Early Cretaceous (ca. 131–111 Ma), Late Cretaceous to Early Cenozoic (ca. 76–64 Ma), and Early Cenozoic (ca. 55–49 Ma) [11,34]. Notably, the Early Cenozoic (55–49 Ma) magmatic episode may correlate with slab break-off following the initial India–Asia collision [11,35].

Therefore, establishing further constraints on the Cretaceous tectono-magmatic events of this region is crucial, and as an important component of the southeastern ETTD, the Tengchong terrane consists of various Early and Late Cretaceous magmatic rocks and offers valuable clues to constrain the Early and Late Cretaceous tectono-magmatic events. This study focuses on rhyolites from the Qushi area of the Tengchong terrane, and an integrated investigation, including field geological mapping, petrographic observations, and zircon LA-ICP-MS U-Pb geochronology, was conducted. These multidisciplinary approaches aim to provide more data on Early Cretaceous magmatic events and advance our understanding of southeastern ETTD's Mesozoic tectono-magmatic evolution.

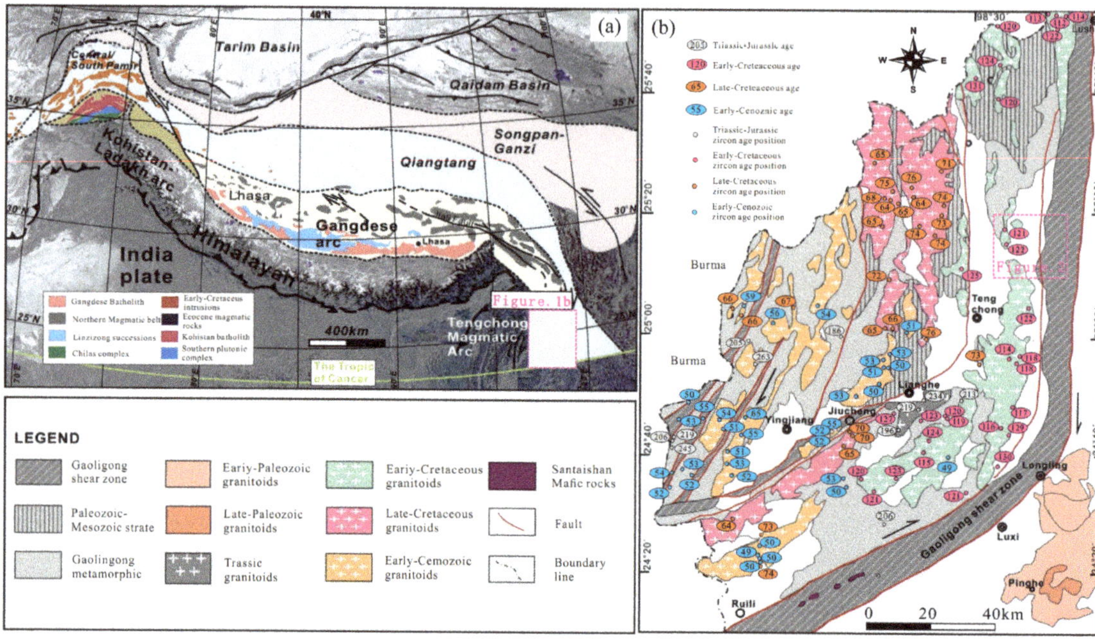

Figure 1. (**a**) Generalized main continental blocks in the ETTD and the positions of the Tengchong terrane (modified after [11,36,37]). (**b**) Schematic regional geological map of the Tengchong area illustrating the distribution of the Mesozoic igneous rocks (modified after [5,6,11,13,34,37,38]).

2. Geological Setting

The study area is located in Qushi town, north of Tengchong city, which is part of the eastern Tengchong terrane, with the Gaoligong shear zone to the east (Figure 1b). The stratigraphic sequence in the study area mainly comprises the Neoproterozoic Meijiashan Group, Permian, and Triassic strata (Figure 2). The Meijiashan Group is subdivided into four formations: the Jiuduhe Formation, Danlonghe Formation, Baohuashan Formation, and Erdaohe Formation. The Permian Bangdu Formation consists of silty slate, while the Triassic Hewanjie Formation is characterized by dolomite and dolomitic limestone. The Neogene Mangbang Formation is composed of sandstone, conglomerate, and claystone. Quaternary basalts are extensively distributed along the western margin of the study area, representing the southwestern extension of the Tengchong Volcanic Belt. Intrusive rocks are predominantly Late Jurassic to Early Cretaceous granitoids. NNE-trending fault structures are well developed in the eastern part of the study area, transecting distribution zone of the Meijiashan Group and inducing mylonitization of the host rocks.

This study focused on rhyolites (originally assigned to the Erdaohe Formation of the Meijiashan Rock Group, [39]) from the Sunjiaying and Xinqiao villages in the Qushi area of the Tengchong terrane (Figures 1b, 2 and 3). Five representative samples were collected and used for comprehensive petrographic, geochemical, and isotopic geochronological analyses. Detailed sampling information such as rock types, GPS locations, and mineral assemblages are systematically listed in Table S1.

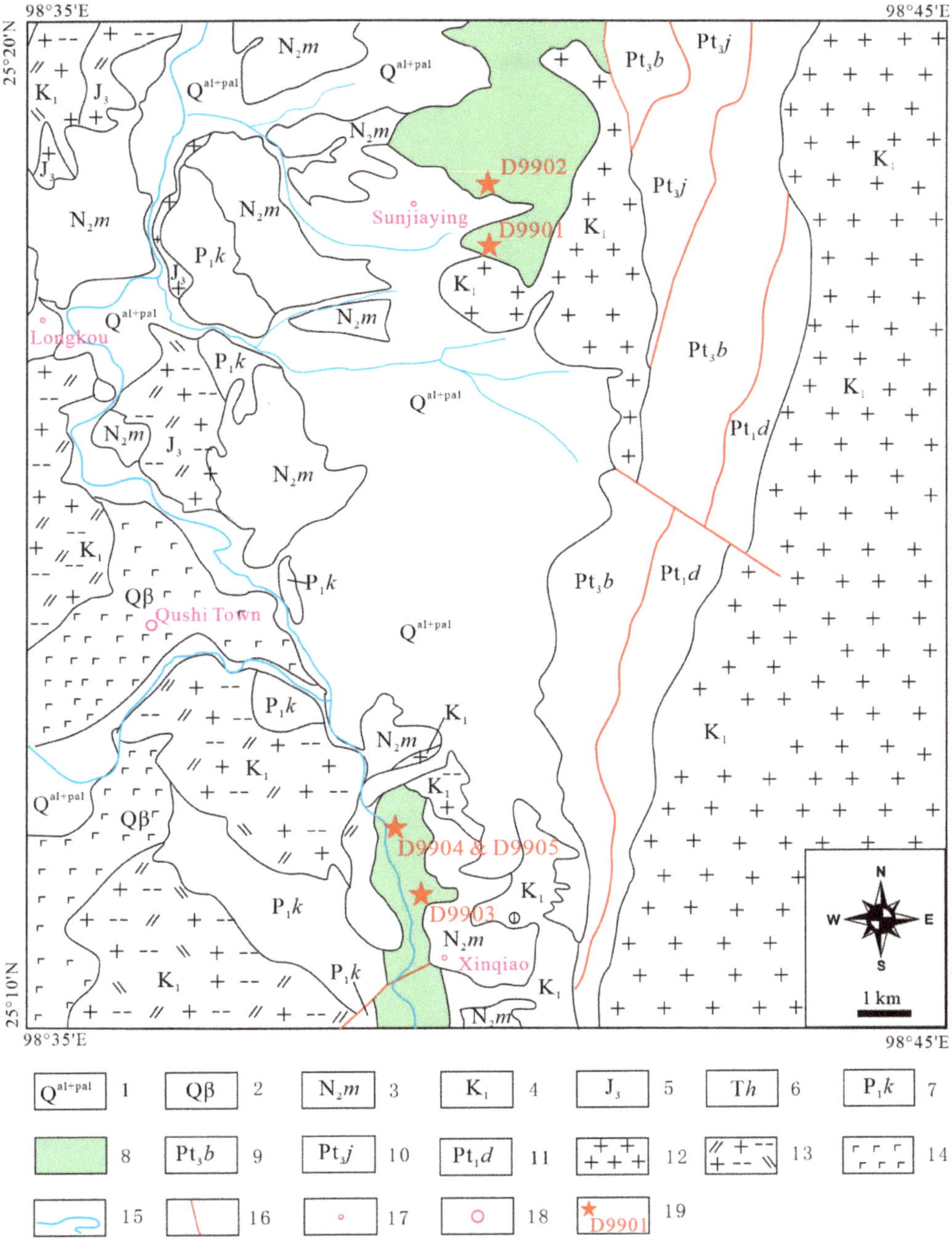

Figure 2. Geological sketch map of the Qushi area (modified after [39]). 1. Quaternary alluvial and proluvial deposits; 2. Quaternary basalt; 3. Mangbang Formation; 4. Early Cretaceous granite; 5. Late Jurassic monzogranite; 6. Triassic Hewanjie Formation; 7. Permian Kongshuhe Formation; 8. Rhyolites; 9. Baohuashan Formation; 10. Jiuduhe Formation; 11. Shanlonghe Formation; 12. granite; 13. biotite monzogranite; 14. basalt; 15. geological boundary; 16. fault; 17. village; 18. town; 19. sampling locations.

Figure 3. Field photos of the rhyolites for sample D9901 (**a**), D9902 (**b**), D9903 (**c**), D9904 (**d**), and D9905 (**e**,**f**) in the Qushi area.

3. Samples and Analytical Methods

3.1. Geochemical Analysis

Geochemical analyses were executed at the Yunnan Institute of Geological Survey to establish the major and trace element compositions of rhyolites. Major elements were measured via X-ray fluorescence (XRF), while trace elements (including REEs) were analyzed with inductively coupled plasma mass spectrometry (ICP-MS). Trace element concentrations were adjusted relative to chondritic and primitive mantle values following [40]'s method. The analytical precision obtained was within ±5% for major elements and within ±10% for trace elements.

3.2. Zircon U-Pb Dating and Trace Element Analysis

Zircon U-Pb dating via LA-ICP-MS (laser ablation inductively coupled plasma mass spectrometry) was executed at Wuhan Sample Solution Analytical Technology Co., Ltd. in Wuhan, China. The detailed operational conditions for the laser ablation system and ICP-MS instrument, as well as data reduction procedures, are presented in [41]. Laser sampling utilized a GeolasPro laser ablation system, which included a COMPexPro 102 ArF excimer laser (Lambda Physik, Göttingen, Germany; 193 nm wavelength, 200 mJ maximum energy) and a MicroLas optical system. An Agilent 7900 ICP-MS instrument was employed to measure ion signal intensities. Helium was applied as the carrier gas, and argon functioned as the make-up gas, combined with the carrier gas via a T-connector prior to entering the ICP torch. A signal smoothing device, known as the "wire" device, was incorporated into the laser ablation system to enhance signal stability [42]. In this study, the laser spot size was set to 24 μm, and the repetition rate was 6 Hz. For U-Pb dating, zircon 91,500 was used as the external calibration standard [43]. Each analysis included a background acquisition period of approximately 20–30 s to measure the gas blank, followed by 50 s of data acquisition from the sample. The trace element compositions of zircon were calibrated using NIST 610 glass (National Institute of Standards & Technology, Gaithersburg, MD, USA) as an external calibration. The Excel-based software ICPMSDataCal (version 9.5) was utilized for offline data reduction, enabling the selection and integration of background and analytical signals, time-drift correction, and quantitative calibration for U-Pb dating and trace element analysis [44,45]. Weighted mean age calculations and concordia diagrams were generated utilizing Isoplot/Ex version 3.0 [46]. A total of 93 spots were analyzed across three samples (D9901, D9902, D9904) with $^{206}Pb/^{238}U$ ages (Table S4). Ti-in-zircon thermometry was calculated using the calibration of [47] (Table S5).

4. Results

4.1. Petrography Characteristics

The rhyolites exhibited prominent flow banding structures, characterized by strongly oriented microcrystals within the matrix, which may be associated with lava flow during rapid cooling processes (Figure 4a–c), and they exhibited a homogeneous porphyritic texture with a glassy to cryptocrystalline groundmass (Figure 4a–f). These rhyolites were mainly composed of mineral assemblages of quartz (Qz, 30–50 vol.%, 200–1500 μm in length), potassium feldspar (Kfs, 20–35 vol.%, euhedral-to-subhedral crystals, 500–3000 μm in length), and minor plagioclase (Pl, 5–15 vol.%). biotite (Bt, 2–6 vol.%) and magnetite (Mag, 1–3 vol.%) were also present, along with accessory minerals such as apatite (Ap) and zircon (Zr) (Table S1; Figure 4). Under plane-polarized light, the quartz phenocrysts displayed resorbed boundaries, indicating magmatic resorption processes.

4.2. Geochemical Features

In the Al_2O_3 vs. TiO_2 discrimination diagram, these samples fell within the field of acidic igneous rocks (Figure 5a) [48,49]. In the Nb/Y vs. Zr/TiO_2 diagram, they were plotted in the rhyodacite/dacite or trachyte–andesite fields (Figure 5b) [50]. The A/NK ratios of the Qushi rhyolites ranged from 1.61 to 1.75, with occasional values up to 2.91. All A/CNK values exceeded 1.0 (Table S2), indicating weakly peraluminous to peraluminous characteristics. The Rittmann index (σ43) varied between 0.96 and 1.72 (Table S2), classifying these rocks as calc-alkaline rhyolites.

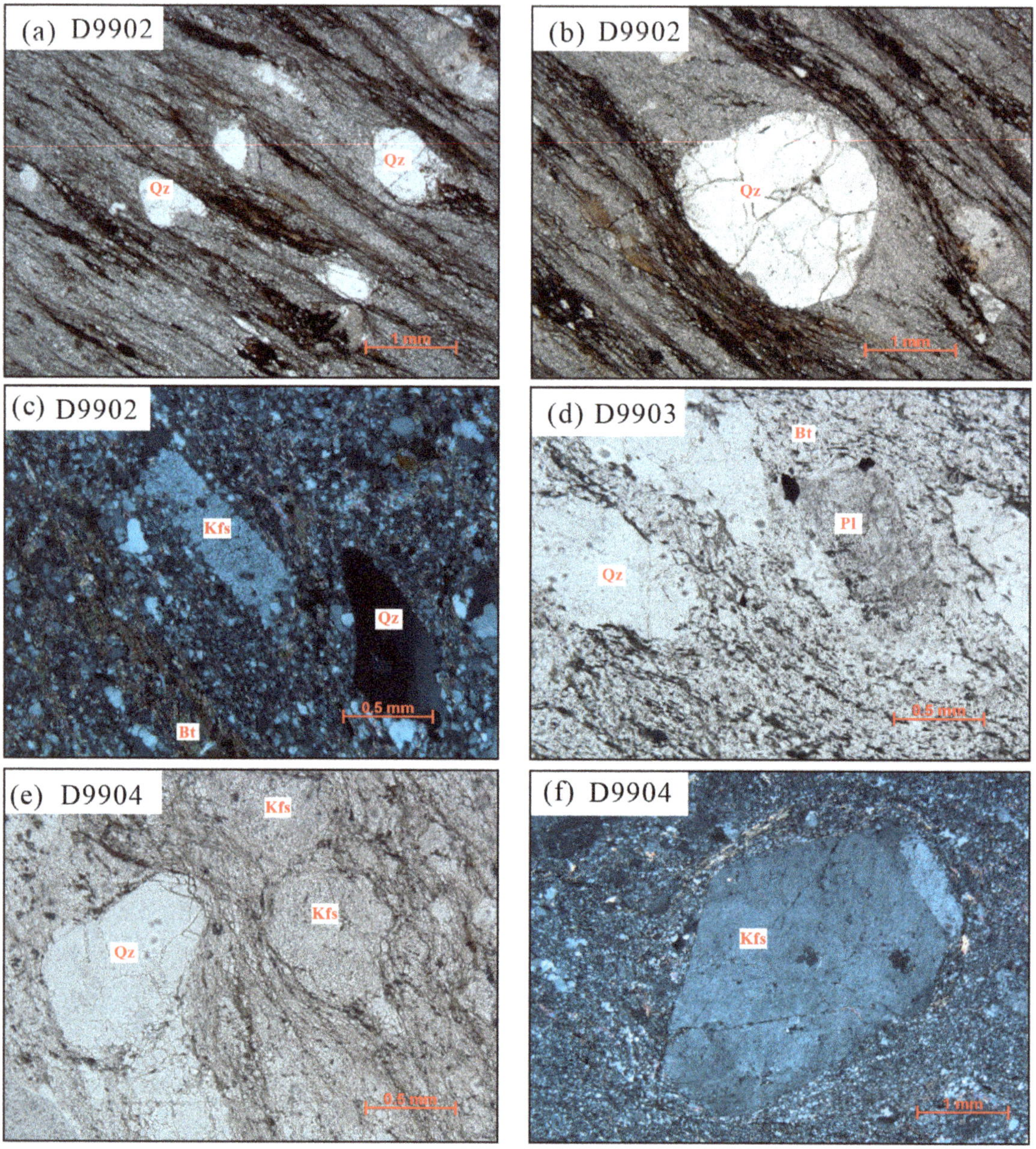

Figure 4. Representative photomicrographs of rhyolites for samples D9902 (**a–c**), D9903 (**d**), and D9904 (**e**,**f**) from the Qushi area. (**a–e**) Photomicrographs taken under plane-polarized light; (**c**,**f**) photomicrographs taken under cross-polarized light. Qz—quartz, Kfs—potassium feldspar, Bt—biotite, Pl—plagioclase.

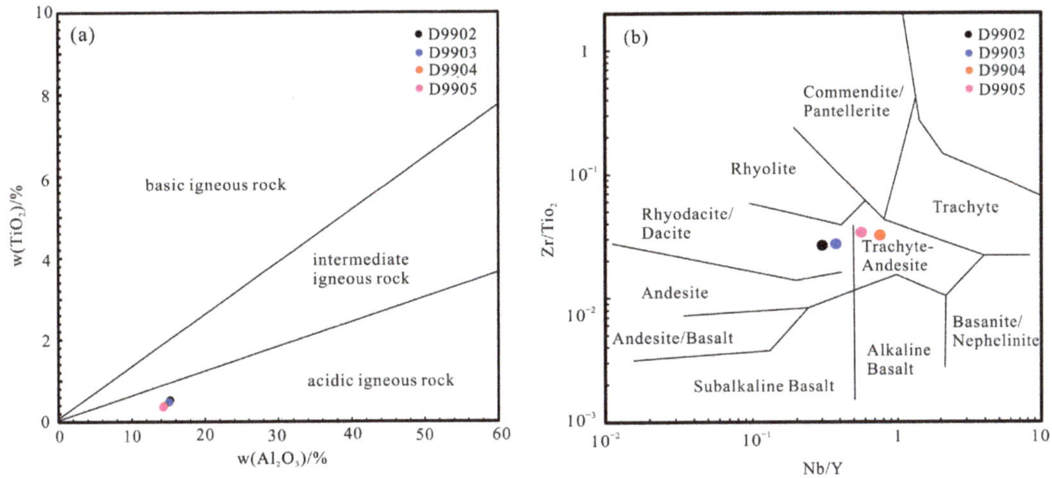

Figure 5. Discrimination diagram of rhyolites in the Qushi area. (**a**) TiO_2 versus Al_2O_3 diagram (modified after [48,49]); (**b**) Zr/TiO_2 versus Nb/Y diagram (modified after [50]).

The total rare earth element (ΣREE) concentrations of the Qushi rhyolites ranged from 173 to 195 ppm. Their LREE/HREE ratios varied between 7.6 and 14.9, while the La_N/Yb_N ratios ranged from 9.8 to 18.5. The δEu and δCe values ranged from 0.69 to 0.76 and 0.48 to 1.01, respectively (Table S3). Chondrite-normalized REE patterns exhibited uniformly right-sloping trends with mid-REE "V"-shaped troughs (Figure 6a).

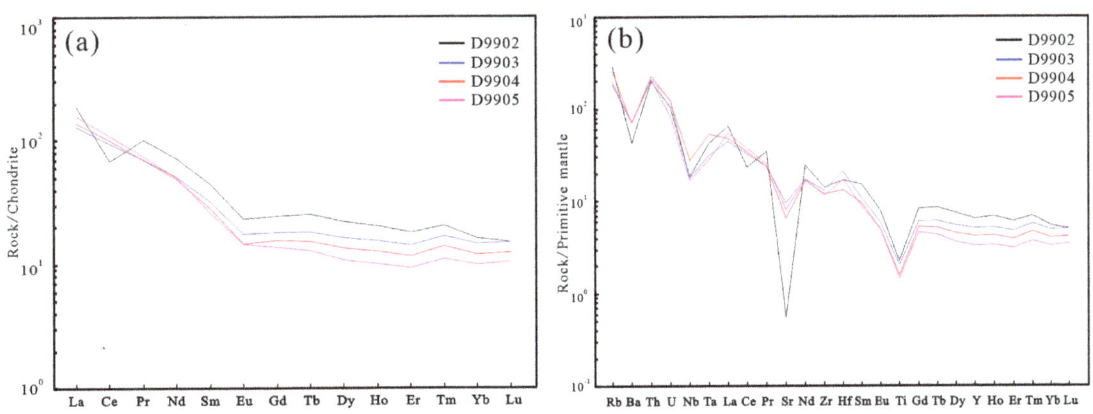

Figure 6. (**a**) Chondrite-normalized REE for the rhyolites from the Qushi area; (**b**) Primitive-mantle-normalized trace element patterns for the rhyolites from the Qushi area. The values for chondrite and primitive mantle are from [40].

Primitive-mantle-normalized trace element spider diagrams displayed a right-sloping "M"-type multi-peak pattern. This pattern shows significant enrichment in large-ion lithophile elements (LILEs: Rb, K), radiogenic elements (Th, U), and magmaphile elements (Ce, La, Zr, Hf). There were also marked negative anomalies in high-field-strength elements (HFSEs: Nb, Ta, P, Ti) and some LILEs (Sr, Ba) (Figure 6b).

4.3. Zircon U-Pb Geochronology

Zircon grains from sample D9901 were predominantly euhedral-to-subhedral prismatic, with rare subangular or fragmented grains. Their lengths ranged from 50 to 100 μm (rarely exceeding 100 μm), and their aspect ratios ranged from 1:1 to 2:1. Cathodoluminescence (CL) images revealed distinct oscillatory zoning in all zircons (Figure 7a). Due to significant radiogenic Pb loss, spots 1, 4, 5, 6, 14, 18, 20, 22, and 26 were excluded (Table S4). The remaining analytical spots yielded Th and U concentrations of 125–751 ppm and 100–1027 ppm, respectively, with Th/U ratios of 0.73–1.47 (Table S4). These zircons displayed a chondrite-normalized rare earth element (REE) pattern characterized by light REE (LREE) depletion, heavy REE (HREE) enrichment, and pronounced negative δEu anomalies (Figure 8a). The magmatic nature of these zircons was confirmed by their concordant ($^{206}U/^{208}Pb$) ages of 118–124 Ma, with a weighted mean age of 120.5 ± 1.2 Ma (MSWD = 2.2, n = 17; Table S4, Figure 8a,b). This age was interpreted as the crystallization age of the rhyolite.

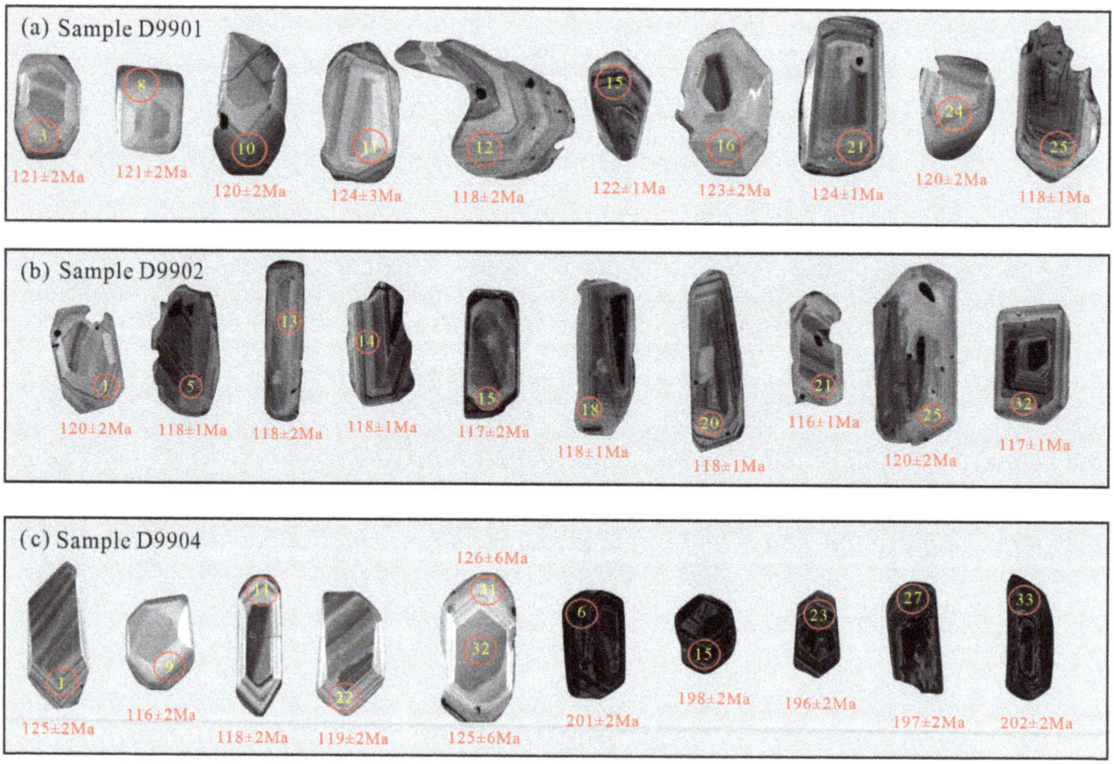

Figure 7. Cathodoluminescence images labeled with $^{206}Pb/^{238}U$ ages of the zircons separated from rhyolites (**a**–**c**). The yellow numbers are the analytical spot symbols, the red numbers are the respective ages ($^{206}U/^{208}Pb$ ages were used when the zircon age was less than 1000 Ma), and the red circle stands for 24 μm.

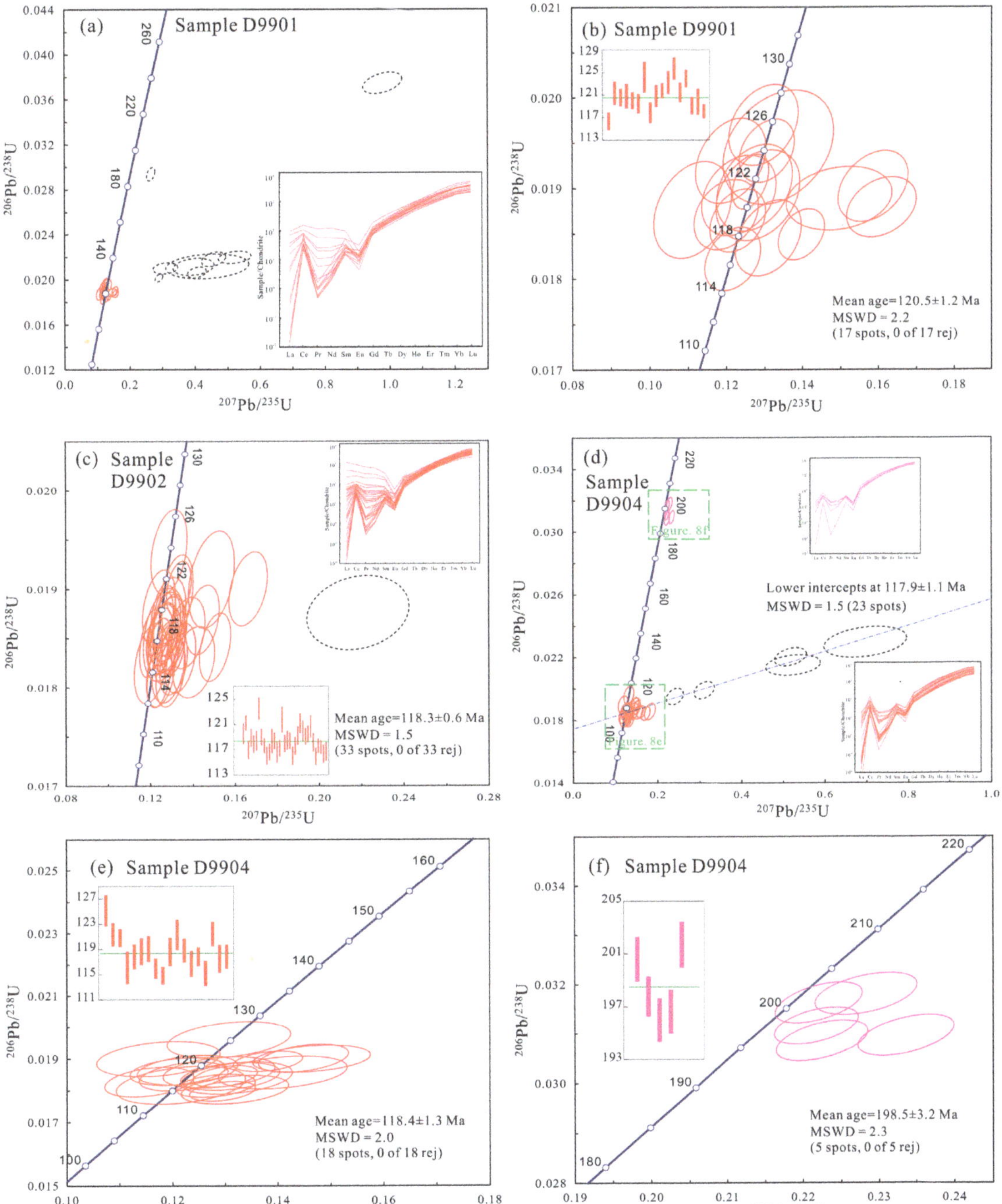

Figure 8. Concordia diagrams, chondrite-normalized rare-earth element patterns, and histograms of Cenozoic ages for the magmatic zircons of rhyolites. (The red circles denote individual zircon analyses from the ca. 120 Ma sample, while the pink circles represent zircon analyses from the ca. 200 Ma sample. The red bar delineates the ca. 120 Ma age range of the younger sample, whereas the pink bar illustrates the ca. 200Ma span of the older sample.) Chondrite values are from [40]. (**a**) Concordia diagram for Sample D9901, showing the distribution of 206Pb/238U and 207Pb/235U

ratios. (**b**) Concordia diagram for Sample D9901 with age histograms and mean age calculation. (**c**) Concordia diagram for Sample D9902. (**d**) Concordia diagram for Sample D9904 with age histograms and mean age calculation. (**e**) Concordia diagram for Sample D9904 with age histograms and mean age calculation. (**f**) Concordia diagram for Sample D9904 with age histograms and mean age calculation.

Zircons from sample D9902 were predominantly euhedral-to-subhedral prismatic or pyramidal crystals, with rare irregular grains. The grain lengths ranged from 100 to 150 μm (rarely exceeding 150 μm), with aspect ratios of 2:1 to 5:1. CL imaging showed well-developed oscillatory zoning (Figure 7b). Excluding spot 7 due to radiogenic Pb loss (Table S4), the remaining spots yielded Th and U concentrations of 129–738 ppm and 167–1165 ppm, respectively, with Th/U ratios of 0.44–1.09 (Table S4). Their REE patterns exhibited LREE depletion, HREE enrichment, and marked negative δEu anomalies (Figure 8c). The magmatic zircons yielded $^{206}U/^{208}Pb$ ages of 116–120 Ma, with a weighted mean age of 118.3 ± 0.6 Ma (MSWD = 1.5, n = 33; Table S4, Figure 8c), representing the crystallization age of the rhyolite.

Zircons from sample D9904 were divided into two groups based on CL characteristics. Group 1 (including spots 6, 15, 23, 27, and 33) were predominantly elongated prisms (50–100 μm in length, with aspect ratios of 1:1 to 3:1, Figure 7c). Their REE patterns showed LREE depletion, HREE enrichment, and strong negative δEu anomalies (Figure 8d). The Th and U concentrations ranged from 792.7 to 1922.2 ppm and 2228 to 3786 ppm, respectively, with Th/U ratios of 0.29 to 0.63. These zircons yielded magmatic U-Pb ages of 183–205 Ma, with a weighted mean age of 198.5 ± 3.2 Ma (MSWD = 2.3, n = 5; Table S4, Figure 8f), interpreted as inherited/captured zircons from an older magmatic source. Group 2 (excluding spots 2, 3, 7, 10, 13, 16, 17, 24, and 28 due to radiogenic Pb loss; Table S4) displayed light CL images, with grain lengths of 50 to 150 μm (rarely exceeding 150 μm) and aspect ratios of 1:1 to 4:1. Their REE patterns also showed LREE depletion, HREE enrichment, and negative δEu anomalies (Figure 8d). The Th and U concentrations ranged from 83.9 to 1089.9 ppm and 121.7 to 530.8 ppm, respectively, with Th/U ratios of 0.29 to 1.69. A lower intercept age of 117.9 ± 1.1 Ma was obtained (Figure 8d), with magmatic U-Pb ages clustering at 112–128 Ma and a weighted mean age of 118.4 ± 1.3 Ma (MSWD = 2, n = 18; Table S4, Figure 8e). This age was interpreted as the crystallization age of the rhyolite.

5. Discussion

5.1. Cretaceous Magmatism in Tengchong Terrane

The Cretaceous magmatic activity in the Tengchong terrane is a significant stage in the Mesozoic tectono-magmatic evolution of the southeastern ETTD. Zircon LA-ICP-MS U-Pb dating of the Qushi rhyolite gave crystallization ages ranging from 120.5 ± 1.2 Ma to 118.3 ± 0.6 Ma, in agreement with Early Cretaceous magmatic events documented in the Tengchong terrane [6,11,25,51,52]. Contemporaneous intrusive and extrusive rocks have been identified throughout the Tengchong terrane.

The study area and its surrounding regions contain monzogranite, syenite, and rhyolite, with ages spanning 104–125 Ma. Specifically, (1) the monzogranites exhibit zircon U-Pb ages of 120–122 Ma [6]; (2) the syenites yield ages of 104–120 Ma [53]; and (3) the rhyolites display ages of 116–125 Ma [54].

The northern region of Tengchong terrane (north of the study area) contains granodiorites, granites, and monzogranites with distinct age clusters. (1) The granodiorites at 120 km north of the study area are 117–125 Ma [30]; (2) the granites at 100 km north of the study area are 115–125 Ma [55]; (3) the northeastern granites (140 km of the study area) are

122–128 Ma [56]; (4) the northern granites (130 km of the study area) are 115–129 Ma [57]; and (5) the monzogranite porphyries (130 km north of the study area) are 120–123 Ma [58].

The western and southern areas of the Tengchong terrane include diverse lithologies. (1) The monzogranites (a) 20 km west of the study area are 113–128 Ma, and (b) those 50 km south of the study area are 114–122 Ma [59]; (2) the quartz diorites (70 km southwest of the study area) are 123.7–130.2 Ma [60]; (3) the quartz monzonites are 121–129 Ma [61]; (4) the granites (60 km southwest of the study area) are 117–131 Ma [55]; (5) the monzonites and syenogranites (70 km south of the study area) are 113–126 Ma [57]; and (6) the granodiorites (80 km south of the study area) are 110–126 Ma [62].

5.2. Rhyolite and Its Petrogenesis

Early Cretaceous magmatic activities were extensively distributed across the Tengchong terrane (130–111Ma, peak at ca. 120 Ma) [25]. In the northern part of the Tengchong terrane, Early Cretaceous granites are mainly high-K calc-alkaline, whereas those in the southern part are medium-K calc-alkaline. These Early Cretaceous granites of the Tengchong terrane exhibit metaluminous-to-peraluminous characteristics, displaying high LREE/HREE and $(La/Yb)_N$ ratios, along with negative Eu anomalies ([25] and its internal references). Compared with Early Cretaceous granites reported by other studies, the Qushi rhyolites are classified as acidic high-K calc-alkaline igneous rocks with REE patterns of strong LREE enrichment and heavy REE depletion with moderately negative Eu anomalies, suggesting crustal melting with plagioclase fractional crystallization. These geochemical features, including strong LREE enrichment, HREE depletion, moderate negative Eu anomalies, and localized weak negative Ce anomalies, suggest a crustal melting-dominated origin with plagioclase fractional crystallization. Coupled with their calc-alkaline and peraluminous characteristics, these data imply magma generation through crustal melting or crust–mantle interaction in a subduction-related tectonic setting. Thus, the geochemical characteristics of the Qushi rhyolite may be closer to the granites in the northern part of the Tengchong terrane.

Prior studies indicate that the dominant rocks of the Early Cretaceous granites of the Tengchong terrane exhibit I-type or highly fractionated I-type granites, with minor S-type granites [6,25,63]. In I-type and S-type granite discrimination diagrams, the Qushi rhyolites span both I-type and S-type granite fields (Figure 9) [52]. Sample D9902 exhibited an A/CNK value of 2.86 (significantly >1.1) along with an extremely low Sr content of 12.4 ppm (significantly <100 ppm) and lower Mg# value of 0.279. In contrast, samples D9903, D9904, and D9905 showed A/CNK ratios ranging from 1.07 to 1.18, with elevated Sr concentrations between 163.8 and 244.7 ppm (all >100 ppm) and higher Mg# values (0.306–0.353) (Table S2). These distinct geochemical signatures suggest that sample D9902 likely represents S-type granite, while samples D9903, D9904, and D9905 may belong to I-type granite [64].

Most of the Early Cretaceous granitoids in northern region of the Tengchong terrane may originated from basaltic magmas, and the minor S-type granitoids in northern region exhibit sedimentary affinities, predominantly originating from argillaceous protoliths [6,25,30,58]. Multiple petrogenetic studies propose that the Early Cretaceous magmatic systems in Tengchong terrane may represent hybrid crust–mantle sources, as evidenced by isotopic signatures and mineralogical assemblages [27,29,54,59,63,65–68]. In the southern part of the Tengchong terrane, the Early Cretaceous granitoids display features typical of a volcanic arc origin, whereas the northern region's granitoids demonstrate collisional origin properties [25]. The occurrence of mafic enclaves in both the southern and northern areas indicates a potential gradual rise in mantle-derived magma during the Early Cretaceous period [25,59,62,63,65–67,69].

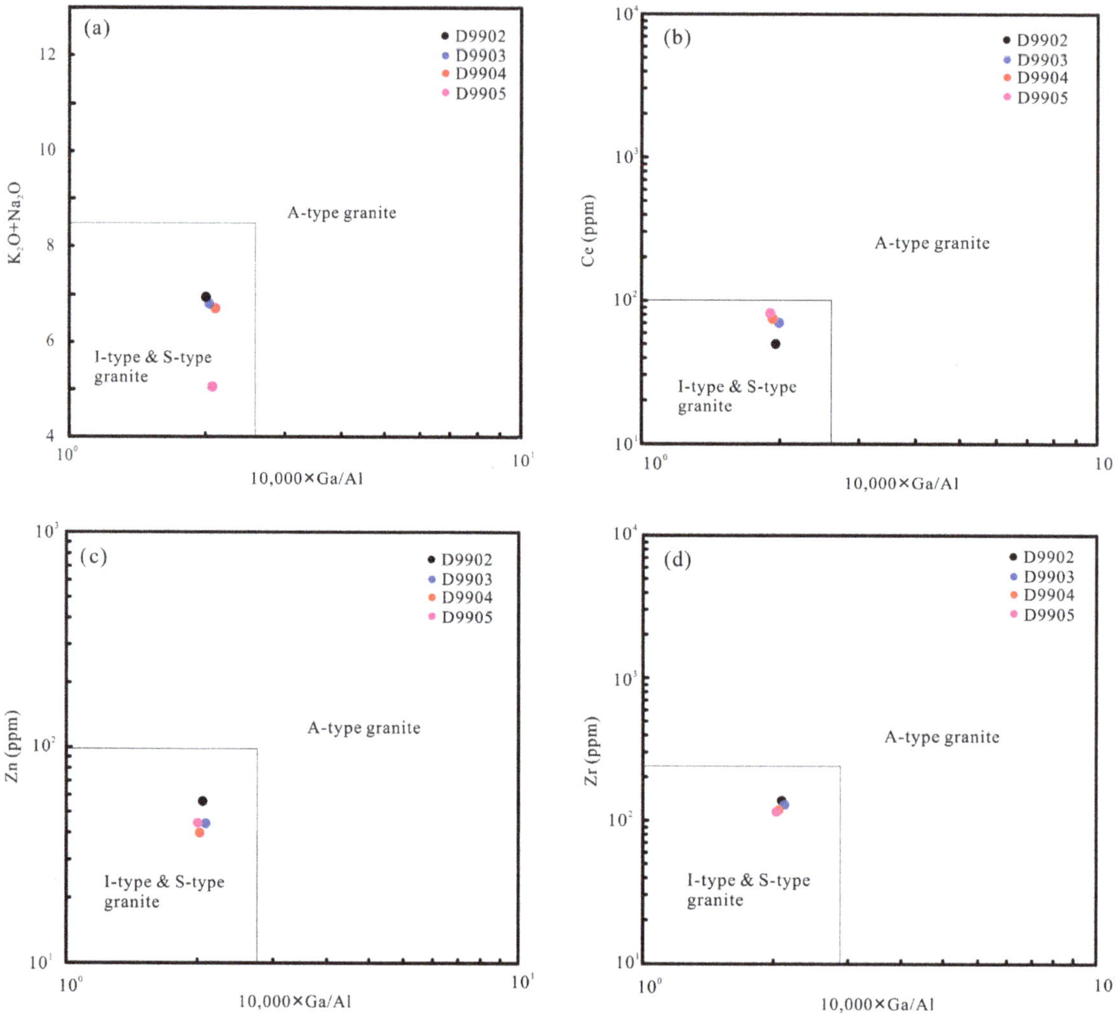

Figure 9. I-type and S-type granite discrimination diagram for the rhyolites from the Qushi area (after [51]). Geochemical classification diagrams for granitic samples. (**a**) K_2O + Na_2O vs. 10,000 × Ga/Al diagram, showing the distribution of samples D9902, D9903, D9904, and D9905. The diagram distinguishes A-type granite from I-type and S-type granite. (**b**) Ce vs. 10,000 × Ga/Al diagram, illustrating the geochemical characteristics of the samples and their classification into A-type or I-type & S-type granite. (**c**) Zr vs. 10,000 × Ga/Al diagram, further differentiating the granitic samples based on their geochemical properties. (**d**) Another Zr vs. 10,000 × Ga/Al diagram, providing additional insights into the classification and geochemical features of the samples.

The A/MF vs. C/MF diagram (Figure 10d) shows that the Qushi rhyolite samples of D9903, D9904, and D9905 may have originated from the partial melt of meta-tonalitic or meta-greywacke, while sample D9902 may have originated from a meta-pelitic source [70]. In the Rb/Sr vs. Rb/Ba diagram (Figure 10e), the samples D9903, D9904, and D9905 of the rhyolites were plotted in a shale or greywacke source, and sample D9902 was plotted in a pelite-derived melt [71]. The igneous source of the Qushi rhyolites align with those reported by Qi et al. (2020) from the neighboring region (Figure 10d,e) [54]. The presence of inherited zircon grains of ca. 198.5 Ma in the Qushi region (Figure 8f) suggests a polygenetic origin for the parent magmas of the rhyolites. This includes contributions from lower-crustal mafic meta-igneous components, meta-sedimentary materials, and minor inputs from mantle-derived sources, with potential contributions from some arc-related materials.

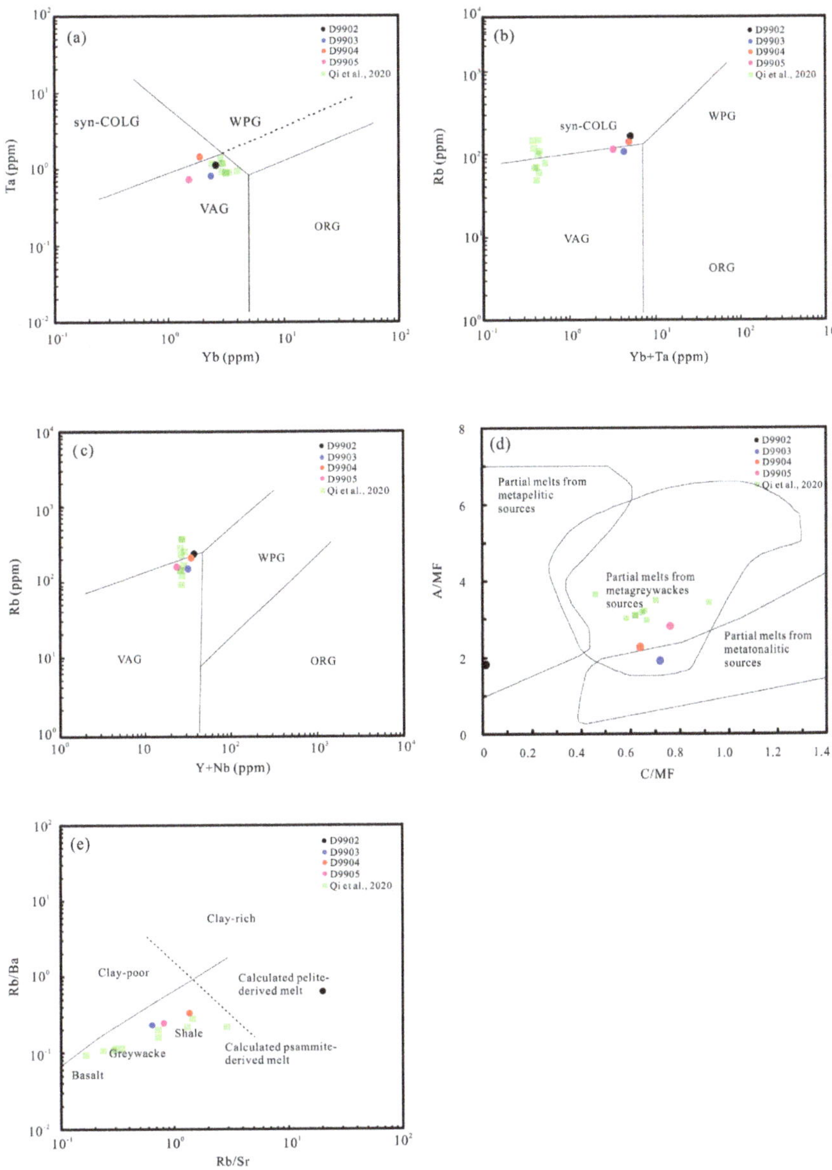

Figure 10. The discrimination diagrams of tectonic setting (after [72]). (**a**) Ta vs. Yb, (**b**) Rb vs. Yb + Ta, and (**c**) Rb vs. Y + Nb discrimination diagrams of tectonic setting. (**d**) A/MF vs. C/MF discrimination diagrams of magma sources (after [70]), A/MF = molar $Al_2O_3/(MgO + TFeO)$, C/MF: molar $CaO/(MgO + TFeO)$. (**e**) Rb/Ba vs. Rb/Sr discrimination diagram for the rhyolites from the Qushi area and [54] (after [71]).

5.3. Tectonic Significance

The Qushi rhyolites were formed at 118.3–120.5 Ma with Ti-in-Zr temperatures of 641–816 °C (average = 716 °C), representing Early Cretaceous magmatism in the Tengchong terrane (Figures 8, 11c and 12). The inherited zircon ages of ca. 198.5 Ma (615–699 °C, average = 657 °C) (Figures 8, 11c and 12) suggest that Early Jurassic igneous units may be present in the basement of the Qushi region. Previous studies indicate that the zircon saturation temperatures for magmatic activities of 198–180 Ma (Tzr < 750 °C) are significantly lower than those for magmatic activities of 130–111 Ma (Tzr > 800 °C) in the Tengchong terrane [25], which is roughly consistent with this study. The Early Jurassic magmatism of S-type granites in the Tengchong terrane may have formed in the tectonic background of a syn-collision or post-collision environment linked to the collision between the Tengchong terrane and the Eurasian continent [6,25,26,53].

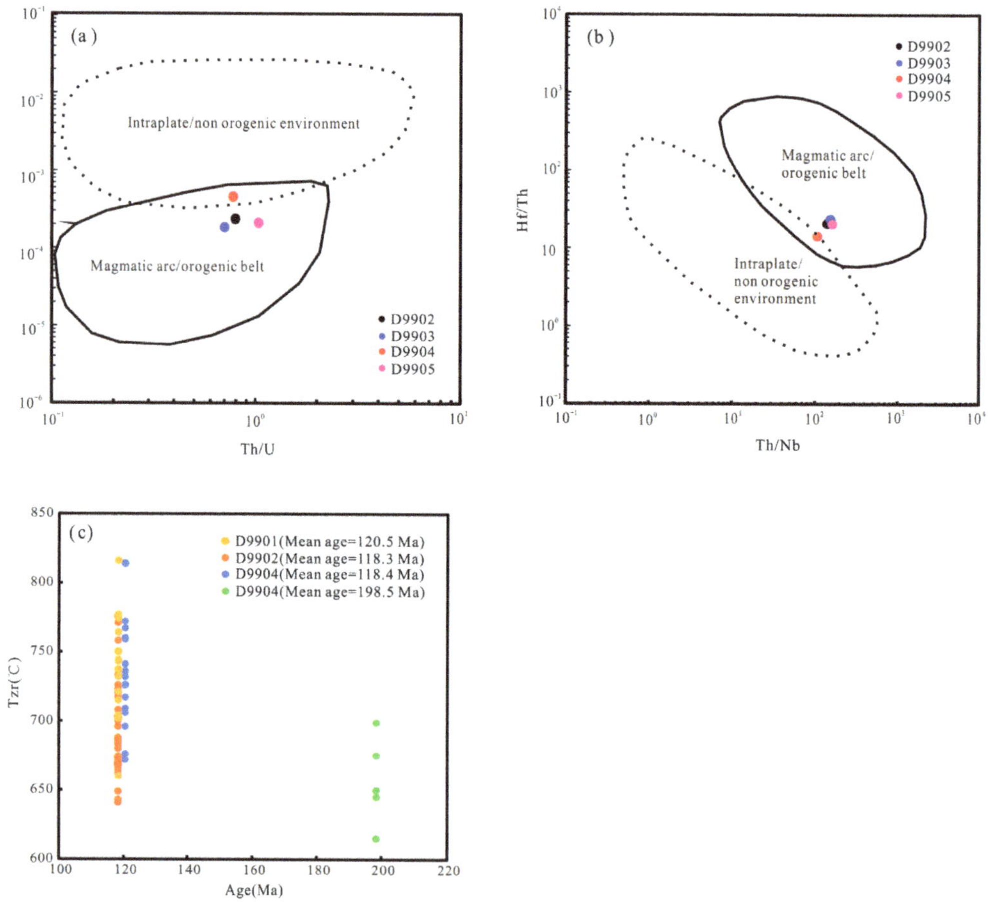

Figure 11. Magmatic zircon trace element discriminating diagrams of rhyolites in the Qushi area (after [73,74]). (**a**) Th/U vs. Nb/Hf, (**b**) Th/Nb vs. Hf/Th discrimination diagrams of tectonic setting. (**c**) Rhyolites Ti-in-Zr temperatures [47]. WPG = within-plate granite, VAG = volcanic arc granite, post-COLG = post-collision granite, syn-COLG = syn-collisional granite, ORG = ocean ridge granite.

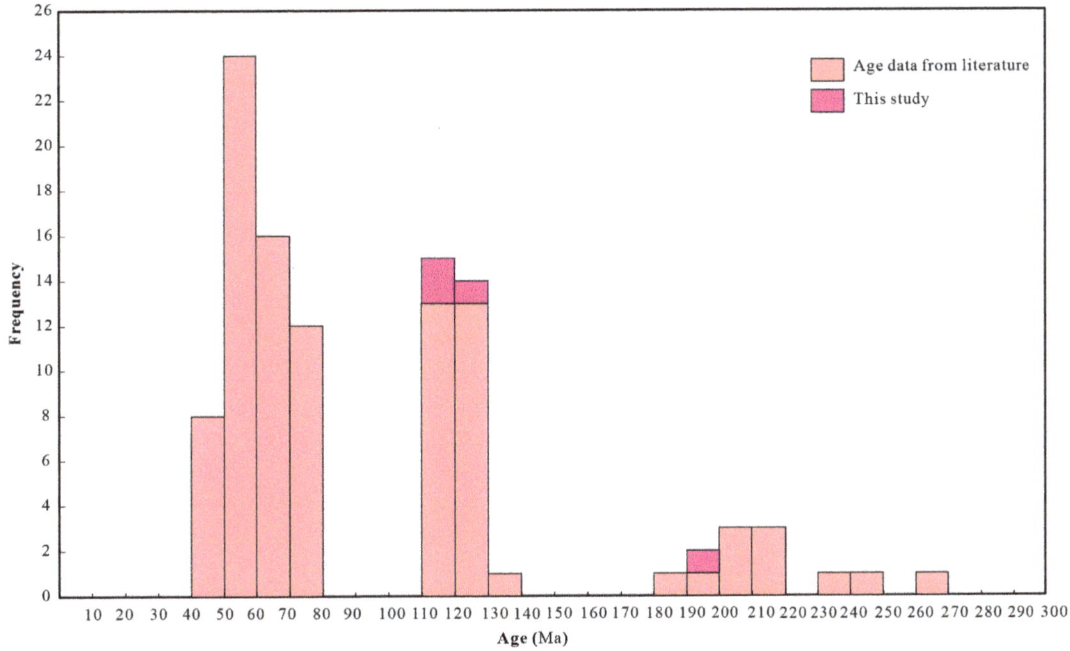

Figure 12. Histogram of reported magmatic zircon ages from the Tengchong terrane. Data sourced from the same as Figure 1b.

In relation to the Early Cretaceous magmatic activities, six unique tectonic models have been proposed for the Tengchong terrane, including (1) an arc setting related to the NTO flat subducted eastward beneath the Tengchong terrane, followed by rollback of the subducted slab [62,69,75–79]; (2) a magmatic arc setting linked to northward (eastward) subduction of the MTO crust beneath the Tengchong terrane [6,27–29,63,66,80–83]; (3) a post-collisional tectonic setting of crustal thickening related to the closure of the MTO [30,58]; (4) southward subduction of the MTO beneath the Tengchong terrane [28,29,53,65,67,68]; (5) the slab break-off of the MTO beneath the Tengchong terrane linked to the closure of the Meso-Tethyan Ocean [28,59,66]; and (6) the north-to-south scissor-shaped closure of the MTO [25].

On the tectonic discrimination diagrams (Figure 10a–c), the rhyolites were plotted within the volcanic arc granite (VAG) and syn-collisional granite (syn-COLG) fields, indicating a subduction-related setting. In addition, on the tectonic discrimination diagrams (Figure 11a–b), the rhyolites were plotted within the magmatic arc and orogenic belt, indicating a subduction-related compression setting. Considering the MTO was situated between the Tengchong and Burma terranes and still open between 130–111 Ma [6,63,81], it is unlikely that the Early Cretaceous magmatism was triggered by the subduction of the NTO or affected by the southward (westward) subduction of the MTO. Bimodal volcanic rocks of ca. 120 Ma formed in a back-arc basin system have been discovered in Baoshan terrane and indicate that the magmatic arc may be present in the Tengchong terrane [78]. Furthermore, previous studies suggest that the collision between the Burma terrane and the Tengchong terrane may have occurred during the Cretaceous period [6,63,81,84]. Yang et al. (2024) proposed that the MTO's north-to-south scissor-shaped closure may have triggered the Early Cretaceous magmatism of the Tengchong terrane [25]; however, due to limited regional data, additional evidence is required to validate this hypothesis.

In conclusion, the formation of the Qushi rhyolite may be related to the subduction, slab rollback, slab break-off, and upwelling of the asthenosphere associated with the MTO, and it may also be linked with the final closure of the MTO (the collision and malgamation of the Burma terrane and the Tengchong terrane) (Figure 13).

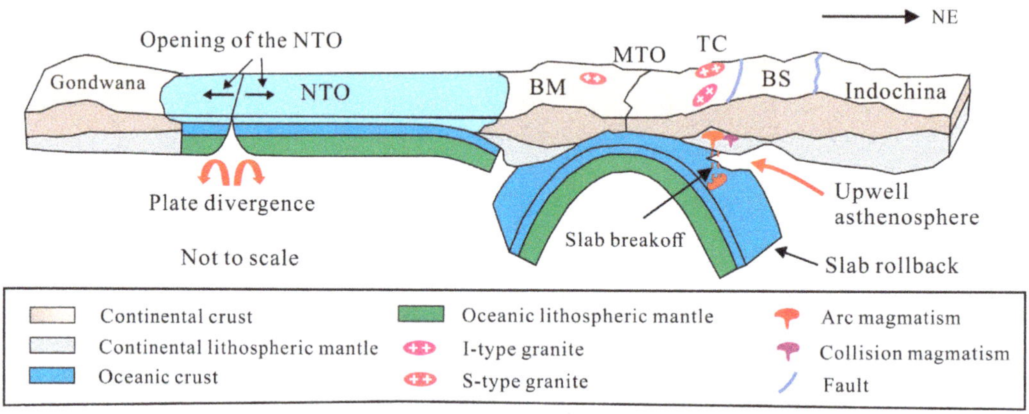

Figure 13. Conceptual frameworks illustrating the Early Cretaceous geodynamic evolution of the Tengchong terrane (modified after [25]). TC: Tengchong terrane, BS: Baoshan terrane, PTO: Paleo-Tethys Ocean, BM: Burma terrane, NTO: Neo-Tethys Ocean, MTO: Meso-Tethys Ocean.

(1) Zircon LA-ICP-MS U-Pb dating indicates that the Qushi rhyolites were crystallized from 118.3 Ma to 120.5 Ma, representing Early Cretaceous magmatic activity in the Tengchong terrane. Additionally, the rhyolites contain inherited zircon ages of approximately 198.5 Ma, suggesting the potential presence of Early Jurassic igneous units in the basement of the Qushi region.
(2) The Qushi rhyolites are classified as acidic, calc-alkaline rhyolites with weakly peraluminous-to-peraluminous characteristics and exhibit petrogeochemical characteristics of enrichment in LREE, Rb, K, Th, U, Ce, La, Zr and Hf and depletion of HREE, Nb, Ta, P, Ti, Sr and Ba.
(3) The A/CNK values of the Qushi rhyolites range from 1.07 to 2.86, alongside Sr content variations from 12.4 to 244.7 ppm, suggest both I-type and S-type granite affinities, indicative of crustal melting with plagioclase fractionation.
(4) The Qushi rhyolites probably formed in syn-COLG and VAG tectonic settings related to the combined effects of the subduction and the closure of the MTO.

Supplementary Materials: The following supporting information can be downloaded at: https://www.mdpi.com/article/10.3390/min15030315/s1, Table S1: Sample numbers, rock types, GPS locations, and mineral assemblages of the rocks in this study. The mineral abbreviations are derived from [85]; Table S2: Analysis results of major elements of rhyolites from the Qushi area. A/NK = Al_2O_3/(Na_2O + K_2O) (mol); A/CNK = Al_2O_3/(CaO + Na_2O + K_2O) (mol); σ43 = $(Na_2O + K_2O)^2/(SiO_2\text{-}43)$(Wt%); A/MF = Al_2O_3/(TFeO + MgO) (mol); C/MF = CaO/(TFeO + MgO) (mol); Mg# = MgO/(MgO + TFeO) (mol); Table S3: Analysis results of trace elements of rhyolites from the Qushi area; Table S4: LA-ICP-MS U–Pb isotope data for zircons of rhyolites from the Qushi area; Table S5: Trace element data for zircons of rhyolites from the Qushi area. Ti-in-Zr temperatures were calculated based on [47].

Author Contributions: X.M. and C.G. equally contributed to the work, and they are joint first authors. Conceptualization, X.K.; methodology, X.M. and C.G.; validation, X.M., C.G. and X.K.; formal analysis, X.M.; investigation, X.M., J.W. (Jialin Wu), R.Q., X.W. and Q.G.; resources, X.M.; data curation, X.M., C.G. and X.K.; writing—original draft preparation, X.M. and C.G.; writing—review and editing, X.K., Y.S. and J.W. (Jinglong Wu); supervision, X.K.; funding acquisition, X.K., Y.S. and X.M. All authors have read and agreed to the published version of the manuscript.

Funding: This work was supported by the High-level Talent Scientific Research Start-Up Fund of Ordos (grant number: DC2400001462), Ordos Mining Area Geohazard Prevention and Geoenvironmental Protection Engineering Research Center (grant number: RZ2300001544), Planning and Management of Paleontological Fossil Conservation in Yunnan Province (project number: 530000210000000021416), and 1:50,000 Scale Regional Geological Survey of six geological maps including Dongpengyang in Yunnan Province (project number: D2017012).

Data Availability Statement: Data are contained within the article or Supplementary Materials.

Acknowledgments: We extend our sincere gratitude to Jing Li from the Yunnan Geological Survey for their enthusiastic help and assistance during the fieldwork of this study. We sincerely thank Ming Wang for their helpful discussions in improving the quality of this paper. We also thank two anonymous reviewers for their constructive comments.

Conflicts of Interest: The authors declare no conflicts of interest.

References

1. Metcalfe, I. Gondwana dispersion and Asian accretion: Tectonic and palaeogeographic evolution of eastern Tethys. *J. Asian Earth Sci.* **2013**, *66*, 1–33. [CrossRef]
2. Zhu, D.C.; Mo, X.; Niu, Y.; Zhao, Z.; Wang, L.; Liu, Y.; Wu, F. Geochemical investigation of Early Cretaceous igneous rocks along an east–west traverse throughout the central Lhasa Terrane, Tibet. *Chem. Geol.* **2009**, *268*, 298–312. [CrossRef]
3. Hou, Z.; Zhang, H. Geodynamics and metallogeny of the eastern Tethyan metallogenic domain. *Ore Geol. Rev.* **2015**, *70*, 346–384. [CrossRef]
4. Hu, X.; Garzanti, E.; Wang, J.; Huang, W.; An, W.; Webb, A.A. The timing of India-Asia collision onset–Facts, theories, controversies. *Earth-Sci. Rev.* **2016**, *160*, 264–299. [CrossRef]
5. Cao, H.-W.; Zhang, Y.-H.; Santosh, M.; Zhang, S.-T.; Tang, L.; Pei, Q.-M. Mineralogy, zircon U–Pb–Hf isotopes, and whole-rock geochemistry of Late Cretaceous–Eocene granites from the Tengchong terrane, western Yunnan, China: Record of the closure of the Neo-Tethyan Ocean. *Geol. J.* **2017**, *52*, 1423–1441. [CrossRef]
6. Cao, H.; Zhang, Y.; Tang, L.; Hollis, S.P.; Zhang, S.; Pei, Q.; Yang, C.; Zhu, X. Geochemistry, zircon U–Pb geochronology and Hf isotopes of Jurassic-Cretaceous granites in the Tengchong terrane, SW China: Implications for the Mesozoic tectono-magmatic evolution of the Eastern Tethyan Tectonic Domain. *Int. Geol. Rev.* **2019**, *61*, 257–279. [CrossRef]
7. Kong, X.; Mi, W.; Zhu, L.; Yang, W. Zircon U-Pb chronology and provenance of the Paleogene sandstones in the Nima basin, Tibet: Implication for coeval paleogeography. *Arab. J. Geosci.* **2019**, *12*, 692. [CrossRef]
8. Zhu, D.C.; Wang, Q.; Zhao, Z.D.; Chung, S.L.; Cawood, P.A.; Niu, Y.; Mo, X.X. Magmatic record of India-Asia collision. *Sci. Rep.* **2015**, *5*, 14289. [CrossRef] [PubMed]
9. Qi, X.X.; Sheng, H.; Wei, C.; Ren, Y.F.; Cai, Z.H.; Ji, F.B.; Liang, F.; Liu, X. Longlin-Ruili subduction-accretionary complex belt in the southeastern Gaoligong orogen, and its relationship with the evolution of the Meso-Tethyan Ocean. *Acta Petrol. Sin.* **2021**, *37*, 3067–3094. (In Chinese with English abstract)
10. Yin, A.; Harrison, T. Geologic Evolution of the Himalayan-Tibetan Orogen. *Annu. Rev. Earth Planet. Sci.* **2000**, *28*, 211–280. [CrossRef]
11. Zhu, R.Z.; Lai, S.; Paterson, S.R.; Luffi, P.; Zhang, B.; Pompe, L.R. Westward migration of high-magma addition rate events in SE Tibet. *Tectonophysics* **2022**, *807*, 229096. [CrossRef]
12. Kornfeld, D.; Eckert, S.; Appel, E.; Ratschbacher, L.; Sonntag, B.; Pfänder, J.A.; Ding, L.; Liu, D. Cenozoic clockwise rotation of the Tengchong block, southeastern Tibetan Plateau: A paleomagnetic and geochronologic study. *Tectonophysics* **2014**, *628*, 105–122. [CrossRef]
13. Xu, Z.Q.; Wang, Q.; Cai, Z.H.; Dong, H.W.; Li, H.Q.; Chen, X.J.; Duan, X.D.; Cao, H.; Li, J.; Burg, J.-P. Kinematics of the Tengchong terrane in SE Tibet from the Late Eocene to Early Miocene: Insights from coeval mid-crustal detachments and strike-slip shear zones. *Tectonophysics* **2015**, *665*, 127–148. [CrossRef]

14. Qu, X.X.; Wang, R.; Xin, H.; Jiang, J.; Chen, H. Age and petrogenesis of A-type granites in the middle segment of the Bangonghu–Nujiang suture, Tibetan plateau. *Lithos* **2012**, *146–147*, 264–275. [CrossRef]
15. Chen, Y.; Zhu, D.; Zhao, Z.; Meng, F.; Wang, Q.; Santosh, M.; Wang, L.; Dong, G.; Mo, X. Slab breakoff triggered ca. 113 Ma magmatism around Xainza area of the Lhasa Terrane, Tibet. *Gondwana Res.* **2014**, *26*, 449–463.
16. Li, F.; Tang, J.; Wang, N.; Liu, Z.; Song, Y.; Zhang, J.; Ma, X.; Fu, B.; Li, H.; Han, S. Petrogenesis and Geodynamic Implications of Early Cretaceous (~130 Ma) Magmatism in the Baingoin Batholith, Central Tibet: Products of Subducting Slab Rollback. *Acta Geol. Sin. (Engl. Ed.)* **2022**, *96*, 1960–1978. [CrossRef]
17. Zhu, D.C.; Zhao, Z.; Niu, Y.; Mo, X.; Chung, S.; Hou, Z.; Wang, L.; Wu, F. The Lhasa Terrane: Record of a microcontinent and its histories of drift and growth. Earth Planet. *Sci. Lett.* **2011**, *301*, 241–255.
18. Mi, W.T.; Kong, X.; Zhang, D.; Hu, J.; Wan, Y. Zircon LA-ICP-MS dating and geochemical characteristics of I-type granitoids from the Yanhu area, west segment of the Bangongco-Nujiang suture (western Tibet): Petrogenesis and implications for the southward subduction of the Tethyan Ocean. *J. Geol. Soc. India* **2017**, *90*, 335–346. [CrossRef]
19. Ma, X.; Song, Y.; Tang, J.; Chen, W. Newly identified rhyolite-biotite monzogranite (A2-type granite)-norite belt from the Bangong-Nujiang collision zone in Tibet Plateau: Evidence for the slab break-off beneath the Lhasa Terrane. *Lithos* **2020**, *366–367*, 105565. [CrossRef]
20. Li, Y.; He, J.; Wang, C.; Santosh, M.; Dai, J.; Zhang, Y.; Wei, Y.; Wang, J. Late Cretaceous K-rich magmatism in central Tibet: Evidence for early elevation of the Tibetan plateau? *Lithos* **2013**, *160*, 1–13. [CrossRef]
21. Wang, Q.; Zhu, D.; Zhao, Z.; Liu, S.; Chung, S.; Li, S.; Liu, D.; Dai, J.; Wang, L.; Mo, X. Origin of the ca. 90 Ma magnesia-rich volcanic rocks in SE Nyima, central Tibet: Products of lithospheric delamination beneath the Lhasa-Qiangtang collision zone. *Lithos* **2014**, *198*, 24–37. [CrossRef]
22. Chen, J.; Xu, J.; Yu, H.; Wang, B.; Wu, J.; Feng, Y. Late Cretaceous high-Mg# granitoids in southern Tibet: Implications for the early crustal thickening and tectonic evolution of the Tibetan Plateau? *Lithos* **2015**, *232*, 12–22.
23. Yi, J.; Wang, Q.; Zhu, D.; Li, S.; Liu, S.; Wang, R.; Zhang, L.L.; Zhao, Z. Westward-younging high-Mg adakitic magmatism in central Tibet: Record of a westward-migrating lithospheric foundering beneath the Lhasa–Qiangtang collision zone during the Late Cretaceous. *Lithos* **2018**, *316*, 92–103. [CrossRef]
24. Sun, M.; Tang, J.; Chen, W.; Ma, X.; Qu, X.; Song, Y.; Li, X.; Ding, J. Process of lithospheric delamination beneath the Lhasa–Qiangtang collision orogen: Constraints from the geochronology and geochemistry of Late Cretaceous volcanic rocks in the Lhasa terrane, central Tibet. *Lithos* **2020**, *356*, 10529. [CrossRef]
25. Yang, X.; Zhang, Z.; Zhou, Y.; Yang, J. Spatio-temporal analysis of Permian-Cretaceous magmatic activities in the Tengchong block: Implications for its tectono-magmatic evolution. *Geosci. Front.* **2024**, *15*, 101920. [CrossRef]
26. Wang, G.; Liu, Z.; Tan, S.; He, X.; Wang, Y.; Qi, C.; Ye, H. Petrogenesis of the Early Jurassic Longtang and Menglong Peraluminous Granites in Tengchong Terrane, and their Tectonic Implication. *Acta Geol. Sin. (Engl. Ed.)* **2022**, *96*, 1079–1990. [CrossRef]
27. Zhu, R.Z.; Lai, S.; Qin, J.; Zhao, S. Early-Cretaceous highly fractionated I-type granites from the northern Tengchong block, western Yunnan, SW China: Petrogenesis and tectonic implications. *J. Asian Earth Sci.* **2015**, *100*, 145–163. [CrossRef]
28. Zhu, R.Z.; Lai, S.; Qin, J.; Zhao, S. Petrogenesis of late Paleozoic-to-early Mesozoic granitoids and metagabbroic rocks of the Tengchong Block, SW China: Implications for the evolution of the eastern Paleo-Tethys. *Int. J. Earth Sci.* **2017**, *107*, 431–457. [CrossRef]
29. Zhao, S.; Lai, S.; Qin, J.; Zhu, R. Tectono-magmatic evolution of the Gaoligong belt, southeastern margin of the Tibetan plateau: Constraints from granitic gneisses and granitoid intrusions. *Gondwana Res.* **2016**, *35*, 238–256. [CrossRef]
30. Xu, Y.G.; Yang, Q.J.; Lan, J.B.; Luo, Z.Y.; Huang, X.L.; Shi, Y.R.; Xie, L.W. Temporal-spatial distribution and tectonic implications of the batholiths in the Gaoligong-Tengliang-Yingjiang area, western Yunnan: Constraints from zircon U-Pb ages and Hf isotopes. *J. Asian Earth Sci.* **2012**, *53*, 151–175. [CrossRef]
31. Qi, X.; Zhu, L.; Grimmer, J.C.; Hu, Z. Tracing the Transhimalayan magmatic belt and the Lhasa block southward using zircon U–Pb, Lu–Hf isotopic and geochemical data: Cretaceous—Cenozoic granitoids in the Tengchong block, Yunnan, China. *J. Asian Earth Sci.* **2015**, *110*, 170–188. [CrossRef]
32. Chen, X.; Hu, R.; Bi, X.; Zhong, H.; Lan, J.; Zhao, C.; Zhu, J. Petrogenesis of metaluminous A-type granitoids in the Tengchong–Lianghe tin belt of southwestern China: Evidences from zircon U–Pb ages and Hf–O isotopes, and whole-rock Sr–Nd isotopes. *Lithos* **2015**, *212*, 93–110. [CrossRef]
33. Zhao, S.; Lai, S.; Qin, J.; Zhu, R.; Wang, J. Geochemical and geochronological characteristics of Late Cretaceous to Early Paleocene granitoids in the Tengchong Block, Southwestern China: Implications for crustal anatexis and thickness variations along the eastern Neo-Tethys subduction zone. *Tectonophysics* **2017**, *694*, 87–100. [CrossRef]
34. Cao, H.; Zou, H.; Zhang, Y.H.; Zhang, S.T.; Zheng, L.; Zhang, L.K.; Tang, L.; Pei, Q.-M. Late Cretaceous magmatism and related metallogeny in the Tengchong area: Evidence from geochronological, isotopic and geochemical data from the Xiaolonghe Sn deposit, western Yunnan, China. *Ore Geol. Rev.* **2016**, *78*, 196–212. [CrossRef]

35. Bouilhol, P.; Jagoutz, O.; Hanchar, J.M.; Dudas, F.O. Dating the India–Eurasia collision through arc magmatic records. *Earth Planet. Sci. Lett.* **2013**, *366*, 163–175. [CrossRef]
36. Chung, S.; Chu, M.F.; Zhang, Y.; Xie, Y.; Lo, C.H.; Lee, T.Y.; Lan, C.Y.; Li, X.; Zhang, Q.R.; Wang, Y. Tibetan tectonic evolution inferred from spatial and temporal variations in post-collisional magmatism. *Earth-Sci. Rev.* **2005**, *68*, 173–196. [CrossRef]
37. Chapman, J.B.; Scoggin, S.H.; Kapp, P.A.; Carrapa, B.; Ducea, M.; Worthington, J.R.; Oimahmadov, I.; Gadoev, M. Mesozoic to Cenozoic magmatic history of the Pamir. *Earth Planet. Sci. Lett.* **2018**, *482*, 181–192. [CrossRef]
38. Zhao, S.; Fang, X.; Tai, W.; Lai, S.; Pei, X.; Li, Z. Origin of granitic magma within continental margin arc magmatic lull: Constraints by the Late Cretaceous granites in Tengchong Block, SE Tibet. *Acta Petrol. Sin.* **2025**, *41*, 91–111, (In Chinese with English abstract).
39. Yunnan Institute of Geological Sciences (YIGS). *Regional Geological Survey Report of the Qushijie, Dajujie, and Dadongjie Areas at 1:50,000 Scale, Yunnan Province*; Yunnan Institute of Geological Sciences: Kunming, China, 2004. (In Chinese)
40. Sun, S.S.; McDonough, W.F. Chemical and isotopic systematics of oceanic basalts: Implications for mantle composition and processes. *Geol. Soc. Spec. Pub.* **1989**, *42*, 313–345. [CrossRef]
41. Zong, K.; Klemd, R.; Yuan, Y.; He, Z.Y.; Guo, J.L.; Shi, X.L.; Liu, Y.S.; Hu, Z.C.; Zhang, Z.M. The assembly of Rodinia: The correlation of early Neoproterozoic (ca. 900 Ma) high-grade metamorphism and continental arc formation in the southern Beishan Orogen, southern Central Asian Orogenic Belt (CAOB). *Precambrian Res.* **2017**, *290*, 32–48. [CrossRef]
42. Hu, Z.C.; Zhang, W.; Liu, Y.S.; Gao, S.; Li, M.; Zong, K.Q.; Chen, H.H.; Hu, S.H. "Wave" signal-smoothing and mercury-removing device for laser ablation quadrupole and multiple collector ICPMS analysis: Application to lead isotope analysis. *Anal. Chem.* **2015**, *87*, 1152–1157. [CrossRef] [PubMed]
43. Hu, Z.C.; Li, X.H.; Luo, T.; Zhang, W.; Crowley, J.; Li, Q.L.; Ling, X.X.; Yang, C.; Li, Y.; Feng, L.P.; et al. Tanz zircon megacrysts: A new zircon reference material for the microbeam determination of U–Pb ages and Zr–O isotopes. *J. Anal. At. Spectrom.* **2021**, *36*, 2715–2734. [CrossRef]
44. Liu, Y.S.; Hu, Z.C.; Gao, S.; Günther, D.; Xu, J.; Gao, C.G.; Chen, H.H. In situ analysis of major and trace elements of anhydrous minerals by LA-ICP-MS without applying an internal standard. *Chem. Geol.* **2008**, *257*, 34–43. [CrossRef]
45. Liu, Y.S.; Gao, S.; Hu, Z.C.; Gao, C.G.; Zong, K.Q.; Wang, D.B. Continental and oceanic crust recycling-induced melt-peridotite interactions in the Trans-North China Orogen: U-Pb dating, Hf isotopes and trace elements in zircons of mantle xenoliths. *J. Petrol.* **2010**, *51*, 537–571. [CrossRef]
46. Ludwig, K.R. *ISOPLOT 3.00: A Geochronological Toolkit for Microsoft Excel*; Berkeley Geochronology Center: Berkeley, CA, USA, 2003; Volume 4, pp. 1–70.
47. Ferry, J.M.; Watson, E.B. New thermodynamic models and revised calibrations for the Ti-in-zircon and Zr-in-rutile thermometers. *Contr. Mineral. Petrol.* **2007**, *154*, 429–437. [CrossRef]
48. Hayashi, K.; Fujisawa, H.; Holland, H.D.; Ohmoto, H. Geochemistry of ~1.9 Ga sedimentary rocks from northeastern Labrador, Canada. *Geochim. Cosmochim. Acta* **1997**, *61*, 4115–4137. [CrossRef]
49. He, B.; Zhong, Y.; Xu, Y.; Li, X. Triggers of Permo-Triassic boundary mass extinction in South China: The Siberian Traps or Paleo-Tethys ignimbrite flare-up? *Lithos* **2014**, *204*, 258–267. [CrossRef]
50. Winchester, J.A.; Floyd, P.A. Geochemical discrimination of different magma series and their differentiation products using immobile elements. *Chem. Geol.* **1977**, *20*, 325–343. [CrossRef]
51. Cao, H.; Zhang, Y.H.; Pei, Q.M.; Zhang, R.Q.; Tang, L.; Lin, B.; Cai, G.J. U–Pb dating of zircon and cassiterite from the Early Cretaceous Jiaojiguan iron-tin polymetallic deposit, implications for magmatism and metallogeny of the Tengchong area, western Yunnan, China. *Int. Geol. Rev.* **2017**, *59*, 234. [CrossRef]
52. Whalen, J.B.; Currie, K.L.; Chappell, B.W. A-type granites: Geochemical characteristics, discrimination and petrogenesis. *Contr. Mineral. Petrol.* **1987**, *95*, 407–419. [CrossRef]
53. Zhu, R.Z.; Lai, S.C.; Qin, J.; Zhao, S. Early-Cretaceous Syenites and Granites in the Northeastern Tengchong Block, SW China: Petrogenesis and Tectonic Implications. *Acta Geol. Sin. Engl. Ed.* **2018**, *92*, 1349–1365. [CrossRef]
54. Qi, X.; Shen, H.; Ren, Y.; Wei, C.; Cai, Z.; Zhang, C.; Ji, F.; Liang, F. Early Cretaceous volcanism in the northwestern Gaoligong orogen and its relationship with subduction of the Nujiang Ocean: Evidence from geochemistry, zircon U-Pb dating and Hf isotopic compositions of rhyolites. *Acta Petrol. Sin.* **2020**, *36*, 2946–2962. (In Chinese with English abstract)
55. Yu, L. Genesis and Tectonic Significance of the Mesozoic Granitoids in the Tengchong-Baoshan Block, Sanjiang Area. Ph.D. Thesis, China University of Geosciences, Beijing, China, 2016; 137p.
56. Ma, X.; Shu, L.; Meert, J.G.; Li, J. The paleozoic evolution of central tianshan: Geochemical and geochronological evidence. *Gondwana Res.* **2014**, *25*, 797–819. [CrossRef]
57. Zhou, S.M. Petrogenesis of the Late Triassic-Early Eocene Granitoids in the Tengchong Block. Ph.D. Thesis, China University of Geosciences, Beijing, China, 2019; 168p. (In Chinese).
58. Cao, H.; Zhang, S.T.; Lin, J.Z.; Zheng, L.; Wu, J.D.; Li, D. Geology, geochemistry and geochronology of the Jiaojiguanliangzi Fe-polymetallic deposit, Tengchong County, Western Yunnan (China): Regional tectonic implications. *J. Asian Earth Sci.* **2014**, *81*, 142–152. [CrossRef]

59. Xie, J.C.; Zhu, D.C.; Dong, G.C.; Zhao, Z.D.; Wang, Q.; Mo, X.X. Linking the Tengchong Terrane in SW Yunnan with the Lhasa Terrane in southern Tibet through magmatic correlation. *Gondwana Res.* **2016**, *39*, 217–229. [CrossRef]
60. Li, Z.H.; Lin, S.L.; Cong, F.; Xie, T.; Zou, G.F. U-Pb ages of zircon from metamorphic rocks of the Gaoligongshan Group in western Yunnan and its tectonic significance. *Acta Petrol. Sin.* **2012**, *28*, 1529–1541. (In Chinese with English abstract)
61. Qi, X.X.; Zhu, L.H.; Hu, Z.C.; Li, Z.Q. Zircon SHRIMP U-Pb dating and Lu-Hf isotopic composition for Early Cretaceous plutonic rocks in Tengchong block, southeastern Tibet, and its tectonic implications. *Acta Petrol. Sin.* **2011**, *27*, 3409–3421. (In Chinese with English abstract)
62. Cong, F.; Lin, S.; Zou, G.; Li, Z.; Xie, T.; Peng, Z.; Liang, T. Magma mixing of granites at Lianghe: In-situ zircon analysis for trace elements, U-Pb ages and Hf isotopes. *Sci. China Earth Sci.* **2011**, *54*, 1346–1359. [CrossRef]
63. Fang, Y.; Zhang, Y.; Zhang, S.; Cao, H.; Zou, H.; Dong, J. Early Cretaceous I-type granites in the Tengchong terrane: New constraints on the late Mesozoic tectonic evolution of southwestern China. *Geosci. Front.* **2018**, *9*, 459–470. [CrossRef]
64. Chappell, B.W.; White, A.J. Two contrasting granite types. *J. Geol. Soc. Aust.* **1974**, *21*, 353–369.
65. Zhu, R.Z.; Lai, S.; Santosh, M.; Qin, J.; Zhao, S. Early Cretaceous Na-rich granitoids and their enclaves in the Tengchong Block, SW China: Magmatism in relation to subduction of the Bangong-Nujiang Tethys ocean. *Lithos* **2017**, *286*, 175–190. [CrossRef]
66. Zhang, J.; Peng, T.; Fan, W.; Zhao, G.; Dong, X.; Gao, J.; Peng, B.; Wei, C.; Xia, X.; Chen, L.; et al. Petrogenesis of the Early Cretaceous granitoids and its mafic enclaves in the Northern Tengchong Terrane, southern margin of the Tibetan Plateau and its tectonic implications. *Lithos* **2018**, *314–315*, 283–298. [CrossRef]
67. Qi, X.; Wei, C.; Zhang, C.; Zhang, S.; Hu, Z.; Ji, F. Southward extension of the Bangonghu–Nujiang Suture: Evidence from Early Cretaceous intermediate and felsic magmatism in the Gaoligong Orogen, China. *J. Asian Earth Sci.* **2019**, *182*, 1–25. [CrossRef]
68. Chen, Y.; Li, G.; Qin, L.; Lu, Y.; Jiang, C. Geochronology and geochemistry of Cretaceous–Eocene granites, Tengchong Block (SW China): Petrogenesis and implications for Mesozoic-Cenozoic tectonic evolution of Eastern Tethys. *Geosci. Front.* **2022**, *13*, 101338. [CrossRef]
69. Zhang, J.; Fan, W.; Xu, X.; Peng, T.; Ratschbacher, L.; Wang, B. Early Cretaceous magmatism in Yunnan traces the tectonic transition from the Tengchong-Baoshan continental collision to Neo-Tethyan slab rollback. *Lithos* **2023**, *456*, 106717. [CrossRef]
70. Altherr, R.; Holl, A.; Hegner, E.; Langer, C.; Kreuzer, H. High-potassium, calc-alkaline I-type plutonism in the European Variscides: Northern Vosges (France) and northern Schwarzwald (Germany). *Lithos* **2000**, *50*, 51–73. [CrossRef]
71. Sylvester, P.J. Post-collisional strongly peraluminous granites. *Lithos* **1998**, *45*, 29–44. [CrossRef]
72. Pearce, J.A.; Harris, N.B.W.; Tindle, A.G. Trace element discrimination diagrams for the tectonic interpretation of granitic rocks. *J. Petrol.* **1984**, *25*, 956–983. [CrossRef]
73. Yang, J.; Cawood, P.A.; Du, Y.; Huang, H.; Hu, L. Detrital record of Indosinian mountain building in SW China: Provenance of the Middle Triassic turbidites in the Youjiang Basin. *Tectonophysics* **2012**, *574*, 105–117. [CrossRef]
74. Yang, J.; Cawood, P.A.; Du, Y.; Huang, H.; Huang, H.; Tao, P. Large Igneous Province and magmatic arc sourced Permian–Triassic volcanogenic sediments in China. *Sediment. Geol.* **2012**, *261*, 120–131. [CrossRef]
75. He, X.; Liu, Z.; Wang, G.; Leonard, N.D.; Tao, W.; Tan, S. Petrogenesis and tectonic setting of the Early Cretaceous granitoids in the eastern Tengchong terrane, SW China: Constraint on the evolution of Meso-Tethys. *Lithosphere* **2020**, *12*, 150–165. [CrossRef]
76. Chen, X.; Hu, R.; Bi, X.; Li, H.; Lan, J.; Zhao, C.; Zhu, J. Cassiterite LA-MC-ICP-MS U/Pb and muscovite $^{40}Ar/^{39}Ar$ dating of tin deposits in the Tengchong-Lianghe tin district, NW Yunnan, China. *Miner. Depos.* **2014**, *49*, 843–860. [CrossRef]
77. Zhang, J.; Fan, W.; Peng, T.; Ratschbacher, L. Southeastern continuation of the Bangong-Nujiang suture zone: Constraints from Middle Jurassic–Early Cretaceous sedimentary rocks in the western Baoshan block, SW China. *J. Asian Earth Sci.* **2021**, *212*, 104944. [CrossRef]
78. Zhang, J.; Fan, W.; Peng, T.; Ratschbacher, L. An early cretaceous arc–back-arc system in western Yunnan, SW China: Constraints from U-Pb zircon ages and geochemistry of volcanic rocks in the western Baoshan block. *Lithos* **2022**, *412*, 106555. [CrossRef]
79. Cong, F.; Lin, S.; Zou, G.; Xie, T.; Li, Z.; Tang, F.; Peng, Z. Geochronology and Petrogenesis for the Protolith of Biotite Plagioclase Gneiss at Lianghe, Western Yunnan. *Acta Geol. Sin. (Engl. Ed.)* **2011**, *85*, 891–904. [CrossRef]
80. Deng, J.; Wang, Q.; Li, G.; Li, C.; Wang, C. Tethys tectonic evolution and its bearing on the distribution of important mineral deposits in the Sanjiang region, SW China. *Gondwana Res.* **2014**, *26*, 419–437. [CrossRef]
81. Li, J.; Zhang, L.; Fan, W.; Ding, L.; Sun, Y.; Peng, T.; Li, G.; Sein, K. Mesozoic-Cenozoic tectonic evolution and metallogeny in Myanmar: Evidence from zircon/cassiterite U–Pb and molybdenite Re–Os geochronology. *Ore Geol. Rev.* **2018**, *97*, 829–845. [CrossRef]
82. Mao, W.; Zhong, H.; Yang, J.; Liu, L.; Fu, Y.; Zhang, X.; Doh, S.H. Geochronology of Sn mineralization in Myanmar: Metallogenic implications. *Econ. Geol.* **2022**, *117*, 1387–1403. [CrossRef]
83. Mao, W.; Zhong, H.; Yang, J.; Tang, Y.; Liu, L.; Fu, Y.; Zhang, X.; Sein, K.; Aung, S.M.; Li, J.; et al. Combined Zircon, Molybdenite, and Cassiterite Geochronology and Cassiterite Geochemistry of the Kuntabin Tin-Tungsten Deposit in Myanmar. *Econ. Geol.* **2020**, *115*, 603–625. [CrossRef]

84. Liu, C.; Chung, S.; Wu, F.; Zhang, C.; Xu, Y.; Wang, J.; Chen, Y.; Guo, S. Tethyan suturing in Southeast Asia: Zircon U-Pb and Hf-O isotopic constraints from Myanmar ophiolites. *Geology* **2016**, *44*, 311–314. [CrossRef]
85. Warr, L.N. IMA-CNMNC approved mineral symbols. *Mineral. Mag.* **2021**, *85*, 291–320. [CrossRef]

Disclaimer/Publisher's Note: The statements, opinions and data contained in all publications are solely those of the individual author(s) and contributor(s) and not of MDPI and/or the editor(s). MDPI and/or the editor(s) disclaim responsibility for any injury to people or property resulting from any ideas, methods, instructions or products referred to in the content.

minerals

MDPI

Article

S-Type Granites from the Guomang-Co Area in Central Tibet: A Response to Early Paleozoic Andean-Type Orogeny Along the Northern Margin of East Gondwana

Yuhe Zhang [1], Ming Wang [1,2,*], Changsheng Yu [1] and Zhenglong Li [1]

[1] College of Earth Sciences, Jilin University, Changchun 130061, China; zhangyh2220@mails.jlu.edu.cn (Y.Z.); yucs22@mails.jlu.edu.cn (C.Y.); 19180829141@163.com (Z.L.)

[2] Key Laboratory of Mineral Resources Evaluation in Northeast Asia, Ministry of Nature Resources, Changchun 130061, China

* Correspondence: wm609@163.com

Abstract: The Proto-Tethys Ocean existed between Gondwana and Laurussia during the late Neoproterozoic to Early Paleozoic. As part of the northern margin of East Gondwana, the Lhasa terrane records subduction-related processes of the Proto-Tethys Ocean. This study analyzes mylonitized granites from the Guomang-Co area in the central Lhasa terrane, focusing on their major and trace elements, U-Pb age values, and Sr-Nd-Pb-Hf isotopes. Geochemical and isotopic data consistently indicate S-type affinity derived from Paleoproterozoic metasedimentary sources, and likely formed in a syn-collisional setting. Combined with previous studies, the granites are interpreted as products of the Early Paleozoic Andean-type orogeny along the northern margin of East Gondwana, which indicate southward subduction of the Proto-Tethys Ocean during the Cambrian–Ordovician.

Keywords: Proto-Tethys; Lhasa terrane; Early Paleozoic; S-type granite; Andean-type orogeny; U-Pb; Sr-Nd-Pb-Hf

1. Introduction

The Proto-Tethys Ocean, which originated from the breakup of the Rodinia supercontinent during the late Neoproterozoic, existed between the northern margin of Gondwana and Laurussia from the late Neoproterozoic to the late Paleozoic. Its evolution profoundly influenced the global tectonic framework of the Early Paleozoic [1–5] and is closely linked to the early formation and evolution of the Tibetan Plateau.

The Proto-Tethys Ocean underwent a general evolution from extensional rifting and expansion during the late Neoproterozoic, to the initiation of subduction in the Early Paleozoic, and finally to the closure of the ocean basin by the late Silurian. However, extensive later magmatic activity and tectonic overprinting have left many issues concerning the Proto-Tethys Ocean unresolved, with the debate over its subduction polarity being one of them. Some researchers advocate a unidirectional northward subduction model [6–9], whereas others propose a bidirectional model involving both northward and southward subduction [10–14]. This debate is closely linked to whether the northern margin of Gondwana during the Early Paleozoic functioned as a passive continental margin or experienced Andean-type orogenesis.

In recent years, numerous Neoproterozoic and Early Paleozoic magmatic events along the northern margin of East Gondwana have been dated by researchers (Table 1). The relationship between these magmatic activities, the Pan-African orogeny, and Proto-Tethys subduction has become a focal point of debate. Currently, there are three prevailing viewpoints regarding these magmatic events:

(1) The traditional perspective is on the basis of the understanding that the Pan-African orogeny occurred between 550~450 Ma [15]; it posits that the Early Paleozoic magmatism in the Tibetan Plateau is part of the Pan-African event; e.g., [16–23].

(2) Another perspective posits that during the late Neoproterozoic, the northern margin of Gondwana functioned as a passive continental margin (e.g., [9,24–27]), and that Early Paleozoic magmatic rocks were generated in a post-collisional extensional setting following the Pan-African orogeny (e.g., [7,28–33]).

(3) The third perspective argues that an Andean-type orogenic belt developed along the northern margin of Gondwana during the Early Paleozoic, and that these rocks were associated with subduction-related processes (e.g., [10,34–38]).

The primary reasons behind these debates are as follows: (1) Differences in understanding the spatiotemporal extent of the Pan-African orogeny. (2) These reports are colloreted to primarily acidic rocks, with few mafic rocks and fewer intermediate rocks reported, and records of Andean-type orogenic events on the northern margin of Australia are limited. (3) The northern margin of Gondwana is considered as a passive continental margin during the Neoproterozoic, the mechanism through which a passive continental margin transitioned into a subduction zone remains unclear.

Table 1. Early Paleozoic statistical age data reported along the northern margin of East Gondwana.

Location	Terrane	Lithology	Genesis Type	Method	Age	Ref.
Mabjia	Himalaya	Granitic gneiss	\	SIMS	530	[39]
Kampa	Himalaya	Granitic gneiss	\	SHRIMP	527, 506	[40]
Kangmar	Himalaya	Granitic gneiss	\	U-Pb	508	[41]
Kangmar	Himalaya	Granitic gneiss	\	SHRIMP	515	[16]
Kangmar	Himalaya	Granitic gneiss	S-type	LA-ICP-MS	500~478	[42]
Kangmar	Himalaya	Granite	S-type	SHRIMP	499~498	[43]
Gyirong	Himalaya	Granitic gneiss	\	SHRIMP	514	[16]
Gyirong	Himalaya	Granitic gneiss	\	SHRIMP	499	[44]
Gyirong	Himalaya	Granitic gneiss	S-type	LA-ICP-MS	486~474	[42]
Gyirong	Himalaya	Granite	S-type	LA-ICP-MS	486~429	[43]
Laguigangri	Himalaya	Granitic gneiss	S-type	LA-ICP-MS	514	[29]
Yadong	Himalaya	Granite	\	SHRIMP	512	[16]
Yadong	Himalaya	Granite	S-type	SHRIMP	508	[43]
Yadong	Himalaya	Granite	S-type	LA-ICP-MS	499	[45]
Namche Bawa	Himalaya	Granite	S-type	SHRIMP	506	[43]
Namche Bawa	Himalaya	Granitic gneiss	I-type	LA-ICP-MS	503~490	[46]
Dinggye	Himalaya	Granitic gneiss	S-type	SHRIMP	504~482	[47]
Nielamu	Himalaya	Granitic gneiss	\	SHRIMP	501	[16]
Yalaxiangbo	Himalaya	Granitic gneiss	S-type	LA-ICP-MS	496~488	[42]
Xiaru	Himalaya	Granite	S-type	MC-ICP-MS	486~446	[43]
Xiaru	Himalaya	Othogneiss	I-type or S-type	SIMS	480~470	[48]
Xiaru	Himalaya	Granite	S-type	LA-ICP-MS	478, 475	[49]
Kathmandu, Dadeldhura	Himalaya	Granite	\	MC-ICP-MS	484~476	[34,50]
Kathmandu	Himalaya	Granite	\	LA-ICP-MS	478, 477	[10]
Dadeldhura	Himalaya	Granite	\	MC-ICP-MS	482~474	[51]
Dadeldhura	Himalaya	Granite	\	LA-ICP-MS	478	[52]
Rupshu	Himalaya	Granite	Peral. I-type	IDMS	482	[53]
Tso Morari	Himalaya	Granitic gneiss	S-type	IDMS	479	[53]
Lhagoi Kangri	Himalaya	Granite	S-type	SHRIMP	477	[43]

Table 1. *Cont.*

Location	Terrane	Lithology	Genesis Type	Method	Age	Ref.
Mabjia	Himalaya	Granite	S-type	SHRIMP	475	[43]
Simchar	Himalaya	Granite	\	LA-ICP-MS	471	[54]
Duguer	South Qiangtang	Granitic gneiss	S-type	LA-ICP-MS	502~492	[55]
Duguer	South Qiangtang	Granitic gneiss	S-type	MC-ICP-MS	490~473	[56]
Duguer	South Qiangtang	Gabbro	E-MORB	MC-ICP-MS	490	[57]
Duguer	South Qiangtang	Basalt	E-MORB	MC-ICP-MS	483	[57]
Duguer	South Qiangtang	Andesite	OIB	MC-ICP-MS	477~473	[57]
Duguer	South Qiangtang	Granitic gneiss	\	LA-ICP-MS	476~471	[58]
Duguer	South Qiangtang	Granitic gneiss	A-type	MC-ICP-MS	454	[57]
Bensong-Co	South Qiangtang	Granitic gneiss	S-type	LA-ICP-MS	497, 496	[59]
Bensong-Co	South Qiangtang	Granitic gneiss	S-type	LA-ICP-MS	486, 481	[60]
Gemuri	South Qiangtang	Granite	\	MC-ICP-MS	489, 480	[61]
Gemuri	South Qiangtang	Granite	S-type	LA-ICP-MS	486, 480	[60]
Gemuri	South Qiangtang	Granite	S-type	MC-ICP-MS	469~455	[56]
Gemuri	South Qiangtang	Granite	\	MC-ICP-MS	439	[61]
Ejiumai	South Qiangtang	Granite	\	SHRIMP	476~463	[62]
Moon-Co	South Qiangtang	Metabasalt	OIB	LA-ICP-MS	476	[63]
Moon-Co	South Qiangtang	Metarhyolite	\	LA-ICP-MS	470, 461	[63]
Moon-Co	South Qiangtang	Metarhyolite	\	LA-ICP-MS	458, 454	[64]
Heibailing	South Qiangtang	Metarhyolite	\	LA-ICP-MS	467	[63]
Wugongshan	South Qiangtang	Granitic gneiss	\	LA-ICP-MS	465	[19]
Jitang	South Qiangtang	Granite	S-type	LA-ICP-MS	455	[65]
Guomang-Co	Lhasa	Granite	S-type	MC-ICP-MS	530, 480	this study
Banglei	Lhasa	Metarhyolite	\	LA-ICP-MS	536	[28]
Zhaqian	Lhasa	Metarhyolite	S-type	LA-ICP-MS	525, 511	[36]
Zhakang	Lhasa	Metarhyolite	Peral. A-type	MC-ICP-MS	512	[38]
Zhakang	Lhasa	Metarhyolite	S-type	LA-ICP-MS	510	[36]
Zhakang	Lhasa	Metarhyolite	\	LA-ICP-MS	501	[66]
Xainza	Lhasa	Granite	\	LA-ICP-MS	510	[67]
Tongka	Lhasa	Granite	\	SHRIMP	507	[18]
Jili	Lhasa	Granitic gneiss	Peral. I-type	LA-ICP-MS	504, 495	[68]
Milin	Lhasa	Granite	S-type	LA-ICP-MS	501~496	[69]
Nyingchi	Lhasa	Granitic gneiss	\	LA-ICP-MS	496	[70]
Banglei	Lhasa	Metabasalt	\	LA-ICP-MS	492	[11]
Banglei	Lhasa	Metarhyolite	S-type	LA-ICP-MS	492	[11]
Amdo	Amdo	Granitic gneiss	\	LA-ICP-MS	532~483	[71]
Amdo	Amdo	Granitic gneiss	\	U–Pb	531	[72]
Amdo	Amdo	Granitic gneiss	I-type	LA-ICP-MS	517~505	[37]
Amdo	Amdo	Granitic gneiss	\	SHRIMP	502~483	[46]
Amdo	Amdo	Granitic gneiss	\	LA-ICP-MS	488	[21]

The data are primarily concentrated between 530~460 Ma.

The Lhasa terrane, one of the principal tectonic blocks forming the Tibetan Plateau, is generally regarded as a microcontinent that developed along the northern margin of Gondwana (i.e., southern margin of the Proto-Tethys Ocean) during the Early Paleozoic [11,73]. This terrane holds significant importance for understanding the Early Paleozoic evolution of the northern Gondwana margin. In this study, we report a suite of Early Paleozoic mylonitized granites from the Guomang-Co area in Xainza County, Lhasa terrane. On the basis of geochemical data, U-Pb geochronological results, and Sr-Nd-Pb-Hf isotopic analyses, this paper investigates the petrogenesis and tectonic setting of these granites, providing new constraints on the Early Paleozoic evolution of the Proto-Tethys Ocean along the northern margin of Gondwana.

2. Geological Setting and Samples

On the basis of the current plate tectonic model for the Tibetan Plateau and the Early Paleozoic tectonic framework [73,74], five major plate sutures with a nearly east-west orientation have been identified (Figure 1a). From north to south, these sutures are the Kunlun Mountain Suture Zone (KSZ), the Jinsha River Suture Zone (JSSZ), the Longmu Co-Shuanghu Suture Zone (LSSZ), the Bangong Lake-Nujiang Suture Zone (BNSZ), and the India-Yarlung Zangbo River Suture Zone (IYZSZ). The plates defined by these sutures are, in turn: the Bayan Har-Ganzi terrane, the North Qiangtang terrane, the South Qiangtang terrane, the Lhasa terrane, and the Himalaya terrane (Figure 1a).

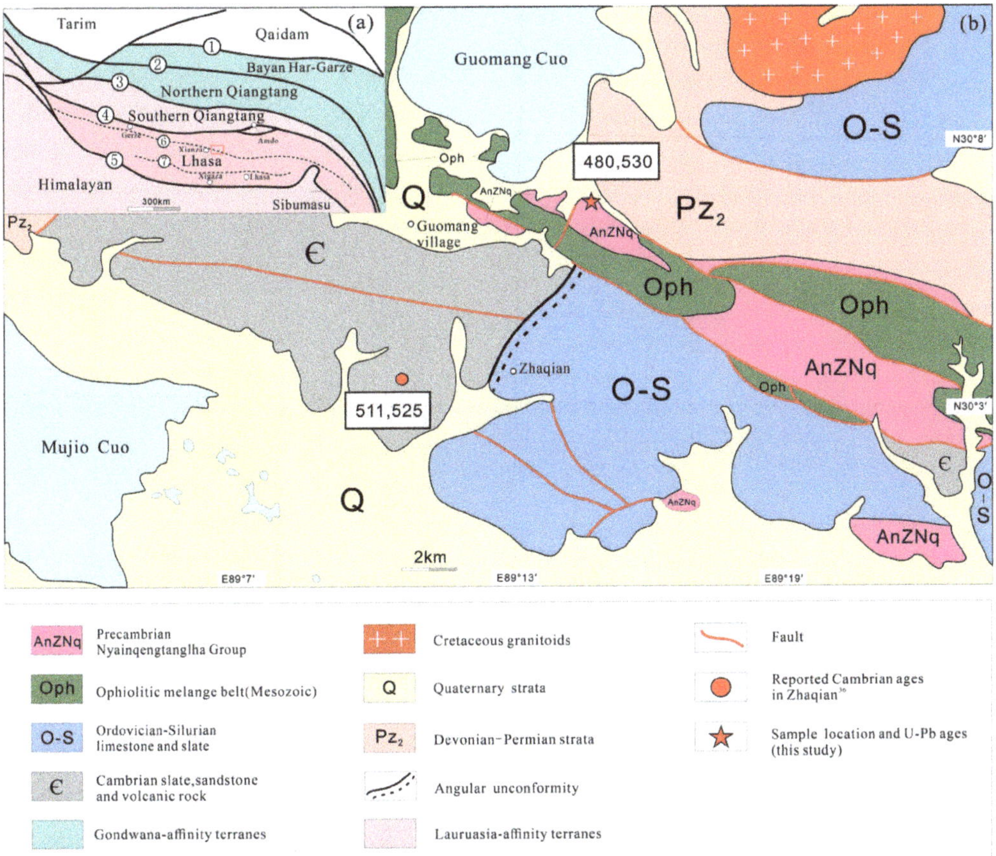

Figure 1. (**a**) Geotectonic map of Tibetan Plateau (revised by [73,74]); (**b**) geological map of Guomang-Co area (revised by [36,75]). 1: Kunlun Mountain Suture Zone (KSZ); 2: Jinsha River Suture Zone (JSSZ); 3: Longmu Co-Shuanghu Suture Zone (LSSZ); 4: Bangong Lake-Nujiang Suture Zone (BNSZ); 5: India-Yarlung Zangbo River Suture Zone (IYZSZ); 6: Shiquanhe-Nam Co Ophiolitic Mélange Zone (SNMZ); 7: Luobadui-Mila Mountain Fault (LMF).

The Lhasa terrane is situated between the Bangong Lake-Nujiang Suture Zone (BNSZ) and the Yarlung Zangbo River Suture Zone (IYZSZ), and exhibits characteristics akin to the northern margin of Gondwana. It can be further subdivided into three regions by the Shiquan River-Nam Co Mélange Zone (SNMZ) and Luobadui-Mila Moutain Fault (LMF)—northern, central, and southern regions. However, the formation of the northern and southern Lhasa terrane during the Meso-Cenozoic mainly occurred during the Mesozoic and Cenozoic, which has little relation to the Early Paleozoic evolutionary processes [74]. Therefore, for the purpose of this study, the Lhasa terrane is considered as a unified whole.

The granitic mylonite of Guomang-Co is exposed approximately 6 km east of Guomang Village in Xainza County, central Tibet (Figure 1b). Tectonically, it is situated near the Shiquanhe-Nam Co Ophiolitic Mélange Zone (SNMZ), within the central part of the Lhasa terrane. Mesozoic strata dominate to the north of Guomang-Co, while Paleozoic strata are predominantly exposed to the south (Figure 1b). In the Zhaqian area, about 12 km south of Guomang-Co, a well-preserved angular unconformity between Cambrian and Ordovician strata has been identified [76].

The coordinates of the Guomang-Co granite sampling site are E: 89°13′48.68″, N: 31°7′44.78″, and it is located within the Precambrian Nyainqengtanglha Group (AnZNq) previously defined by other researchers (Figure 1b). The Nyainqentanglha Group is distributed as a belt-like manner in the NWW direction in the area. It is a rock group with basement characteristics in Lhasa terrane, which is formed by the oldest sedimentary processes and generally has undergone metamorphism within the amphibolite facies-granulite facies. Its lithology includes biotite granodiorite gneiss, biotite plagiogneiss, greenstone plagioclase amphibolite, diopside marble, feldspar-quartzite, quartzite, quartz schist, granitic gneiss, slate, and phyllite [77,78]. Ji et al. [45] first identified an Early Cambrian rhyolite (501 Ma) separated from the Nyainqengtanglha Group in the Shenzha area and suggested that at least a portion of the upper strata belongs to the Cambrian. Subsequently, several Early Paleozoic reports have emerged from the originally defined Nyainqengtanglha Group region [11,28,36,38,69,70]. The Guomang-Co granite samples collected in this study are also part of the Nyainqengtanglha group.

The outcrop conditions at the sampling site are poor, and the contact relationship with adjacent rock bodies is unknown (Figure 2a). Field observations show that the Guomang-Co granite consists of a suite of dark-colored granitoid mylonite (Figure 2b,c), locally experiencing intense ductile deformation, with the development of sheath folds, stretching lineation, and shear bands (Figure 2b,c). Under the microscope, the samples are severely mylonitized, exhibiting a porphyroclastic texture and mylonitic structure (Figure 2d,e). The primary constituents are quartz and alkali feldspar; the porphyroblasts constitute about 50%, with 80% alkali feldspar and 20% quartz. The matrix shows a fine-grained granoblastic texture, making up about 47%, with a certain degree of preferred orientation (Figure 2f). White mica fills mineral gaps and constitutes about 3%, with mica fish structures formed due to ductile deformation (Figure 2f,g).

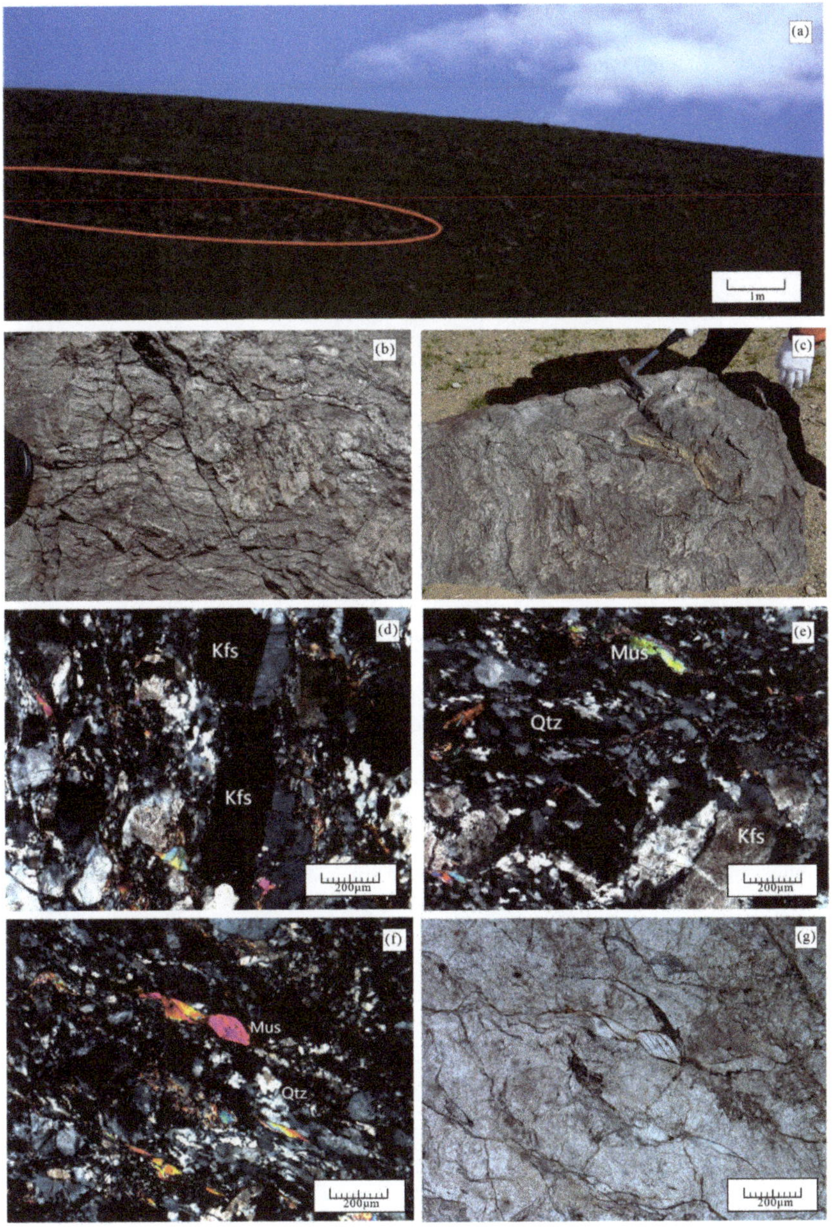

Figure 2. (**a**) Outcrop of map of mylonitized granites from Guomang-Co area in central Lhasa terrane. (**b**,**c**) Field features of the sample of Guomang-Co mylonitized granites. (**d**–**g**) Photomicrographs of mylonitized granites from Guomang-Co. Qzt—quart; Kfs—K-feldspar; Mus—muscovite.

3. Materials and Methods

3.1. Whole-Rock Geochemical Data

Four fresh samples (Q22T4H1–Q22T4H4) were processed for geochemical characterization. The samples were crushed to a grain size <200 mesh at the Regional Geological Survey Center of Hebei Province, China. Major and trace element testing were conducted at the Key Laboratory of Mineral Resources Evaluation in Northeast Asia of Jilin University, Changchun, China. Major elements were analyzed with a Leeman Prodigy ICP-OES and Rigaku ZSX Primus II X-ray fluorescence spectrometer achieving analytical precision of 1%–3%. Trace elements were determined using an Agilent 7500a ICP-MS, with elemental uncertainties generally <5%. Loss on ignition (LOI) was measured by heating ~1 g powder at 1000 °C. Analytical protocols followed established procedures in references [79,80].

3.2. Zircon U-Pb Geochronological Investigations

Two samples (numbered Q22T4-1 and Q22T4-2) of Guomang-Co granite were selected for zircon U-Pb dating. Zircon grains were separated using conventional heavy liquid and magnetic techniques at the Regional Geological Survey Center of Hebei Province, China, with cathodoluminescence (CL) images taken in the same place. Zircon U-Pb dating was performed at the Key Laboratory of Mineral Resources Evaluation in Northeast Asia of Jilin University, Changchun, China, using an Agilent 7500a inductively coupled plasma mass spectrometer (ICP-MS) with 193 nm laser ablation system (UP193SS). In this study, 91,500 was used as the zircon standard [81]. NIST 610 was used as the external standard, and ^{29}Si was used as the internal standard for trace element calculations, the GJ-1 reference zircon was analyzed as an unknown sample, yielding an average $^{206}Pb/^{238}U$ age of 609.5 ± 2.4 Ma, matching its certified value (608.5 ± 0.37 Ma) [82]. Methodology details can be founded in [83]. Data reduction utilized ICPMS Data Cal v10.2 [84] and Isoplot v4.15 [85].

3.3. Zircon Hf Isotope

In situ zircon Hf isotope analysis was conducted at Beijing Createch Testing Technology Co. Ltd., Beijing, China, by use of a Neptune Plus Multi-collector Inductively Coupled Plasma Mass Spectrometer (MC-ICP-MS) equipped with a New Wave 213 laser-ablation system (30 or 38 μm diameter laser beam). The laboratory temperature was maintained between 18 °C and 22 °C, with relative humidity kept below 65%. The laser ablation beam used for testing had an energy density of 6 J/cm^2, a pulse frequency of 6 Hz, and a diameter of 55 μm. Signal acquisition time was approximately 1 min. The international zircon standard GJ-1 was used as the external standard during the analysis. The material produced by laser ablation was transported to the Neptune Plus MC-ICP-MS for analysis, with high-purity helium (He) used as the carrier gas. Hf isotope data were obtained in low-resolution static mode. Detailed analytical procedures and parameters can be found in [86]. The decay constant for ^{176}Lu used to correct the results was 1.867×10^{-11} a^{-1}, following [87]. Additionally, Hf model age values and εHf(t) values were calculated on the basis of the methods of [88,89].

3.4. Whole-Rock Sr-Nd-Pb Isotopes

Two samples were selected for Sr-Nd-Pb isotope analysis (numbered Q22T4H1 and Q22T4H2) at Beijing Createch Testing Technology Co., Ltd., by use of a Thermo Scientific Neptune Plus Multi-Collector Inductively Coupled Plasma Mass Spectrometer (MC-ICP-MS). The analytical procedure is briefly outlined as follows: first, the samples were digested by use of a mixture of nitric acid and HF acid in Teflon beakers, followed by separation and analysis by use of a dedicated cation exchange column. Detailed testing methods and related parameters can be found in [90]. Sr and Nd isotopes were normalized using $^{86}Sr/^{88}Sr$ = 0.7219 and $^{146}Nd/^{144}Nd$ = 0.7219, respectively. The international standards NBS 987 ($^{87}Sr/^{86}Sr$ = 0.710248 ± 14 (2SD, n = 33)) and GSB Nd ($^{143}Nd/^{144}Nd$ = 0.512190 ± 10 (2SD, n = 10)) were used for external calibration.

4. Results

4.1. Whole-Rock Geochemical Characteristics

The whole-rock major and trace elements of the samples show good consistency, and the analytical results are presented in Table S1. The mineral assemblage observed under the microscope shows orientation, indicating that the samples underwent late-stage deformation. However, no significant mineral phase transformations were observed (Figure 2d–g). The loss on ignition (LOI) of the samples has a range of 0.63%~0.80%, showing relatively low values and small variation. The K_2O/Na_2O ratios are stable (0.97~1.05), and no significant changes in Sr and Ba content were observed, suggesting that these elements did not experience significant loss or enrichment [91]. The high field strength elements (HFSEs) and rare earth elements (REEs) distribution patterns are consistent with those of granite rocks (Figure 3a,b) [92], indicating that the samples were not significantly affected by late-stage alteration. Therefore, the whole-rock major and trace element data of the samples are considered reliable for geochemical analysis.

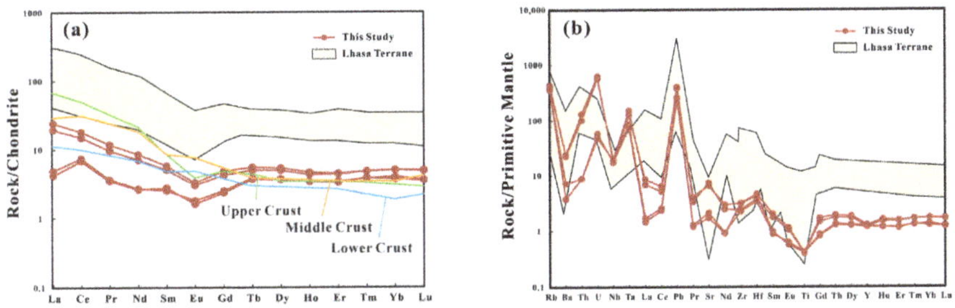

Figure 3. (**a**) Chondrite-normalized REE patterns of the Guomang-Co granites. (**b**) Primitive mantle-normalized trace elements spider diagrams of the Guomang-Co granites. Coeval granites data of Lhasa terrane are from [11,28,36,38,66,68]. Normalizing values are from [93]. The REE and trace element contents for the upper, middle, and lower crust are from [94].

The samples are rich in SiO_2 (75.18%~75.76%) and Al_2O_3 (13.32%~13.63%), with a Na_2O content range of 4.02%~4.15% and a K_2O content range of 3.98%~4.23%. They are low in TiO_2 (0.08%~0.09%), CaO (0.51%~0.59%), P_2O_5 (~0.17%), MgO (0.20%~0.25%), MnO (0.01%~0.03%), and FeO^T (0.68%~0.79%), with Mg# values ranging within 38.5~39.7. In the TAS diagram, the samples fall within the granite field (Figure 4a) and belong to the high-potassium calc-alkaline series (Figure 4b). Mineral calculations using the CIPW norm [95] show that the samples primarily consist of quartz (34.4%~35.2%) and orthoclase (58.8%~59.5%), with minor amounts of plagioclase (1.4%~1.8%) and corundum (1.5%~2.0%). The QAP diagram places the samples in the orthoclase granite field (Figure 4c). The A/CNK values of the samples have a range of 1.09~1.13, classifying them as weakly peraluminous granites (Figure 4d). These geochemical features are consistent with those of S-type granites [96,97].

In the whole-rock REE chondrite-normalized diagram (Figure 3a), the samples exhibit a pronounced rightward inclination. These rocks have REE patterns characterized by flat to fractionated LREE [$(La/Sm)_N$ = (1.48~3.77)] and flat to depleted HREE [$(Gd/Yb)_N$ = 0.61~0.95], with moderate negative Eu anomalies (δEu = 0.61~0.72) and positive Ce anomalies (δCe = 1.07~1.80). The REE distribution pattern is similar to that of arc-related magmatic rocks [98]. Two of the samples exhibited a relatively pronounced tetrad effect, which is often observed in highly evolved peraluminous S-type granite rocks [99].

In the trace element primitive mantle-normalized diagram (Figure 3b), the samples display enrichment in large ion lithophile elements (LILEs) such as Rb, Th, U, and Pb, depletion in high field strength elements (HFSEs) such as Nb and Ti, and depletion in Sr and Ba. Compared with early Paleozoic granites reported from the Lhasa terrane, the Guomangcu granite exhibits overall lower elemental abundances (REE = 10.3~27.8 ppm) and a more pronounced depletion of heavy rare earth elements. This may be attributed to garnet-bearing residue, indicating the derivation from the lower crust and a higher degree of fractional crystallization [98].

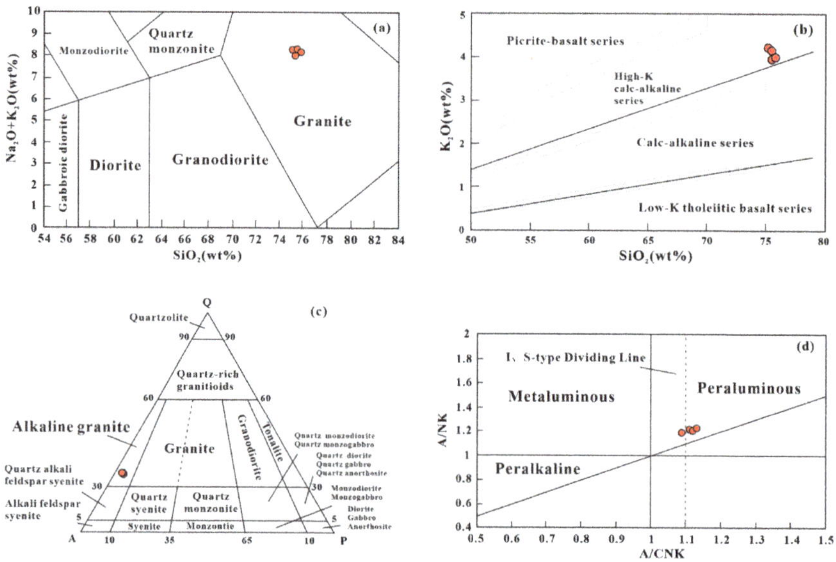

Figure 4. (**a**) TAS diagram [100]; (**b**) SiO_2-K_2O diagram [101]; (**c**) QAP diagram [102], mineral composition derived from the CIPW normative mineral calculation; (**d**) A/NK-A/CNK diagram [103].

4.2. Zircon U-Pb Geochronological Investigations

In this study, 30 zircon grains from samples Q22T41 and Q22T42 were selected for LA-ICP-MS dating. Under cathodoluminescence (CL) images (Figure 5), the zircon grains are transparent to semi-transparent and range in color from colorless to light brown, predominantly exhibiting sub-euhedral columnar or granular morphologies, with some grains showing damaged, incomplete crystal forms. The grain size has a range of 50~100 μm, and the length-to-width ratio is 1:1~1:2. The Th/U ratios of these zircon grains are significantly greater than 0.5, with a range of 1.22~2.38, which is consistent with a magmatic origin [104]. Given that these zircon grains possess clear cores and most exhibit distinct and regular oscillatory zoning, a feature not typical of hydrothermal zircons, the zircon grains' cores are most likely of magmatic origin and can be used to constrain the formation age of the primary granite [105]. Moreover, the narrow widths of the oscillatory zones indicate relatively low crystallization temperatures, which concurs with the zircon saturation temperature calculations presented later [104].

The results of the zircon U-Pb geochronological investigations are shown in Table S1, and the concordia diagrams for the 21 zircon analysis points are presented in Figure 5. The $^{206}Pb/^{238}U$ ages of the nine analysis points from sample Q22T41 have a range of 528~531 Ma, with a weighted mean age of 529.4 ± 3.0 Ma (MSWD = 0.074, n = 9). The $^{206}Pb/^{238}U$ ages of the 12 analysis points from sample Q22T42 range of 478~482 Ma, with a weighted mean age of 479.8 ± 2.5 Ma (MSWD = 0.800, n = 12). Both samples exhibit good concordance and represent two distinct episodes of magmatic activity.

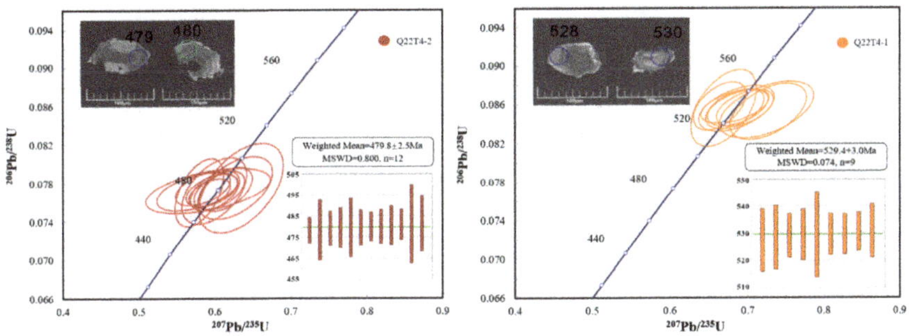

Figure 5. Cathodoluminescence (CL) images of representative zircon grains and concordia diagrams of the Early Paleozoic granites from Guomang-Co.

4.3. Zircon Hf Isotope

Zircon in situ Lu-Hf isotope analysis was performed on 19 concordant test points from the two samples (Figure 6), with the analytical data presented in Table S1. For sample Q22T41, the $^{176}Yb/^{177}Hf$ ratio has a range of 0.045832~0.093184, and the $^{176}Lu/^{177}Hf$ ratio has a range 0.001548~0.003017. The $f_{Lu/Hf}$ values have a range of −0.91~−0.95, with an average value of −0.93. The $\varepsilon_{Hf}(t)$ values have a range of −4.10~−8.36, and the two-stage model age (T_{DM2}) has a range of 1745~2010 Ma. For sample Q22T42, the $^{176}Yb/^{177}Hf$ ratio has a range of 0.044872~0.160229, and the $^{176}Lu/^{177}Hf$ ratio has a range of 0.001543~0.004988. The $f_{Lu/Hf}$ values have a range of −0.85~−0.95, with an average value of −0.93. The $\varepsilon_{Hf}(t)$ values have a range of −2.20~−9.33, and the two-stage model age (T_{DM2}) has a range of 1587~2033 Ma.

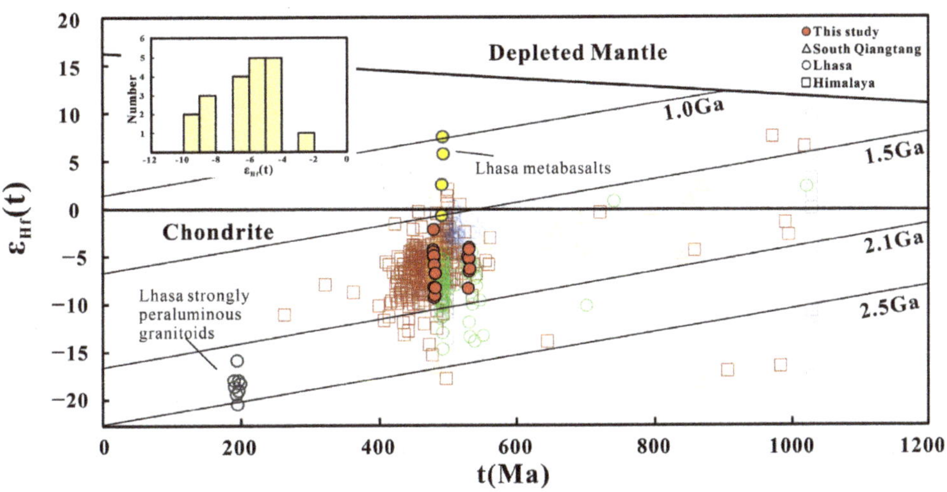

Figure 6. $\varepsilon_{Hf}(t)$-t diagram and $\varepsilon_{Hf}(t)$ histogram for zircon grains from the Guomang-Co granites. Data of Banglei metabasalts and metarhyolites [11], Early Jurassic strongly peraluminous granitoids [106], Zhaqian metarhyolites [36] in the Lhasa terrane, Gemuri and Bensong Co granites [60] in the South Qiangtang terrane, Kangmar and Gyirong Granitic Gneiss [42] in the Himalaya terrane are shown for comparison.

4.4. Whole-Rock Sr–Nd–Pb Isotopes

The Sr-Nd-Pb isotopic test results for the Guomang-Co granite samples Q22T41 and Q22T42 are shown in Table S1. The samples exhibit relatively high initial $^{87}Sr/^{86}Sr$ ratios (0.706955~0.707155) and modern values (0.744868~0.747049), indicating a crustal origin. The $(^{143}Nd/^{144}Nd)t$ ratios have a range of 0.511896~0.511933, with negative $\varepsilon_{Nd}(t)$ values of −10.72 to −9.99 (Figure 7). The relatively low initial $^{143}Nd/^{144}Nd$ ratios (0.511406~0.511408) also suggest that their magma source region has undergone long-term radioactive decay. The Pb isotopic ratios are as follows: $^{206}Pb/^{204}Pb$, $^{207}Pb/^{204}Pb$, and $^{208}Pb/^{204}Pb$ with a range of 19.4131~19.4225, 15.8121~15.8164, and 39.9654~40.0308, respectively; $(^{206}Pb/^{204}Pb)t$, $(^{207}Pb/^{204}Pb)t$, and $(^{208}Pb/^{204}Pb)t$ ratios with a range of 17.0398~17.1831, 15.6817~15.6821, and 39.3308~39.5131, respectively, all of which are consistent with a crustal source composition.

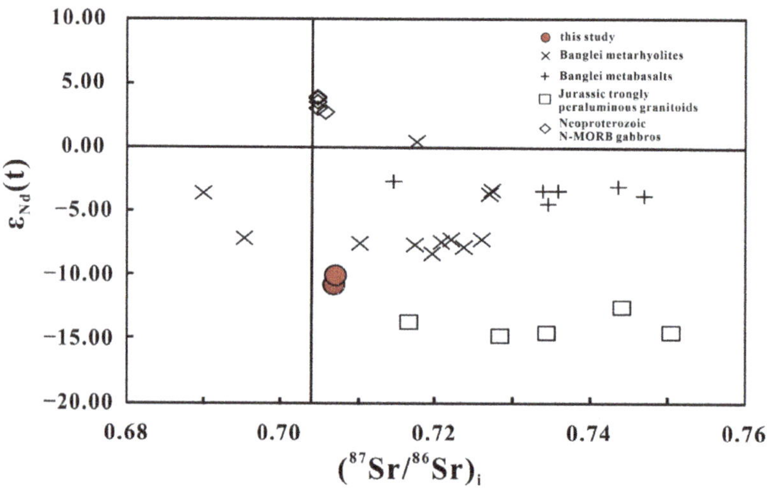

Figure 7. Initial Sr-Nd isotopic composition of Guomang-Co granites. Data of Cambrian metabasalts [28] and metarhyolites [11], Neoproterozoic N-MORB gabbros [107], Jurrassic strongly peraluminous [108] in the Lhasa terrane terrane are shown for comparison.

5. Discussion

5.1. Petrogenesis and Magma Source

The general granite classification scheme divides granites into four types: A-type (anorogenic), S-type (sedimentary source-modified), I-type (crust-mantle co-melting), and M-type (mantle source) [109].

The granite samples exhibit S-type granite characteristics, as evidenced by the following observations: (1) The samples possess high SiO_2 and K_2O contents and display peraluminous features, with CIPW normative mineral calculations indicating the presence of 1.5~2.0% corundum [96,97]. (2) The REE distribution pattern is analogous to that of arc-related magmatic rocks, with relatively low Sr and Ba contents implying dehydration melting of mica [110,111]. Furthermore, the samples exhibit low 10,000 Ga/Al and FeO^T/MgO ratios and a depletion of HFSEs (Zr + Nb + Ce + Y), positioning them within the non-A-type granite field (Figure 8b,c) [84]. (3) Using the zircon saturation temperature formula $T_{Zr} = 12{,}900/[2.95 + 0.85M + \ln(496{,}000/Zr_{melt})]$ [112], the formation temperature of the granite is calculated to the range of 653 °C~679 °C, consistent with the low formation temperature characteristic of S-type granites. (4) The samples exhibit relatively high initial $^{87}Sr/^{86}Sr$ ratios (0.706955~0.707155) and very low $\varepsilon_{Nd}(t)$ values (−10.72~−9.99), which effectively distinguish them from I-type granites [97]. Moreover, in the ACF diagram, the samples plot within the S-type granite field (Figure 8a).

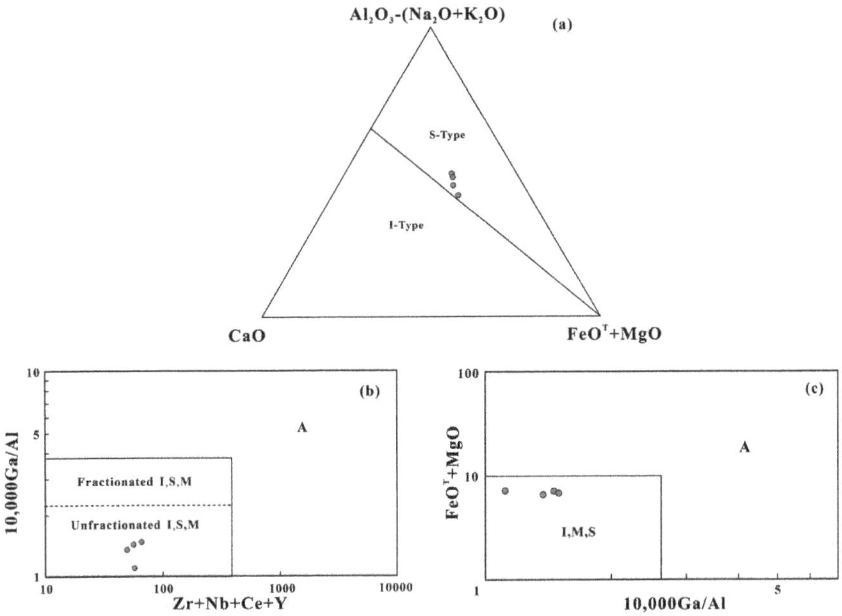

Figure 8. (**a**) CaO-FeO^T + MgO-Al_2O_3-(Na_2O + K_2O) diagram [96]; (**b**) FeO^T/MgO-10,000 Ga/Al diagram [109]; (**c**) 10,000Ga/Al-(Zr + Nb + Ce + Y) diagram [109].

S-type granites can originate either from the fractional crystallization of basaltic magmas or from the deep melting of metamorphosed sedimentary rocks in the crust [113,114]. During fractional crystallization, the isotopic composition of the magma remains constant in a closed system, meaning that acid rocks formed through crystallization differentiation would exhibit isotopic compositions similar to those of the mafic parent magma. However, the Hf isotopic signature of samples ($\varepsilon_{Hf}(t)$ = −9.3~−2.2) is significantly lower than that of nearly coeval Banglei metamorphic basalts from the adjacent area (ca. 492 Ma, $\varepsilon_{Hf}(t)$ = −0.7~+7.5) (Figure 6). Therefore, the most likely origin is the partial melting of metamorphosed sedimentary rocks in the crust.

Additionally, the εHf(t) values of the samples are significantly higher than those of the strongly peraluminous Early Jurassic granites from central Lhasa ($\varepsilon_{Hf}(t)$ = −20.5~−16.0) (Figure 6) [106], indicating the involvement of mantle-derived materials. Compared with other contemporaneous S-type granites (Figure 6), the Guomangcuo granite generally exhibits higher εHf(t) values than those in other regions, such as Banglei in the Lhasa terrane (ca. 492 Ma, 538 Ma; $\varepsilon_{Hf}(t)$ = −14.7~−4.6) [11], Kangmar in the Himalaya terrane (ca. 478~500 Ma, $\varepsilon_{Hf}(t)$ = −17.8~1.99) [42], and Gyirong (ca. 486~474 Ma, $\varepsilon_{Hf}(t)$ = −13.6~−4.6) [42], but lower than those of the adjacent Zhaqian area (ca. 515 Ma, $\varepsilon_{Hf}(t)$ = −4.89~−0.3) [36] (Figure 1b). It can be comparable to those of Gemuri and Bensong Co in the southern Qiangtang terrane (ca. 486 Ma, $\varepsilon_{Hf}(t)$ = −8.5~−4.0) [60]. The two-stage Hf isotopic model age values, with a range of 1587~2033 Ma, suggest that the samples were derived from the ancient Mesoproterozoic crust. Notably, the Hf isotopic composition of sample Q22T4-2 (ca. 480 Ma) is more variable and heterogeneous than that of sample Q22T4-1, which may be attributed to a greater contribution of mantle-derived components. In conjunction with Nd isotopic data, the samples exhibit relatively low $\varepsilon_{Nd}(t)$ values (−10.72~−9.99), consistent with a crustal origin, but slightly higher than those of the strongly peraluminous Early Jurassic granites in the Lhasa terrane ($\varepsilon_{Nd}(t)$ = −14.79~−12.57) (Figure 7), further supporting the involvement of mantle-derived materials. The samples also display relatively high lead isotope ratios, and when plotted on the $^{07}Pb/^{204}Pb$-$^{206}Pb/^{204}P$ diagram (Figure 9) [115], the data points fall near the lower crustal field, which can also be corroborated by their depleted REE distribution patterns (Figure 3a).

The aluminous S-type granites formed by the crustal partial melting of metasedimentary rocks can be identified by whole-rock Rb/Sr, Rb/Ba, and CaO/Na_2O ratios, which reflect the characteristics of their source region. In the Rb/Sr-Rb/Ba diagram (Figure 10a), most of our samples fall into the source region of the clay-rich, plagioclase-poor mudstone-derived melt. The CaO/Na_2O ratio in granites is primarily controlled by plagioclase to clay ratio in the source region. Strongly peraluminous granites derived from clay-rich rocks with low plagioclase content exhibit $CaO/Na_2O < 0.3$, while peraluminous granites derived from sandstones exhibit $CaO/Na_2O > 0.3$ [114,116]. The CaO/Na_2O ratio of the Guomang-Co granites have a range of 0.12~0.15, significantly lower than 0.3, indicating that their protolith was a metamorphosed mudstone. In the A/MF-C/MF diagram (Figure 10b), samples match with the model of partial melting of metamorphosed mudstone as well.

In summary, the Guomangcuo granite was most likely derived from the partial melting of Mesoproterozoic lower crustal metapelitic rocks, with a certain degree of mantle-derived material involvement.

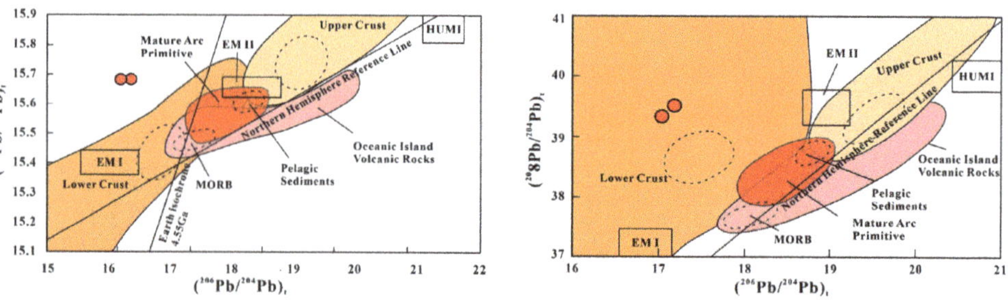

Figure 9. The $^{207}Pb/^{204}Pb$-$^{206}Pb/^{204}Pb$ diagrams [115]. EMI, EMII, HIMU, earth isochrone, and north hemisphere reference line after [117]; lower crust, mature arc, and upper crust after [115]. The area within the dashed lines represents the region of data concentration in each area.

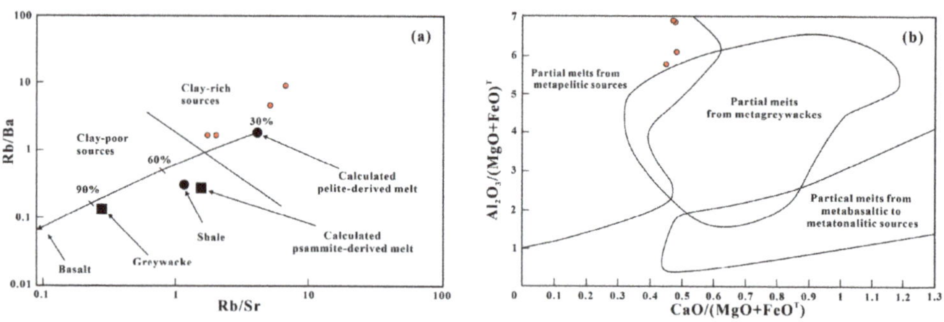

Figure 10. Diagram for magma source region discrimination. (**a**) Rb/Sr-Rb/Ba diagram [114]; (**b**) A/MF-C/MF diagram [118].

5.2. Tectonic Setting and Significance

In orogenic belts around the world, S-type granites are closely associated with the partial melting of the thickened crust [96]. They typically form during the continental–continental collision stage of mountain building, while they can also form in continental arcs, back arcs, and post-collision settings [113,114,119]. However, discussing tectonic settings must first focus on the regional geological context.

The samples have a high SiO_2 content, exhibiting peraluminous characteristics, and belong to the typical high-potassium calc-alkaline series. It is enriched in LILEs, depleted in HFSEs and shows negative δEu. These geochemical features are consistent with granites from active continental margins [98,120]. In the granite tectonic discrimination diagrams related to Nb, Ta, Rb (Figure 11a–d), the Guomangcuo granite samples fall into the syn-collision region. Notably, the sample data also fall within the region of granites related to slab break off [121]. In the Hf-Rb-Ta diagram (Figure 11e), the sample data fall within the late-collision to post-collision tectonic setting. Similarly, in the R1–R2 diagram (Figure 11f), the sample data also fall near the syn-collision region, post-collision. Taken together, the formation of the Guomangcuo granite aligns with a syn-collision to post-collision tectonic background.

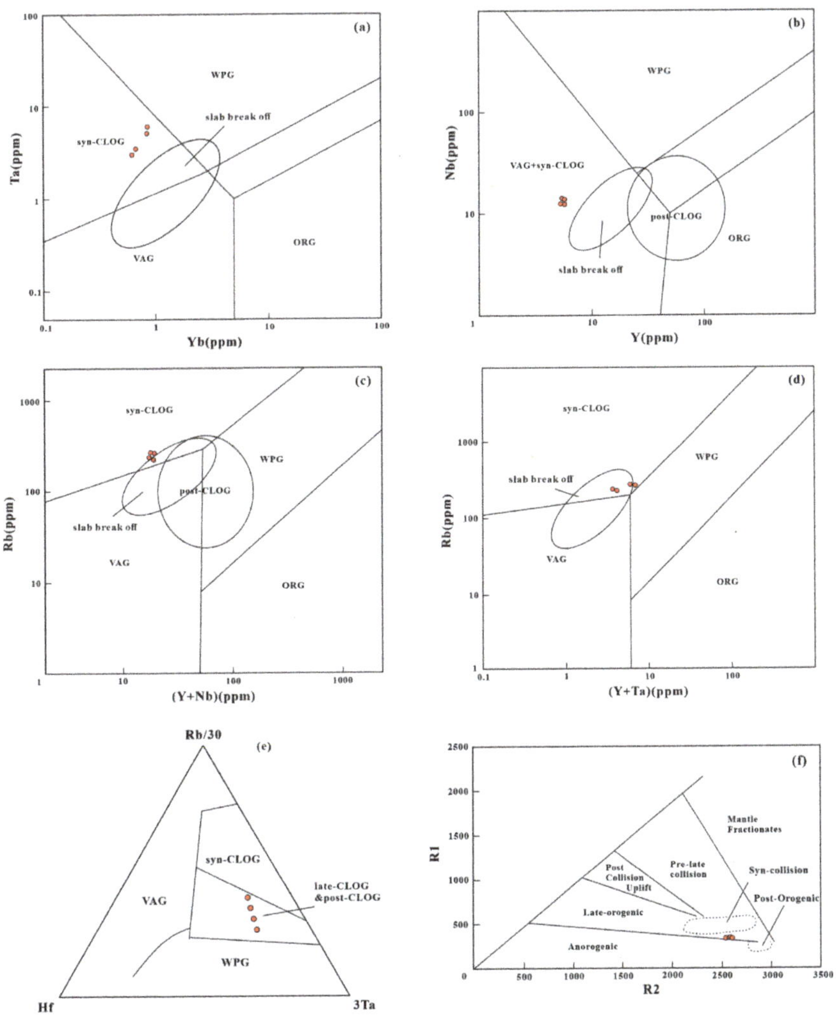

Figure 11. Granite tectonic discrimination diagrams. (**a–d**) Nb, Ta, Rb revalant diagrams [122,123]; (**e**) Hf-Rb-Ta diagram [124]; (**f**) R1-R2 dagram [125]. R1 = 4Si − 11(Na + K) − 2(Fe + Ti), R2 = 6Ca + 2Mg + Al. The granite data area related to slab break off comes from [121].

Collins and Pisarevsky [126] summarized the spatiotemporal distribution patterns of the Pan-African orogeny on the East Gondwana continent, dividing it into two tectonic events. The early orogenic event occurred only in the eastern part of Africa and is referred to as the East African Orogeny (650~620 Ma), while the later orogenic event had a broader influence and is known as the Pan-African Orogeny. The latter includes the Kuunga Orogen (570~500 Ma) between southeast Antarctica and India, and the Malagasy Orogen (550~520 Ma) between East and West Gondwana. Subsequently, more researchers have supported this view, asserting that the Pan-African orogeny refers to a series of collisional orogenic events between different continental blocks within the Gondwana continent, with a timeframe concentrated at 570~520 Ma [10,31,33,35,36,127]. In contrast, the various blocks of the Tibetan Plateau and its surrounding regions (such as the Himalayan terrane, Lhasa terrane, and South Qiangtang terrane) were located along the northern margin of the Gondwana during the Early Paleozoic. Through our statistics of Early Paleozoic age from these regions (Table 1), it is not hard to find out that the tectonic movements mainly occurred at 530~460 Ma, which is generally 40~60 Ma later than the timing of the Pan-African orogeny. Therefore, we do not consider these magmatic events to be part of the Pan-African orogeny. It is worth noting that most of the granites reported in these records are S-type granites. Although an extensional setting can also produce S-type granites, it does not provide the geological conditions for the formation of large-scale S-type granites.

On the basis of the principle of global volume balance, Cawood et al. [10] proposed that after the end of the Pan-African orogeny, a new subduction zone formed along the northern margin of the East Gondwana, leading to Andean-type accretionary orogeny. This viewpoint fully considers the global tectonic evolution and the delayed tectonic movements in the Early Paleozoic of the Tibetan Plateau and its surrounding regions, and may be consistent with the Ediacaran-Cambrian Cadomian orogeny on the northern margin of West Gondwana [128–130]. It has since been widely accepted by many scholars [11,12,35,36,49,131]. The Early Paleozoic magmatic rocks are believed to be related to the subduction and consumption of oceanic crust [36,38], and micro-continental block collisions occurred at the edge of the Andes magmatic arc [11,44].

Through the study of the Early Paleozoic bimodal magmatism in the Lhasa terrane, Zhu et al. [11] proposed an Andean-type subduction-slab break-off evolution model. According to this model, magmatic activity at the margin of the Gondwana continent has exsited around 536 Ma. As the subduction depth of the Proto-Tethys oceanic crust increased, the oceanic plate sank under gravity. Around 510 Ma, the slab began to retreat, and subsequently, the back-arc basin underwent extension, accompanied by asthenosphere upwelling, triggering deep melting of the continental crust and resulting in extensive magmatic activity. Around 492 Ma, slab break-off occurred, and the upwelling asthenosphere initiated large-scale magmatic activity, which also marked the onset of post-collision orogeny.

The Guomangcuo granite supports the hypothesis of Andean-type subduction, aligning well with the aforementioned evolutionary model. Notably, in the adjacent Zhaqian area (Figure 1b), a Cambrian–Ordovician angular unconformity is present. The overlying undeformed Lower Ordovician strata are nearly orthogonal (~90°) to the underlying strongly deformed Cambrian strata [76]. A volcanic interlayer within the underlying sequence has yielded an age of ca. 501 Ma, thereby constraining the termination of the collisional event to after this time. Thus, the syn-collisional phase likely occurred during the Cambrian, corresponding to the ca. 530 Ma age of the samples. By the early Ordovician (ca. 480 Ma), slab break-off triggered partial melting of the mantle, which in turn facilitated deep crustal anatexis. Some mantle-derived material mixed with the metasedimentary-derived melts, modifying the isotopic composition of the magma (Figure 12). This transition marked the shift of the Andean-type orogenic process into its post-orogenic stage, which also explains why the sample data plot near the post-orogenic field in tectonic discrimination diagrams (Figure 11c,e,f).

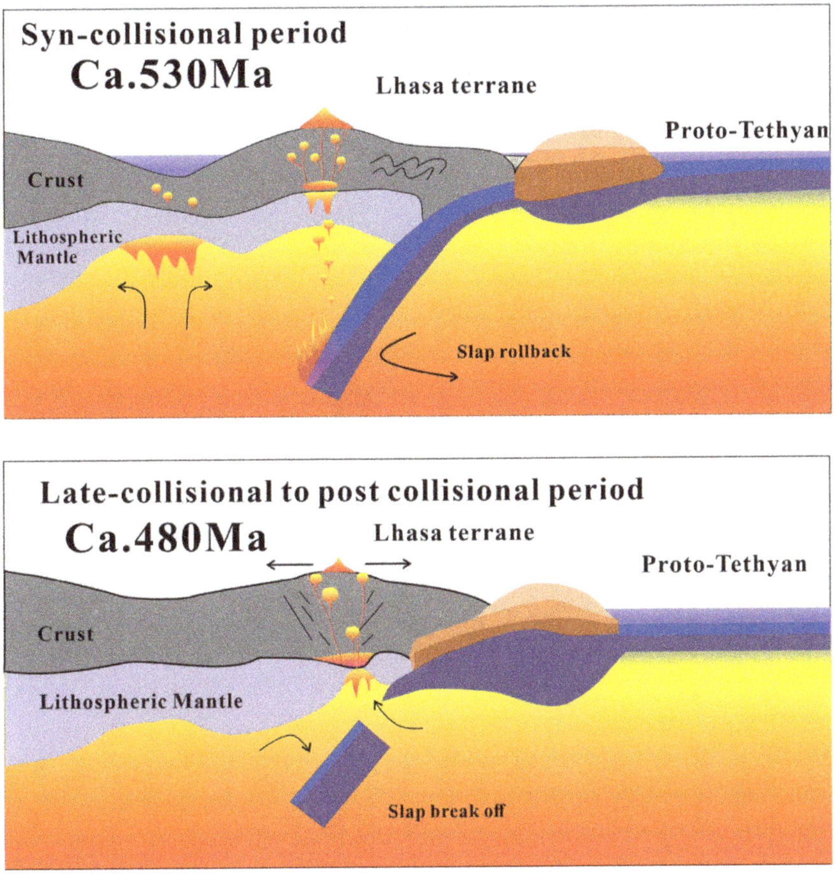

Figure 12. Schematic diagram of the Early Paleozoic tectonic evolution of the Lhasa terrane. (revised by [11]).

Our research provides new constraints for the study of the Early Paleozoic evolution of the Proto-Tethys Ocean and supplements the Early Paleozoic data of the Lhasa terrane.

6. Conclusions

The Guomang-Co granites are a set of dark mylonitized granite newly defined from the originally defined Nyainqengtanglha Group (AnZNq). Its main components are quartz and alkaline feldspar, with no significant mineralogical phase changes. The protolith is considered to be an alkaline granite.

The zircon U-Pb age results for the two samples of Guomangcuo granite are 529.4 ± 3.0 Ma and 479.8 ± 2.5 Ma, corresponding to the Early Cambrian and Early Ordovician, respectively, indicating two magmatic events during the Early Paleozoic in this region.

The Guomang-Co granites exhibit classic S-type geochemical signatures, including peraluminous composition (A/CNK = 1.09~1.13), high-potassium calc-alkaline, negative Nd, Hf systematics ($\varepsilon_{Nd}(t) = -10.72 \sim -9.99$, $\varepsilon_{Hf}(t) = -2.20 \sim -9.33$). It may originate from partial melting of lower crustal Paleo-Proterozoic mudstone influenced by mantle-derived components, likely a product of the Andean-type orogeny along the northern margin of

Gondwana in the Early Paleozoic, indicating the existence of the Proto-Tethys Ocean's southward subduction during the Cambrian–Ordovician period.

Supplementary Materials: The following supporting information can be downloaded at https://www.mdpi.com/article/10.3390/min15030284/s1; Table S1: Geochemistry and Sr-Nd-Pb-Hf isotope data of Guomang-Co granite samples.

Author Contributions: Conceptualization, Y.Z.; formal analysis, Y.Z.; investigation, Z.L.; writing—original draft preparation, Y.Z.; writing—review and editing, M.W. and C.Y.; supervision, M.W.; funding acquisition, M.W. and C.Y. All authors have read and agreed to the published version of the manuscript.

Funding: This work was supported by the Deep Earth Probe and Mineral Resources Exploration-National Science and Technology Major Project (Grant No.2024ZD1000100), the National Key R&D Program of China (Grant No. 2022YFC3080200), the Scientific Research Project of the Education Department of Jilin Province (Grant No. JJKH20241254KJ), and Graduate Innovation Fund of Jilin University (Grant No. 2024CX102).

Data Availability Statement: The data presented in this study are available in this paper.

Acknowledgments: We acknowledge anonymous reviewers of their constructive reviews and valuable suggestions that led to great improvement in the presentation of the paper. We also thank Teng Wang for his help in data collection in this paper.

Conflicts of Interest: The authors declare that they have no competing interests.

References

1. Stampfli, G.M.; Borel, G.D. A plate tectonic model for the Paleozoic and Mesozoic constrained by dynamic plate boundaries and restored synthetic oceanic isochrons. *Earth Planet. Sci. Lett.* **2002**, *196*, 17–33. [CrossRef]
2. Stampfli, G.M.; Hochard, C.; Vérard, C. The formation of Pangea. *Tectonophysics* **2013**, *593*, 1–19. [CrossRef]
3. Cocks, L.R.M.; Torsvik, T.H. The Palaeozoic geography of Laurentia and western Laurussia: A stable craton with mobile margins. *Earth Sci. Rev.* **2011**, *106*, 1–51. [CrossRef]
4. Wu, F.-Y.; Wan, B.; Zhao, L.; Xiao, W.-J.; Zhu, R.-X. Tethyan geodynamics. *Acta Petrol. Sin.* **2020**, *36*, 1627–1674.
5. Zhao, G.-C.; Wang, Y.-J.; Huang, B.-C.; Dong, Y.-P.; Li, S.-Z.; Zhang, G.-W.; Yu, S. Geological reconstructions of the East Asian blocks: From the breakup of Rodinia to the assembly of Pangea. *Earth Sci. Rev.* **2018**, *186*, 262–286. [CrossRef]
6. Wilhem, C.; Windley, B.F.; Stampfli, G.M. The Altaids of Central Asia: A tectonic and evolutionary innovative review. *Earth Sci. Rev.* **2012**, *113*, 303–341. [CrossRef]
7. Miller, C.; Thöni, M.; Frank, W. The early Palaeozoic magmatic event in the Northwest Himalaya, India: Source, tectonic setting and age of emplacement. *Geol. Mag.* **2001**, *138*, 237–251. [CrossRef]
8. Ji, W.-Q.; Wu, F.-Y.; Chung, S.-L. Zircon U–Pb geochronology and Hf isotopic constraints on petrogenesis of the Gangdese batholith, southern Tibet. *Chem. Geol.* **2009**, *262*, 229–245. [CrossRef]
9. Pan, G.-T.; Wang, L.-Q.; Zhang, W.-P.; Wang, B.-D.; Pan, G.T.; Wang, L.Q.; Li, R.S.; Yuan, S.H.; Ji, W.H.; Yin, F.G.; et al. Tectonic evolution of the Qinghai-Tibet plateau. *Asian Earth Sci.* **2012**, *53*, 314. [CrossRef]
10. Cawood, P.A.; Johnson, M.R.W.; Nemchin, A.A. Early Palaeozoic orogenesis along the Indian margin of Gondwana: Tectonic response to Gondwana assembly. *Earth Planet. Sci. Lett.* **2007**, *255*, 70–84. [CrossRef]
11. Zhu, D.-C.; Zhao, Z.-D.; Niu, Y.-L.; Dilek, Y.; Wang, Q.; Ji, W.-H.; Dong, G.-C.; Sui, Q.-L.; Liu, Y.-S.; Yuan, H.-L.; et al. Cambrian bimodal volcanism in the Lhasa Terrane, southern Tibet: Record of an early Paleozoic Andean-type magmatic arc in the Australian proto-Tethyan margin. *Chem. Geol.* **2012**, *328*, 290–308. [CrossRef]
12. Wang, Y.-J.; Xing, X.-W.; Cawood, P.A.; Lai, S.-C.; Xia, X.-P.; Fan, W.-M.; Liu, H.-C.; Zhang, F.-F. Petrogenesis of early Paleozoic peraluminous granite in the Sibumasu Block of SW Yunnan and diachronous accretionary orogenesis along the northern margin of Gondwana. *Lithos* **2013**, *182–183*, 67–85. [CrossRef]
13. Liu, G.-C. Petrology Component and Geochronology of Early Paleozoic Proto-Tethys Ophiolitemélange in SW Yunnan. Ph.D. Thesis, China University of Geosciences, Wuhan, China, 2020.
14. Liu, G.-C.; Zi, J.-W.; Nie, X.-M.; Santosh, M.; Zhu, C.; Wang, W.; Zhao, T.-Y.; Chen, G.-Y.; Feng, Q.-L. Late Cambrian Magmatic Events in SW Yunnan and Implications for the Tectonic Reconstruction of Northern Gondwana. *J. Earth Sci.* **2024**, *35*, 1407–1425. [CrossRef]
15. Kennedy, W.Q. The structural differentiation of Africa in the Pan-Africa (±500 my) tectonic episode. *Univ. Leeds Res. Inst. Afr. Geol. Ann. Rept.* **1964**, *8*, 48–49.

16. Xu, Z.-Q.; Yang, J.-S.; Liang, F.-H.; Qi, X.-X.; Liu, F.-C.; Zeng, L.-S.; Liu, M.-Y.; Li, H.-B.; Wu, C.-L.; Shi, R.-D.; et al. Pan-African and Early Paleozoic orogenic events in the Himalaya terrane: Inference from SHRIMP U-Ph zircon ages. *Acta Petrol.* **2005**, *21*, 3–14.
17. Liu, W.-C.; Zhou, Z.-G.; Zhang, X.-X.; Zhao, X.-G. SHRIMP zircon geochronological constraints on a Pan-African orogeny in the Yadong Area, Southern Tibet. *Geochim. Cosmochim. Acta Suppl.* **2006**, *70*, A365. [CrossRef]
18. Li, C.; Xie, Y.-W.; Sha, S.-L.; Dong, Y.-S. SHRIMP U-Pb zircon dating of the Pan-African granite in Baxoi County, eastern Tibet, China. *Geol. Bull. China* **2008**, *27*, 64–68.
19. Hu, P.-Y.; Li, C.; Su, L.; Li, C.-B.; Yu, H. Zircon U-Pb dating of granitic gneiss in Wugong Mountain area, central Qiangtang, Qinghai-Tibet Plateau: Age records of Pan-African movement and Indo-China movement. *Geol. China* **2010**, *37*, 1050–1061.
20. Xie, C.-M.; Li, C.; Su, L.; Wu, Y.-W.; Wang, M.; Yu, H. LA-ICP-MS U-Pb dating of zircon from granite-gneiss in the Amdo area, northern Tibet, China. *Geol. Bull. China* **2010**, *29*, 1737–1744.
21. Liu, Q.-S.; Ye, P.-S.; Wu, Z.-H. SHRIMP zircon U-Pb dating and petrogeochemistry of Ordovician granite bodies in the southern segment of Gaoligong Mountain, western Yunnan Province. *Geol. Bull. China* **2012**, *31*, 250–257.
22. Lin, S.-L.; Cong, F.; Gao, Y.-J.; Zou, G.-F. LA-ICP-MS zircon U-Pb age of gneiss from Gaoligong Mountain Group on the southeastern margin of Tengchong block in western Yunnan Province. *Geol. Bull. China* **2012**, *31*, 258–263.
23. Wang, Y.-W.; Zhou, Q.; Xie, Q.-X.; Sun, P.; Zhang, J.-R.; Wang, H.; Xiao, L. Pan-African and Early Paleozoic Geological Orogenic Events in the Tethyan Himalaya: From Chronological Evidence of Gneissic Monzonitic Granite, Kangma, Tibet. *Bull. Geol. Sci. Technol.* **2016**, *35*, 30.
24. Murphy, J.B.; Nance, R.D. Supercontinent model for the contrasting character of Late Proterozoic orogenic belts. *Geology* **1991**, *19*, 469–472. [CrossRef]
25. Brookfield, M.E. The Himalayan passive margin from Precambrian to Cretaceous times. *Sediment. Geol.* **1993**, *84*, 1–35. [CrossRef]
26. Wang, M.; Li, C.; Fan, J.-J. Geochronology and geochemistry of the Dabure basalts, central Qiangtang, Tibet: Evidence for~550 Ma rifting of Gondwana. *Int. Geol. Rev.* **2015**, *57*, 1791–1805. [CrossRef]
27. Wang, M.; Li, C.; Xie, C.-M. Dating of detrital zircons from the Dabure clastic rocks: The discovery of Neoproterozoic strata in southern Qiangtang, Tibet. *Int. Geol. Rev.* **2016**, *58*, 216–227. [CrossRef]
28. Pan, X.-P.; Li, R.-S.; Wang, C.; Yu, P.-S.; Gu, P.-Y.; Zha, X.-F. Geochemical characteristics of the Cambrian volcanic rocks in Banglecun area on the northern margin of Gangdise, Nyima County, Tibet. *Geol. Bull. China* **2012**, *31*, 63–74.
29. Gu, P.-Y.; He, S.-P.; Li, R.-S.; Shi, C.; Dong, Z.-C.; Zha, X.-F.; Wu, J.-L.; Wang, Y. Geochemical features and tectonic significance of granitic gneiss of Laguigangri metamorphic core complexes in southern Tibet. *Acta Petrolog. Sin.* **2013**, *29*, 756–768.
30. Yang, X.-J.; Jia, X.-C.; Xiong, C.-L.; Bai, X.-Z.; Huang, B.-X.; Luo, G.; Yang, C.-B. LA-ICP-MS zircon U-Pb age of metamorphic basic volcanic rock in Gongyanghe Group of southern Gaoligong Mountain, western Yunnan Province, and its geological significance. *Geol. Bull. China* **2012**, *31*, 264–276.
31. Liu, Y.-M. Magmatism at 550~450 Ma Within the SouthQiangtang-Baoshan Plate of the Tibetan Plateau: Constraintson the Tectonic Evolution of the Basement of the northernGondwana Margin. Ph.D. Thesis, Jinlin University, Changchun, China, 2017.
32. Lv, P.-R.; Zhang, H.-D.; Luo, Y.-J.; Yao, W.-G. LA-ICP-MS Zircon U-Pb Dating of Granitic Gneiss from Yadong Area in South Tibet and Its Geochemical Characteristics. *Earth Sci.* **2018**, *43*, 4459–4474.
33. Zhang, T.-Y. Early Paleozoic Tectonic Movement on the Tibetan Plateauand Its Adjacent Areas: A Case Study of Cambrian-Ordovician Unconformity. Ph.D. Thesis, Jinlin University, Changchun, China, 2018.
34. Gehrels, G.E.; DeCelles, P.G.; Martin, A.; Ojha, T.P.; Pinhassi, G. Initiation of the Himalayan orogen as an early Paleozoic thin-skinned thrust belt. *GSA Today* **2003**, *13*, 4–9. [CrossRef]
35. Zhang, Z.-M.; Wang, J.-L.; Shen, K.; Shi, C. Paleozoic circum-Gondwana orogens: Petrology and geochronology of the Namche Barwa Complex in the eastern Himalayan syntaxis, Tibet. *Acta Petrolog. Sin.* **2008**, *24*, 1627–1637.
36. Hu, P.-Y.; Li, C.; Wang, M.; Xie, C.-M.; Wu, Y.-W. Cambrian volcanism in the Lhasa terrane, southern Tibet: Record of an early Paleozoic Andean-type magmatic arc along the Gondwana proto-Tethyan margin. *J. Asian Earth Sci.* **2013**, *77*, 91–107. [CrossRef]
37. Xie, C.-M.; Li, C.; Su, L.; Wu, Y.-W.; Xie, Y.-W. Pan-African and early Paleozoic tectonothermal events in the Nyainrong microcontinent: Constraints from geochronology and geochemistry. *Sci. China Earth Sci.* **2014**, *56*, 2066–2079. [CrossRef]
38. Ding, H.-X.; Zhang, Z.-M.; Dong, X.; Yan, R.; Lin, Y.-H.; Jiang, H.-Y. Cambrian ultrapotassic rhyolites from the Lhasa terrane, south Tibet: Evidence for Andean-type magmatism along the northern active margin of Gondwana. *Gondwana Res.* **2014**, *27*, 1616–1629. [CrossRef]
39. Lee, J.; Whitehouse, M.J. Onset of mid-crustal extensional flow in southern Tibet: Evidence from U/Pb zircon ages. *Geology* **2007**, *35*, 45–48. [CrossRef]
40. Quigley, M.C.; Liangjun, Y.; Gregory, C.; Corvino, A.; Sandiford, M.; Wilson, C.J.L.; Xiaohan, L. U–Pb SHRIMP zircon geochronology and T–t–d history of the Kampa Dome, southern Tibet. *Tectonophysics* **2008**, *446*, 97–113. [CrossRef]
41. Lee, J.; Hacker, B.R.; Dinklage, W.S.; Wang, Y.; Gans, P.; Calvert, A.; Wan, J.-L.; Chen, W.-J.; Blythe, A.E.; McClelland, W. Evolution of the Kangmar Dome, southern Tibet: Structural, petrologic, and thermochronologic constraints. *Tectonics* **2000**, *19*, 872–895. [CrossRef]

42. Wang, X.-X.; Zhang, J.-J.; Santosh, M.; Liu, J.; Yan, S.-Y.; Lei, G. Andean-type orogeny in the Himalayas of south Tibet: Implications for early Paleozoic tectonics along the Indian margin of Gondwana. *Lithos* **2012**, *154*, 248–262. [CrossRef]
43. Gao, L.-E.; Zeng, L.-S.; Hu, G.-Y.; Wang, Y.-Y.; Wang, Q.; Guo, C.-L.; Hou, K.-J. Early Paleozoic magmatism along the northern margin of East Gondwana (Review). *Lithos* **2019**, *334–335*, 25–41. [CrossRef]
44. Wang, X.-X.; Zhang, J.-J.; Yang, X.-Y.; Zhang, B. Zircon SHRIMP U-Pb ages, Hf isotopie features and their geologicalsiqnificance of the Greater Himalayan Crystalline Complex augen qneiss in Gyirong Area, south Tibet. *Earth Sci. Front.* **2011**, *18*, 127–139.
45. Shi, C.; Li, R.-S.; He, S.-P.; Wang, C.; Pan, S.-J.; Liu, Y.; Gu, P.-Y. LA-ICP-MS zircon U-Pb dating for gneissic garnet-bearing bi otite granodiorite in the Yadong area, southern Tibet, China and its geological significance. *Geol. Bull. China* **2010**, *29*, 1745–1753.
46. Zhang, Z.-M.; Dong, X.; Santosh, M.; Liu, F.; Wang, W.; Yiu, F.; He, Z.-Y.; Shen, K. Petrology and geochronology of the Namche Barwa Complex in the eastern Himalayan syntaxis, Tibet: Constraints on the origin and evolution of the north-eastern margin of the Indian Craton. *Gondwana Res.* **2012**, *21*, 123–137. [CrossRef]
47. Yang, D.-T. Geochronology, Geochemistry and Tectonic Implication of EarlyPaleozoic Granitic Gneisses in the Dinggyê Area, Southern Tibet. Master's Thesis, Chinese Academy of Geological Science, Beijing, China, 2015.
48. Liu, Z.-C.; Wu, F.-Y.; Ding, L.; Liu, X.-C.; Wang, J.-G.; Ji, W.-Q. Highly fractionated Late Eocene (~35 Ma) leucogranite in the Xiaru Dome, Tethyan Himalaya, South Tibet. *Lithos* **2016**, *240–243*, 337–354. [CrossRef]
49. Zhang, S.-Z.; Li, F.-Q.; Li, Y.; Liu, W.; Qin, Y.-D. Early Ordovician strongly peraluminous granite in the middle section of the Yarlung Zangbo junction zone and its geological significance. *Sci. China Earth Sci.* **2014**, *57*, 630–643. [CrossRef]
50. Gehrels, G.E.; DeCelles, P.G.; Ojha, T.P.; Upreti, B.N. Geologic and U-Th-Pb geochronologic evidence for early Paleozoic tectonism in the Kathmandu thrust sheet, central Nepal Himalaya. *Geol. Soc. Am. Bull.* **2006**, *118*, 185–198. [CrossRef]
51. Gehrels, G.E.; DeCelles, P.G.; Ojha, T.P.; Upreti, B.N. Geologic and U–Pb geochronologic evidence for early Paleozoic tectonism in the Dadeldhura thrust sheet, far-west Nepal Himalaya. *J. Asian Earth Sci.* **2006**, *28*, 385–408. [CrossRef]
52. DeCelles, P.G.; Gehrels, G.E.; Quade, J.; LaReau, B.; Spurlin, M. Tectonic implications of U-Pb zircon ages of the Himalayan Orogenic Belt in Nepal. *Science* **2000**, *288*, 497–499. [CrossRef]
53. Girard, M.; Bussy, F. Late Pan-African magmatism in the Himalaya: New geochronological and geochemical data from the Ordovician Tso Morari metagranites (Ladakh, NW India). *Schweiz. Mineral. Petrogr. Mitt.* **1999**, *79*, 399–418.
54. Johnson, M.R.W.; Oliver, G.J.H.; Parrish, R.R.; Johnson, S.P. Synthrusting metamor phism, cooling and erosion of the Himalayan Kathmandu complex, Nepal. *Tectonics* **2001**, *20*, 394–415. [CrossRef]
55. Liu, Y.-M.; Li, C.; Xie, C.-M.; Fan, J.-J.; Wu, H.; Jiang, Q.-Y.; Li, X. Cambrian granitic gneiss within the central Qiangtang terrane, Tibetan Plateau: Implications for the early Palaeozoic tectonic evolution of the Gondwanan margin. *Int. Geol. Rev.* **2016**, *58*, 1043–1063. [CrossRef]
56. Wang, H.-T.; Zhai, Q.-G.; Hu, P.-Y.; Zeng, L.-S.; Tang, Y.; Zhu, Z.-C. Early Paleozoic granitic rocks of the South Qiangtang Terrane, northern Tibetan Plateau: Implications for subduction of the Proto- (Paleo-) Tethys Ocean. *J. Asian Earth Sci.* **2020**, *204*, 104579. [CrossRef]
57. Liu, Y.-M.; Xie, C.-M.; Li, C.; Li, S.-Z.; Santosh, M.; Wang, M.; Fan, J.-J. Breakup of the northern margin of Gondwana through lithospheric delamination: Evidence from the Tibetan Plateau. *Geol. Soc. Am. Bull.* **2019**, *131*, 675–694. [CrossRef]
58. Pullen, A.; Kapp, P.; Gehrels, G.E.; Ding, L.; Zhang, Q. Metamorphic rocks in central Tibet: Lateral variations and implications for crustal structure. *Bulletin* **2011**, *123*, 585–600. [CrossRef]
59. Peng, Z.M.; Geng, Q.R.; Wang, L.Q.; Zhang, Z.; Guan, J.L.; Cong, F.; Liu, S.S. Zircon U-Pb ages and Hf isotopic characteristics of granitic gneiss from Bunsumco, central Qiangtang, Qinghai-Tibet Plateau. *Chin. Sci. Bull.* **2014**, *59*, 2621–2629. (In Chinese)
60. Hu, P.-Y.; Zhai, Q.-G.; Jahn, B. Early Ordovician granites from the South Qiangtang terrane, northern Tibet: Implications for the early Paleozoic tectonic evolution along the Gondwanan proto-Tethyan margin. *Lithos* **2015**, *220*, 318–338. [CrossRef]
61. Wang, H.-T.; Zhai, Q.-G.; Hu, P.-Y.; Zeng, L.-S.; Tang, Y.; Zhu, Z.-C. Late Cambrian to Early Silurian Granitic Rocks of the Gemuri Area, Central Qiangtang, North Tibet: New Constraints on the Tectonic Evolution of the Northern Margin of Gondwana. *Acta Geol. Sin.* **2020**, *94*, 1007–1019. [CrossRef]
62. Zheng, Y.-L.; Wang, G.-H.; Guo, Z.-W.; Liang, X.; Yuan, G.-L.; Wang, H.-D.; Huang, B.; He, Y.-D. The record of the Pan African and the Indosinian tectono-thermal event in Qiangtang terrane, northern Tibet: Evidence from geochemical characteristics and U-Pb geochronology of the metamorphic complex in Ejiumai area. *Acta Petrol. Sin.* **2015**, *31*, 1137–1152.
63. Xie, C.-M.; Li, C.; Fan, J.-J.; Su, L. Ordovician sedimentation and bimodal volcanism in the Southern Qiangtang terrane of northern Tibet: Implications for the evolution of the northern Gondwana margin. *Int. Geol. Rev.* **2017**, *59*, 2078–2105. [CrossRef]
64. Xie, C.-M.; Li, C.; Fan, J.-J.; Wang, M.; Liu, Y.-M.; Zhang, H.-Y. Establishment of Middle-Upper Ordovician Dawashan Formationin central Oiangtang, Tibetan Plateau, and its significance. *Geol. Bull. China* **2015**, *34*, 1812–1820.
65. Ren, F.; Yin, F.-G.; Xu, B.; Liu, H.-L.; Fan, B.-L.; Xu, C.-H.; Bai, J.-G. Zircon U-Pb age and Hf isotope of Early Paleozoic granite from the Jitang area in eastern Tibet and its insight into the evolution of the Proto-Tethys Ocean. *Geol. Bull. China* **2021**, *40*, 1865–1876.
66. Ji, W.-H.; Chen, S.-J.; Zhao, Z.-M.; Li, R.-S.; He, S.-P.; Wang, C. Discovery of the Cambiran volcanic rocks in the Xainza area, Gangdese orogenic belt, Tibet, China and its significance. *Geol. Bull. China* **2009**, *28*, 1350–1354.

67. Gehrels, G.; Kapp, P.; DeCelles, P.; Pullen, A.; Blakey, R.; Weislogel, A.; Ding, L.; Guynn, J.; Martin, A.; McQuarrie, N.; et al. Detrital zircon geochronology of pre-Tertiary strata in the Tibetan-Himalayan orogen. *Tectonics* **2011**, *30*, TC5016. [CrossRef]
68. Kang, C.; Dai, K.; Li, H.; Liang, C.; Chen, X.; Zha, X.; Wang, T.; Li, J. The Cambrian metamorphic granites in the Jili area, Baxoi, Xizang: U-Pb dating, geochemical signatures and geological significance. *Sediment. Geol.* **2019**, *39*, 1–13.
69. Dong, X.; Zhang, Z.-M. Cambrian granitoids from the southeastern Tibetan Plateau: Research on petrology and zircon Hf isotope. *Acta Petrolog. Sin.* **2015**, *31*, 1183–1199.
70. Dong, X.; Zhang, Z.-M.; Wang, J.-L.; Zhao, G.-C.; Liu, F.; Wang, W.; Yu, F. Provenance and formation age of the Nyingchi Group in the southern Lhasa terrane, Tibetan Plateau: Petrology and zircon U-Pb geochronology. *Acta Petrolog. Sin.* **2009**, *25*, 1678–1694.
71. Guynn, J.; Kapp, P.; Gehrels, G.E.; Ding, L. U-Pb geochronology of basement rocks in central Tibet and paleogeographic implications. *J. Asian Earth Sci.* **2012**, *43*, 23–50. [CrossRef]
72. Xu, R.-H.; Schärer, U.; Claude, J. Allègre.Magmatism and Metamorphism in the Lhasa Block (Tibet): A Geochronological Study. *J. Geol.* **1985**, *93*, 41–57. [CrossRef]
73. Li, C.; Huang, X.-P.; Zhai, Q.-G.; Zhu, T.-X.; Yu, Y.-S.; Wang, G.-H.; Zeng, Q.-G. The Longmu Co-Shuanghu-Jitang plate suture andthe northern boundary of Gondwanaland in the Qinghai-Tibet plateau. *Earth Sci. Front.* **2006**, *13*, 136–147.
74. Zhu, D.-C.; Zhao, Z.-D.; Niu, Y.-L.; Wang, Q.; Dilek, Y.; Dong, G.-C.; Mo, X.-X. Origin and Paleozoic Tectonic Evolution of the Lhasa Terrane. *Geol. J. China Univ.* **2012**, *18*, 1–15.
75. Xu, M.-J.; Li, C.; Wu, Y.-W.; Xie, C.-M. Geochemical characteristics and sedimentary environments of siliceous rocks in Guomang-Co ophiolitic mélange of Tibet. *Geol. Bull. China* **2014**, *33*, 1061–1066.
76. Li, C.; Wu, Y.-W.; Wang, M.; Yang, H.-T. Significant progress on Pan-African and Early Paleozoic orogenic events in Qinghai-Tibet Plateau—Discovery of Pan-African orogenic unconformity and Cambrian System in the Gangdise area, Tibet, China. *Geol. Bull. China* **2010**, *29*, 1733–1736.
77. Hu, D.; Wu, Z.; Jiang, W.; Shi, Y.; Ye, P.; Liu, Q. SHRIMP zircon U-Pb age and Nd isotope study of the Nianqing Tanggula Group in Tibet. *Sci. China Ser. D Earth Sci.* **2005**, *48*, 1377–1386. [CrossRef]
78. Wu, Y.; Ma, X.-X.; Zhsng, Z.-P.; Jiao, S.-W.; Duan, K.; Dong, H.; Wang, X.-W.; Ma, T.; Li, P.-J.; Liang, Z.-Y.; et al. Geochemical Features of the Nyainqentanglha Group in the Western Lhasa Terrane, Western Tibet and Their Tectonic Significance. *Acta Geol. Sin.* **2016**, *90*, 3081–3098.
79. Yu, H. Mineral Geochemical Characteristics and Genetic Mechanism of Olivine Rocks in Shangnan, Shanxi. Master's Thesis, China University of Geosciences, Beijing, China, 2011.
80. Li, W.-Q. Testing accuracy of 10 major elements in diabase by borate melting sample preparation method. *World Geol.* **2019**, *38*, 843–851.
81. Wiedenbeck, M.; Alle, P.; Corfu, F.Y. Three natural zircon standards for U-Th-Pb, Lu-Hf, trace element and REE analyses. *Geostand. Newslett.* **1995**, *19*, 1–23. [CrossRef]
82. Jackson, S.E.; Pearson, N.J.; Griffin, W.L. The application of laser ablation-inductively coupled plasma-mass spectrometry to in situ U–Pb zircon geochronology. *Chem. Geol.* **2004**, *211*, 47–69. [CrossRef]
83. Hao, Y.-J.; Ren, Y.-S.; Zhao, H.-L.; Lai, K.; Zhao, X.; Ma, Y.-P. Metallogenic mechanism and tectonic setting of tungsten mineralization in the Yangbishan deposit in northeastern China. *Acta Geol. Sin.* **2018**, *92*, 241–267. [CrossRef]
84. Liu, J.-F.; Chi, X.G.; Dong, C.-Y.; Zhao, Z.-L.; Guan, R.; Zhao, Y.-D. Discovery of Early Paleozoic granites in the eastern Xiao Hinggan Mountains, northeastern China and their tectonic significance. *Geol. Bull. China* **2008**, *27*, 534–544.
85. Ludwig, K.R. *Isoplot version 4.15: A geochronological toolkit for microsoft Excel*; Berkeley Geochronology Center, Special Publication: Berkeley, CA, USA, 2008; Volume 4, pp. 247–270.
86. Hou, K.-J. Laser ablation-MC-ICP-MS technique for Hf isotope microanalysis of zircon and its geological applications. *Acta Petrol. Sin.* **2007**, *23*, 2595–2604.
87. Söderlund, U.; Patchett, P.J.; Vervoort, J.D. The 176Lu decay constant determined by Lu–Hf and U–Pb isotope systematics of Precambrian mafic intrusions. *Earth Planet. Sci. Lett.* **2004**, *219*, 311–324. [CrossRef]
88. Bouvier, A.; Vervoort, J.D.; Patchett, P.J. The Lu–Hf and Sm–Nd isotopic composition of CHUR: Constraints from unequilibrated chondrites and implications for the bulk composition of terrestrial planets. *Earth Planet. Sci. Lett.* **2008**, *273*, 48–57. [CrossRef]
89. Griffin, W.L.; Wang, X.; Jackson, S.E. Zircon chemistry and magma mixing, SE China: In-situ analysis of Hf isotopes, Tonglu and Pingtan igneous complexes. *Lithos* **2002**, *61*, 237–269. [CrossRef]
90. Liang, X.-R. Precise measurement of 143Nd/144Nd and Sm/Na ratios using multiple collectors-inductively coupled plasma mass spectrometer (MC-ICPMS). *Geochimica* **2003**, *32*, 91–96.
91. Rollinson, H.R. *Using Geochemical Data: Evaluation, Presentation, Interpretation*; Routledge: London, UK, 2014.
92. Taylor, S.R.; McLennan, S.M. The Continental Crust: Its Composition and Evolution. 1985. Available online: https://www.osti.gov/biblio/6582885 (accessed on 3 January 2025).
93. Sun, S.S.; McDonough, W.F. Chemical and isotopic systematics of oceanic basalt: Implication for mantle composition and processes. *Geol. Soc. Lond. Spec. Publ.* **1989**, *42*, 313–345. [CrossRef]

94. Rudnick, R.L.; Fountain, D.M. Nature and composition of the continental crust: A lower crustal perspective. *Rev. Geophys.* **1995**, *33*, 267–309. [CrossRef]
95. Kelsey, C.H. Calculation of the CIPW norm. *Mineral. Mag.* **1965**, *34*, 276–282.
96. Chappell, B.W.; White, A.J.R. I-and S-type granites in the Lachlan Fold Belt. *Earth Environ. Sci. Trans. R. Soc. Edinb.* **1992**, *83*, 1–26.
97. Chappell, B.W.; White, A.J. Two contrasting granite types: 25 years later. *Aust. J. Earth Sci.* **2001**, *48*, 489–499. [CrossRef]
98. Hou, Q.-L. *Advanced Structural Geology*; Science Press: Beijing, China, 2018; Volume 1.
99. Irber, W. The lanthanide tetrad effect and its correlation with K/Rb, Eu/Eu∗, Sr/Eu, Y/Ho, and Zr/Hf of evolving peraluminous granite suites. *Geochim. Cosmochim. Acta* **1999**, *63*, 489–508. [CrossRef]
100. Middlemost, E.A.K. Naming materials in the magma/igneous rock system. *Earth Sci. Rev.* **1994**, *37*, 215–224. [CrossRef]
101. Rickwood, P.C. Boundary lines within petrologic diagrams which use oxides of major and minor elements. *Lithos* **1989**, *22*, 247–263. [CrossRef]
102. Maniar, P.D.; Piccoli, P.M. Tectonic discrimination of granitoids. *Geol. Soc. Am. Bull.* **1989**, *101*, 635–643. [CrossRef]
103. Streckeisen, A.L. Classification of the common igneous rocks by means of their chemical composition: A provisional attempt. *N. Jb. Miner. Mon.* **1976**, *1*, 1–15.
104. Hoskin, P.W.; Schaltegger, U. The composition of zircon and igneous and metamorphic petrogenesis. *Rev. Mineral. Geochem.* **2003**, *53*, 27–62. [CrossRef]
105. Vavra, G.; Schmid, R.; Gebauer, D. Internal morphology, habit and U-Th-Pb microanalysis of amphibolite-to-granulite facies zircons: Geochronology of the Ivrea Zone (Southern Alps). *Contrib. Mineral. Petrol.* **1999**, *134*, 380–404. [CrossRef]
106. Zhu, D.-C.; Zhao, Z.-D.; Niu, Y.; Mo, X.-X.; Chung, S.-L.; Hou, Z.-Q.; Wang, L.-Q.; Wu, F.-Y. The Lhasa Terrane: Record of a microcontinent and its histories of drift and growth. *Earth Planet. Sci. Lett.* **2011**, *301*, 241–255. [CrossRef]
107. Hu, P.-Y.; Zhai, Q.-G.; Wang, J.; Tang, Y.; Wang, H.-T.; Zhu, Z.-C.; Wu, H. Middle Neoproterozoic (ca. 760 Ma) arc and back-arc system in the North Lhasa terrane, Tibet, inferred from coeval N-MORB-and arc-type gabbros. *Precambrian Res.* **2018**, *316*, 275–290. [CrossRef]
108. Liu, Q.-S.; Jiang, W.; Jian, P.; Ye, P.-S.; Wu, Z.-H.; Hu, D.-G. Zircon SHRIMP U-Pb age and petrochemical and geochemicalfeatures of Mesozoic muscovite monzonitic granite at Ningzhong, Tibet. *Acta Petrol. Sin.* **2006**, *22*, 643–652.
109. Whalen, J.B.; Currie, K.L.; Chappell, B.W. A-type granites: Geochemical characteristics, discrimination and petrogenesis. *Contrib. Mineral. Petrol.* **1987**, *95*, 407–419. [CrossRef]
110. Inger, S.; Harris, N. Geochemical constraints on leucogranite magmatism in the Langtang Valley, Nepal Himalaya. *J. Petrol.* **1993**, *34*, 345–368. [CrossRef]
111. Zhang, Z.-M.; Kang, D.-Y.; Ding, H.-X.; Tian, Z.-L.; Dong, X.; Qin, S.-K.; Mu, H.-C.; Li, M.-M. Partial meltig of Himalayan Orogen and formation mechanism of leucogranites. *Earth Sci.* **2018**, *43*, 82–98.
112. Watson, E.B.; Harrison, T.M. Zircon saturation revisited: Temperature and composition effects in a variety of crustal magma types. *Earth Planet. Sci. Lett.* **1983**, *64*, 295–304. [CrossRef]
113. Clemens, J.D. S-type granitic magmas–petrogenetic issues, models and evidence. *Earth Sci. Rev.* **2003**, *61*, 1–18. [CrossRef]
114. Sylvester, P.J. Post-collisional strongly peraluminous granites. *Lithos* **1998**, *45*, 29–44. [CrossRef]
115. Zartman, R.E.; Doe, B.R. Plumbotectonics—The model. *Tectonophysics* **1981**, *75*, 135–162. [CrossRef]
116. Zindler, A.; Hart, S.R. Chemical geodynamics. *Annu. Rev. Earth Planet. Sci.* **1986**, *14*, 493–571. [CrossRef]
117. Patiño Douce, A.E.; Johnston, A.D. Phase equilibria and melt productivity in the pelitic system: Implications for the origin of peraluminous granitoids and aluminous granulites. *Contrib. Mineral. Petrol.* **1991**, *107*, 202–218. [CrossRef]
118. Altherr, R.; Holl, A.; Hegner, E.; Langer, C.; Kreuzer, H. High-potassium, calc-alkaline I-type plutonism in the European Variscides: Northern Vosges (France) and northern Schwarzwald (Germany). *Lithos* **2000**, *50*, 51–73. [CrossRef]
119. Collins, W.J.; Richards, S.W. Geodynamic significance of S-type granites in circum-Pacific orogens. *Geology* **2008**, *36*, 559–562. [CrossRef]
120. Barbarin, B. A review of the relationships between granitoid types, their origins and their geodynamic environments. *Lithos* **1999**, *46*, 605–626. [CrossRef]
121. Whalen, J.B.; Hildebrand, R.S. Trace element discrimination of arc, slab failure, and A-type granitic rocks. *Lithos* **2019**, *348*, 105–179.
122. Pearce, J.A.; Harris, N.B.W.; Tindle, A.G. Trace element discrimination diagrams for the tectonic interp retation of granitic rocks. *J. Petrol.* **1984**, *25*, 956–983. [CrossRef]
123. Pearce, J. Sources and settings of granitic rocks. *Episodes* **1996**, *19*, 120–125. [CrossRef]
124. Harris, N.B.W.; Pearce, J.A.; Tindle, A.G. Geochemical characteristics of collision-zone magmatism. *Geol. Soc. Lond. Spec. Publ.* **1986**, *19*, 67–81. [CrossRef]
125. Batchelor, R.A.; Bowden, P. Petrogenetic interpretation of granitoid rock series using multicationic parameters. *Chem. Geol.* **1985**, *48*, 43–55. [CrossRef]

126. Collins, A.S.; Pisarevsky, S.A. Amalgamating eastern Gondwana: The evolution of the Circum-Indian Orogen. *Earth Sci. Rev.* **2005**, *71*, 229–270. [CrossRef]
127. Zhang, Z.-M.; Wang, J.-L.; Zhao, G.-C.; Shi, C. Geochronology and Precambrian tectonic evolution of the Namche Barwacomplex from the eastern Himalayan syntaxis, Tibet. *Acta Petrolog. Sin.* **2008**, *24*, 1477–1487.
128. Rojo-Pérez, E.; Fuenlabrada, J.M.; Díez Fernández, R.; Arenas, R. Origin and evolution of Cadomian magmatism in SW Iberia: From subduction onset and arc building to a tectonic switching. *Int. Geol. Rev.* **2024**, *66*, 1885–1909. [CrossRef]
129. Moreno-Martín, D.; Fernández, R.D.; Albert, R.; Martínez, S.S.; Rojo-Pérez, E.; Gerdes, A.; Arenas, R. Cuartel Ophiolite: Structure, timing and exhumation mechanisms for a Cadomian suture zone in the peri-Gondwanan Realm (SW Iberia). *Gondwana Res.* **2025**, *137*, 255–273. [CrossRef]
130. Fuenlabrada, J.M.; Arenas, R.; Martínez, S.S.; Fernández, R.D.; Pieren, A.P.; Pereira, M.F.; Chichorro, M.; Silva, J.B. Geochemical and isotopic (Sm-Nd) provenance of Ediacaran-Cambrian metasedimentary series from the Iberian Massif. Paleoreconstruction of the North Gondwana margin. *Earth-Sci. Rev.* **2020**, *201*, 103079. [CrossRef]
131. Cai, Z.-H.; Xu, Z.-Q.; Duan, X.-D.; Li, H.-Q.; Cao, H.; Huang, X.-M. Early stage of Farly Paleozoie orogenic event in western Yunnan Province, south eastern margin of Tibet Plateau. *Acta Petrol. Sin.* **2013**, *29*, 2123–2140.

minerals

MDPI

Article

Mafic Intrusions in the Cuona Area, Eastern Tethyan Himalaya: Early Kerguelen Mantle Plume Activity and East Gondwana Rifting

Chunxi Shan [1], Zhiqiang Kang [1,2,*], Feng Yang [1,2], Chengyou Ma [1], Zedong Qiao [1], Zonghao Liu [1], Jizhong Mu [1], Lingling Wu [1] and Lu Zhou [1]

[1] School of Earth Sciences, Guilin University of Technology, Guilin 541004, China; shanchunxi1997@163.com (C.S.)
[2] Collaborative Innovation Center for Exploration of Concealed Deposits and Development of Materials in Nonferrous Metals, Guilin University of Technology, Guilin 541004, China
* Correspondence: zk99201@163.com

Abstract: The widespread occurrence of Mesozoic ocean island basalt (OIB)-like igneous rocks in the Southern Tibetan Himalayan Belt provides important constraints on the rifting of East Gondwana. This study undertook a petrological, geochronological, and geochemical investigation of mafic intrusive rocks in the Cuona area of the eastern Tethyan Himalayan Belt. The mafic intrusions have OIB-type geochemical signatures, including diabase porphyrite, gabbro, and diabase. Zircon U–Pb dating indicates that the diabase porphyrite formed at 135.0 $\pm$ 1.6 Ma. The diabase porphyrite and gabbro are enriched in high-field-strength elements (Nb and Ti) and large-ion lithophile elements (Sr and Pb) and experienced negligible lithospheric mantle or crustal contamination. The diabase is enriched in large-ion lithophile elements (LILEs, e.g., La and Ce) and depleted in high-field-strength elements (HFSEs, e.g., Ru, Zr and Ti). In general, the mafic intrusions exhibit significant light REE enrichment and heavy REE depletion and have no Eu anomalies. Whole-rock neodymium ($\varepsilon_{Nd}(t)$ = 1.55) and zircon Hf ($\varepsilon_{Hf}(t)$ = 0.60–3.73) isotopic compositions indicate derivation of the magma from enriched type I mantle. We propose that the diabase porphyrite and diabase formed in a continental margin rift setting, influenced by the Kerguelen mantle plume, and represent magmatism related to the breakup of East Gondwana. However, the gabbro formed in a relatively stable continental intraplate environment, likely derived from deep magmatic processes associated with the Kerguelen mantle plume. Our results provide new constraints on the early activity of the Kerguelen mantle plume and offer insights into the breakup and tectonic evolution of East Gondwana.

Keywords: East Gondwana; Tethyan Himalaya; Cuona area; Kerguelen mantle plume; mafic intrusive rock

1. Introduction

The Tethyan Himalayan Belt is a key geological unit of the Tibetan Plateau and records continental rifting, lithospheric thinning, and mantle plume activity [1–11]. Mantle plume activity links large igneous provinces (LIPs) and the rifting of (super)continents [12]. The Kerguelen mantle plume was one of the largest mantle plumes in Earth history and had a significant role in volcanic activity in the central Indian Ocean and rifting of East Gondwana [13–18] (Figure 1a). As such, the Kerguelen mantle plume has been a focus of research. Some studies have suggested that during the Early Cretaceous, the Kerguelen mantle plume may have accelerated the rifting of East Gondwana, leading to the subsequent separation of the Indian, Australian, and Antarctic plates [13,16].

The Tethyan Himalayan Belt contains extensive Late Jurassic to Early Cretaceous mafic igneous rocks, which can be broadly classified into three types: ocean island basalts (OIBs), normal-type mid-ocean ridge basalts (N-MORBs), and transitional rocks. These rocks generally formed in a tectonic setting characterized by lithospheric extension and thinning along the northern margin of Gondwana [19–24]. OIB-type mafic rocks are considered to be the products of rifting of the East Gondwana continent and provide key insights into the regional magmatism and tectonic evolution [25–29]. Previous studies have suggested that an extensional tectonic setting existed during the Early Cretaceous in both the eastern and western segments of the Tethyan Himalayan tectonic belt [30]. However, the petrogenesis and tectonic setting of igneous rocks in this region are debated. Some studies have proposed that Early Cretaceous mafic magmatism in the eastern Tethyan Himalayan Belt was closely associated with the Kerguelen mantle plume, suggesting the plume was the primary driver of rifting in East Gondwana [15,25,27,30–33]. In contrast, other studies have proposed that the Early Cretaceous magmatism resulted from partial melting of enriched asthenospheric mantle, with no direct connection to the Kerguelen mantle plume. In this case, rifting of East Gondwana was driven primarily by passive rifting between the Indian and Australian–Antarctic plates [7,26–34].

To better understand the formation of mafic intrusions in this region and the relationship between the Kerguelen mantle plume and rifting of East Gondwana, we undertook a petrological, geochronological, and geochemical study of mafic intrusions in the Cuona area of the eastern Tethyan Himalayan Belt. By determining the ages, sources, and tectonic settings of the intrusions, we provide new constraints on the early activity of the Kerguelen mantle plume and the rifting and tectonic evolution of East Gondwana.

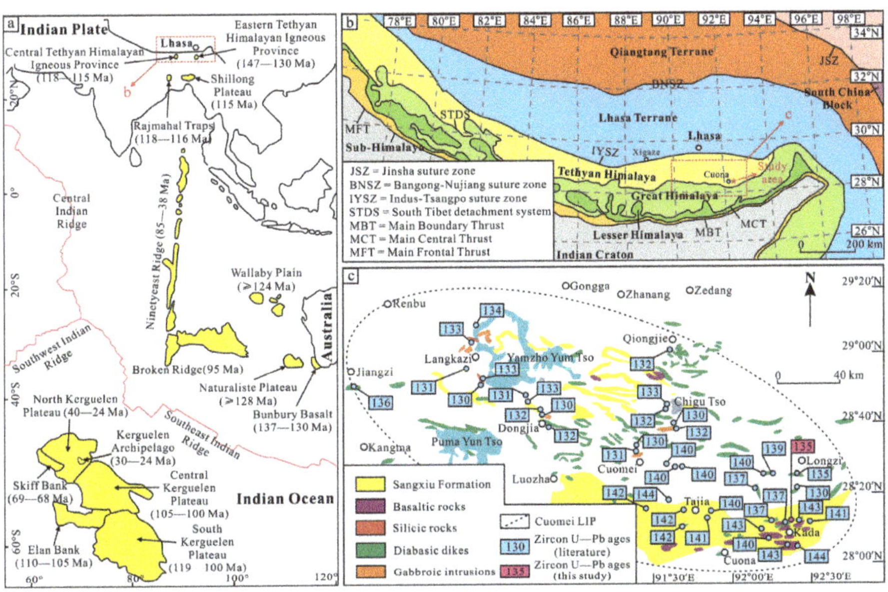

Figure 1. (**a**) Distribution of igneous rocks related to the Kerguelen mantle plume in the Indian Ocean and adjacent continental margins [2,4,31]. (**b**) Geological map of the Tethyan Himalaya and adjacent areas [35,36]. (**c**) Geological map of the Cuomei large igneous province showing the distribution of igneous rocks with ages of >130 Ma [10,16].

2. Geological Setting and Sample Descriptions

The Himalayan Orogenic Belt was part of the Rodinia and Gondwana supercontinents and has experienced both convergent and passive continental margin tectonism [37,38]. The Himalayan Orogenic Belt is divided into the Tethyan Himalaya, Great Himalaya, and Lesser Himalaya [35]. The Tethyan Himalayan Belt is the northernmost tectonic unit of the Himalayan Orogenic Belt and is located south of the Yarlung Zangbo Suture Zone and north of the High Himalayan Belt. The Tethyan Himalayan Belt is E–W-trending (Figure 1b). The Tethyan Himalaya consists primarily of Paleozoic–Mesozoic marine sedimentary rocks, within which mafic volcanic rocks are widely developed. The magmatism in this region occurred mainly from the Permian to Cretaceous [39,40]. The Tethyan Himalayan Belt can be divided into southern and northern zones. The southern zone consists of weakly metamorphosed sedimentary rocks from the Indian continental margin, while the northern zone is characterized by deep-water sedimentary rocks, including flysch and ophiolitic fragments.

Igneous rocks are widely distributed in the Tethyan Himalayan Belt and mainly intrude Triassic, Jurassic, and Lower Cretaceous sedimentary rocks. The igneous rocks consist mainly of basaltic volcanic rocks, diabase dikes/sills, and gabbroic intrusions, which are typically exposed in an E–W-trending belt. The study area is located in the Cuona region of the eastern Tethyan Himalaya (Figure 1c). Jurassic and Cretaceous sedimentary rocks interlayered with volcanic rocks are widely distributed in this region, occurring in the Middle to Upper Jurassic Zhela, Lure, and Sangxiu formations. The Zhela Formation consists mainly of siltstone, black shale, and graywacke interlayered with chert, limestone, and volcanic rocks. The Lure Formation is dominated by dark gray muddy (micritic) limestone, which is locally interlayered with basalt, as well as dark gray calcareous silty slate interbedded with dark gray, muddy, fine-grained limestone.

Seven samples were collected from near Juela Township in Cuona County, within the eastern Tethyan Himalayan Belt; the main rock types are diabase porphyrite (21CN-01, 21CN-02, 21CN-04, 21CN-08), gabbro (21CN-06), and diabase (21CN-07-1, 21CN-07-2) (Figure 2). The samples were collected from mafic dikes in the Lure and Zhela formations. Field investigations revealed that these intrusions have clear intrusive contacts and are extensively exposed at the surface (Figure 3a–c).

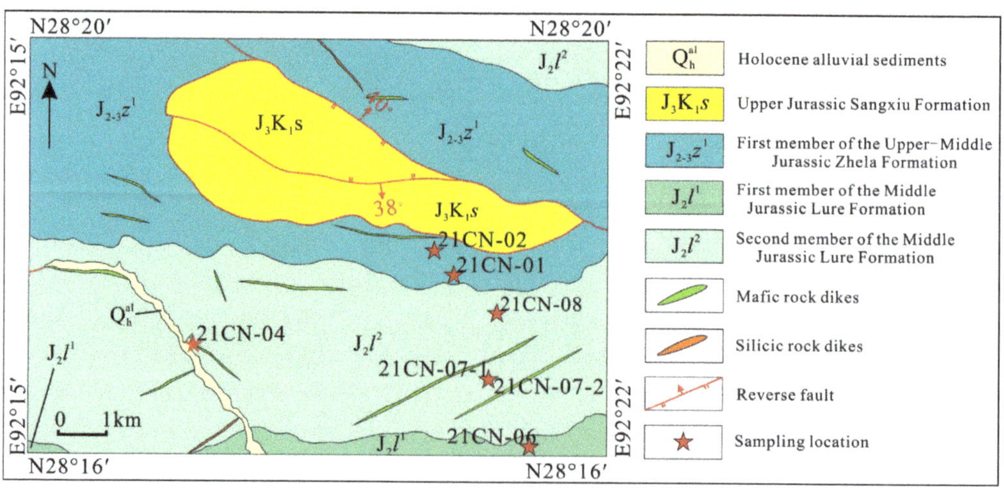

Figure 2. Simplified geological map of the Cuona area in the eastern segment of the Tethyan Himalayan Belt (modified from the 1:50,000 Juela Geological Map).

Figure 3. (**a–c**) Field photographs, (**d–f**) hand specimens, and (**g–i**) microscopic characteristics of mafic intrusions in the Cuona area. Pl = plagioclase; Cpx = clinopyroxene; Mt = magnetite; Bt = biotite; Kfs = K-feldspar; Cal = calcite; Chl = chlorite.

The diabase porphyrite has gray–green fresh surfaces and weathers to a light brown color (Figure 3d). It is characterized by a diabasic texture and massive structure. The primary minerals are plagioclase (~50 vol.%) and clinopyroxene (~35 vol.%). Plagioclase occurs mainly as subhedral–euhedral lath-like and tabular crystals that exhibit significant alteration, with some grains being opaque due to sericitization and chloritization. These crystals display oscillatory zoning visible under polarized light, suggesting rapid crystallization from a magmatic melt. Some plagioclase grains show fine-grained sericite filling the interstitial spaces between crystals, indicating low-temperature alteration. Clinopyroxene forms anhedral granular to short prismatic crystals that exhibit chloritization. The groundmass consists mainly of microcrysts of plagioclase, pyroxene, and biotite. Minor minerals include quartz and amphibole, with quartz appearing as small, subhedral grains often found in the interstitial spaces. Accessory minerals include magnetite, which forms opaque grains, and apatite, which appears as fine, needle-like crystals (Figure 3g).

The gabbro has dark gray fresh surfaces and weathers to a yellow–brown color (Figure 3e). It is characterized by a gabbroic texture and massive structure. The primary minerals are plagioclase (~45 vol.%) and clinopyroxene (~40 vol.%). Plagioclase is commonly subhedral, altered to clay minerals, and has a patchy texture indicative of replacement by K-feldspar. Clinopyroxene typically occurs as subhedral–anhedral, short-prismatic crystals that are interstitial to plagioclase and exhibit partial chloritization. The accessory minerals are magnetite, titanite, and apatite, which occur as irregular or acicular crystals that are randomly distributed. Micro-cracks are well developed and infilled with crystalline quartz and carbonate veinlets (Figure 3h). Additionally, some clinopyroxene crystals display distinct edge alteration features, which may be related to compositional exchange during the crystal growth process. The plagioclase shows some degree of fractures and altered textures, which may suggest that the rock has undergone an alteration process, such as weathering or hydrothermal alteration. Some titanite crystals exhibit a subtle metallic luster, which may reflect the high-temperature environment in which they formed.

The diabase has grayish blue fresh surfaces and weathers to a yellow–brown color (Figure 3f). It has a diabasic texture and massive structure. The primary mineral is plagioclase (~55 vol.%), which is subhedral–euhedral and exhibits significant alteration, including chloritization, carbonatization, and silicification. Some plagioclase microcrysts are irregularly arranged, indicative of variable degrees of alteration. These crystals display oscillatory zoning, visible under polarized light, suggesting rapid crystallization from a magmatic melt. Secondary minerals include chlorite and calcite, and accessory minerals are magnetite and apatite (Figure 3i). Magnetite occurs as opaque, subhedral to anhedral grains, while apatite appears as fine, needle-like crystals interspersed in the groundmass. The rock also contains minor quartz, which appears as small, subhedral grains, commonly located in the interstitial spaces between plagioclase crystals.

3. Methods

3.1. Zircon U–Pb Dating

Two samples (21CN-02) was selected for zircon U–Pb dating by laser ablation–inductively coupled plasma–mass spectrometry (LA–ICP–MS). Zircon separation was conducted by the Shangyi Rock and Mineral Testing Technology Service Company Limited, Langfang, Hebei Province, China. Zircon mounting and cathodoluminescence (CL) imaging were undertaken by the Yujin Technology Company Limited in Chongqing, China. The analyses were carried out in the Guangxi Key Laboratory of Hidden Metallic Ore Exploration, Guilin University of Technology, Guilin, Guangxi, China. The LA–ICP–MS instrument comprised a New Wave 193 nm LA system coupled to an Agilent 7500 ICP–MS instrument. The laser was operated at a frequency of 5 Hz, and high-purity He was used as the carrier gas. A laser spot size of 32 μm was used for the analyses. Zircon ages were calculated by use of the standard zircon Plesovice (PL) as an external standard, and elemental concentrations were calibrated by use of the synthetic silicate glass standard NIST610 from the United States National Institute of Standards and Technology (NIST). Data processing was performed by use of ICP–MS DataCal 7.2 software [41], with common Pb corrected following [42]. Zircon U–Pb ages were calculated by use of concordia diagrams and weighted-mean age calculations, both using Isoplot 3.2 software [43]. The detailed instrumental operating conditions and data-processing methods were described by Wang et al., (2006) [44] and Yuan et al., (2004) [45].

3.2. Major and Trace Elements

The contents of major and trace elements in seven samples of diabase porphyry (21CN-01, 21CN-02, 21CN-04, 21CN-08), gabbro (21CN-06), and diabase (21CN-07-1, 21CN-07-2) were determined in the Key Laboratory of Guangxi Concealed Metal Ore Exploration of Guilin University of Technology. The rock samples were washed and dried, and weathered surfaces were removed. The samples were then crushed to <5 mm and ground into powder finer than 200 mesh. Major elements were analyzed by use of an alkali fusion glass bead method, with a ZSX Primus II X-ray fluorescence spectrometer. The precision of the major element analyses was better than ±10%. We determined trace elements with an Agilent 7500cx ICP–MS instrument after acid digestion, achieving a precision better than ±5%. Our analytical procedures followed Li et al., (2016) [46].

3.3. Sr–Nd Isotopes

The Sr–Nd isotopic compositions of sample (21CN-02) was determined at the Guangxi Key Laboratory of Hidden Metallic Ore Exploration, Guilin University of Technology. The samples were fully dissolved by acid digestion. Rubidium, Sr, and the rare earth elements (REEs) were separated by use of cation exchange resin, and Sm and Nd were further separated by use of HDEHP resin. The Sr and Nd fractions were analyzed with a Neptune Plus multi-collector–ICP–MS instrument. The $^{87}Sr/^{86}Sr$ and $^{143}Nd/^{144}Nd$ ratios were corrected for instrumental mass bias by normalization to $^{86}Sr/^{88}Sr$ = 0.1194 and $^{146}Nd/^{144}Nd$ = 0.7219, respectively. The details of the Sr–Nd isotopic analyses were described by Liang et al., (2003) [47].

3.4. Zircon Hf Isotopes

The Zircon Hf isotopic analysis was conducted on one sample (21CN-02) at the Guangxi Key Laboratory of Hidden Metallic Ore Exploration, Guilin University of Technology. The Hf isotopic analyses were undertaken near the zircon U–Pb dating sites. The Hf isotopic analyses were undertaken with a Neptune Plus multi-collector–ICP–MS instrument coupled to a GeoLas HD excimer ArF LA system. The laser spot diameter was set to 32 μm, and the laser frequency was 8 Hz. Crustal-model Hf isotopic ages were calculated by assuming the parental magma was derived from average continental crust. The $^{176}Lu/^{177}Hf$ ratio of 0.015 represents that of a depleted mantle source [48].

4. Results

4.1. Zircon U–Pb Ages

The zircon U–Pb dating results are presented in Table 1. The zircon grains from both samples are primarily euhedral–subhedral, with lengths of 50–100 μm, aspect ratios of 1:1 to 2:1, and prismatic shapes (Figure 4a). The U contents in the zircons from the diabase porphyrite sample (21CN-02) are 494–1050 ppm, and the Th contents are 462–1432 ppm. The Th/U ratios are 0.94–1.59, with an average of 1.17. These values are consistent with those of typical magmatic zircon.

Table 1. Zircon U-Th-Pb isotopic dating results of mafic intrusive rocks in the Cuona region.

Sample Point	Content/10^{-6}			Th/U	Isotopic Ratio				Isotopic Age/Ma				
	Pb	Th	U		$^{207}Pb/^{235}U$	1σ	$^{206}Pb/^{238}U$	1σ	$^{207}Pb/^{235}U$	1σ	$^{206}Pb/^{238}U$	1σ	Con*
21CN-02-06	21.77792	755.0443	739.7702	1.02	0.14585	0.00607	0.02159	0.00023	138.2	5.4	137.7	1.4	99%
21CN-02-11	32.31679	1302.509	1049.637	1.24	0.14207	0.00551	0.02125	0.00025	134.9	4.9	135.5	1.6	99%
21CN-02-14	14.03134	462.2057	493.6375	0.94	0.14318	0.00859	0.02134	0.00030	135.9	7.6	136.1	1.9	99%
21CN-02-16	18.90271	701.7517	629.2832	1.12	0.14162	0.00695	0.02179	0.00029	134.5	6.2	138.9	1.8	96%
21CN-02-17	16.85502	588.0218	581.0979	1.01	0.13800	0.00694	0.02094	0.00030	131.3	6.2	133.6	1.9	98%
21CN-02-30	15.79062	511.8948	516.887	0.99	0.15149	0.01039	0.02152	0.00037	143.2	9.2	137.2	2.3	95%
21CN-02-31	30.23075	1272.806	971.9639	1.31	0.14006	0.00910	0.02047	0.00025	133.1	8.1	130.6	1.6	98%
21CN-02-33	30.33958	1412.224	916.9356	1.54	0.13682	0.00604	0.02101	0.00024	130.2	5.4	134.1	1.5	97%
21CN-02-34	24.13773	930.2534	796.9868	1.17	0.13195	0.00561	0.02113	0.00025	125.9	5.0	134.8	1.6	93%
21CN-02-37	23.66793	838.1252	767.2604	1.09	0.13738	0.00566	0.02177	0.00025	130.7	5.1	138.8	1.6	93%
21CN-02-38	29.24066	1148.345	976.7321	1.18	0.13456	0.00550	0.02063	0.00026	128.2	4.9	131.6	1.6	97%
21CN-02-39	29.71245	1432.31	901.637	1.59	0.13925	0.00604	0.02096	0.00024	132.4	5.4	133.7	1.5	98%
21CN-02-40	16.04238	557.4163	553.4698	1.01	0.13057	0.00650	0.02103	0.00029	124.6	5.8	134.1	1.8	92%

Con*: Concordance.

After excluding discordant ages and those with large errors, the analyses were used to construct concordia diagrams. The weighted-mean $^{206}Pb/^{238}U$ age for sample 21CN-02 is 135.0 ± 1.6 Ma (MSWD = 0.68; n = 13) (Figure 4b).

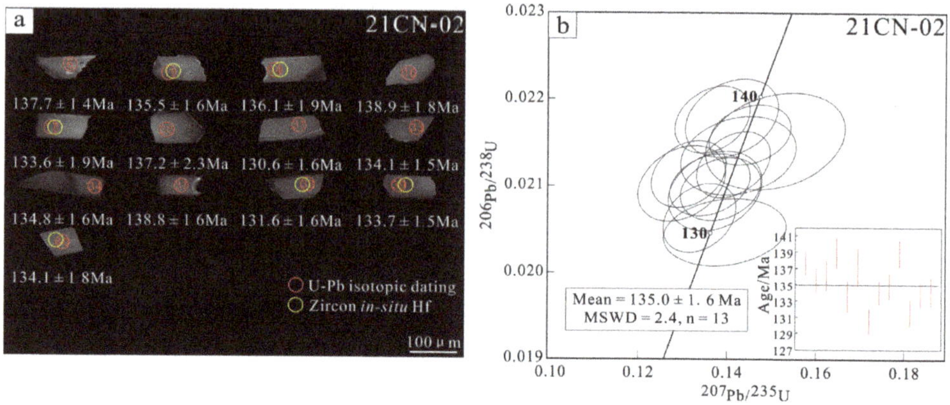

Figure 4. (**a**) Cathodoluminescence (CL) images from mafic intrusions in the Cuona area, (**b**) U–Pb concordia diagrams of zircon grains from mafic intrusions in the Cuona area.

4.2. Major and Trace Elements

The major element data are presented in Table 2. The SiO_2 contents of the mafic intrusive rock samples are 49.24–51.95 wt.%. The Al_2O_3 contents are 10.51–15.58 wt.%, and the Na_2O contents are 2.18–5.38 wt.%. The average K_2O content is 0.61 wt.%, and the total alkali (K_2O + Na_2O) contents are 2.20–5.40 wt.%. $Mg^{\#}$ values vary from 44.2 to 66.2. The loss-on-ignition (LOI) values are 1.26–5.42 wt.%, indicating these mafic intrusive rocks have undergone variable degrees of alteration. In a Nb/Y–Zr/TiO_2 diagram, the samples plot in the subalkaline and alkaline basalt fields (Figure 5a), indicating transitional geochemical signatures between subalkaline and alkaline affinities. In an A/CNK–A/NK diagram, the samples plot in the metaluminous field (Figure 5b). Due to their relatively high alumina saturation index (A/CNK; 0.52–0.92) and Laitman index (σ; 0.76–3.77) values, the samples plot in the calc-alkaline to metaluminous fields on an AR–SiO_2 classification diagram (Figure 5c), showing a trend in transition from sub-alkaline to alkaline components. In a SiO_2–K_2O diagram, the samples plot in the calc-alkaline to low-K (tholeiitic) fields (Figure 5d).

Table 2. Major (wt.%), trace (ppm), and rare earth element (ppm) analysis results of mafic intrusive rocks from the Cuona region.

Sample	21CN-01	21CN-02	21CN-04	21CN-06	21CN-07-1	21CN-07-2	21CN-08
Rock type	Diabase porphyrite	Diabase porphyrite	Diabase porphyrite	Gabbro	Diabase	Diabase	Diabase porphyrite
Al_2O_3	13.06	13.31	13.70	12.82	13.76	14.68	9.99
CaO	7.28	9.67	10.77	6.67	5.79	4.20	8.68
$Fe_2O_3^T$	11.52	10.04	11.00	12.80	13.10	12.05	13.17
K_2O	0.80	0.98	0.87	1.40	0.03	0.02	0.02
MgO	6.12	6.62	6.49	4.28	4.60	4.09	11.06
MnO	0.16	0.15	0.16	0.20	0.19	0.18	0.17
Na_2O	3.98	3.20	2.23	3.15	4.54	5.07	2.07
P_2O_5	0.40	0.28	0.30	0.45	1.56	1.43	0.29
SiO_2	49.05	48.95	49.55	50.89	46.38	48.94	46.94
TiO_2	3.99	2.74	2.85	4.58	4.25	3.55	2.68
LOI	2.74	3.17	1.26	2.62	5.42	4.89	4.90
Total	99.11	99.12	99.19	99.87	99.62	99.10	99.93

Table 2. *Cont.*

Sample	21CN-01	21CN-02	21CN-04	21CN-06	21CN-07-1	21CN-07-2	21CN-08
Li	18.69	9.94	14.80	13.74	25.97	21.01	31.60
Sc	35.47	33.39	38.37	14.39	18.57	13.82	19.99
Ti	24,816.87	17,116.76	17,443.96	28,228.77	27,042.95	22,586.42	16,904.02
V	272.22	231.41	247.41	251.02	243.71	171.81	242.08
Cr	298.87	440.10	179.86	14.81	98.47	14.96	451.08
Mn	1104.18	1066.03	1110.47	1429.79	1355.65	1174.42	1171.26
Co	40.97	39.17	42.95	34.39	29.58	23.20	63.07
Ni	124.74	154.79	67.24	17.70	42.62	6.51	259.45
Cu	36.88	38.13	26.41	13.99	19.85	10.83	32.69
Zn	128.99	109.63	113.31	127.48	135.54	150.72	124.56
Ga	25.89	14.11	14.08	16.11	16.94	14.28	9.41
Rb	17.93	29.81	29.64	18.02	1.67	1.21	0.38
Sr	548.45	437.22	578.08	672.44	1046.50	587.62	79.02
Y	30.22	23.25	25.58	31.12	38.76	40.01	25.63
Zr	240.84	165.96	177.32	275.20	215.56	239.69	162.32
Nb	20.79	15.30	15.40	23.31	29.46	33.46	12.94
Mo	1.78	2.24	0.65	1.53	1.75	1.58	0.72
Cs	0.18	0.32	0.60	0.31	1.12	0.48	0.72
Ba	538.48	195.03	166.93	236.48	214.15	100.33	22.52
La	24.41	16.89	18.18	21.48	44.31	48.41	13.21
Ce	52.31	36.36	38.76	52.73	109.73	116.73	32.96
Pr	6.96	4.87	5.20	6.55	13.67	14.22	4.58
Nd	31.51	22.21	23.54	33.38	62.53	72.87	21.13
Sm	7.84	5.57	6.09	8.59	15.84	15.83	5.67
Eu	2.51	1.87	1.99	2.57	5.35	5.15	1.93
Gd	7.62	5.59	6.08	8.33	14.31	14.20	5.74
Tb	1.19	0.89	0.95	1.33	1.94	1.94	0.94
Dy	6.64	5.00	5.46	7.45	9.71	9.76	5.47
Ho	1.25	0.96	1.02	1.40	1.66	1.68	1.03
Er	3.15	2.40	2.58	3.49	3.81	3.91	2.66
Tm	0.41	0.32	0.35	0.47	0.45	0.47	0.37
Yb	2.52	1.93	2.08	2.83	2.50	2.65	2.23
Lu	0.35	0.27	0.29	0.39	0.33	0.35	0.31
Hf	6.32	4.34	4.59	7.22	5.54	5.92	4.24
Ta	1.00	0.93	0.79	1.27	1.70	1.76	0.61
Tl	0.07	0.05	0.07	0.11	0.01	0.01	0.14
Pb	2.41	1.53	2.25	3.01	3.19	2.75	2.15
Th	1.88	1.40	1.89	1.83	3.29	3.72	1.34
U	0.48	0.33	0.42	0.60	0.74	0.84	0.33
A/CNK	0.63	0.56	0.57	0.68	0.76	0.92	0.52
A/NK	1.76	2.10	2.97	1.91	1.83	1.76	2.92
AR	1.61	1.44	1.29	1.25	1.61	1.61	1.74
$Mg^{\#}$	55.32	60.58	57.89	43.80	45.00	44.17	66.18
FeO^T	10.76	9.42	10.11	11.84	12.51	11.51	12.47
∑REE	148.67	105.14	112.56	150.98	286.14	308.16	98.23
LREE	133.16	93.37	99.84	133.62	265.73	287.40	85.22
HREE	15.52	11.77	12.73	17.36	20.40	20.76	13.01
LREE/HREE	8.58	7.93	7.85	7.70	13.02	13.84	6.55
Eu*	0.99	1.03	1.00	0.93	1.09	1.05	1.04
$(La/Yb)_N$	6.96	6.26	6.28	5.45	12.70	13.09	4.25
$(Sm/Yb)_N$	3.46	3.20	3.25	3.37	7.03	6.62	2.83
$(La/Sm)_N$	2.01	1.96	1.93	1.62	1.81	1.98	1.51
$(Tb/Yb)_N$	2.16	2.10	2.10	2.14	3.54	3.34	1.92

Notes: AR = $(Al_2O_3 + CaO + Na_2O + K_2O)/(Al_2O_3 + CaO\text{-}Na_2O\text{-}K_2O)$.

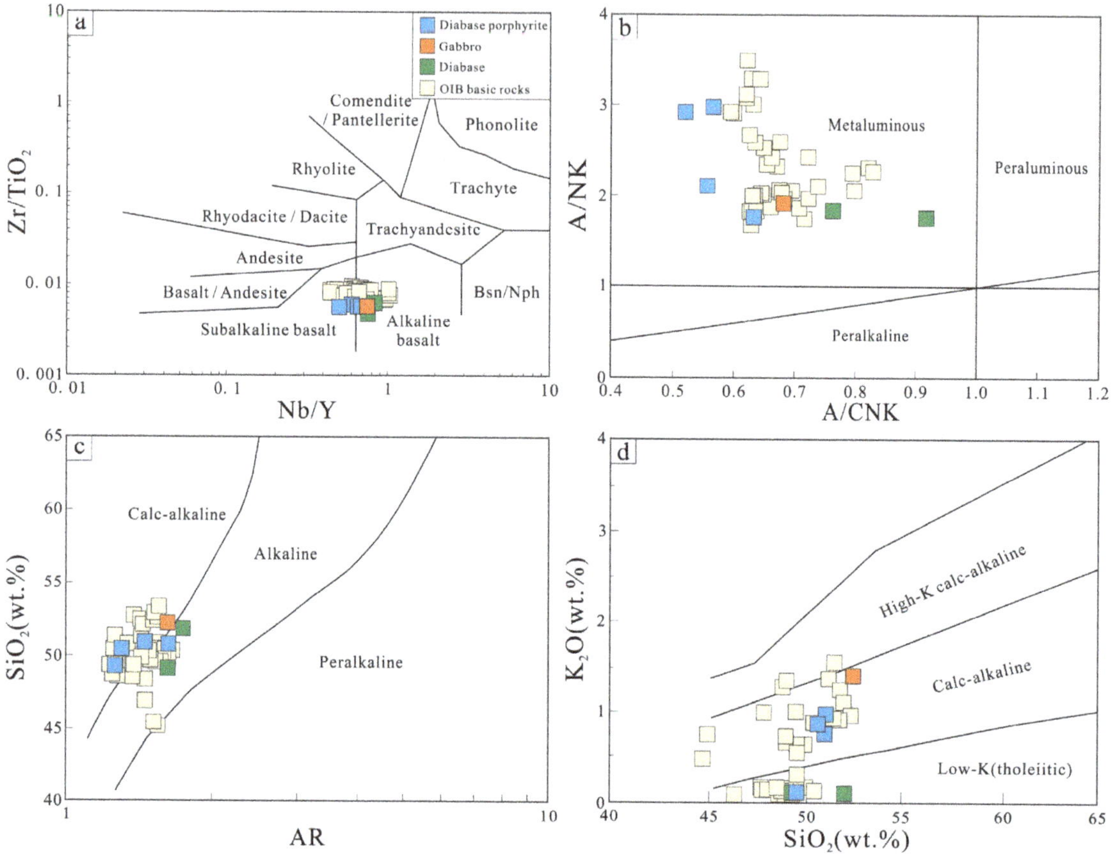

Figure 5. Geochemical classification diagrams for mafic intrusive rocks in the Cuona area and related rocks [49–52]. (**a**) Nb/Y–Zr/TiO_2 diagram; (**b**) A/CNK–A/NK diagram; (**c**) AR–SiO_2 classification diagram; (**d**) SiO_2–K_2O diagram. Data for regional OIB-type mafic rocks are from Zhu et al., (2008), Chen et al., (2021), and Yang et al., (2022) [2,4,29].

The trace element data are listed in Table 2. The total REE contents of the mafic intrusive rock samples range from 98.2 to 308.2 ppm. The REE patterns are light-REE-enriched (Figure 6a). The light/heavy REE ratios vary from 6.55 to 13.84, and $(La/Yb)_N$ = 4.25–13.09. δEu values range from 0.93 to 1.09, with an average of 1.02, indicating there are no significant Eu anomalies. In a primitive-mantle-normalized trace element diagram (Figure 6b), the diabase porphyrite and gabbro have similar trace element characteristics, with enrichments in high-field-strength elements (HFSEs, e.g., Nb and Ti) and large-ion lithophile elements (LILEs, e.g., Sr and Pb), similar to OIBs. The diabase is enriched in large-ion lithophile elements (LILEs, e.g., La and Ce) and depleted in high-field-strength elements (HFSEs, e.g., Ru, Zr, and Ti).

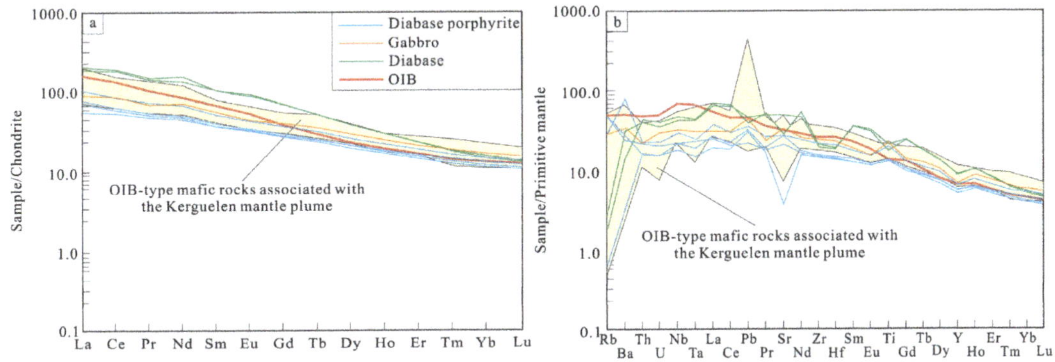

Figure 6. (**a**) Chondrite-normalized rare earth element (REE) patterns and (**b**) primitive-mantle-normalized trace element diagram for mafic intrusive rocks in the Cuona area and related rocks (chondrite, primitive mantle, OIB, N-MORB, and E-MORB data are from Sun and McDonough, 1989 [53]; OIB-type mafic rock data associated with the Kerguelen mantle plume are from Zhu et al., 2008; Chen et al., 2021; Yang et al., 2022) [2,4,29].

4.3. Sr–Nd Isotopes

The Sr–Nd isotopic data are listed in Table 3. The diabase has $(^{87}Sr/^{86}Sr)_i = 0.705583$, $(^{143}Nd/^{144}Nd)_i = 0.512541$, and $\varepsilon_{Nd}(t) = 1.51$. The diabase porphyrite has relatively high $(^{87}Sr/^{86}Sr)_i$ and low $\varepsilon_{Nd}(t)$ values, with T_{DM} ages of 1168 Ma (Figure 7).

Table 3. Sr-Nd isotope analysis results of mafic intrusive rocks in the Cuona region.

Sample	Sm (ppm)	Nd (ppm)	Rb (ppm)	Sr (ppm)	$^{147}Sm/^{144}Nd$	$^{143}Nd/^{144}Nd$	$(^{143}Nd/^{144}Nd)_i$	$\varepsilon_{Nd}(t)$	$^{87}Rb/^{86}Sr$	$^{87}Sr/^{86}Sr$	$(^{87}Sr/^{86}Sr)_i$	T_{DM}(Ga)
21CN-02	5.57	22.21	29.81	437.22	0.15171	0.51268	0.51254	1.55	0.19725	0.70597	0.70558	1.1668

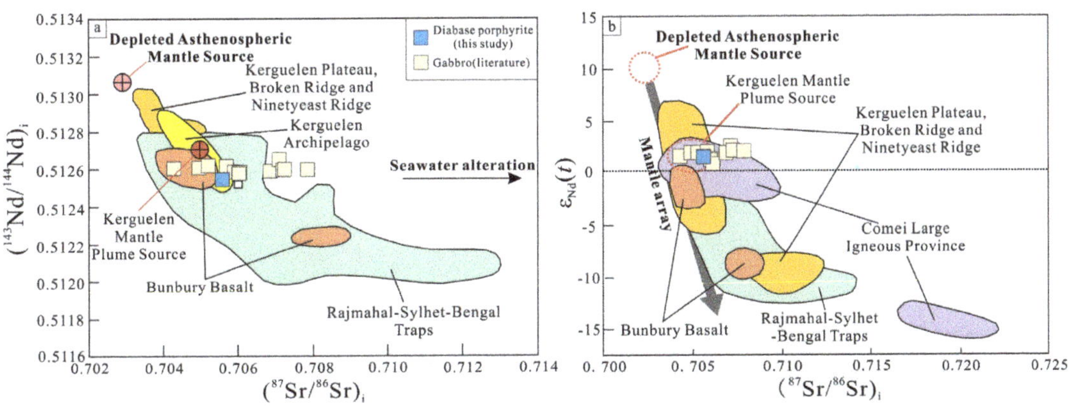

Figure 7. (**a**) $(^{87}Sr/^{86}Sr)_i$ versus $(^{143}Nd/^{144}Nd)_i$ and (**b**) $(^{87}Sr/^{86}Sr)_i$ versus $\varepsilon_{Nd}(t)$ diagrams for mafic intrusive rocks in the Cuona area and related rocks (Kerguelen mantle plume, Kerguelen Islands, Kerguelen Plateau, Ninetyeast Ridge, and Rajmahal–Bengal–Sylhet Traps data are from Olierook et al., 2019; Peng et al., 2022; gabbro data are from Wang et al., 2022) [8,18,26].

4.4. Zircon Hf Isotopes

The in situ zircon Hf isotopic data are listed in Table 4. The diabase porphyrite has an average zircon $^{176}Lu/^{177}Hf$ value of 0.000803 (1σ; n = 14) and an average zircon $^{176}Hf/^{177}Hf$ value of 0.2828746 (1σ; n = 14), which represents the initial $^{176}Hf/^{177}Hf$ ratio, given the low

Lu/Hf ratio. The $\varepsilon_{Hf}(t)$ values range from 0.60 to 3.73 (Figure 8), with T_{DM} Hf model ages of 641–776 Ma.

Table 4. Lu-Hf isotope analysis results of zircons from mafic intrusive rocks in the Cuona region.

Sample Point	Age (Ma)	$^{176}Yb/^{177}Hf$	$^{176}Lu/^{177}Hf$	$^{176}Hf/^{177}Hf$	$\pm 2\sigma$	$\varepsilon_{Hf}(0)$	$\varepsilon_{Hf}(t)$	$(^{176}Hf/^{177}Hf)_i$	$T_{DM}(Hf)$	T_{DM} (Ma)	$f_{Lu/Hf}$
21CN-02-02	143	0.02742	0.00063	0.28272	0.000016	−1.83040	1.30543	0.28272	745	1113	−0.981
21CN-02-03	142	0.03072	0.00073	0.28273	0.000017	−1.65096	1.46297	0.28273	739	1102	−0.978
21CN-02-04	149	0.03014	0.00073	0.28271	0.000016	−2.25922	1.00827	0.28271	763	1137	−0.978
21CN-02-11	136	0.02601	0.00061	0.28278	0.000018	0.20608	3.18877	0.28278	664	988	−0.982
21CN-02-12	140	0.02556	0.00059	0.28277	0.000016	−0.16955	2.90089	0.28277	678	1009	−0.982
21CN-02-14	136	0.02367	0.00055	0.28274	0.000018	−1.20716	1.77511	0.28274	719	1078	−0.983
21CN-02-17	134	0.02746	0.00065	0.28280	0.000016	0.79427	3.73320	0.28279	641	951	−0.980
21CN-02-19	148	0.02574	0.00062	0.28274	0.000018	−1.10972	2.13618	0.28274	716	1064	−0.981
21CN-02-20	141	0.03373	0.00081	0.28271	0.000017	−2.39056	0.70116	0.28270	770	1150	−0.976
21CN-02-31	131	0.05426	0.00139	0.28276	0.000035	−0.42895	2.44373	0.28276	701	1031	−0.958
21CN-02-36	139	0.04103	0.00097	0.28276	0.000017	−0.61020	2.43815	0.28275	701	1038	−0.971
21CN-02-38	132	0.03465	0.00084	0.28276	0.000016	−0.41738	2.47727	0.28276	692	1030	−0.975
21CN-02-39	134	0.04543	0.00110	0.28277	0.000017	−0.29332	2.64530	0.28276	691	1021	−0.967
21CN-02-40	140	0.03928	0.00102	0.28270	0.000020	−2.47044	0.59930	0.28270	776	1156	−0.969

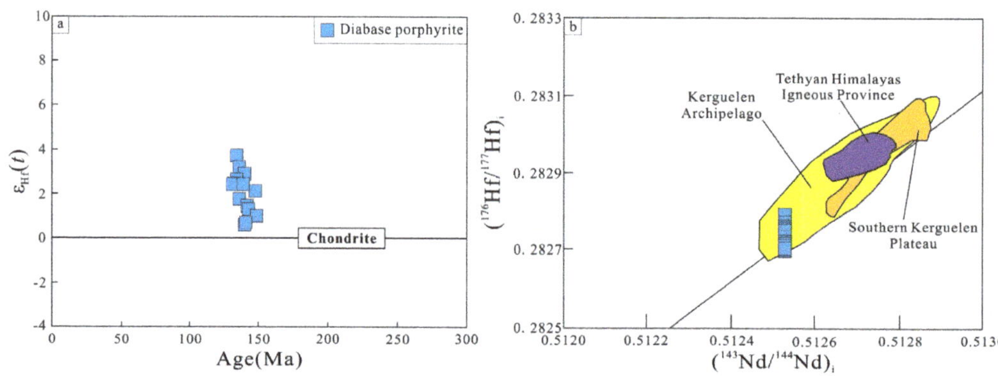

Figure 8. (**a**) Zircon $\varepsilon_{Hf}(t)$ and (**b**) $(^{143}Nd/^{144}Nd)_i$–$(^{176}Hf/^{177}Hf)_i$ diagrams for mafic intrusive rocks in the Cuona area and related rocks (Kerguelen Islands and Tethyan Himalayan Igneous Province data are from Chen et al., 2018 and Chen et al., 2021) [3,4].

5. Discussion

5.1. Ages

Zircon U–Pb dating was undertaken on two samples of mafic intrusive rock from the Cuona area in the eastern Tethyan Himalayan Belt of southern Tibet. The diabase porphyrite sample (21CN-02) has a crystallization age of 135.0 ± 1.6 Ma. The age of diabase porphyrite indicates that there is Early Cretaceous mafic magmatism in this area.

We compiled crystallization ages for Late Jurassic to Early Cretaceous mafic rocks in the Tethyan Himalayan Belt and constructed an age histogram (Figure 9). On the basis of previous studies, Late Jurassic to Early Cretaceous mafic magmatism in the Tethyan Himalayan Belt occurred mainly during 147–129 Ma. The crystallization age of the diabase porphyrite of this study (135.0 ± 1.6 Ma) is generally consistent with diabase crystallization ages (137–131 Ma) previously reported from the Cuona area [2,4]. It is also consistent with the Late Jurassic to Early Cretaceous magmatism in the Tethyan Himalayan Belt (Figure 9). Therefore, the studied mafic intrusive rocks may be the products of two distinct magmatic events. The diabase porphyrite corresponds to Late Jurassic to Early Cretaceous magmatism in the Tethyan Himalayan Belt (147–129 Ma), whereas the gabbro may represent the initial activity of the Kerguelen mantle plume during the Late Jurassic to Early Cretaceous.

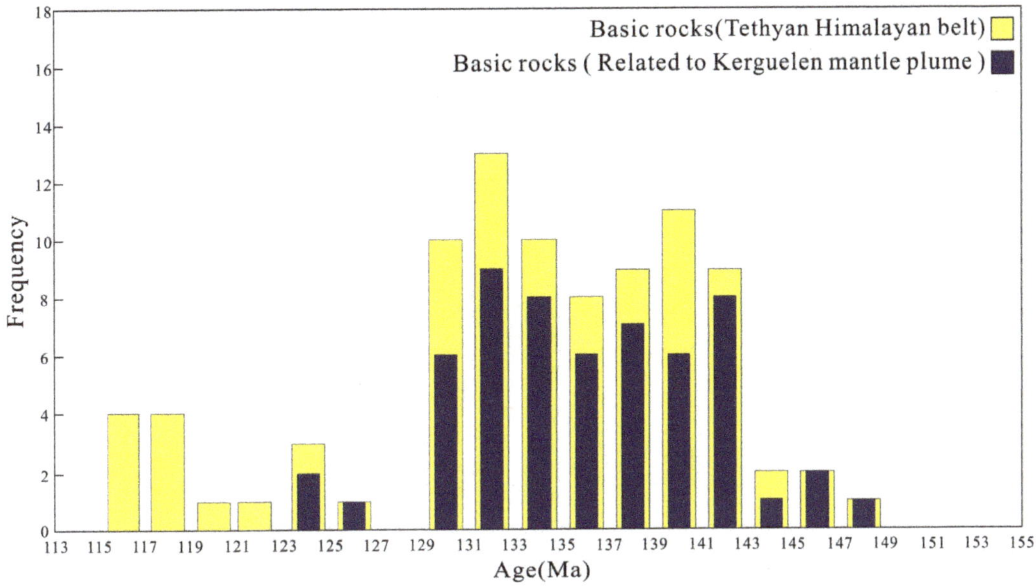

Figure 9. Histogram of ages for Late Jurassic to Early Cretaceous mafic rocks in the Tethyan Himalaya (age data are from Zhu et al., 2009; Hou Chenyang, 2017; Shi et al., 2017; Chen et al., 2021; Deng Jin, 2021; Liang et al., 2023) [4,16,36,54].

5.2. Fractional Crystallization and Crustal Contamination

Alteration can complicate the use of geochemical data to constrain the petrogenesis of igneous rocks. Some trace elements in mafic rocks, such as REEs, HFSEs, and transition metals (e.g., V, Cr, and Ni), are barely affected by low-temperature alteration or weathering [55–57]. The mafic rock samples from the Cuona area have LOI values of 1.26–5.42 wt.%, indicating they have undergone variable degrees of alteration. Therefore, elements that are relatively immobile during alteration are used in the following discussion.

Primary magmas from a mantle source typically have high $Mg^{\#}$ values (73–81) and elevated Cr (>1000 ppm) and Ni (>400 ppm) concentrations [58]. The studied samples have $Mg^{\#}$ = 43.80–66.18, Cr = 14.8–451 ppm, and Ni = 6.5–259 ppm, which are lower than the values for primary mantle-derived magmas. This indicates that the samples have undergone fractional crystallization. Both Cr and Ni exhibit positive correlations with MgO (Figure 10d–e), indicative of the early fractional crystallization of olivine or clinopyroxene. δEu values of the mafic rock samples are 0.93–1.09, with an average of 1.02, indicative of no significant plagioclase fractionation. Al_2O_3 exhibits a negative correlation with MgO (Figure 10a), which is also indicative of little plagioclase fractionation. These results suggest the magma chemistry was affected mainly by clinopyroxene fractionation. Moreover, CaO exhibits a positive correlation with MgO and a negative correlation with SiO_2 (Figure 10b,f), which is consistent with clinopyroxene fractionation being dominant. TiO_2 exhibits a negative correlation with MgO (Figure 10c) because TiO_2 was an incompatible element that was progressively enriched in the magma during early fractionation. As the magma evolved further, late fractional crystallization of magnetite occurred, which caused a decrease in TiO_2 concentrations.

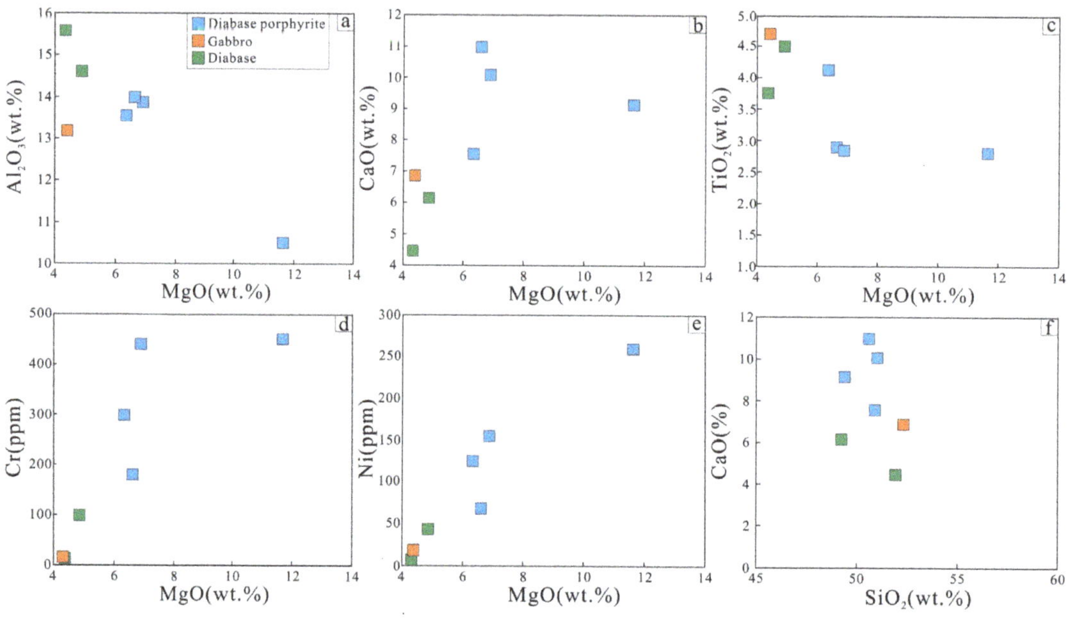

Figure 10. Major elements variation diagram for mafic rocks from the Cuona area. (**a**) MgO-Al_2O_3 diagram; (**b**) MgO-CaO diagram; (**c**) MgO-TiO_2 diagram; (**d**) MgO-Cr diagram; (**e**) MgO-Ni diagram; (**f**) SiO_2-CaO diagram.

The diabase porphyrite samples have La/Ta = 18.2–24.4 and La/Sm = 2.33–3.12, while the gabbro samples have La/Ta = 17.0 and La/Sm = 2.50, indicating these rocks were not significantly contaminated by the crust or lithospheric mantle [59]. Moreover, the diabase porphyrite samples have Zr/Hf = 38.08–38.62 and Th/Ta = 1.5–2.40, while the gabbro samples have Zr/Hf = 38.13 and Th/Ta = 1.45, which are both different from typical crustal values [53,60]. The diabase samples have La/Ta = 26.1–27.6 and La/Sm = 2.80–3.06, indicative of no significant crustal contamination but some degree of lithospheric mantle contamination. In addition, the diabase samples have Zr/Hf = 38.91–40.51 and Th/Ta = 1.94–2.12, which are also markedly different from typical crustal values. In a La/Sm–La/Nb diagram (Figure 11a), all samples plot far from the crustal contamination trend line, indicative of insignificant crustal contamination. In a $(Th/Ta)_{PM}$–$(La/Nb)_{PM}$ diagram (Figure 11b), the diabase porphyrite and gabbro samples plot near the primitive mantle field and away from the crustal field. Some diabase porphyrite and gabbro samples plot in the OIB-type igneous rock field associated with the Kerguelen mantle plume. The diabase samples plot farther away from the middle and upper crustal fields and closer to the lower crustal field. In a Nb/Yb–Th/Yb diagram (Figure 11c), the diabase porphyrite, gabbro, and diabase samples all exhibit characteristics similar to those of rocks from the Kerguelen Islands. They exhibit an overall trend towards the intraplate enrichment field, with no involvement of subduction-related or crustal materials. In a $(La/Nb)_{PM}$–$(Th/Nb)_{PM}$ diagram (Figure 11d), the diabase porphyrite and gabbro samples plot in the Cuomei LIP field, close to the primitive mantle values. In contrast, the diabase samples plot closer to the lower crustal field, indicative of possible interactions with the lower crust.

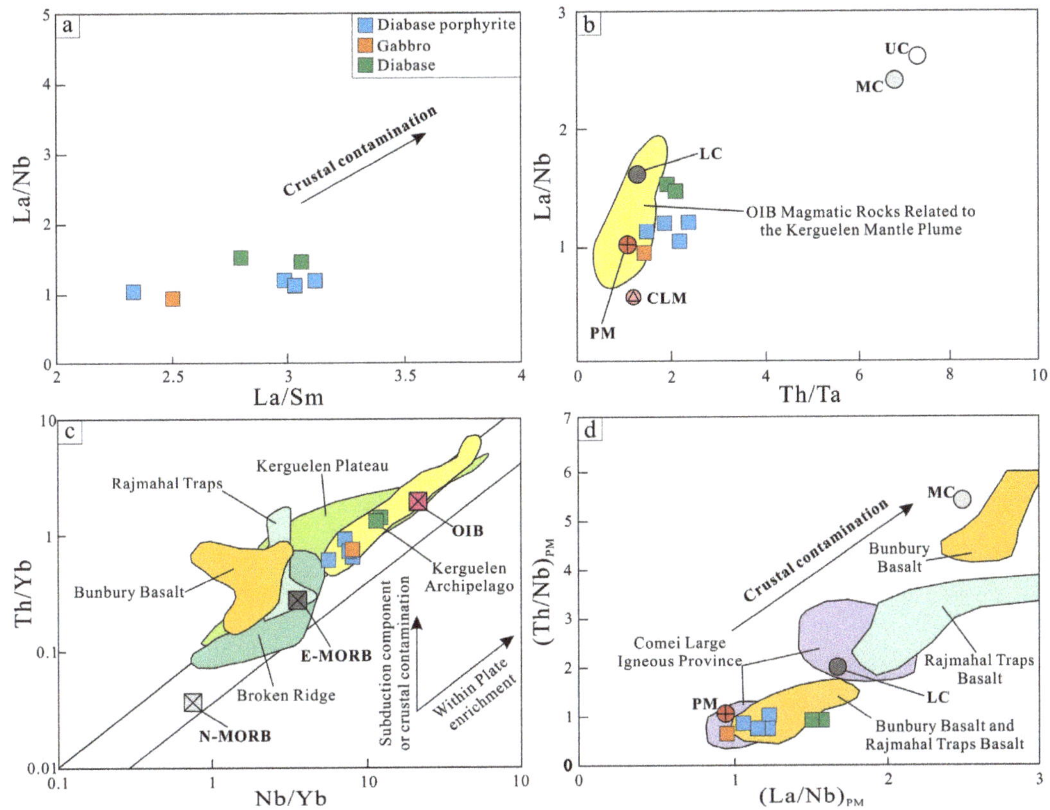

Figure 11. Plots used to identify crustal contamination of mafic intrusions in the Cuona area: (**a**) La/Sm versus La/Nb; (**b**) Th/Ta versus La/Nb; (**c**) Nb/Yb versus Th/Yb; (**d**) $(La/Nb)_{PM}$ versus $(Th/Nb)_{PM}$ [1,2,34]. Data sources: upper, middle, and lower crust [61]; primitive mantle [62]; lithospheric mantle [63]; OIB-type igneous rocks related to the Kerguelen mantle plume [1]. OIB = ocean island basalt; E-MORB = enriched mid-ocean ridge basalt; N-MORB = normal mid-ocean ridge basalt; UC = upper crust; MC = middle crust; LC = lower crust; PM = primitive mantle; CLM = continental lithospheric mantle.

In summary, the diabase porphyrite and gabbro show no features of crustal or lithospheric mantle contamination, and the diabase is contaminated by a certain degree of lithospheric mantle.

5.3. Magma Sources and Tectonic Setting

In a $(^{87}Sr/^{86}Sr)_i$–$(^{143}Nd/^{144}Nd)_i$ diagram (Figure 7a), the samples plot near the field for the Kerguelen Islands and Kerguelen mantle plume. In a $(^{87}Sr/^{86}Sr)_i$–$\varepsilon_{Nd}(t)$ diagram (Figure 7b), the samples plot within the fields for the Cuomei LIP and Kerguelen mantle plume and have similar isotopic compositions as the gabbro of the Lakang Formation in the southeastern Tethyan Himalayan Belt (Wang et al., 2022) [18]. In a $(^{143}Nd/^{144}Nd)_i$–$(^{176}Hf/^{177}Hf)_i$ diagram (Figure 8b), the diabase porphyrite plots within the field for the Kerguelen Islands, and its isotopic composition is similar to that of the Cuomei LIP (Xia et al., 2012) [22], with $(^{87}Sr/^{86}Sr)_i$ = 0.7047 and $\varepsilon_{Nd}(t)$ = 1.5. These characteristics suggest the diabase porphyrite has similar isotopic compositions as other Early Cretaceous OIB-type mafic rocks in the Tethyan Himalayan Belt, and were associated with the Kerguelen mantle plume.

The average Zr/Nb and La/Nb ratios of the studied mafic intrusive rocks are 9.42 and 1.28, respectively, which are close to the values for EM I (Zr/Nb = 4.2–11.5 and La/Nb = 0.86–1.19). The geochemical characteristics of the samples are relatively uniform. In La–La/Nb and Nb–Nb/Th diagrams (Figure 12a,b), all samples plot in the OIB field. In a Zr/Nb–Y/Nb diagram (Figure 12c), all samples plot in the enriched mantle region. In a Ta/Yb–Th/Yb diagram (Figure 12d), the samples plot in the enriched mantle field and close to that for active continental margins. The OIB-type mafic intrusions in this region have EM I geochemical features, similar to other such rocks in several locations within the Tethyan Himalayan Belt in southern Tibet [19,24,29].

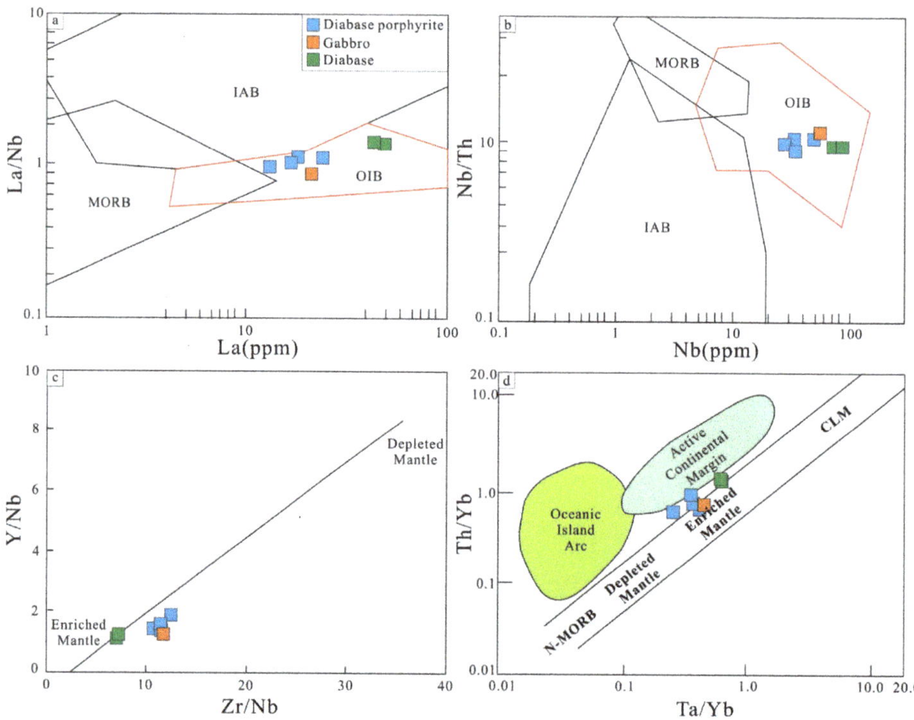

Figure 12. (**a**) La versus La/Nb ((**a**) is from Shuguang Li, 1993) [64], (**b**) Nb versus Nb/Th ((**b**) is from Shuguang Li, 1993) [64], (**c**) Zr/Nb versus Y/Nb ((**c**) is from Wilson, 1997) [58], and (**d**) Ta/Yb versus Th/Yb diagrams ((**d**) is from Meschede, 1986) [65] for mafic intrusive rocks in the Cuona area. OIB = ocean island basalt; MORB = mid-ocean ridge basalt; IAB = island arc basalt; N-MORB = normal mid-ocean ridge basalt; CLM = continental lithospheric mantle.

In Zr–Zr/Y and Zr–TiO_2 diagrams, all the samples plot in the intraplate basalt field (Figure 13a,b). In a Ta/Hf–Th/Hf tectonic discrimination diagram, the diabase porphyrite samples plot in the initial rift basalt field, the gabbro samples plot in the oceanic intraplate basalt field (including OIBs, seamounts, transitional-type MORBs, and enriched-type MORBs), and the diabase samples plot in the boundary region between the initial and intracontinental rift fields (Figure 13c). In a 2Nb–Zr/4–Y tectonic discrimination diagram, the diabase porphyrite and diabase samples plot in the intraplate alkaline basalt + intraplate tholeiitic basalt field, while the gabbro samples plot in the intraplate alkaline basalt field (Figure 13d). Compared with previous studies, the diabase porphyrite and diabase have similar geochemical characteristics to volcanic rocks in the Sangxiu, Weimei, and Zhela formations in the eastern Tethyan Himalayan Belt [1,8]. This similarity suggests that the diabase porphyrite and diabase may have formed in a continental rift environment, while the gabbro, with its distinct composition, could potentially represent an continental intraplate environment. However, the age differences between these rock types have not been clearly established, and further investigations are needed to support these interpretations.

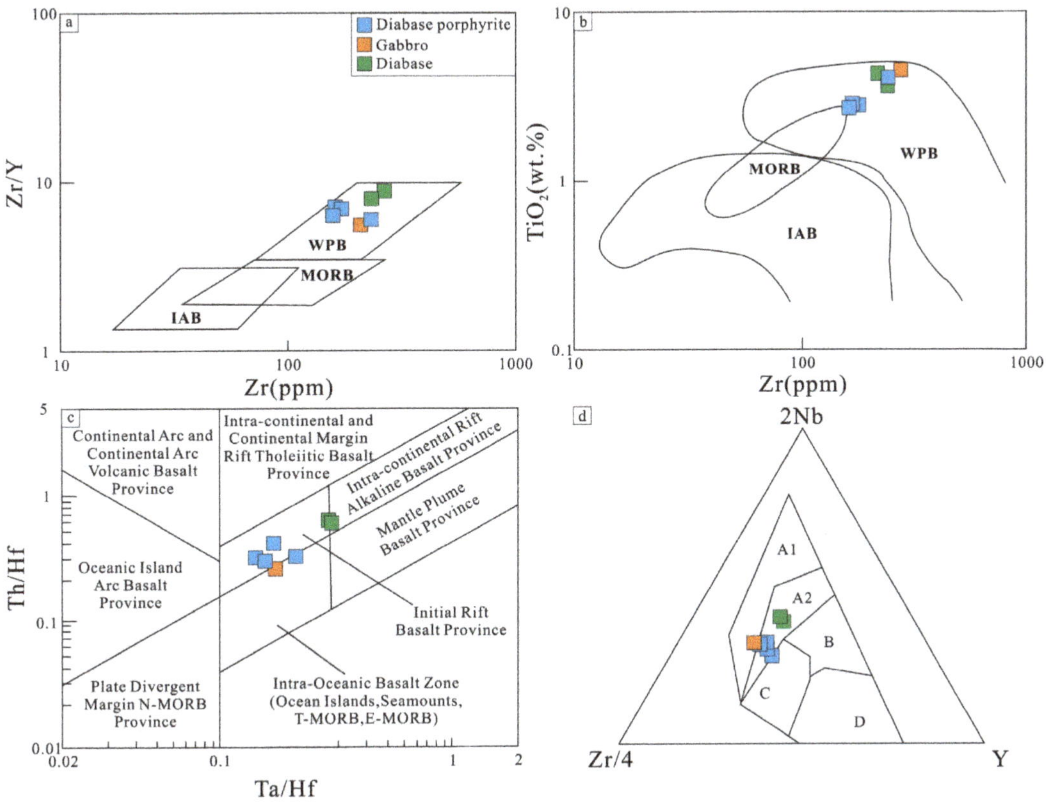

Figure 13. (**a**) Zr/Y versus Zr ((**a**) is from Pearce and Norry, 1979) [66], (**b**) Zr versus Ti ((**b**) is from Pearce, 1982) [67], (**c**) Th/Hf versus Ta/Hf ((**c**) is from Wang et al., 2001) [68], and (**d**) 2Nb–Zr/4–Y tectonic discrimination diagrams ((**d**) are from Meschede, 1986) [65] for mafic intrusive rocks in the Cuona area. WPB = within-plate basalt; IAB = island arc basalt; MORB = mid-ocean ridge basalt; A1 = within-plate alkaline basalts; A2 = within-plate alkaline basalts + within-plate tholeiites; B = E-MORBs; C = within-plate tholeiites + volcanic arc basalts; D = volcanic arc basalts + E-MORBs.

5.4. Geological Significance

Although most Cretaceous magmatism in East Gondwana was unrelated to the Kerguelen mantle plume, the Early Cretaceous magmatism in the Tethyan Himalayan Belt was closely associated with the plume [26,34]. Previous studies have shown that the OIB-type mafic magmatism in this tectonic belt began at ca. 147 Ma and ended at 125 Ma [2,4,5,24,29,69]. According to an Early Cretaceous paleogeographic reconstruction of Gondwana, the Tethyan Himalayan Belt was located along a passive continental margin at the time. There is a lack of arc-related igneous rocks that coexist with the Early Cretaceous OIB-type mafic rocks, which rules out the possibility of their formation in a subduction setting [70,71]. During the Late Jurassic to Early Cretaceous, the Cuona area was located along the passive continental margin of eastern Gondwana, adjacent to the Neo-Tethys Ocean. The rapid expansion of the Neo-Tethys Ocean caused intense extensional tectonism south of the mid-ocean ridge, leading to the formation of rift structures [8,27,72]. OIB-type igneous rocks are generally considered to be the products of a mantle plume or hotspot activity [73]. The Early Cretaceous OIB-type mafic rocks in the Tethyan Himalayan Belt were closely related to the Kerguelen mantle plume, which had a key role in the rifting of eastern Gondwana [16,25,28,30].

The diabase porphyrite formed at 135.0 ± 1.6 Ma, which is slightly earlier than the peak of Kerguelen mantle plume activity (ca. 132 Ma; Zhu et al., 2008; Zhu et al., 2009; Ma et al., 2016) [2,16,74], and was coeval with the Bunbury basalts (137–136 Ma) and the oceanic magnetic anomaly (136 Ma) that record the rifting of eastern Gondwana [34,75]. Furthermore, the diabase porphyrite has distinct OIB-type geochemical characteristics, with enrichments in light REEs and HFSEs, indicative of a mantle plume source. Its age coincides with the peak of mafic magmatism during the Late Jurassic–Early Cretaceous (147–129 Ma) in the Tethyan Himalayan Belt (Figure 9). Therefore, the diabase porphyrite was formed in a continental rift setting associated with the Kerguelen mantle plume, and represents a product of the rifting of eastern Gondwana.

5.5. *Kerguelen Mantle Plume and Evolution of Eastern Gondwana*

The initial stages of continental rifting are typically associated with LIPs [76,77]. There are differing views on the relationship between the rifting of eastern Gondwana and the Kerguelen mantle plume. Some studies have suggested there is no direct relationship between the two [26]. However, there is some evidence that the Kerguelen mantle plume had a key role in the rifting of East Gondwana. For example, the bimodal magmatism during 147–130 Ma, the formation of A-type granites [6,25], and the development of the Indian Ocean Ridge at 136 Ma were all associated with mantle plume activity. Some studies have suggested that the small magmatic provinces on East Gondwana were associated with the activity of the Kerguelen mantle plume [16,18,26,34]. The magmatism associated with the plume lasted for an extended period, with intermittent eruptions for potentially >100 Myr on both sides of the Southeast Indian Ridge [78]. These lines of evidence suggest that the prolonged magmatic activity of the Kerguelen mantle plume played a key role in the rifting of East Gondwana and potentially caused the rifting.

As East Gondwana began to rift, the Indian Plate gradually moved northward, and the Kerguelen mantle plume activity became increasingly focused in the Tethyan Himalayan region. At ca. 147 Ma, East Gondwana underwent initial rifting and the Cuona area in the eastern Tethyan Himalayan Belt was located in the central part of the Kerguelen mantle plume head (Figure 14a; Zhang et al., 2023) [10]. As the rifting of East Gondwana proceeded, the Cuona area began to drift towards the edge of the Kerguelen mantle plume, and the mantle plume magmatism peaked, which formed a large volume of OIB-type basalts at ca. 135 Ma (Figure 14b). With the further expansion of the mantle plume head, younger OIB-type volcanic rocks (125 Ma) formed (Dong et al., 2019) [70]. The bimodal magmatism (118–115 Ma) and N-MORB-type basalts (ca. 120 Ma) in the eastern Tethyan Himalaya are the products of extension rather than Kerguelen mantle plume activity [2,3,5,16]. Furthermore, the Australia–Antarctic plate had a relatively stable paleo-latitude during 140–120 Ma (Torsvik et al., 2012) [79], indicating the Indian Plate had completely separated from East Gondwana by ca. 125 Ma (Figure 14c).

Our results indicate that early activity of the Kerguelen mantle plume not only facilitated the rifting of East Gondwana, but also affected the tectonic and magmatic evolution of the Tethyan Himalaya. Our study has provided new constraints on the early activity of the Kerguelen mantle plume, which had a key role in the geological evolution of the Tethyan Himalaya, and the origins of basaltic magmatism in continental margin rift and intraplate settings during the rifting of East Gondwana.

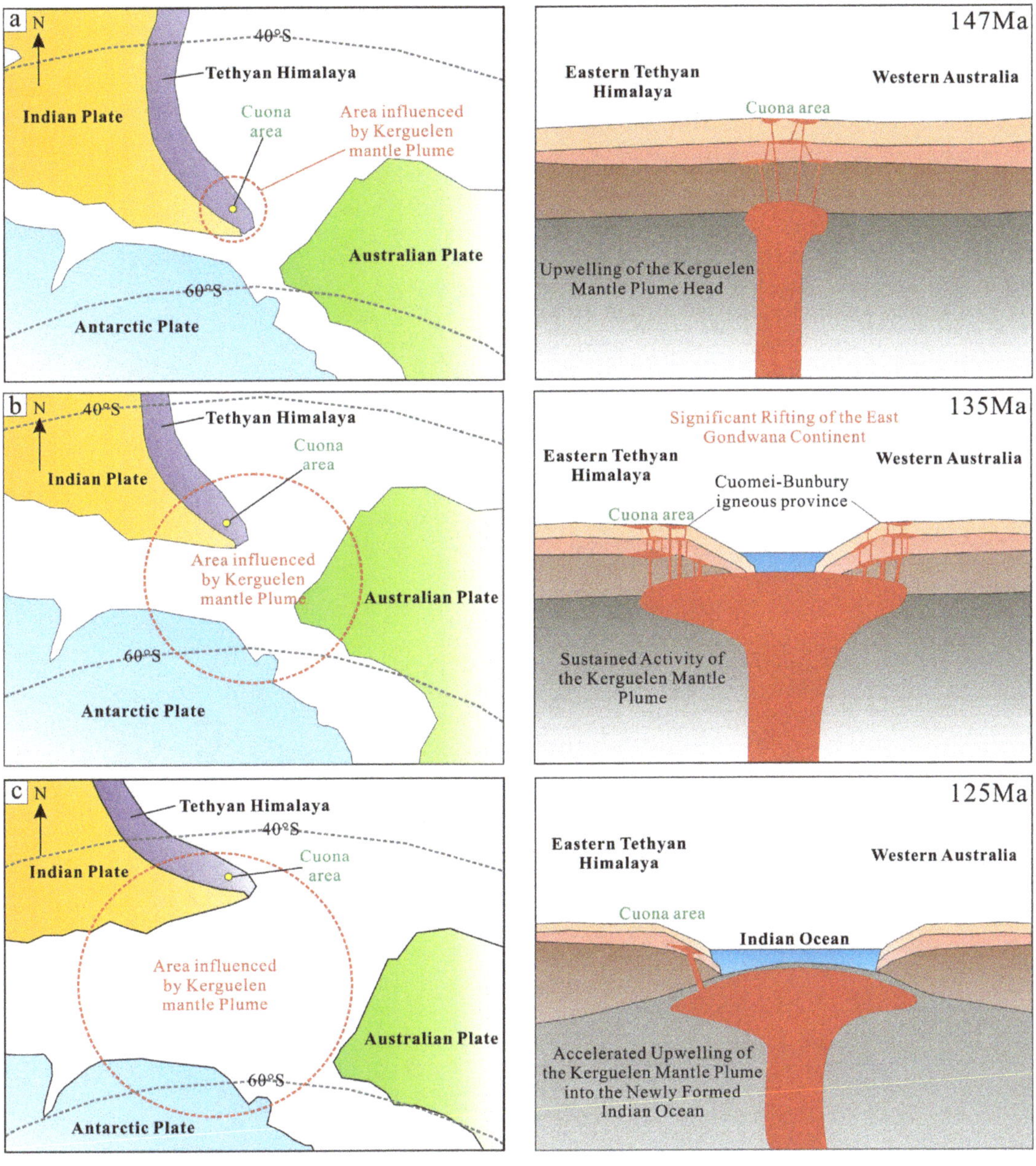

Figure 14. Model of the evolution of the Kerguelen mantle plume and East Gondwana (modified from Gibbons et al., 2013; Olierook et al., 2019; Wang et al., 2022; Liang et al., 2023) [9,18,26,77]. (**a**) At ca. 147 Ma, East Gondwana underwent initial rifting and the Cuona area in the eastern Tethyan Himalayan Belt was located in the central part of the Kerguelen mantle plume head. (**b**) At ca. 135 Ma, as the rifting of East Gondwana proceeded, the Cuona area began to drift towards the edge of the Kerguelen mantle plume, and the mantle plume magmatism peaked, which formed a large volume of OIB-type basalts. (**c**) By ca. 125 Ma, the Indian Plate had completely separated from East Gondwana, as indicated by the paleo-latitude stability of the Australia–Antarctic Plate.

6. Conclusions

We investigated the petrology, geochronology, and geochemistry of mafic intrusive rocks in the Cuona region of the eastern Tethyan Himalaya. Our main conclusions are as follows:

(1) The diabase porphyrite formed in the Early Cretaceous (135.0 ± 1.6 Ma). The diabase porphyrite and gabbro have a similar petrogenesis to the numerous OIB-type mafic igneous rocks in the Cuomei LIP, displaying typical OIB-type geochemical signatures without contamination by lithospheric mantle or crust. These rocks were likely derived from EMI-like enriched mantle. The diabase porphyrite formed in a continental rift setting related to the Kerguelen mantle plume, whereas the gabbro should be the product of the deep magma of the Kerguelen mantle plume in a stable continental intraplate environment.

(2) The mafic intrusive rocks in the Cuona region are subalkaline to alkaline. There is significant fractionation between the light (enriched) and heavy (depleted) REEs and no significant Eu anomalies. The trace element characteristics of the diabase porphyrite and gabbro are similar, with both being enriched in HFSEs and LILEs. The diabase is enriched in large-ion lithophile elements (LILEs, e.g., La and Ce) and depleted in high-field-strength elements (HFSEs, e.g., Ru, Zr and Ti).

(3) The age of the diabase porphyrite (135 Ma) is slightly older than the peak of Kerguelen mantle plume activity (132 Ma), indicating it was the product of East Gondwana continental rifting associated with the effects of the Kerguelen mantle plume.

Author Contributions: Conceptualization, F.Y. and C.M.; methodology, Z.Q.; software, Z.L.; validation, F.Y.; formal analysis, J.M.; investigation, Z.L.; resources, L.W.; data curation, Z.Q.; writing—original draft preparation, C.S.; writing—review and editing, Z.K.; visualization, L.Z.; supervision, F.Y.; project administration, C.S.; funding acquisition, Z.K. All authors have read and agreed to the published version of the manuscript.

Funding: This research was funded by the following projects: The project of National Natural Science Foundation of China "The genesis of mafic rocks in the middle and eastern parts of the Tethys Himalayan belt and its constraints on the early evolution history of the Neo-Tethys Ocean", grant number: 42173048; National Key R & D Program "Study on Comprehensive Application of Refractory Elements and Isotope Analysis Technology", grant number: 2020YFA0714804; and the Innovation Project of Guangxi Graduate Education "Determination of the initial breakup time of the Neo-Tethys Ocean in Tibet", grant number: YCSW2021206.

Data Availability Statement: Data are contained within the article.

Conflicts of Interest: The authors declare no conflict of interest.

References

1. Zhu, D.; Pan, G.; Mo, X.; Liao, Z.; Jiang, X.; Wang, L.; Zhao, Z. Petrogenesis of volcanic rocks in the Sangxiu Formation, central segment of Tethyan Himalaya: A probable example of plume–lithosphere interaction. *J. Asian Earth Sci.* **2007**, *29*, 320–335. [CrossRef]
2. Zhu, D.; Mo, X.; Pan, G.; Zhao, Z.; Dong, G.; Shi, Y.; Liao, Z.; Wang, L.; Zhou, C. Petrogenesis of the Earliest Early Cretaceous Mafic Rocks from the Cona Area of the Eastern Tethyan Himalaya in South Tibet: Interaction Between the Incubating Kerguelen Plume and the Eastern Greater India Lithosphere? *Lithos* **2008**, *100*, 147–173. [CrossRef]
3. Chen, S.; Fan, W.; Shi, R.; Liu, X.; Zhou, X. 118–115 Ma Magmatism in the Tethyan Himalaya Igneous Province: Constraints on Early Cretaceous Rifting of the Northern Margin of Greater India. *Earth Planet. Sci. Lett.* **2018**, *491*, 21–33. [CrossRef]
4. Chen, S.; Fan, W.; Shi, R.; Xu, J.; Liu, Y. The Tethyan Himalaya Igneous Province: Early Melting Products of the Kerguelen Mantle Plume. *J. Petrol.* **2021**, *62*, 301–320. [CrossRef]

5. Shi, Y.; Hou, C.; Anderson, J.L.; Yang, T.; Bian, W.; Jin, J. Zircon SHRIMP U–Pb age of Late Jurassic OIB-type volcanic rocks from the Tethyan Himalaya: Constraints on the initial activity time of the Kerguelen mantle plume. *Acta Geochim.* **2018**, *37*, 441–455. [CrossRef]
6. Tian, Y.; Gong, J.; Chen, H.; Guo, L.; Xu, Q.; Chen, L.; Lin, X.; Cheng, X.; Yang, R.; Zhao, L.; et al. Early Cretaceous bimodal magmatism in the eastern Tethyan Himalayas, Tibet: Indicative of records on precursory continental rifting and initial breakup of eastern Gondwana. *Lithos* **2018**, *324*, 699–715. [CrossRef]
7. Zeng, Y.; Xu, J.; Chen, J.; Wang, B.; Huang, F.; Yu, H.; Chen, X.; Zhao, P. Breakup of Eastern Gondwana as Inferred from the Lower Cretaceous Charong Dolerites in the Central Tethyan Himalaya, Southern Tibet. *Palaeogeogr. Palaeoclimatol. Palaeoecol.* **2019**, *515*, 70–82. [CrossRef]
8. Peng, W.; Yang, T.; Shi, Y.; Bian, W.; Ma, Y.; Wang, S.; Jiao, X.; Ma, J.; Kang, Y.; Zhang, S.; et al. Role of the Kerguelen Mantle Plume in Breakup of Eastern Gondwana: Evidence from Early Cretaceous Volcanic Rocks in the Eastern Tethyan Himalaya. *Palaeogeogr. Palaeoclimatol. Palaeoecol.* **2022**, *588*, 110823. [CrossRef]
9. Liang, J.; Bian, W.; Jiao, X.; Peng, W.; Ma, J.; Wang, S.; Ma, Y.; Zhang, S.; Wu, H.; Li, H.; et al. Geochronological Results from the Zhela Formation Volcanics of the Tethyan Himalaya and Their Implications for the Breakup of Eastern Gondwana. *Sci. Rep.* **2023**, *13*, 20035. [CrossRef]
10. Zhang, Z.; Li, G.; He, X.; Zhang, L.; Dong, S.; Liang, W.; Han, S. The evolution of Kerguelen mantle plume and breakup of eastern Gondwana: New insights from multistage Cretaceous magmatism in the Tethyan Himalaya. *Gondwana Res.* **2023**, *119*, 68–85. [CrossRef]
11. Meng, F.; Meng, Y.; Wei, Y. Identification of the EMII-type Mantle End-member of the Kerguelen Mantle Plume and Its Implications for the Initial Breakup of Eastern Gondwana: New Constraints from the Cretaceous Diabase in the Tethyan Himalaya. *J. Asian Earth Sci.* **2024**, *272*, 106250. [CrossRef]
12. Duncan, R.A.; Hooper, P.R.; Rehacek, J.; Marsh, J.S.; Duncan, A.R. The timing and duration of the Karoo igneous event, southern Gondwana. *J. Geophys. Res. Solid Earth* **1997**, *102*, 18127–18138. [CrossRef]
13. Pirajno, F.; Santosh, M. Mantle plumes, supercontinents, intracontinental rifting and mineral systems. *Precambrian Res.* **2015**, *259*, 243–261. [CrossRef]
14. Müller, R.D.; Royer, J.Y.; Lawver, L.A. Revised plate motions relative to the hotspots from combined Atlantic and Indian Ocean hotspot tracks. *Geology* **1993**, *21*, 275–278. [CrossRef]
15. Bryan, S.E.; Constantine, A.E.; Stephens, C.J.; Ewart, A.; Schön, R.W.; Parianos, J. Early Cretaceous volcano-sedimentary successions along the eastern Australian continental margin: Implications for the break-up of eastern Gondwana. *Earth Planet. Sci. Lett.* **1997**, *153*, 85–102. [CrossRef]
16. Zhu, D.; Chung, S.; Mo, X.; Zhao, Z.; Niu, Y.; Song, B.; Yang, Y. The 132 Ma Comei-Bunbury Large Igneous Province: Remnants Identified in Present-day Southeastern Tibet and Southwestern Australia. *Geology* **2009**, *37*, 583–586. [CrossRef]
17. Wei, Y.; Liang, W.; Shang, Y.; Zhang, B.; Pan, W. Petrogenesis and Tectonic Implications of ~130 Ma Diabase Dikes in the Western Tethyan Himalaya (Western Tibet). *J. Asian Earth Sci.* **2017**, *143*, 236–248. [CrossRef]
18. Wang, Z.; Shi, Y.; Yang, T.; Lawford Anderson, j.; Zhao, J.; Peng, X.; Hou, C.; Sun, H.; Bian, W.; Ma, Y. Early Activity of the Kerguelen Mantle Plume: Geochronology, Geochemistry and Sr-Nd-Pb Isotopes of Mafic Dykes and Sills from the Tethyan Himalaya. *Int. Geol. Rev.* **2022**, *65*, 512–526. [CrossRef]
19. Jiang, S.; Nie, F.; Hu, P.; Liu, Y.; Lai, X. Geochemical Characteristics of Mafic Dykes in Southern Tibet. *Acta Geol. Sin.* **2007**, *81*, 60–71. (In Chinese)
20. Tong, J.; Liu, J.; Zhong, H.; Xia, J.; Lu, R.; Li, Y. Zircon U-Pb Dating, Geochemistry, and Tectonic Significance of Mafic Dykes in the Luozha Area of Southern Tibet. *Geol. Bull. China* **2007**, *26*, 1654–1664. (In Chinese)
21. Qiu, B.; Zhu, D.; Zhao, Z.; Wang, L. Western Extension and Significance of the Cuomei Large Igneous Province in Southern Tibet. *Acta Petrol. Sin.* **2010**, *26*, 2207–2216. (In Chinese)
22. Xia, Y.; Zhu, D.; Zhao, Z.; Wang, Q.; Yuan, S.; Chen, Y.; Mo, X. Whole-Rock Geochemistry and Zircon Hf Isotopes of OIB-Type Mafic Rocks in the Cuomei Large Igneous Province, Southeastern Tibet. *Acta Petrol. Sin.* **2012**, *28*, 1588–1602. (In Chinese)
23. Zhu, D.; Xia, Y.; Qiu, B.; Wang, Q.; Zhao, Z. Why Propose the Early Cretaceous Cuomei Large Igneous Province in Southeastern Tibet? *Acta Petrol. Sin.* **2013**, *29*, 3659–3670. (In Chinese)
24. Ding, F.; Gao, J.; Xu, K. Geochemical and Geochronological Characteristics of Mafic Intrusions in the Rongbu Area of Southern Tibet and Their Geological Significance. *Acta Petrol. Sin.* **2020**, *36*, 391–408. (In Chinese)
25. Zhou, Q.; Liu, Z.; Lai, Y.; Wang, G.; Liao, Z. Petrogenesis of Mafic and Felsic Rocks from the Comei Large Igneous Province, South Tibet: Implications for the Initial Activity of the Kerguelen Plume. *Geol. Soc. Am. Bull.* **2018**, *130*, 811–824. [CrossRef]
26. Olierook, H.K.; Jiang, Q.; Jourdan, F.; Chiaradia, M. Greater Kerguelen Large Igneous Province Reveals No Role for Kerguelen Mantle Plume in the Continental Breakup of Eastern Gondwana. *Earth Planet. Sci. Lett.* **2019**, *511*, 244–255. [CrossRef]
27. Tian, Y. Early Cretaceous Bimodal Magmatic Activity in the Eastern Tethyan Himalayas: A Marker for Continental Rifting and the Initial Breakup of East Gondwana. *Lithos* **2019**, *324–325*, 699–715. (In Chinese) [CrossRef]

28. Lian, D.; Yang, J.; Liu, F.; Wu, Q.; Xu, Z. Fingerprints of the Kerguelen Mantle Plume in Southern Tibet: Evidence from Early Cretaceous Magmatism in the Tethyan Himalaya. *J. Geol.* **2021**, *129*, 207–231. [CrossRef]
29. Yang, C.; Li, Y.; Zhang, J.; Zhang, G. Age, Genesis, and Tectonic Significance of the Karlong Gabbro in the Langkazi Area of Southern Tibet. *Geol. Sci.* **2022**, *57*, 399–414. (In Chinese)
30. Liu, Z.; Zhou, Q.; Lai, Y.; Qing, C.; Li, X.; Wu, J.; Xia, X. Petrogenesis of the Early Cretaceous Laguila Bimodal Intrusive Rocks from the Tethyan Himalaya: Implications for the Break-up of Eastern Gondwana. *Lithos* **2015**, *236–237*, 190–202. [CrossRef]
31. Coffin, M.F.; Pringle, M.S.; Duncan, R.A.; Gladczenko, T.P.; Storey, M.; Müller, R.D. Kerguelen Hotspot Magma Output since 130 Ma. *J. Petrol.* **2002**, *43*, 1121–1139. [CrossRef]
32. Ingle, S.; Weis, D.; Doucet, S.; Mattielli, N. Hf isotope constraints on mantle sources and shallow-level contaminants during Kerguelen hot spot activity since ~120 Ma. *Geochem. Geophys. Geosystems* **2003**, *4*, 767–782. [CrossRef]
33. Liu, Z.; Kang, Z.; Yang, F.; Qiao, Z.; Ma, C.; Shen, Y. Geochronology, Geochemistry, and Petrogenesis of Volcanic Rocks from the Eastern Tethyan Himalayas. *Geol. Explor. J.* **2024**, *39*, 219–231. (In Chinese)
34. Olierook, H.K.; Jourdan, F.; Merle, R.E.; Timms, N.E.; Kusznir, N.; Muhling, J.R. Bunbury Basalt: Gondwana breakup products or earliest vestiges of the Kerguelen mantle plume? *Earth Planet. Sci. Lett.* **2016**, *440*, 20–32. [CrossRef]
35. Yin, A.; Harrison, T.M. Geologic Evolution of the Himalayan-Tibetan Orogen. *Annu. Rev. Earth Planet. Sci.* **2000**, *28*, 211–280. [CrossRef]
36. Hou, C. Zircon Geochronology, Geochemistry, and Geological Significance of the Lacon Formation Volcanic Rocks in the Tethyan Himalayas. Master's Thesis., China University of Geosciences, Beijing, China, 2017. (In Chinese)
37. Boger, S.D. Antarctic Geology: Extending a Gondwanan Perspective. *Gondwana Res.* **2011**, *19*, 548–566.
38. Merdith, A.S.; Collins, A.S.; Williams, S.E.; Pisarevsky, S.; Foden, J.D.; Archibald, D.B.; Blades, M.L.; Alessio, B.L.; Armistead, S.; Plavsa, D.; et al. A Full-Plate Global Reconstruction of the Neoproterozoic. *Gondwana Res.* **2017**, *50*, 84–134. [CrossRef]
39. Huang, Y.; Liang, W.; Zhang, L.; Li, G.; Huang, C.; Xia, X.; Dong, S.; Wu, J. Initial Rifting of the Tethyan Himalayas-Indo-Asian Crust: Evidence from Late Cretaceous OIB-Type Basalts in Southern Tibet. *Earth Sci.* **2018**, *43*, 2651–2663. (In Chinese)
40. Pan, G.; Wang, L.; Yin, F.; Geng, Q.; Li, G.; Zhu, D. Review, Progress, and Prospects of the Formation and Evolution of the Tibetan Plateau. *Sediment. Tethys Geol.* **2022**, *42*, 151–175. (In Chinese)
41. Liu, Y.; Gao, S.; Hu, Z.; Gao, C.; Zong, K.; Wang, D. Continental and Oceanic Crust Recycling-induced Melt-Peridotite Interactions in the Trans-North China Orogen: U-Pb Dating, Hf Isotopes and Trace Elements in Zircons from Mantle Xenoliths. *J. Petrol.* **2009**, *51*, 392–399. [CrossRef]
42. Andersen, T. Correction of common lead in U-Pb analyses that do not report ^{204}Pb. *Chem. Geol.* **2002**, *192*, 59–79. [CrossRef]
43. Ludwig, K. *A Plotting and Regression Program for Radiogenic Isotope Data*; Open-File Report; 91-445; U.S. Geological Survey: Sunrise Valley Drive Reston, VA, USA, 1991.
44. Wang, X.; Zhou, J.; Qiu, J.; Zhang, W.; LIU, X.; Zhang, G. LA-ICP-MS U-Pb zircon geochronology of the Neoproterozoic igneous rocks from northern Guangxi, south China: Implications for tectonic evolution. *Precambrian Res.* **2006**, *145*, 111–130. [CrossRef]
45. Yuan, H.; Gao, S.; Liu, X.; Li, H.; Günther, D.; Wu, F. Accurate U-Pb Age and Trace Element Determinations of Zircon by Laser Ablation-Inductively Coupled Plasma-Mass Spectrometry. *Geostand. Geoanal. Res.* **2004**, *28*, 353–370. [CrossRef]
46. Li, X.Y.; Wang, H.Y.; Fan, G.Q. Accurate determination of more than 40 trace elements in rock samples by ICP-MS. *Priv. Sci. Technol.* **2016**, 18. (In Chinese)
47. Liang, X.; Wei, G.; Li, X.; Liu, Y. Precise Determination of ^{143}Nd/^{144}Nd and Sm/Nd Ratios Using MC-ICPMS. *Geochimica* **2003**, *32*, 91–96. (In Chinese)
48. Griffin, W.L.; Pearson, N.J.; Belousova, E.; Jackson, S.E.; Achterbergh, V.E.; O'Reilly, S.Y.; Shee, S.R. The Hf isotope composition of cratonic mantle: LAM-MC-ICP-MS analysis of zircon megacrysts in kimberlites. *Geochim. Cosmochim. Acta* **2000**, *64*, 133–147. [CrossRef]
49. Pearce, J. A User's Guide to Basalt Discrimination Diagrams. In *Trace Element Geochemistry of Volcanic Rocks: Applications for Massive Sulphide Exploration*; Geological Association of Canada Short Course Notes; Geological Association of Canada: Toronto, ON, Canada, 1996; Volume 12, pp. 79–113.
50. Maniar, P.D.; Piccoli, P.M. Tectonic Discrimination of Granitoids. *Geol. Soc. Am. Bull.* **1989**, *101*, 635–643. [CrossRef]
51. Wright, J.B. A Simple Alkalinity Ratio and Its Application to Questions of Non-Orogenic Granite Genesis. *Geol. Mag.* **1969**, *106*, 370–384. [CrossRef]
52. Winchester, J.A.; Floyd, P.A. Geochemical discrimination of different magma series and their differentiation products using immobile elements. *Chem. Geol.* **1977**, *20*, 325–343. [CrossRef]
53. Sun, S.S.; McDonough, W.F. Chemical and isotopic systematics of oceanic basalts: Implications for mantle composition and processes. *Geol. Soc. Lond. Spec. Publ.* **1989**, *42*, 313–345. [CrossRef]
54. Deng, J. Paleomagnetic and Zircon U-Pb Geochronology of the Zhela Formation Volcanic Rocks in the Tethyan Himalayas. Master's Thesis, China University of Geosciences, Beijing, China, 2021. (In Chinese)

55. Ludden, J.N.; Thompson, G. Behaviour of Rare Earth Elements During Submarine Weathering of Tholeiitic Basalt. *Nature* **1978**, *274*, 147–149. [CrossRef]
56. Bienvenu, P.; Bougault, H.; Joron, J.L.; Treuil, M.; Dmitriev, L. MORB Alteration: Rare-Earth Element/Non-Rare-Earth Hygromagmaphile Element Fractionation. *Chem. Geol.* **1990**, *82*, 1–14. [CrossRef]
57. Staudigel, H.; Plank, T.; White, B.; Schmincke, H. Geochemical Fluxes During Seafloor Alteration of the Basaltic Upper Oceanic Crust: DSDP Sites 417 and 418. In *Subduction: Top to Bottom*; Geophysical Monograph Series; Advancing Earth and Space Sciences: Washington, DC, USA, 2013; Volume 96, pp. 19–38.
58. Wilson, M.; Hyman, U. *Igneous Petrogenesis*; Springer: Dordrecht, The Netherlands, 1997.
59. Lassiter, J.C.; Depaolo, D.J. *Plume/Lithosphere Interaction in the Generation of Continental and Oceanic Flood Basalts: Chemical and Isotopic Constraints*; American Geophysical Union (AGU): Washington, DC, USA, 2013.
60. Weaver, B.L. The origin of ocean island basalt end-member compositions: Trace element and isotopic constraints. *Earth Planet. Sci. Lett.* **1991**, *104*, 381–397. [CrossRef]
61. Rudnick, R.L.; Gao, S. Composition of the Continental Crust. *Treatise Geochem.* **2003**, *3*, 1–64.
62. Taylor, S.R.; Mclennan, S.M. The continental crust: Its composition and evolution. *J. Geol.* **1985**, *94*, 57–72. [CrossRef]
63. McDonough, W.F. Contraints on the composition of the continental lithospheric mantle. *Earth Planet. Sci. Lett.* **1990**, *101*, 1–18. [CrossRef]
64. Shuguang, L. Ba-Nb-Th-La diagrams used to identify tectonic environments of ophiolite. *Acta Petrol. Sin.* **1993**, *9*, 146.
65. Meschede, M. A method of discriminating between different type of mid-ocean ridge basalts and continental tholeiites with the Nb-Zr-Y diagram. *Chem. Geol.* **1998**, *56*, 207–218. [CrossRef]
66. Pearce, J.A.; Norry, M.J. Petrogenetic Implications of Ti, Zr, Y, and Nb Variations in Volcanic Rocks. *Contrib. Miner. Petrol.* **1979**, *69*, 33–47. [CrossRef]
67. Pearce, J. Trace element characteristics of lavas from destructive plate boundaries. In *Andesites: Orogenic Andesites and Related Rocks*; John and Wiley and Sons: Hoboken, NJ, USA, 1982.
68. Wang, Y.; Zhang, C.; Xiu, S. Th/Hf-Ta/Hf Discrimination Diagram for Tectonic Environments of Basalt Formation. *Acta Petrol. Sin.* **2001**, *17*, 413–421. (In Chinese)
69. Dong, L.; Li, G.; Zhang, Z.; Huang, Y.; Cao, H. Zircon U-Pb Ages, Geochemistry, and Sr-Nd-Pb Isotopes of Diabase Dykes from the Xuesha Area, Southern Tibet. *J. Chengdu Univ. Technol. (Sci. Technol. Ed.)* **2019**, *46*, 734–745. (In Chinese)
70. Yang, G.; Li, Y.; Xiao, W.; Tong, L. OIB-type rocks within West Junggar ophiolitic melanges: Evidence for the accretion of seamounts. *Earth-Sci. Rev.* **2015**, *150*, 477–496. [CrossRef]
71. Liu, H.; Wang, Y.; Cawood, P.A.; Guo, X. Episodic slab rollback and back-arc extension in the Yunnan-Burma region: Insights from Cretaceous Nb-enriched and oceanic-island basalt-like mafic rocks. *Geol. Soc. Am. Bull.* **2017**, *129*, B31604. [CrossRef]
72. Ren, C. Geological Geochronology, Geochemistry, and Geological Significance of Early Cretaceous Magmatic Rocks in Cuomei, Tibet. Bachelor's Thesis, China University of Geosciences, Beijing, China, 2015. (In Chinese)
73. Zhu, D.; Pan, G.; Mo, X.; Liao, Z.; Jiang, X.; Wang, L.; Zhao, Z. Geochemistry and Petrogenesis of the Basalts from the Sanxiu Formation in the Central Tethyan Himalayas. *Geochimica* **2005**, 7–19. (In Chinese)
74. Ma, Y.; Yang, T.; Bian, W.; Zhang, S.; Wu, H.; Li, H. Early Cretaceous paleomagnetic and geochronologic results from the Tethyan Himalaya: Insights into the Neotethyan paleogeography and the India-Asia collision. *Sci. Rep.* **2016**, *6*, 21605. [CrossRef]
75. Davis, J.K.; Lawver, L.A.; Norton, I.O.; Gahagan, L.M. New Somali Basin magnetic anomalies and a plate model for the early Indian Ocean. *Gondwana Res.* **2016**, *34*, 16–28. [CrossRef]
76. Storey, B.C. The role of mantle plumes in continental breakup: Case histories from Gondwanaland. *Nature* **1995**, *377*, 301–308. [CrossRef]
77. Gibbons, A.D. The breakup of East Gondwana: Assimilating constraints from Cretaceous ocean basins around India into a best-fit tectonic model. *J. Geophys. Res. Solid Earth* **2013**, *118*, 808–822. [CrossRef]
78. Jiang, Q.; Jourdan, F.; Olierook, H.K. Longest continuously erupting large igneous province driven by plume-ridge interaction. *Geology* **2020**, *49*, 206–210. [CrossRef]
79. Torsvik, T.H.; Van der Voo, R.; Preeden, U.; Niocaill, C.M.; Steinberger, B.; Doubrovine, P.V.; Hinsbergen, D.; Domeier, M.; Gaina, C.; Tohver, E. Phanerozoic polar wander, palaeogeography and dynamics. *Earth-Sci. Rev.* **2012**, *114*, 325–368. [CrossRef]

minerals

MDPI

Article

The Northernmost Effects of the Neo-Tethys Oceanic Slab Subduction Under the Lhasa Terrane: Evidence from the Mazin Rhyolite Porphyry

Zhuosheng Wang [1], Nan Wang [2,*], Zhibo Liu [2] and Xudong Ma [2]

1 School of Earth Sciences and Resources, China University of Geosciences, Beijing 100083, China; 2001220106@email.cugb.edu.cn

2 Institute of Mineral Resources, Chinese Academy of Geological Sciences, Beijing 100037, China; geoleo@163.com (Z.L.); maxudong2011@126.com (X.M.)

* Correspondence: wangnan_0912@sina.com

Abstract: The India–Asia collision represents the most significant geological event in the formation of the Tibetan plateau. The subsidence of the Neo-Tethys oceanic slab and the closure of the ocean basin were precursors of the India–Asia collision. The Linzizong volcanic formations, which range in age from the late Cretaceous to early Cenozoic (70–40 Ma), are widely distributed across the Lhasa terrane and are considered products of the closure of the Neo-Tethys oceanic basin and the India–Asia collision. Here, we report a newly identified series of rhyolite porphyries, which share similar age and geochemical features with typical Linzizong volcanic formations. These porphyries are the northernmost extension of Linzizong volcanic formations discovered to date. Zircon U-Pb dating suggests that they formed between 58.8 and 56.1 Ma. These porphyries are characterized by high SiO_2 (75.04%–77.82%), total alkali (K_2O: 4.71%–5.03%), and Na_2O (2.54%–3.63%) values; relatively low Al_2O_3 (12.30%–13.62%) and MgO (0.13%–0.33%) values; and low Mg# values (15.8–25.7). They also exhibit strong enrichment in light rare earth elements ($[La/Yb]_N$ = 3.76–11.08); negative Eu anomalies (Eu/Eu* = 0.10–0.32); Rb, Ba, Th, U, and Pb enrichments; as well as Nb and Ta depletions. The samples have relatively low εNd(t) values (−6.0 to −3.8) and variable zircon εHf(t) values (−6.3 to +3.6). These features suggest they originated from the remelting of the juvenile lower crust of the North Lhasa terrane under high-temperature and extensional conditions. We propose that the Mazin rhyolite porphyries resulted from mantle-derived magma diapirism, triggering juvenile lower crust remelting during Neo-Tethys oceanic slab rollback at the onset of the India–Asia collision. These findings provide new insights into the magmatic processes associated with early collisional tectonics.

Keywords: rhyolite porphyry; North Lhasa terrane; slab rollback; India–Asia initial collision; Linzizong group

1. Introduction

The evolution of the Neo-Tethys Ocean (NTO) and the India–Asia collision represent the most significant geological processes shaping the Tibetan plateau. The Linzizong volcanic formations are widely distributed across the Lhasa terrane, with activity spanning from the late Cretaceous to the early Cenozoic (70–40 Ma). The Linzizong volcanic formations are characterized by multiple phases and sources, and they provide valuable insights into the complex processes associated with the India–Asia collision [1–3]. Previous studies have shown that the Linzizong volcanic formations initially exhibited features of continental margin arc igneous rocks; over time, these characteristics transitioned to those of intracontinental volcanic rocks in the middle stage and potassium-rich basaltic rocks in the late stage. Consequently, the Linzizong volcanic formations are deemed products of the NTO closure and the subsequent India–Asia collision [4,5]. The volcanic formations are mainly distributed along the southern part of the Shiquan River ophiolitic belt in the

Lhasa terrane, including the Linzhou Basin and Junuo, Cuoqin, and Geji-Shiquanhe areas [4,6–20]. The volcanic cycles in the Linzhizong formations exhibit distinct evolutionary characteristics, transitioning from early calc-alkaline continental arc rocks to late-stage high-potassium calc-alkaline and potassium basalt series intraplate volcanic rocks. This progression suggests that the source region and magmatic processes involved are highly complex [11,15,16,20–29].

Recently, we identified a series of rhyolite porphyries in the Baingoin area of the North Lhasa terrane, which shares similar ages and geochemical characteristics with typical Linzizong volcanic formations. This discovery is the northernmost extent of Linzizong volcanic formations identified to date. The emergence of these rocks is crucial for understanding the far-reaching effects of the Neo-Tethys oceanic slab subduction on the deep lithosphere beyond the Lhasa terrane. These rhyolite porphyries serve as valuable natural samples for investigating magma formation processes at the continental margin during the syn-collision period. Additionally, they provide key insights into the activity and dynamic behavior of the oceanic slab during the initial stages of continental collision.

In this study, we present data on new zircon U-Pb ages, Hf isotope, whole-rock major, trace element, and Sr-Nd isotope for these rhyolite porphyries. These data enable us to understand the petrogenesis of the porphyries and provide further insights into the key geodynamic processes shaping the continental lithosphere in response to oceanic closure and continental collision.

2. Geological Background

The Tibetan plateau consists of multiple terranes, namely the Tethys Himalaya, Lhasa, and the Qiangtang, separated by the Yarlung Zangbo, Bangong-Nujiang, Longmucuo-Shuanghu, and Jinshajiang suture zones [30,31]. It is bounded by the Luobadui–Milashan Fault zone and the Shiquan River-Nam Co ophiolite mélange zone, and the Lhasa terrane is further divided into the South Lhasa, the Central Lhasa, and the North Lhasa terranes [32]. The Linzizong volcanic formations, predominantly found in the central and south Lhasa terranes, extend from the Shiquan River in western to eastern Lhasa. This distribution covers approximately 1200 km east to west and 100 km north to south (Figure 1b [21]).

The Linzizong volcanic formations are classified into three stratigraphic units: the Dianzhong Formation, the Nianbo Formation, and the Pana Formation. Comprehensive chronological studies on the Linzizong Group have yielded extensive isotopic age data. Zhou et al. [33] utilized the ^{40}Ar-^{39}Ar dating method to establish a precise chronological framework for the Linzizong volcanic formations in the Linzhou Basin, dating them from 64.4 to 43.9 Ma. This study highlighted an unconformity between the Dianzhong Formation volcanic rocks and the underlying strata, occurring at approximately 65 Ma [6,21,33]. The ages of the Dianzhong Formation vary across the Lhasa terrane. In the eastern section, they range from 70.9 to 58.3 Ma [9–11,13,34–36]. In the central section, they exhibit ages between 70.7 and 57.4 Ma [29,37–39], while in the western section, they are dated from 70.8 to 64.2 Ma [19].

The newly identified Mazin rhyolite porphyry is located southeast of the Baingoin batholith in the central region of the North Lhasa terrane (Figure 1c) and distributed in an east–west direction. The outcrops exhibit eruptive unconformities, overlying the Lagongtang Formation, Langshan Formation, Jingzhushan Formation, and Baingoin batholith. Widespread wallrock breccias are observed within the Mazin rhyolite porphyry (Figure 2c,d). The rocks exhibit a gray-white coloration and display porphyritic, aphanitic, and massive structures. Phenocrysts predominantly range from 1 to 3 mm in size, with the composition dominated by quartz (15%–20%), potassium feldspar (10%–15%), plagioclase (10%–15%), and biotite (1%–2%). The matrix is mainly composed of felsic components such as potassium feldspar (15%–20%), plagioclase (10%–15%), and quartz (10%–15%).

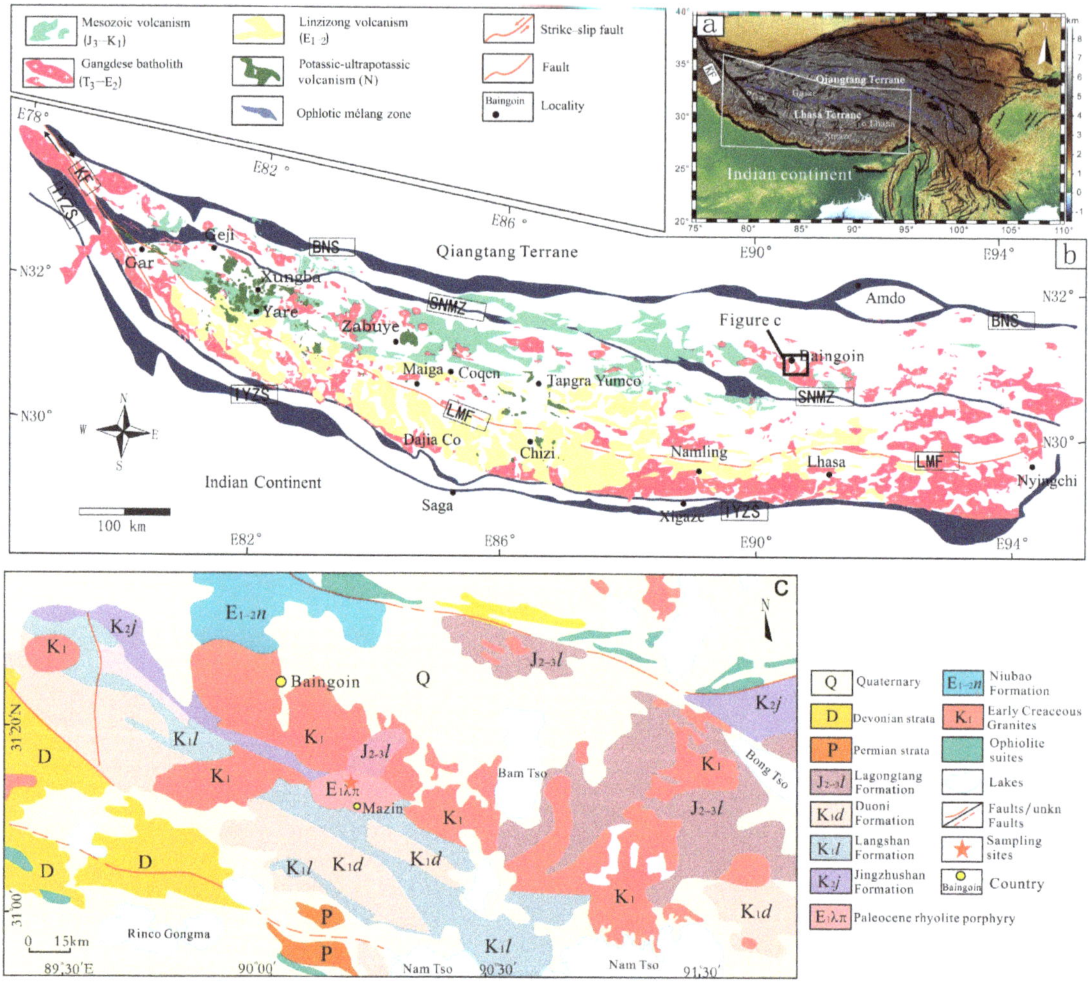

Figure 1. (**a**) Topographic map of Tibetan Plateau. (**b**) Geological map illustrating spatial distribution of Mesozoic-Cenozoic magmatism (modified from [40]). (**c**) Geological map of study area, Baingoin County, northern Lhasa Terrane (modified from [41]). LMF = Luobadui–Milashan Fault; KF = Karakoram Fault; ATF = Altyn Tagh Fault; IYZS = Indus-Yarlung Zangbo suture zone; SNMZ = Shiquan River–Nam Tso mélange zone; JSS = Jinsha suture zone; BNS = Bangong–Nujiang suture zone.

Figure 2. (**a**) Long view of Mazin rhyolite porphyry. (**b**) Photograph of Mazin rhyolite porphyry outcrop. (**c**) Close-up image of Mazin rhyolite porphyry. (**d**) Breccias from wall rock associated with Mazin rhyolite porphyry.

3. Analytical Methods

See Appendix A for the analytical methods.

4. Results

4.1. Zircon U-Pb Ages and Lu-Hf Isotopic Compositions

Two representative samples, BGA38-3 and BGA40-3, were chosen for zircon U-Pb dating and Lu-Hf isotopic analysis, and the results are shown in tables in Appendix B. The cathodoluminescence images reveal a well-preserved morphology of zircons, primarily in the form of short or long columnar shapes with particle sizes ranging from 50 to 250 μm and aspect ratios between 1 and 3. Most zircons exhibit distinct oscillatory zoning patterns, while a few display inherited characteristics.

A zircon analysis of sample BGA38-3 indicated that most zircon grains exhibit elevated and variable concentrations of uranium (257–2176 ppm) and thorium (140–885 ppm), with Th/U ratios ranging from 0.40 to 1.10, indicating a magmatic origin [42]. The $^{206}Pb/^{238}U$ ages range from 79 to 56 Ma, with a weighted mean age of 58.8 ± 0.8 Ma (MSWD = 0.75, n = 19) (Figure 3a). The $^{176}Lu/^{177}Hf$ ratios range from 0.000665 to 0.003827, $^{176}Hf/^{177}Hf$ values range from 0.282742 to 0.282862, and $^{176}Yb/^{177}Hf$ values range from 0.019291 to 0.120023. These data yielded two-stage Hf model ages (T_{DM2}) ranging from 855 to 1122 Ma, with corresponding $\varepsilon Hf_{(t)}$ values ranging from +0.1 to +3.6 based on the in situ age.

Twenty-four zircons from sample BGA40-3 exhibited relatively high thorium (34 to 3495 ppm) and uranium (248 to 6852 ppm) concentrations. The Th/U ratios ranged from 0.06 to 0.73, indicating that most zircons possess characteristics typical of a magmatic origin [42]. The $^{206}Pb/^{238}U$ ages range from 54.0 to 116.0 Ma. A weighted average age was calculated for zircons with higher concordance and more clustered ages, resulting in a $^{206}Pb/^{238}U$ weighted mean age of 56.1 ± 0.8 Ma (MSWD = 1.19, n = 22) (Figure 3b). The $^{176}Lu/^{177}Hf$ ratios ranged from 0.000340 to 0.00408, $^{176}Hf/^{177}Hf$ values ranged from

0.282555 to 0.282804, and $^{176}Yb/^{177}Hf$ values ranged from 0.010517 to 0.122678. These values yielded two-stage Hf model ages (T_{DM2}) in the range of 980–1535 Ma, with corresponding $\varepsilon Hf_{(t)}$ values ranging from −6.3 to +2.2 (Figure 4). The obtained zircon with a $^{206}Pb/^{238}U$ age of 116.0 Ma and an enrichment εHf(t) value of −6.2 is consistent with early Cretaceous magmatic rocks in the North Lhasa terrane (Figure 4), suggesting that it is an inherited zircon.

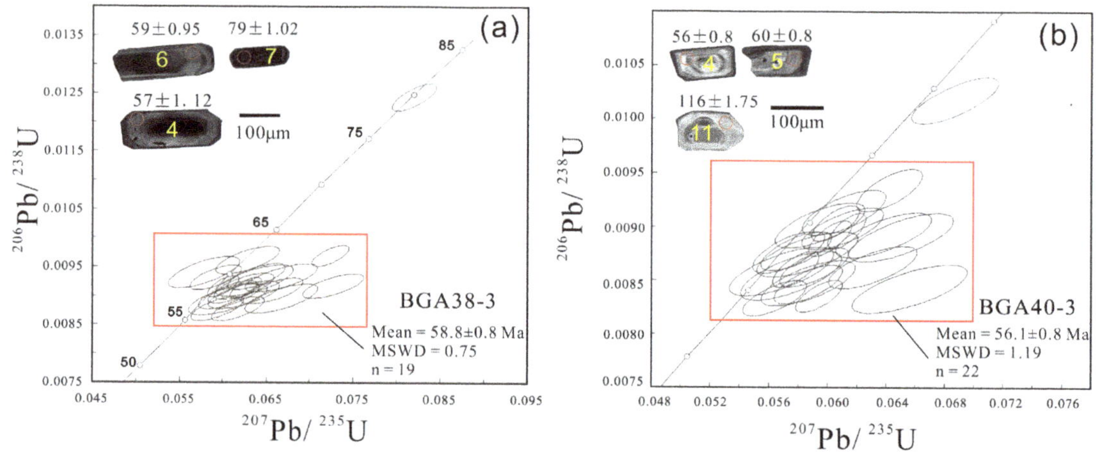

Figure 3. Zircon U-Pb isotope concordia diagrams and representative cathodoluminescence images of zircon from Mazin rhyolite porphyry, Baingoin County, northern Lhasa Terrane. (**a**) BGA38-3; (**b**) BGA40-3; 'n' denotes number of samples; 'MSWD' represents mean square of weighted deviations.

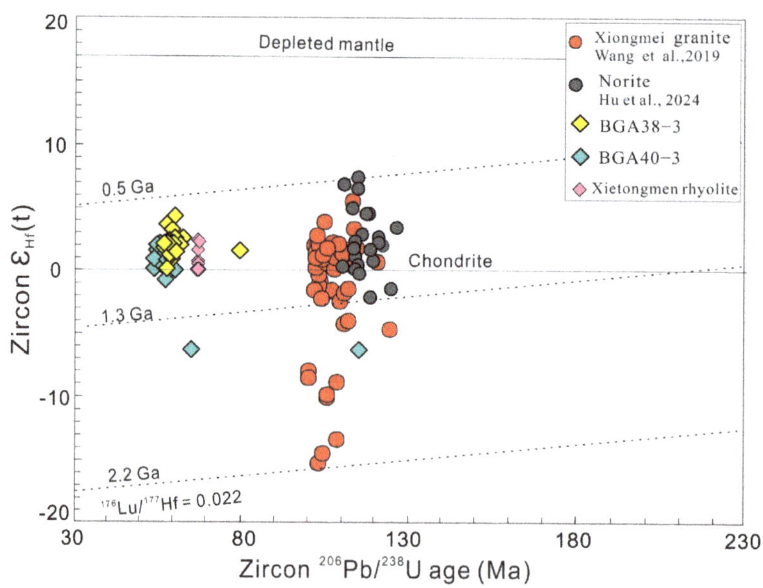

Figure 4. Diagram of zircon εHf(t) versus zircon $^{206}Pb/^{238}U$ age (Ma). Zircon εHf(t) data for Xiongmei granite were compiled from [43]; for Norite, data were compiled from [44]; and for Xietongmen rhyolite, data were compiled from [29].

4.2. Whole-Rock Major, Trace Element, and Sr-Nd Isotopic Compositions

The major and trace element analyses were conducted on seven fresh rhyolite porphyry samples, while the Rb-Sr and Sm-Nd isotope analyses were performed on three samples. The results are presented in Table 1.

All samples are plotted in the rhyolite area in Figure 5a,b. The SiO_2 contents ranged from 75.04% to 77.82%, $Fe_2O_3^T$ from 1.4% to 2.0%, FeO from 0.40% to 1.00%, and MgO from 0.13% to 0.33%, corresponding to Mg# values of 15.8–25.7. The samples showed high K_2O contents (4.71%–5.03%) and Na_2O contents (2.54%–3.63%), placing them within the high-K calc-alkaline series (Figure 5d). The Al_2O_3 contents ranged from 12.30% to 13.62%, while the CaO contents were between 0.7% and 1.08%. The aluminum saturation index (A/CNK) values ranged from 1.03 to 1.22, classifying all samples as peraluminous (Figure 5c). Overall, the Mazin rhyolite porphyry was characterized by high silica and alkali contents, along with relatively low alumina and magnesium levels.

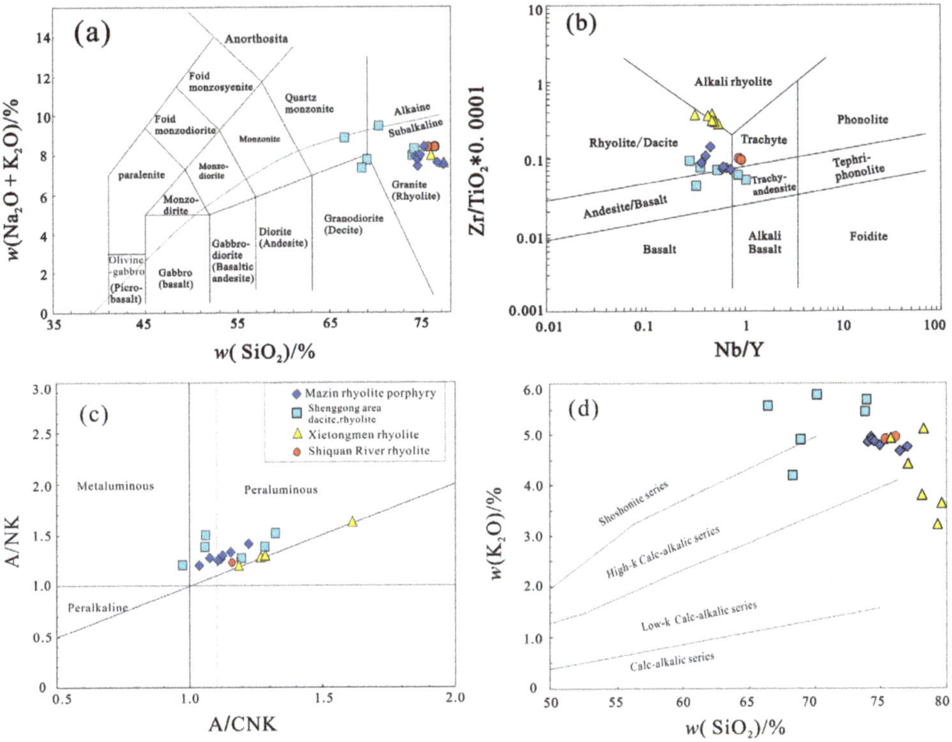

Figure 5. (**a**) Total alkalis versus SiO_2 diagram. (**b**) Diagram of $Zr/TiO_2 \times 0.0001$ versus Nb/Y. (**c**) A/NK versus A/CNK [where A/NK = molar $Al_2O_3/(Na_2O + K_2O)$; A/CNK = molar $Al_2O_3/(CaO + Na_2O + K_2O)$] [45]; (**d**) K_2O versus SiO_2 [46].

The total amount of rare earth elements (∑REE) was relatively high, ranging from 164.22 ppm to 203.29 ppm, with an average value of 181.70 ppm. The chondrite-normalized REE pattern showed a right-dipping curve with fractionated REE (Figure 6a), with $(La/Yb)_N$ ratios ranging from 3.76 to 11.08. Negative Eu anomalies were prominent, and the δEu values ranged from 0.10 to 0.32. The primitive mantle-normalized spider diagram (Figure 6b) shows that the Mazin rhyolite porphyry is characterized by relative depletion in high field strength elements (HFSEs), such as Nb, Ta, P, and Ti; significant enrichment in large ion lithophile elements (LILEs), such as Rb and K as well as Th and U, with distinct negative anomalies of Ba and Sr and positive anomalies of Zr and Hf.

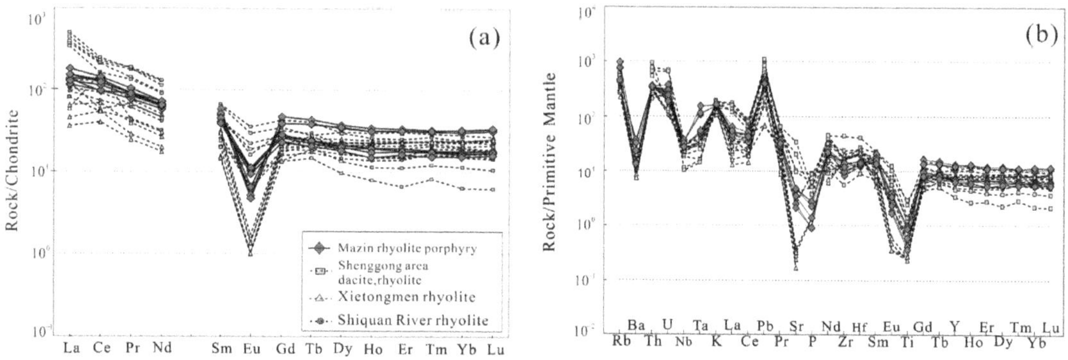

Figure 6. (**a**) Chondrite-normalized rare earth element spidergram and (**b**) primitive mantle-normalized trace element spidergram for Mazin rhyolite porphyry, Baingoin County, North Lhasa Terrane. Normalized values were sourced from [47].

The $^{87}Sr/^{86}Sr$ ratio of the Mazin rhyolite porphyry ranged from 0.7135 to 0.7264, whereas the $^{143}Nd/^{144}Nd$ ratio varied between 0.512315 and 0.512432. The calculated initial $^{87}Sr/^{86}Sr$ ratio ranged from 0.6962 to 0.7069, and the εNd(t) value varied between −6.0 and −3.8. The two-stage Nd isotope model age varied from 1167 to 1348 Ma.

Table 1. Whole-rock major element (%) and trace element (ppm), Rb-Sr and Sm-Nd isotope dates.

Sample No.	17BGA38-3	17BGA38-5	17BGA39-2	17BGA40-3	17BGA41-1	17BGA41-2	17BGA42-1
SiO_2	77.20	76.60	74.41	75.12	74.19	74.46	74.69
TiO_2	0.09	0.08	0.21	0.13	0.21	0.21	0.19
Al_2O_3	12.20	12.30	13.36	13.32	13.35	13.35	13.30
FeO	0.65	0.64	0.40	1.00	0.88	0.88	0.92
$Fe_2O_3{}^T$	1.4	1.5	1.6	1.7	2.0	2.0	2.0
MnO	0.06	0.06	0.07	0.08	0.04	0.04	0.03
MgO	0.13	0.19	0.27	0.20	0.32	0.33	0.30
CaO	0.69	0.69	0.81	0.96	0.88	0.82	1.07
Na_2O	2.76	2.93	2.49	3.62	3.07	2.89	3.14
K_2O	4.74	4.66	4.94	4.78	4.84	4.90	4.86
P_2O_5	0.02	0.02	0.06	0.03	0.06	0.06	0.05
LOI	0.76	0.71	1.23	0.48	0.83	1.04	0.59
Mg#	15.72	20.38	25.65	19.08	24.06	24.63	23.17
A/CNK	1.12	1.11	1.22	1.03	1.12	1.16	1.07
Ga	17.3	18.4	16.9	19.9	17.5	17.4	17.25
Sr	51.90	44.80	94.30	73.00	100.00	97.10	99.00
Hf	3.80	4.20	4.70	4.10	4.70	4.70	4.30
Zr	95.00	113.00	151.00	112.00	155.00	159.00	136.00
Nb	24.30	25.70	19.40	22.50	19.40	19.30	19.30
Ta	4.40	4.60	2.10	6.30	2.00	2.00	2.20
Ba	98.70	93.70	253.00	159.50	250.00	258.00	224.00
Pb	30.60	32.70	27.50	30.00	25.60	28.30	27.60
Th	28.90	30.00	29.50	28.30	29.20	29.00	29.20
U	6.51	4.65	7.70	6.09	5.38	5.13	3.95
Cr	9.00	14.00	8.00	9.00	8.00	8.00	5.00
Ni	0.70	0.90	1.70	1.70	1.60	1.60	1.80
La	32.00	27.60	33.10	30.90	36.80	43.70	36.50
Ce	59.80	60.70	82.90	70.50	82.00	88.00	76.20
Pr	8.82	7.48	8.43	8.07	8.81	9.97	8.67
Nd	34.20	27.90	29.90	30.10	31.40	34.10	30.60
Sm	9.13	7.66	6.40	7.94	6.30	7.08	6.10
Eu	0.31	0.28	0.60	0.39	0.63	0.66	0.56
Gd	9.78	8.18	5.60	8.24	5.40	5.89	4.85
Tb	1.65	1.44	0.94	1.47	0.86	0.91	0.78
Dy	9.70	8.91	5.42	8.99	5.01	5.25	4.71
Ho	1.99	1.83	1.13	1.83	1.11	1.00	0.86
Er	5.68	5.36	3.28	5.30	3.07	3.00	2.74
Tm	0.84	0.81	0.47	0.82	0.46	0.45	0.40

Table 1. *Cont.*

Sample No.	17BGA38-3	17BGA38-5	17BGA39-2	17BGA40-3	17BGA41-1	17BGA41-2	17BGA42-1
Yb	5.61	5.26	3.07	5.62	2.95	2.83	2.68
Lu	0.85	0.81	0.47	0.87	0.44	0.45	0.41
Y	60.10	56.60	32.40	60.60	29.30	31.80	27.00
∑REE	180.36	164.22	181.71	181.04	185.24	203.29	176.06
$(La/Yb)_N$	4.09	3.76	7.73	3.94	8.95	11.08	9.77
δEu	0.10	0.11	0.30	0.15	0.32	0.30	0.30
Tzr(°C)	755	768	800	758	794	799	778
$^{87}Rb/^{86}Sr$	31.3079			24.1950			8.3082
$^{87}Sr/^{86}Sr$	0.726408			0.715432			0.713521
$(^{87}Sr/^{86}Sr)i$	0.700168			0.696184			0.706912
$^{147}Sm/^{144}Nd$	0.1661			0.1596			0.1206
$^{143}Nd/^{144}Nd$	0.512432			0.512315			0.512407
fSm/Nd	−0.16			−0.19			−0.39
$(^{143}Nd/^{144}Nd)i$	0.51237			0.51226			0.51236
$\varepsilon_{Nd}(t)$	−3.8			−6.0			−4.0
T_{DM2}	1167			1348			1184

Note: Major elements (%), trace elements, and rare earth elements (ppm), LOI is loss on ignition; Mg# denotes value of $Mg/(Mg+Fe^{2+})$; A/NK = $Al_2O_3/(Na_2O + K_2O)$ (molar ratio), A/CNK = $Al_2O_3/(CaO + Na_2O + K_2O)$ (molar ratio), Eu = $2 \times Eu_N/(Sm_N + Gd_N)$, the subscript N represents the value normalized by chondrites, and the normalized value is in accordance with [47].

5. Discussion

5.1. Relationships Between the Mazin Rhyolite Porphyry and the Linzizong Group

The Mazin rhyolite porphyry is situated in the central region of the North Lhasa terrane, primarily influenced by the geological processes associated with the evolution of the Bangong–Nujiang Meso-Tethys Ocean (BNO) and potentially the NTO. Previous research suggests that the BNO closed during the late early Cretaceous period (130–110 Ma), followed by delamination after the collision of the Qiangtang and Lhasa terranes in the late Cretaceous period (90–80 Ma). These phases were characterized by intense magmatic activity, after which no significant events occurred [48]. The northward low-angle subduction of the Neo-Tethys oceanic lithosphere may have affected both the Central Lhasa and North Lhasa terranes [7,49–52]. The zircon U-Pb ages of the Mazin rhyolite porphyry (59.8 and 56.1 Ma) align with the formation ages of the Linzizong Dianzhong Formation (70.9–57.4 Ma) [7–11,13,34–36,53]. The Linzizong Group is considered to represent the transitional phase from subduction to the closure of the NTO [15]. Consequently, we deduce that the Mazin rhyolite porphyry shares similarities with the Dianzhong Formation of the Linzizong Group, which is influenced by the evolution of the NTO.

5.2. Magma Source and Melting Condition

The Mazin rhyolite porphyry exhibits geochemical characteristics typical of high-silica, peraluminous composition, and a high-K calc-alkaline series. It is relatively rich in Rb, Th, and U, while it is depleted in Ti, P, Ba, and Sr. Additionally, it shows significant fractionation of REE. These characteristics are similar to those found in continental margin arc volcanic formations within the Dianzhong Formation across different regions [29].

Granitic rocks are commonly divided into three types according to the nature of their protolith and their petrographic and geochemical features: I-type, S-type [54], and A-type granites [55]. However, distinguishing between these types can be challenging, especially for highly fractionated granites. S-type granites are typically strongly peraluminous and contain abundant Al-rich minerals, such as muscovite, garnet, and cordierite [54,56–58]. The Mazin rhyolite porphyry, with a relatively low Al_2O_3 content and an absence of Al-rich minerals, is not classified as S-type. The P_2O_5 vs. SiO_2 contents of the Mazin rhyolite porphyry suggest an affinity with I-type granites (Figure 7a). Most samples fall within the I-type and S-type fields in the plots (Figure 7b,c) and in the fractionated granite field in the (FeO^T/MgO) vs. (Zr+Nb+Ce+Y) diagram (Figure 7d). The petrogenesis of calc-alkaline I-type magmas has been summarized as follows: (1) partial melting of the mafic or intermediate meta-igneous lower crust, with or without interaction with mantle-derived mafic magmas [59–61], (2) assimilation and fractional crystallization (AFC) of mantle-derived basalt or basaltic andesite in shallow crustal magma chambers [62], and (3) reworking of sedimentary materials with variable contributions from mantle-derived magmas [58,63,64].

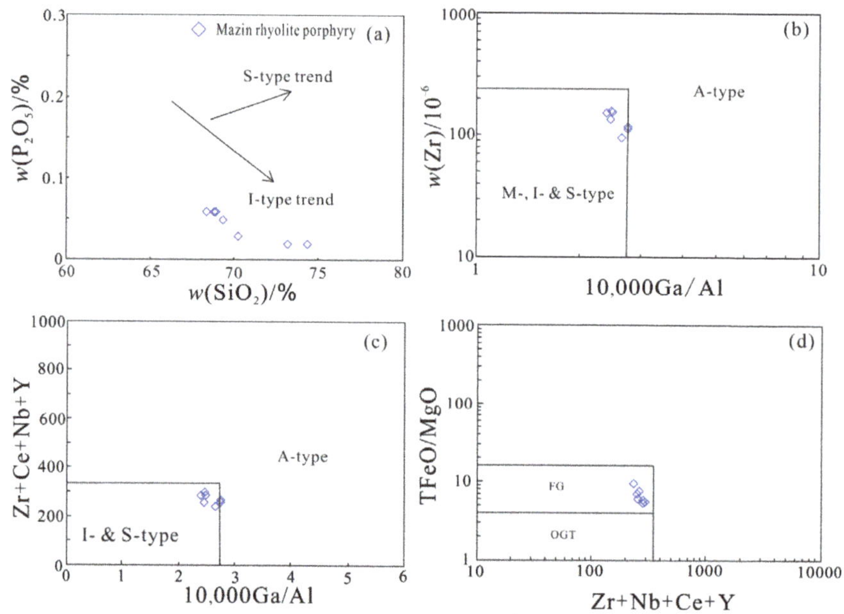

Figure 7. (**a**) $w(SiO_2)/\%$ versus $w(P_2O_5)/\%$; (**b**) 10000Ga/Al versus $w(Zr)/10^{-6}$; (**c**) Zr+Ce+Nb+Y versus10000 Ga/Al; (**d**) TFeO/MgO versus Zr+Nb+Ce+Y (FG = high-grade metamorphic granite, OGT = undifferentiated I-type, S-type, and M-type granites) [65,66].

The wide range of $\varepsilon Hf_{(t)}$ values excludes the AFC model of mantle-derived basalt or basaltic andesite in shallow crustal magma chambers [62]. The positive zircon $\varepsilon Hf_{(t)}$ values, up to +4.3, and the slightly peraluminous nature of these rocks imply partial melting of the mafic or intermediate meta-igneous lower crust. The $\varepsilon Nd_{(t)}$ values varied between −6.0 and −3.8, showing obvious Hf-Nd isotope decoupling. This phenomenon occurs commonly in early Cretaceous igneous rocks from the North Lhasa terrane, which are believed to have originated from the juvenile crust and/or metasomatic mantle wedge during BNO/NTO subduction [43,44]. The Mazin rhyolite porphyry, located within the North Lhasa terrane, shares similarities with typical early Cretaceous igneous rocks within the region, particularly in terms of εHf(t) values and U-Pb ages (Figure 4). Based on these observations, we propose that the juvenile lower crust of the North Lhasa terrane likely served as the source for the Mazin rhyolite porphyry.

During the partial melting of basaltic rocks, the residual phase is influenced by the pressure conditions during melting. Plagioclase is consistently retained in the residual phase under relatively low-pressure conditions, while garnet is retained under relatively high-pressure conditions [46,67]. Additionally, plagioclase preferentially incorporates Sr and Eu, whereas garnet preferentially incorporates heavy rare earth elements (HREEs) [68]. Therefore, the high content of HREEs, negative anomalies in Sr and Eu, and low Sr/Y and La/Yb ratios are commonly employed for estimating the melting pressure [69,70]. The high HREE content, negative Sr and Eu anomalies (Figure 6), and low Sr/Y and La/Yb ratios (Figure 8a) suggest that the source of these felsic rocks likely consisted of plagioclase without garnet, indicating low-pressure melting. This is further supported by a discrimination diagram based on major element compositions (Figure 8b). The significant negative Eu anomaly observed in the Mazin rhyolite porphyry is consistent with its geochemical characteristics. Additionally, the whole-rock zirconium saturation temperature of this rock, calculated to range from 755 to 800 °C, surpasses that of typical rhyolite magma. These findings collectively suggest that the formation of the Mazin rhyolite porphyry occurred within a relatively high-temperature and low-pressure environment.

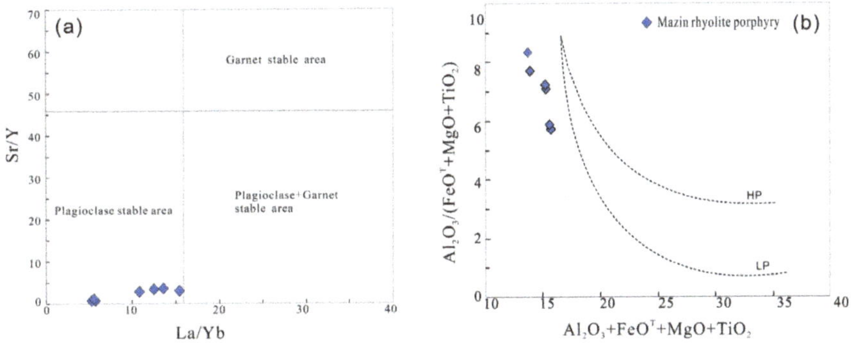

Figure 8. (**a**) Sr\Y versus La/Yb. (**b**) $Al_2O_3\backslash(FeO^T + MgO + TiO_2)$ versus $Al_2O_3 + FeO^T + TiO_2$ (HP means high pressure; LP means low pressure) [71].

5.3. A Formation Model for the Mazin Rhyolite Porphyry

Since the late early Jurassic, the NTO has undergone a series of geological processes, including slab subduction, rollback, and break-off, followed by continent–continent collision. Between 70 and 40 Ma, the oceanic slab of Neo-Tethys transitioned from flat to steep subduction before experiencing rollback or break-off events. During these processes, Linzizhong volcanic rocks were formed [10,25,30,72–76]. Previous studies have indicated that the steep subduction of the Neo-Tethys oceanic lowered the initial melting temperature of the mantle wedge, triggering the melting process in the overlying metasomatic mantle wedge to generate basic magma. This magma then accumulated at the base of the continental crust [25,74]. As subduction deepened, the rollback of the oceanic slab disrupted lithosphere–asthenosphere mantle convection, leading to partial melting of the lower crust and the formation of a small amount of acidic magma mixed with materials derived from the mantle. Subsequently, the acidic magma underwent mixing with the ascending basic magma. This process has been utilized to elucidate the genesis of volcanic rocks within the Dianzhong Formation of the Linzizong Group in various sectors (eastern, western, and central) of the Lhasa terrane [77–79]. The lowest age of the Dianzhong group is approximately 70 Ma [10,11,34,36,39], indicating the onset of steep subduction of the NTO, whereas the uppermost ages in the Dianzhong Formation are predominantly concentrated around 60 Ma, implying that the initial collision between the Indian and Asian continents occurred post−60 Ma [11,39].

The Mazin rhyolite porphyry is located in the North Lhasa terrane, which currently represents the northernmost boundary of the Linzizong Group. It was formed between 58.8 and 56.1 Ma, slightly later than the Dianzhong Formation observed in other regions of the Lhasa terrane. Elements, such as Rb, Y, Zr, Nb, are commonly used to discriminate the tectonic setting of granite formation [80,81].

In terms of the tectonic setting discrimination diagram (Figure 9a,b,d), the Mazin rhyolite porphyry predominantly occurs within a collision-related environment. We also used the R1-R2 tectonic setting discrimination diagram based on major element compositions to analyze the tectonic setting of the Mazin rhyolite porphyry [82]. In Figure 9c, all the samples are plotted in the syn-collision area.

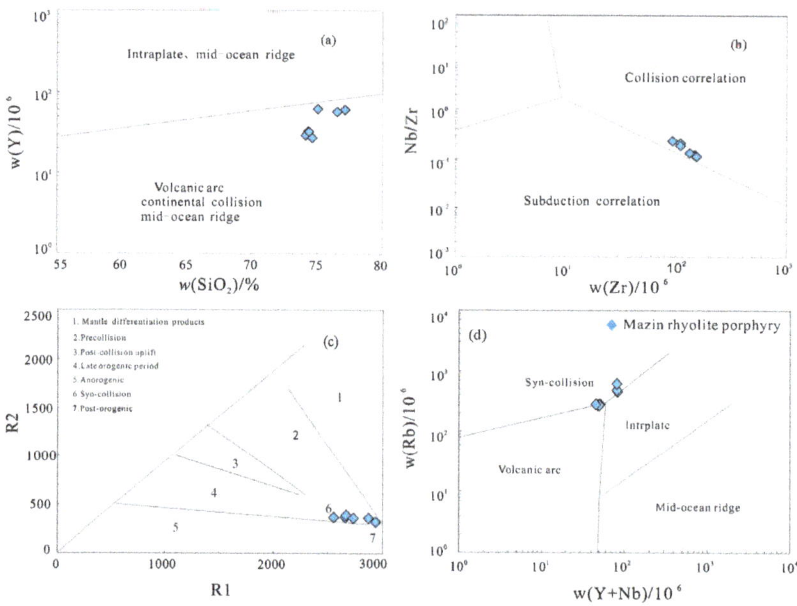

Figure 9. (**a**) Y $\times$ 10^{-6} versus SiO_2; (**b**) Nb/Zr versus Zr $\times$ 10^{-6}; (**c**) R2 versus R1; (**d**) Rb $\times$ 10^{-6} versus (Y + Nb) $\times$ 10^{-6} (modified based on [80–82]).

As discussed in Section 4.2, the Mazin rhyolite porphyry is derived from the melting of the juvenile lower crust of the North Lhasa terrane in a relatively low-pressure, high-temperature environment. Additionally, it is worth noting that there were no significant volcanic or plutonic activities associated with basic magma during this period in the North Lhasa terrane, indicating a scarcity of large-scale basic magma in the region. Therefore, a large-scale melting event is unlikely. Based on these results and observations, we propose a new formation model for the Mazin rhyolite porphyry. Around 60 Ma, the initial collision between India and Asia commenced, leading to the rollback of the slab-subducted Neo-Tethys oceanic plate following earlier NBTO slab steep subduction. During this rollback, a few asthenospheres upwelled due to complementation, triggering diapirism in the overlying lithosphere. This process facilitated heat transfer to the lower crust of the North Lhasa terrane, resulting in partial melting that produced Mazin rhyolite porphyry (Figure 10).

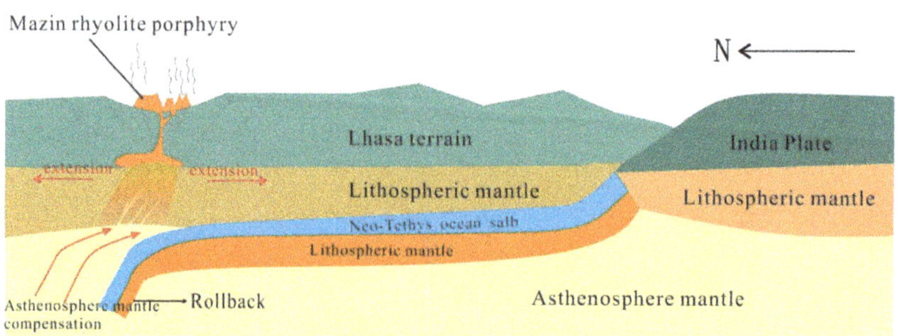

Figure 10. Schematic illustrations depicting the remote effects of the Paleogene geodynamic evolution processes associated with the Neo-Tethys oceanic tectonic regime in the northern Lhasa Terrane.

6. Conclusions

The Mazin rhyolite porphyry, newly identified in the North Lhasa terrane, has zircon U-Pb dating results of 58.8 ± 0.8 Ma and 56.1 ± 0.8 Ma. It belongs to the Dianzhong formation in the Linzizong group and represents the northernmost distribution of the Linzizong group, indicating that the subduction effects of the NTO influenced the North Lhasa terrane during the Paleocene. The Mazin rhyolite porphyry is derived from the melting of the juvenile lower crust of the North Lhasa terrane in a relatively low-pressure, high-temperature environment during slab rollback. These findings provide new insights into the magmatic processes associated with early collisional tectonics.

Author Contributions: Writing—original draft preparation, Z.W.; project administration, Z.L. and X.M.; funding acquisition, N.W. All authors have read and agreed to the published version of the manuscript.

Funding: This research was funded by the Special Funds of Basic Scientific Research of the Chinese Academy of Geological Sciences (grant number JKYQN202331 and KK2321), and the China Geological Survey (grant number DD20230358 and DD20230286).

Data Availability Statement: The original contributions presented in the study are included in the article, further inquiries can be directed to the corresponding author/s.

Conflicts of Interest: The authors declare no conflict of interest.

Appendix A. Analytical Methods

Appendix A.1. LA-ICP-MS Zircon U-Pb Dating

The sorting, target preparation, and cathodoluminescence (CL) imaging of the zircon samples were conducted in the laboratory of Beijing Zircon Linghang Technology Co., Ltd. (Beijing, China). The samples were first crushed to a size of 80–120 mesh followed by dust washing, magnetic mineral removal, and heavy liquid sorting. A number of zircon grains were randomly selected under a binocular stereo microscope, affixed to double-sided adhesive tape, and then subjected to a low-temperature environment. Then, the zircon surfaces were polished using an automatic polishing machine, and the target surface was cleaned sequentially via ultrasonication with alcohol, detergent, and water. After target preparation, transmitted and reflected light images were taken under a polarizing microscope using the widest field of view. The zircon targets were then gold-plated, and CL images were taken with a JEOL JSM6510 electron microscope equipped with a Gatan cathodic fluorescence probe from the United Kingdom.

The tests and analyses were completed at the Institute of Geology, Chinese Academy of Geological Sciences. Zircon U-Pb dating was performed using the latest-generation Neptune Plus multireceiver plasma mass spectrometer (MC-ICP-MS) from Thermo Fisher Company, Waltham, MA, USA, along with the GeoLasPro 193 nm laser ablation system (Coherent, Saxonburg, PA, USA). The laser spot used for ablation had a diameter of 32 μm, a frequency of 10 Hz, and an energy density of approximately 2.5 J/cm^2. Helium was used as the carrier gas. Reflection and transmission photographs of the zircon were taken to avoid cracks and inclusions. The experimental standard sample was 91500, and the error of the weighted mean of the $^{206}Pb/^{238}Pb$ ages was $\pm 1\sigma$. Data processing was performed via the ICPMS DataCal program [83]. During measurements, most analysis points had a $^{206}Pb/^{204}Pb$ ratio greater than 1000, and no common lead correction was applied. ^{204}Pb was detected using an ion counter. Analysis points with abnormally high ^{204}Pb contents, potentially affected by common Pb (e.g., due to inclusions), were excluded from the calculations. The zircon age harmony map was generated via the Isoplot program.

Appendix A.2. Major and Trace Element Analysis

The rock powder crushing and complete chemical analyses of the fresh samples selected for this study were conducted at ALS Analytical and Testing Co., Ltd. (Guangzhou, China). Major element oxides were analyzed using X-ray fluorescence spectrometry. The sample was mixed with fluxes including lithium nitrate, thoroughly blended, and then melted at a high temperature. The melt was poured into a platinum mold to form a flat glass slide, which was then analyzed via an X-ray fluorescence spectrometer. Simultaneously, another sample was weighed, placed in a muffle furnace, and heated at 1000 °C for 1 h. The sample was weighed again after cooling. The weight difference between the pre-heated and post-heated samples represented ignition loss. For the analysis of FeO, the sample was digested with sulfuric acid and hydrofluoric acids and then poured into a beaker containing dilute sulfuric acid, phosphoric acid, or boric acid solution and titrated with a potassium dichromate solution. Two samples for rare earth elements and trace elements were weighed. One sample was digested with perchloric acid, nitric acid, or hydrofluoric acid, evaporated to near dryness, dissolved in dilute hydrochloric acid, and analyzed using plasma optical emission spectrometry and plasma mass spectrometry. The second sample was mixed with a lithium metaborate/lithium tetraborate mixture, melted in a furnace above 1025 °C, cooled, and then diluted with nitric acid, hydrochloric acid, and hydrofluoric acid before being analyzed via a plasma mass spectrometer.

Appendix A.3. Whole-Rock Sr-Nd Isotopic Compositions

The whole-rock Sr-Nd isotope test was performed via MCICP-MS at Beijing Kehui Testing Technology Co., Ltd. (Beijing, China). The static method was used for all Sr-Nd isotopic determinations. Before the samples were tested, the parameters of Neptune Plus, including those for the plasma section (torch tube position and carrier gas flow rate) and ion lens, were optimized via the NIST 987 200 μg/L standard solution and Thermo Fisher conditioning solution to maximize sensitivity. After chemical separation, the sample was introduced into the mass spectrometer in 2% HNO_3 solution, with the signal intensities for ^{88}Sr and ^{146}Nd reaching approximately 8 V (the strontium concentration in the solution was approximately 200 μg/L, and the Nd concentration was 200 μg/L). Sample injection was performed using a nebulizer. After each test, the sampling system was cleaned with 2% HNO_3 solution before measuring the next sample. The $^{87}Sr/^{86}Sr$ ratio and $^{143}Nd/^{144}Nd$ ratios were corrected using exponential normalization with reference values of $^{88}Sr/^{86}Sr$ = 8.375209 and $^{146}Nd/^{144}$ Nd = 0.7219, respectively.

Appendix A.4. Zircon Hf Isotopic Compositions Analysis

The Lu-Hf isotope determination of zircon was performed using LA-MC-ICP-MS (Beijing Kehui Test Technology Co., Ltd.). The analysis was performed on a laser ablation multireceiver inductively coupled plasma mass spectrometer. Zircons were ablated using a Resolution SE 193 nm excimer laser ablation system with a spot diameter of 38 μm, an energy density of 7–8 J/cm^2, and a frequency of 10 Hz. High-purity helium was used as the carrier gas. The gas was directed to a Neptune Plus (MC-ICP MS), and the receiver configuration was identical to that used for solution sampling. The international standard zircon GJ-1 was used as the reference material. Mass discrimination corrections for the Hf isotope ratios were exponentially normalized using $^{179}Hf/^{177}Hf$ = 0.7325, and the Yb isotope ratio was corrected using $^{173}Yb/^{172}Yb$ = 1.35274. The εHf(t) value was calculated using chondrite Hf isotope values of $^{176}Lu/^{177}Hf$ = 0.0336 and $^{176}Hf/^{177}Hf$ = 0.282785 [84]. For calculating Hf mantle model ages, the current depleted mantle value of $^{176}Hf/^{177}Hf$ was 0.28325, the $^{176}Lu/^{177}Hf$ value was 0.0384 [85], and the average crustal $^{176}Lu/^{177}Hf$ ratio of 0.015 was used to calculate the ages of the crustal models [86].

Appendix B

Isotopic data of zircon from the Mazin rhyolite porphyry.

Test Point Number	t (Ma)	$^{176}Yb/^{177}Hf$	1σ	$^{176}Lu/^{177}Hf$	1σ	$^{176}Hf/^{177}Hf$	1σ	εHf(0)	εHf(t)	TDM (Ma)	TDM_2 (Ma)	$f_{Lu/Hf}$
17BGA40-3_1	57	0.015600	0.000019	0.000510	0.000001	0.282799	0.000189	0.9	2.2	634	992	−0.98
17BGA40-3_2	57	0.095442	0.000018	0.002928	0.000042	0.282718	0.000624	−1.9	−0.8	797	1179	−0.91
17BGA40-3_3	58	0.045959	0.000016	0.001456	0.000023	0.282768	0.001203	−0.2	1.1	695	1064	−0.96
17BGA40-3_4	56	0.034152	0.000014	0.001119	0.000006	0.282785	0.000541	0.5	1.7	664	1024	−0.97
17BGA40-3_5	60	0.033350	0.000016	0.001068	0.000006	0.282735	0.000546	−1.3	0.0	734	1135	−0.97
17BGA40-3_6	58	0.039746	0.000016	0.001307	0.000040	0.282804	0.000785	1.1	2.4	640	980	−0.96
17BGA40-3_7	55	0.048725	0.000013	0.001543	0.000036	0.282789	0.000948	0.6	1.7	666	1017	−0.95
17BGA40-3_8	57	0.040171	0.000016	0.001407	0.000022	0.282782	0.000421	0.4	1.6	673	1030	−0.96
17BGA40-3_9	55	0.024348	0.000015	0.000844	0.000003	0.282795	0.000331	0.8	2.0	646	1003	−0.97
~~17BGA40-3_11~~	~~116~~	~~0.026445~~	~~0.000015~~	~~0.000998~~	~~0.000017~~	~~0.282526~~	0.000301	−8.7	−6.2	1027	1572	~~−0.97~~
17BGA40-3_14	54	0.035717	0.000015	0.001232	0.000008	0.282783	0.000106	0.4	1.6	668	1029	−0.96
17BGA40-3_15	54	0.060200	0.000015	0.001966	0.000038	0.282744	0.001093	−1.0	0.1	739	1121	−0.94
17BGA40-3_16	56	0.021108	0.000014	0.000750	0.000003	0.282771	0.000112	−0.1	1.2	678	1056	−0.98
17BGA40-3_17	65	0.010517	0.000013	0.000340	0.000005	0.282555	0.000091	−7.7	−6.3	969	1535	−0.99
17BGA40-3_18	55	0.019575	0.000014	0.000716	0.000021	0.282765	0.000515	−0.2	0.9	685	1069	−0.98
17BGA40-3_19	55	0.018078	0.000014	0.000655	0.000002	0.282797	0.000132	0.9	2.1	639	998	−0.98
17BGA40-3_20	56	0.024329	0.000017	0.000921	0.000015	0.282746	0.000246	−0.9	0.3	715	1111	−0.97
17BGA40-3_21	54	0.036828	0.000017	0.001316	0.000013	0.282794	0.000306	0.8	1.9	654	1005	−0.96
17BGA40-3_22	54	0.039641	0.000015	0.001389	0.000028	0.282765	0.000684	−0.2	0.9	697	1071	−0.96
17BGA40-3_23	58	0.066720	0.000017	0.002219	0.000031	0.282789	0.001260	0.6	1.8	678	1016	−0.93
17BGA40-3_24	58	0.122678	0.000016	0.004008	0.000043	0.282761	0.000970	−0.4	0.7	756	1084	−0.88
17BGA38-3_1	59	0.067176	0.000018	0.002155	0.000037	0.282763	0.001022	−0.3	0.9	716	1076	−0.94
17BGA38-3_2	60	0.120023	0.000018	0.003827	0.000159	0.282862	0.004724	3.2	4.3	598	855	−0.88
17BGA38-3_3	58	0.020785	0.000015	0.000714	0.000001	0.282801	0.000163	1.0	2.3	635	987	−0.98
17BGA38-3_4	57	0.019291	0.000015	0.000665	0.000003	0.282840	0.000234	2.4	3.6	579	899	−0.98
17BGA38-3_5	59	0.034248	0.000015	0.001140	0.000001	0.282769	0.000131	−0.1	1.2	686	1058	−0.97
17BGA38-3_6	59	0.020320	0.000016	0.000696	0.000002	0.282826	0.000197	1.9	3.2	599	930	−0.98
17BGA38-3_7	79	0.041075	0.000016	0.001368	0.000007	0.282767	0.000481	−0.2	1.5	693	1051	−0.96
17BGA38-3_8	56	0.060680	0.000012	0.002002	0.000055	0.282788	0.001468	0.6	1.7	676	1020	−0.94
17BGA38-3_9	62	0.051150	0.000015	0.001619	0.000039	0.282808	0.001701	1.3	2.6	640	971	−0.95
17BGA38-3_10	58	0.034936	0.000015	0.001168	0.000004	0.282770	0.000233	−0.1	1.2	686	1057	−0.96
17BGA38-3_11	58	0.035653	0.000015	0.001195	0.000030	0.282802	0.000703	1.1	2.3	641	985	−0.96
17BGA38-3_12	60	0.022348	0.000013	0.000751	0.000001	0.282808	0.000157	1.3	2.6	625	970	−0.98
17BGA38-3_13	59	0.028763	0.000015	0.000938	0.000004	0.282794	0.000251	0.8	2.0	647	1001	−0.97
17BGA38-3_14	60	0.062997	0.000017	0.002082	0.000026	0.282781	0.001246	0.3	1.5	687	1034	−0.94

Test Point Number	t (Ma)	$^{176}Yb/^{177}Hf$	1σ	$^{176}Lu/^{177}Hf$	1σ	$^{176}Hf/^{177}Hf$	1σ	εHf(0)	εHf(t)	TDM (Ma)	TDM_2 (Ma)	$f_{Lu/Hf}$
17BGA38-3_15	62	0.030634	0.000015	0.001051	0.000004	0.282790	0.000366	0.6	1.9	656	1010	−0.97
17BGA38-3_16	59	0.033109	0.000017	0.001143	0.000042	0.282797	0.001097	0.9	2.2	647	995	−0.97
17BGA38-3_17	58	0.023389	0.000015	0.000822	0.000001	0.282795	0.000176	0.8	2.1	644	1000	−0.98
17BGA38-3_18	60	0.020474	0.000015	0.000700	0.000006	0.282775	0.000159	0.1	1.4	671	1044	−0.98
17BGA38-3_19	57	0.039767	0.000015	0.001351	0.000006	0.282742	0.000151	−1.1	0.1	730	1122	−0.96
17BGA38-3_20	56	0.033664	0.000016	0.001112	0.000011	0.282796	0.000606	0.9	2.0	648	999	−0.97

Zircon U-Pb dating data of the Mazin rhyolite porphyry.

	Isotope Ratios				Isotopic Ratios: 1 Sigma Uncertainty				Age Estimates (Ma)				Age Estimates: 1 Sigma Uncertainty (Ma)			
	$^{207}Pb/^{206}Pb$	$^{206}Pb/^{238}U$	$^{207}Pb/^{235}U$	$^{208}Pb/^{232}Th$	$^{207}Pb/^{206}Pb$	$^{206}Pb/^{238}U$	$^{207}Pb/^{235}U$	$^{208}Pb/^{232}Th$	$^{207}Pb/^{206}Pb$	$^{206}Pb/^{238}U$	$^{207}Pb/^{235}U$	$^{208}Pb/^{232}Th$	$^{207}Pb/^{206}Pb$	$^{206}Pb/^{238}U$	$^{207}Pb/^{235}U$	$^{208}Pb/^{232}Th$
17BGA40-3_1	0.0525	0.00892	0.06381	0.00309	0.00179	0.00014	0.00215	0.00009	307.1	57.2	62.8	62.3	75.76	0.91	2.05	1.79
17BGA40-3_2	0.04817	0.00889	0.06015	0.00308	0.00071	0.00011	0.00095	0.00003	107.7	57	59.3	62.2	34.44	0.7	0.91	0.7
17BGA40-3_3	0.04819	0.00903	0.06052	0.00317	0.0008	0.00011	0.00106	0.00004	108.7	57.9	59.7	64	38.73	0.72	1.02	0.8
17BGA40-3_4	0.04826	0.00876	0.05735	0.00307	0.00136	0.00013	0.00162	0.00007	111.9	56.2	56.6	61.9	65.24	0.8	1.55	1.51
17BGA40-3_5	0.05013	0.00934	0.06419	0.00329	0.00114	0.00013	0.00148	0.00007	201.2	59.9	63.2	66.4	51.86	0.8	1.41	1.35
17BGA40-3_6	0.04892	0.00898	0.06102	0.00313	0.00148	0.00013	0.00183	0.00007	143.9	57.6	60.1	63.1	69.46	0.86	1.76	1.45
17BGA40-3_7	0.04781	0.0086	0.05783	0.0031	0.00083	0.00011	0.00105	0.00005	89.1	55.2	57.1	62.5	41.41	0.69	1.01	0.91
17BGA40-3_8	0.05307	0.0089	0.06499	0.00303	0.00195	0.00015	0.00236	0.00008	331.7	57.1	63.9	61.2	81.27	0.93	2.25	1.58
17BGA40-3_9	0.04967	0.00862	0.05907	0.00299	0.00129	0.00012	0.00154	0.00007	179.6	55.3	58.3	60.4	59.51	0.77	1.48	1.34
17BGA40-3_10	0.04884	0.00913	0.06111	0.00328	0.00151	0.00014	0.00188	0.00008	139.9	58.6	60.2	66.2	71.18	0.88	1.8	1.51
17BGA40-3_11	0.05132	0.01807	0.12852	0.00666	0.00162	0.00028	0.0041	0.00023	255.3	115.5	122.8	134.2	70.89	1.75	3.69	4.64
17BGA40-3_12	0.04821	0.00854	0.05752	0.00313	0.00193	0.00015	0.00226	0.00011	109.7	54.8	56.8	63.2	91.91	0.94	2.17	2.26
17BGA40-3_13	0.05667	0.00842	0.06559	0.00305	0.00232	0.00015	0.00261	0.00012	478.2	54.1	64.5	61.6	88.74	0.99	2.49	2.5
17BGA40-3_14	0.05003	0.00846	0.05868	0.0034	0.00129	0.00012	0.00153	0.00008	196.5	54.3	57.9	68.6	59	0.74	1.47	1.63
17BGA40-3_15	0.05152	0.00835	0.05941	0.00317	0.00086	0.0001	0.00104	0.00005	264.1	53.6	58.6	64.1	38.06	0.66	1	1.1
17BGA40-3_16	0.04854	0.00877	0.05786	0.00307	0.00134	0.00012	0.00159	0.00009	125.7	56.3	57.1	62	63.69	0.8	1.53	1.88
17BGA40-3_17	0.04897	0.01015	0.06862	0.0033	0.00142	0.00015	0.00199	0.00054	146.5	65.1	67.4	66.5	66.61	0.94	1.89	10.88
17BGA40-3_18	0.04794	0.00858	0.05686	0.00325	0.002	0.00015	0.00233	0.00014	95	55.1	56.2	65.6	97.02	0.94	2.24	2.74
17BGA40-3_19	0.05227	0.00862	0.06308	0.00312	0.00168	0.00013	0.002	0.00011	297.2	55.3	62.1	62.9	71.62	0.85	1.91	2.23
17BGA40-3_20	0.04796	0.00874	0.05861	0.00319	0.00188	0.00014	0.00226	0.00011	96.4	56.1	57.8	64.4	91.21	0.92	2.17	2.23
17BGA40-3_21	0.04996	0.00847	0.05866	0.00316	0.00182	0.00014	0.00211	0.00012	192.9	54.4	57.9	63.9	82.77	0.88	2.02	2.42
17BGA40-3_22	0.04884	0.00833	0.05641	0.00329	0.00123	0.00011	0.00143	0.00008	140.1	53.5	55.7	66.4	58.24	0.73	1.38	1.53
17BGA40-3_23	0.04867	0.00901	0.05946	0.00308	0.00165	0.00014	0.00201	0.00008	132.1	57.8	58.6	62.1	78.01	0.88	1.92	1.58
17BGA40-3_24	0.04992	0.00896	0.06128	0.00326	0.00063	0.00011	0.00085	0.00006	191.1	57.5	60.4	65.8	29.09	0.67	0.81	1.13
17BGA38-3_1	0.04739	0.00916	0.06012	0.00274	0.0009	0.00012	0.00118	0.00004	68.5	58.8	59.3	55.3	45.04	0.75	1.13	0.88
17BGA38-3_2	0.05259	0.00931	0.06677	0.00294	0.0022	0.00016	0.00275	0.00009	311.2	59.8	65.6	59.3	92.45	1.04	2.61	1.73
17BGA38-3_3	0.04992	0.00901	0.06244	0.00246	0.00217	0.00016	0.00266	0.0001	191.4	57.8	61.5	49.8	98.23	1.04	2.54	1.95
17BGA38-3_4	0.04919	0.00893	0.06056	0.00251	0.00247	0.00018	0.00298	0.00012	156.7	57.3	59.7	50.8	113.71	1.12	2.85	2.41
17BGA38-3_5	0.04943	0.00913	0.06269	0.00276	0.00089	0.00012	0.00117	0.00004	168.3	58.6	61.7	55.8	41.32	0.74	1.11	0.89
17BGA38-3_6	0.04916	0.00916	0.06245	0.00276	0.00178	0.00015	0.00223	0.00008	155.5	58.8	61.5	55.6	82.59	0.95	2.13	1.7
17BGA38-3_7	0.04839	0.01239	0.08205	0.00383	0.00098	0.00016	0.00172	0.00006	118.4	79.4	80.1	77.3	47.24	1.02	1.61	1.25
17BGA38-3_8	0.05023	0.00877	0.06178	0.00256	0.00163	0.00014	0.00198	0.00008	205.7	56.3	60.9	51.7	73.68	0.86	1.9	1.55
17BGA38-3_9	0.05402	0.00969	0.07241	0.00303	0.00101	0.00012	0.0014	0.00005	371.7	62.2	71	61.2	41.79	0.8	1.33	0.92
17BGA38-3_10	0.05227	0.00908	0.06568	0.00278	0.0019	0.00015	0.00234	0.00008	297.3	58.2	64.6	56.1	80.52	0.96	2.23	1.67
17BGA38-3_11	0.0483	0.00906	0.06117	0.00271	0.00147	0.00013	0.00186	0.00007	113.9	58.1	60.3	54.7	70.51	0.85	1.78	1.36
17BGA38-3_12	0.04641	0.00931	0.05922	0.00273	0.00157	0.00014	0.00198	0.00008	18.9	59.7	58.4	55.1	77.95	0.92	1.89	1.71
17BGA38-3_13	0.05737	0.00917	0.07271	0.00294	0.00185	0.00014	0.00231	0.00008	505.1	58.8	71.3	59.2	69.86	0.92	2.19	1.64
17BGA38-3_14	0.05078	0.00933	0.06387	0.00252	0.00171	0.00015	0.00214	0.00005	231.1	59.9	62.9	50.9	76.07	0.93	2.04	1.02
17BGA38-3_15	0.04794	0.00961	0.06286	0.00263	0.00184	0.00016	0.00238	0.0001	95	61.6	61.9	53.1	89.7	1.03	2.27	1.93
17BGA38-3_16	0.04999	0.00926	0.06367	0.00247	0.00235	0.00018	0.00291	0.00015	194.4	59.4	62.7	49.8	105.85	1.15	2.78	2.96
17BGA38-3_17	0.0492	0.00899	0.06072	0.0028	0.00189	0.00015	0.0023	0.00009	157.4	57.7	59.9	56.5	87.61	0.95	2.2	1.75
17BGA38-3_18	0.04532	0.0094	0.05782	0.00259	0.00224	0.00018	0.00277	0.00014	0.1	60.3	57.1	52.4	76.42	1.16	2.66	2.87
17BGA38-3_19	0.05386	0.00895	0.06678	0.00253	0.00231	0.00017	0.00278	0.00009	364.9	57.4	65.6	51.1	93.79	1.09	2.64	1.86
17BGA38-3_20	0.04988	0.00877	0.05942	0.00249	0.00171	0.00014	0.00201	0.00008	189.3	56.3	58.6	50.3	77.95	0.89	1.93	1.64

References

1. Dickinson, W.R. Potash-depth (K-h) relations in continental margin and intra-oceanic magmatic arcs. *Geology* **1975**, *3*, 53–56. [CrossRef]
2. Gill, J.B. *Orogenic Andesites and Plate Tectonics*; Springer Science & Business Media: Berlin/Heidelberg, Germany, 2012; Volume 16.
3. Schmidt, M.W.; Jagoutz, O. The global systematics of primitive arc melts. *Geochem. Geophys. Geosyst.* **2017**, *18*, 2817–2854. [CrossRef]
4. Mo, X.; Niu, Y.; Dong, G.; Zhao, Z.; Hou, Z.; Zhou, S.; Ke, S. Contribution of syncollisional felsic magmatism to continental crust growth: A case study of the Paleogene Linzizong volcanic Succession in southern Tibet. *Chem. Geol.* **2008**, *250*, 49–67. [CrossRef]
5. Zhu, D.C.; Zhao, Z.D.; Niu, Y.; Dilek, Y.; Hou, Z.Q.; Mo, X.X. The origin and pre-Cenozoic evolution of the Tibetan plateau. *Gondwana Res.* **2013**, *23*, 1429–1454. [CrossRef]
6. Dong, G.C.; Mo, X.X.; Zhao, Z.D.; Wang, L.; Zhou, S. A new understanding of the stratigraphic successions of the Linzizong volcanic rocks in the Linzhou basin, Tibet. *Geol. Bull. China* **2005**, *24*, 549–557. (In Chinese with English Abstract)
7. He, S.; Kapp, P.; DeCelles, P.G.; Gehrels, G.E.; Heizler, M. Cretaceous-Tertiary Geology of the Gangdese Arc in the Linzhou Area, Southern Tibet. *Tectonophysics* **2007**, *433*, 15–37. [CrossRef]
8. Li, H.Y.; Zhong, S.L.; Wang, Y.B.; Zhu, D.C.; Yang, J.H.; Song, B.; Liu, D.Y.; Wu, F.Y. Age, genesis and geological significane of volcanic rocks from Linzizong, Linzhou Basin, Southern Tibet: Zircon U-Pb age and Hf isotope evidenve. *Acta Petrol.* **2007**, *23*, 493–500. (In Chinese with English Abstract)
9. Huang, W.T.; Dupont-Nivet, G.; Lippert, P.C.; van Hinsbergen, D.J.J.; Dekkers, M.J.; Waldrip, R.; Ganerød, M.; Li, X.; Guo, Z.; Kapp, P. What was the Paleogene Latitude of the LhasaTerrane? A Reassessment of the Geochronology and Paleomagnetism of Linzizong Volcanic Rocks (Linzhou Basin, Tibet). *Tectonics* **2015**, *34*, 594–622. [CrossRef]
10. Zhu, D.C.; Wang, Q.; Zhao, Z.D.; Chung, S.L.; Cawood, P.A.; Niu, Y.; Liu, S.; Wu, F.Y.; Mo, X.X. Magmatic Record of India-Asia Collision. *Sci. Rep.* **2015**, *5*, 14289. [CrossRef]
11. Chen, B.; Ding, L.; Xu, Q.; Yue, Y.; Xie, J. U-Pb age framework of the Linzizong volcanic rocks from the Linzhou Basin, Tibet. *Quat. Sci.* **2016**, *36*, 1037–1054. (In Chinese with English Abstract)
12. Liu, A.L.; Wang, Q.; Zhu, D.C.; Zhao, Z.D.; Liu, S.A.; Wang, R.; Dai, J.G.; Zheng, Y.C.; Zhang, L.L. Origin of the Ca. 50 Ma Linzizong Shoshonitic Volcanic Rocks in the Eastern Gangdese Arc, Southern Tibet. *Lithos* **2018**, *304–307*, 374–387. [CrossRef]
13. Yan, H.Y.; Long, X.P.; Li, J.; Wang, Q.; Zhao, B.; Shu, C.; Gou, L.; Zuo, R. Arc Andesitic Rocks Derived from Partial Melts of Mélange Diapir in Subduction Zones: Evidence from Whole-Rock Geochemistry and Sr-Nd-Mo Isotopes of the Paleogene Linzizong Volcanic Succession in Southern Tibet. *J. Geophys. Res. Solid Earth* **2019**, *124*, 456–475. [CrossRef]
14. Chen, J.S.; Huang, B.C.; Sun, L.S. New Constraints to the Onset of the India-Asia Collision: Paleomagnetic Reconnaissance on the Linzizong Group in the Lhasa Block, China. *Tectonophysics* **2010**, *489*, 189–209. [CrossRef]
15. Liang, Y.P.; Zhu, J.; Ci, Q.; He, W.H.; Zhang, K.X. Zircon U-Pb age and geochemical characteristics of volcanic rocks from Linzizong Group, junuo Area, central Gangdisi Belt, Qinghai-Tibet Plateau. *Earth Sci.* **2010**, *35*, 211–223. (In Chinese with English Abstract) [CrossRef]
16. Song, Q.Y. Geochemical features of the volcanic rocks of Linzizong group in Chuoqin Basin. *J. Geomech.* **1999**, *5*, 66–71. (In Chinese with English Abstract)
17. Hu, J.X.; Chen, J.L.; Yao, S. Zircon U-Pb geochronology, genesis for Nianbo Formation volcanic rocks of Linzizong Group in the western part of Gangdese arc and its implication. *Geochimica* **2018**, *47*, 699–711. (In Chinese with English Abstract)
18. Fu, W.C.; Kang, Z.Q.; Pan, H.B. Geochemistry, zircon U-Pb age and implications of the Linzizong Group volcanic rocks in Shiquan River area, western Gangdise belt, Tibet. *Geol. Bull. China* **2014**, *33*, 850–859. (In Chinese with English Abstract)
19. Cao, Y.; Yang, F.; Li, D.X.; Chen, H.; Zhou, T.; Liu, D.M.; Li, Y.X. Geochemical characteristics and tectonic significance of Linzizong volcanic rocks in the west Yare area of Lhasa terrane. *Collect. Geol. Prospect. Pap.* **2020**, *35*, 216–222. (In Chinese with English Abstract)
20. Ding, X.L.; Ding, L.; Wang, C.; Wang, H.Q.; Guo, X.D. Petrogenesis and tectonic implications of Palaeocene (ca.54 Ma) rhyolites in the western Lhasa Terrane, south Tibet: Constraints from geochemistry and Sr-Nd-Hf isotope compositions. *Geol. J.* **2020**, *56*, 494–507. [CrossRef]
21. Mo, X.X. Response of volcanism to the India Asia collision. Earth Sciences. *Frontiers* **2003**, *10*, 135–148. (In Chinese with English Abstract)
22. Zhao, Z.D.; Mo, X.X.; Nomade, S.; Renne, P.R.; Zhou, S.; Dong, G.C.; Wang, L.L.; Zhu, D.C.; Liao, Z.L. Post-collisional ultrapotassic rocks in Lhasa block, Tibetan plateau: Spatial and temporal distribution and its implications. *Acta Petrol. Sin.* **2006**, *22*, 787–794. (In Chinese with English Abstract)
23. Mo, X.X.; Hou, Z.Q.; Niu, Y.L.; Dong, G.C.; Qu, X.M.; Zhao, Z.D.; Yang, Z.M. Mantle contributions to crustal thickening during continental collision: Evidence from Cenozoic igneous rocks in southern Tibet. *Lithos* **2007**, *96*, 225–242. [CrossRef]
24. Zhu, D.C.; Pan, G.T.; Chun, S.L.; Liao, Z.L.; Wang, L.Q.; Li, G.M. SHRIMP zircon age and geochemical constraints on the origin of Early Jurassic volcanic rocks from the Yeba Formation, southern Gangdese in south Tibet. *Int. Geol. Rev.* **2008**, *50*, 442–471. [CrossRef]
25. Lee, H.Y.; Chung, S.L.; Lo, C.H.; Ji, J.Q.; Lee, T.Y.; Qian, Q.; Zhang, Q. Eocene Neotethyan slab breakoff in southern Tibet inferred from the Linzizong volcanic record. *Tectonophysics* **2009**, *477*, 20–35. [CrossRef]

26. Yu, F.; Li, Z.G.; Zhao, Z.D.; Xie, G.G.; Dong, G.C.; Zhou, S.; Zhu, D.C.; Mo, X.X. Geochemistry and implication of the Linzizong volcanic succession in Cuomai area, central-western Gangdese, Tibet. *Acta Petrol. Sin.* **2010**, *26*, 2217–2225. (In Chinese with English Abstract)
27. Xie, B.J.; Zhou, S.; Xie, G.G.; Tian, Z.M.; Liao, Z.L. Zircon SHRIMP U-Pb data and regional contrasts of geochemical characteristics of Linzizong volcanic rocks from Konglong and Dinrenle region, middle Gangdese belt. *Acta Petrol. Sin.* **2013**, *29*, 3803–3814. (In Chinese with English Abstract)
28. Li, Y.; Zhang, S.Z.; Li, F.Q.; Qin, Y.D.; Gong, X.D. Zircon U-Pb ages and implications of the Dianzhong formation in Chazi area, middle Lhasa block, Tibet. *Earth Sci.* **2018**, *43*, 2755–2766. (In Chinese with English Abstract)
29. Zeng, C.; Yan, M.Q.; Shen, Z.Y.; Gao, Q.; Wei, J.H.; Mao, G.Z.; Deng, Y.M. U-Pb age, Hf isotope and geochemical characteristics of zircons from Dianzhong Formation rhyolite in Qieqiang Area, Xietongmen, Tibet. *Geotecton. Metallog.* **2022**, *46*, 154–174. (In Chinese with English Abstract) [CrossRef]
30. Harrison, T.M.; Yin, A.; Grove, M.; Lovera, O.M.; Ryerson, F.J.; Zhou, X.H. The Zedong window: A record of superposed Tertiary convergence in southeastern Tibet. *J. Gephysical Res.* **2000**, *105*, 19211–19230. [CrossRef]
31. Pan, G.T. Spatial temporal framework of the Gangdese Orogenic Belt and its evolution. *Acta Petrol. Sin.* **2006**, *22*, 521–533. (In Chinese with English Abstract)
32. Zhu, D.C.; Zhao, Z.D.; Niu, Y.; Mo, X.X.; Chung, S.L.; Hou, Z.Q.; Wang, L.Q.; Wu, F.Y. The Lhasa Terrane: Record of a microcontinent and its histories of drift and growth. *Earth Planet. Sci. Lett.* **2011**, *301*, 1–2. [CrossRef]
33. Zhou, S.; Mo, X.X.; Zhao, Z.D.; Qiu, R.Z.; Zhang, S.Q.; Guo, T.Y. 40Ar/39Ar isotopic chronology of volcanic rocks in Wuyu Basin, Nanmulin county, Tibet. *China Univ. Geosci.* **2004**, *4*.
34. Tang, P.; Tang, J.X.; Zheng, W.B.; Leng, Q.F.; Lin, B.; Tang, X.Q.; Wang, H. Zircon U-Pb ages, Hf isotopes and geochemistry of the volcanic rocks in Dianzhong Formation from xingaguo area, Tibet. *Acta Petrol. Mineral.* **2018**, *37*, 47–60. (In Chinese with English Abstract)
35. Wei, N.S.; Kang, Z.Q.; Yang, F. Geochronology, geochemical characteristics, and genesis of the Dianzhong Formation volcanic rocks in Changguo area, southeastern Lhasa Block, Tibet. *Geochimica* **2019**, *48*, 30–42. (In Chinese with English Abstract)
36. Zhou, P.; Rong, F.; Li, Q.; Liu, G.X. Zircon U-Pb age and Geological significance of gabbro-diorite veins in Gongting area, Tibet. *Bull. Mineral.* **2019**, *38*, 773–780. (In Chinese with English Abstract) [CrossRef]
37. Fu, Y.G. *The Magmatism and Mineralization Related to the Subduction of Neo-Tethys in the Southern Part of Lhasa Block in Tibet*; Chinese Academy of Geological Sciences: Beijing, China, 2017. (In Chinese with English Abstract)
38. Ding, S.; Chen, Y.C.; Tang, J.X.; Xie, F.W.; Hu, G.Y.; Yang, Z.Y.; Shi, S.; Li, Y.H.; Yang, H.Y. Relationship between Linzizong volcanic rocks and mineralization: A case study of Sinongduo epithermal Ag-Pb-Zn deposit. *J. Mineral Deposits* **2017**, *36*, 1074–1092. (In Chinese with English Abstract)
39. Chen, L.P.; Huang, Z.S.; Jiangba, D.J. Zircon U-Pb age and geochemical characteristics of igneous rocks from the Dianzhong Formation in the Shengong area of Tibet. *J. Geol. Bull. China* **2019**, *38*, 1127–1135. (In Chinese with English Abstract)
40. Liu, D.; Zhao, Z.D.; DePaolo, D.J.; Zhu, D.C.; Meng, F.Y.; Shi, Q.; Wang, Q. Potassic volcanic rocks and adakitic intrusions in southern Tibet: Insights into mantle-crust interaction and mass transfer from Indian plate. *Lithos* **2017**, *268–271*, 48–64. [CrossRef]
41. Tang, W.L.; Huang, F.; Xu, J.F.; Zeng, Y.C.; Liu, X.J. Cretaceous magmatism in the northern Lhasa Terrane:Implications for the tectonic evolution and crustal growth tempos of central Tibet. *GSA Bull.* **2024**, *136*, 3440–3456. [CrossRef]
42. Wu, Y.B.; Zheng, Y.F. Genesis of zircon and its constraints on interpretation of U-Pb age. *Chin. Sci. Bull.* **2004**, *49*, 1554–1569. [CrossRef]
43. Wang, Y.; Ma, X.D.; Chen, W. Geochronology and petrogenesis of the Xiongmei Cu-bearing granodiorite porphyry in North Lhasa subterrane, central Tibet: Implication for the evolution of Bangong-Nujiang metallogenic belt-ScienceDirect. *Ore Geol. Rev.* **2019**, *114*, 103119. [CrossRef]
44. Hu, W.L.; Wang, Q.; Tang, G.J.; Qi, Y.; Wang, J.; Yang, Z.Y.; Sun, P. First identification of Early Cretaceous mafic dikes in the Baingoin area, central Tibet: Implications for crust-mantle interactions and magmatic flare-up. *Geol. Soc. Am. Bull.* **2024**, *136*, 846–860. [CrossRef]
45. Maniar, P.D.; Piccoli, P.M. Tectonic discrimination of granitoids. *Geol. Soc. Am. Bull.* **1989**, *101*, 635–643. [CrossRef]
46. Martin, H.; Smiithies, R.H.; Rapp, R.; Moyen, J.F.; Champion, D. An overview of adakite, tonalite-trondhjemite-granodiorite (TTG), and sanukitoid: Relationships and some implications for crustal evolution. *Lithos* **2005**, *79*, 1–24. [CrossRef]
47. Sun, S.S.; Mcdonough, W.F. Chemical and isotopic systematics of oceanic basalts: Implications for mantle composition and processes. *Geol. Soc. Lond. Spec. Publ.* **1989**, *42*, 313–345. [CrossRef]
48. Yu, S.M.; Ma, X.D.; Hu, Y.C.; Chen, W.; Liu, Q.P.; Song, Y.; Tang, J.X. Post-subdution evolution of the Northern Lhasa Terrane, Tibet: Constraints from geochemical anomalies, chronology and petrogeochemistry. *China Geol.* **2022**, *5*, 84–95. [CrossRef]
49. Coulon, C.; Maluski, H.; Bollinger, C. Mesozoic and Cenozoic Volcanic Rocks from Central and Southern Tibet:39Ar-40Ar Dating, Petrological Characteristics and Geodynamical Significance. *Earth Planet. Sci. Lett.* **1986**, *79*, 281–302. [CrossRef]
50. Pearce, J.A.; Peate, D.W. Tectonic Implications of the Composition of Volcanic ARC Magmas. *Annu. Rev. Earth Planet. Sci.* **1995**, *23*, 251–285. [CrossRef]
51. Marschall, H.R.; Schumacher, J.C. Arc magmas sourced from melange diapirs in subduction zones. *Nat. Geosci.* **2012**, *5*, 862–867. [CrossRef]

52. Wang, N.; Liu, Z.B.; Song, Y.; Li, Z.J.; Yan, X.K.; An, Q. First identification of the Early Cretaceous Suolong Cu-rich porphyry in the Geji area, central Tibet: A new clue for regional porphyry Cu exploration. *J. Asian Earth Sci.* **2024**, 1367–9120. [CrossRef]
53. Yang, H.; Xiang, S.Y.; Wang, X. Age and Tectonic Setting of Dianzhong Formation in the Maxiang Area, Tibet. *China Geol. Sci. Technol. Inf.* **2013**, *32*, 89–96. (In Chinese with English Abstract)
54. Chappell, B.W.; White, A.J.R. Two contrasting granite types. *Pac. Geol.* **1974**, *8*, 173–174.
55. Loiselle, M.C.; Wones, D.R. Characteristics of anorogenic granites. *Geol. Soc. Am. Abstr. Programs* **1979**, *11*, 468.
56. Miller, C.F.; Miller, J.S. Contrasting stratified plutons exposed in tilt blocks, Eldorado Mountains, Colorado River Rift, NV, USA. *Lithos* **2002**, *61*, 209–224. [CrossRef]
57. Frost, C.D.; Frost, B.R.; Bell, J.M.; Chamberlain, K.R. The relationship between A-type granites and residual magmas from anorthosite: Evidence from the northern Sherman batholith, Laramie Mountains, Wyoming, USA. *Precambrian Res.* **2002**, *119*, 45–71. [CrossRef]
58. Clemens, J.D. S-type granitic magmas—Petrogenetic issues, models and evidence. *Earth-Sci. Rev.* **2003**, *61*, 1–18. [CrossRef]
59. Petford, N.; Atherton, M. Na-rich Partial Melts from Newly Underplated Basaltic Crust: The Cordillera Blanca Batholith, Peru. *J. Petrol.* **1996**, *37*, 1491–1521. [CrossRef]
60. Chappell, B.W.; White, A.J.R.; Williams, I.S.; Wyborn, D. Low- and high-temperature granites. *Earth Environ. Sci. Trans. R. Soc. Edinb.* **2004**, *95*, 125–140. [CrossRef]
61. Li, X.H.; Li, W.X.; Li, Z.X. Rediscussion on genetic types and tectonic significance of early Yanshan granites in Nanling. *Chin. Sci. Bull.* **2007**, *52*, 981–991. (In Chinese)
62. Grove, T.L.; Elkins-Tanton, L.T.; Parman, S.W.; Chatterjee, N.; Müntener, O.; Gaetani, G.A. Fractional crystallization and mantle-melting controls on calc-alkaline differentiation trends. *Contrib. Miner. Pet.* **1997**, *145*, 515–533. [CrossRef]
63. Collins, W.J.; Richards, S.W. Geodynamic significance of S-type granites in circum-Pacific orogens. *Geology* **2008**, *36*, 559–562. [CrossRef]
64. Kemp, A.I.S.; Hawkesworth, C.J.; Foster, G.L.; Paterson, B.A.; Woodhead, J.D.; Hergt, J.M.; Gray, C.M.; Whitehouse, M.J. Magmatic and Crustal Differentiation History of Granitic Rocks from Hf-O Isotopes in Zircon. *Science* **2007**, *315*, 980–983. [CrossRef] [PubMed]
65. Eby, G.N. The A-type granitoids: A review of their occurrence and chemical characteristics and speculations on their petrogenesis. *Lithos* **1990**, *26*, 115–134. [CrossRef]
66. Whalen, J.B.; Currie, K.L.; Chappell, B.W. A-type granites: Geochemical characteristics, discrimination and petrogenesis. *Contrib. Mineral. Petrol.* **1987**, *95*, 407–419. [CrossRef]
67. Moyen, J.F.; Stevens, G.; Kisters, A. Record of mid-Archaean subduction from metamorphism in the Barberton terrain, South Africa. *Nature* **2006**, *442*, 559. [CrossRef] [PubMed]
68. Defant, M.J.; Drummond, M.S. Derivation of some modern arc magmas by melting of young subducted lithosphere. *Nature* **1990**, *347*, 662–665. [CrossRef]
69. Miller, C.F.; McDowell, S.M.; Mapes, R.W. Hot and cold granites? Implications of zircon saturation temperatures and preservation of inheritance. *Geology* **2003**, *31*, 529–532. [CrossRef]
70. Bea, F.; Morales, I.; Molina, J.F.; Montero, P.; Cambeses, A. Zircon stability grids in crustal partial melts: Implications for zircon inheritance. *Contrib. Mineral. Petrol.* **2021**, *176*, 1–13. [CrossRef]
71. Rajesh, H.M.; Santosh, M. Charnockitic magmatism in southern India. *Earth Syst. Sci* **2004**, *113*, 565–585. [CrossRef]
72. Ding, L.; Kapp, P.; Zhong, D.L.; Deng, W.M. Cenozoic volcanism in Tibet: Evidence for a transition from oceanic to continental subduction. *J. Petrol.* **2003**, *44*, 1833–1865. [CrossRef]
73. Wen, D.R.; Liu, D.Y.; Chung, S.L.; Chu, M.F.; Ji, J.Q.; Zhang, Q.; Song, B.; Lee, T.Y.; Yeh, M.W.; Lo, C.H. Zircon SHRIMP U-Pb ages of the Gangdese batholith and implications for Neotethyan subduction in southern Tibet. *Chem. Geol.* **2008**, *252*, 191–201. [CrossRef]
74. Lee, H.Y.; Chung, S.L.; Ji, J.Q.; Qian, Q.; Gallet, S.; Lo, C.H.; Lee, T.Y.; Zhang, Q. Geochemical and Sr–Nd isotopic constraints on the genesis of the Cenozoic Linzizong volcanic successions, southern Tibet. *J. Asian Earth Sci.* **2012**, *53*, 96–114. [CrossRef]
75. Ji, W.Q.; Wu, F.Y.; Chung, S.L.; Wang, X.C.; Liu, C.Z.; Li, Q.L.; Liu, Z.C.; Liu, X.C.; Wang, J.G. Eocene neo-Tethyan slab breakoff constrained by 45 Ma oceanic island basalt–type magmatism in southern Tibet. *Geology* **2016**, *44*, 283–286. [CrossRef]
76. Zhang, Y.T.; Huang, F.; Xu, J.F.; Zeng, Y.C.; Wang, B.D.; Lv, M.D.; Zhang, L.; Li, M.J.; Zhang, Z.; Tian, Y.; et al. Origin of the volcanic rocks in Dianzhong Formation, central Lhasa Terrane, Tibet: Implication for the genesis of syn-collision magmatism and Neo-Tethyan slab roll-back. *Int. Geol. Rev.* **2022**, *65*, 21–39. [CrossRef]
77. Chung, S.; Chu, M.W.; Zhang, Y.; Xie, Y.; Lo, C.H.; Lee, T.Y.; Lan, C.Y.; Li, X.H.; Zhang, Q.; Wang, Y.Z. Tibetan tectonic evolution inferred from spatial and temporal variations in post-collisional magmatism. *J. Earth-Sci. Rev.* **2005**, *68*, 173–196. [CrossRef]
78. Hou, Z.Q.; Mo, X.X.; Yang, Z.M.; Gao, Y.F.; Ding, L.; Dong, G.C. Early processes and tectonic model for the Indian-Asian continental collision: Evidence from the Cenozoic Gangdese igneous rocks in Tibet. *J. Acta Geol. Sin.* **2006**, *80*, 1233–1248.
79. Zhang, X.Q. Petrogenesis of Linzizong Group Volcanic Rocks in Zexue Area, Tibet and Its Relation to Mineralization. Ph.D. Thesis, China University of Geosciences, Wuhan, China, 2013; pp. 41–44. (In Chinese with English Abstract)
80. Pearce, J.A.; Harris, N.B.W.; Tindle, A.G. Trace elements discrimination diagrams for the tectonic interpretation of granitic rock. *J. Petrol.* **1984**, *25*, 956–983. [CrossRef]

81. Pearce, J.A.; Norry, M.J. Petrogenetic implications of Ti, Zr, Y and Nb variations in volcanic rocks. *Contrib. Mineral. Petrol.* **1979**, *69*, 33–47. [CrossRef]
82. Batchelor, R.A.; Bowden, P. Petrogenetic interpretation of granitiod rock series using multicationic parameters. *J. Chem. Geol.* **1985**, *48*, 43–55. [CrossRef]
83. Liu, Y.S.; Gao, S.; Hu, Z.C.; Gao, C.G.; Zong, K.Q.; Wang, D.B. Continental and oceanic crust recycling-induced melt-peridotite interactions in the Trans-North China Orogen: U-Pb dating, Hf isotopes and trace elements in zircons of mantle xenoliths. *Petrol* **2010**, *51*, 537–571. [CrossRef]
84. Bouvier, A.; Vervoort, J.D.; Patchett, P.J. The Lu-Hf and Sm-Nd isotopic composition of CHUR: Constraints from unequilibrated chondrites and implications for the bulk composition for terrestrial planets. *Earth Planet. Sci. Lett.* **2008**, *273*, 48–57. [CrossRef]
85. Griffin, W.L.; Pearson, N.J.; Belousova, E.; Jackson, S.E.; Van Achterbergh, E.; O'Reilly, S.Y.; Shee, S.R. The Hf isotope composition of cratonic mantle: LA-MC-ICPMS analysis of zircon megacrysts in kimberlites. *Geochim. Cosmochim. Acta* **2000**, *64*, 133–147. [CrossRef]
86. Griffin, W.L.; Wang, X.; Jackson, S.E.; Pearson, N.J.; O'Reilly, S.Y.; Xu, X.; Zhou, X. Zircon chemistry and magma mixing, SE China: In-situ analysis of Hf isotopes, Tonglu and Pingtan igneous complexes. *Lithos* **2002**, *61*, 237–269. [CrossRef]

MDPI AG
Grosspeteranlage 5
4052 Basel
Switzerland
Tel.: +41 61 683 77 34

Minerals Editorial Office
E-mail: minerals@mdpi.com
www.mdpi.com/journal/minerals

www.ingramcontent.com/pod-product-compliance
Lightning Source LLC
Chambersburg PA
CBHW040820250926
51850CB00007BA/1217
9783725868162